ISBN 978-0-484-41353-4
PIBN 10753199

1890.

PROCEEDINGS

OF THE

UNITED STATES

NAVAL INSTITUTE.

VOLUME XVI.

PUBLISHED QUARTERLY BY THE INSTITUTE.

ANNAPOLIS, MD.

PRESS OF ISAAC FRIEDENWALD CO.
BALTIMORE, MD.

CONTENTS.

THE PROCEEDINGS

OF THE

UNITED STATES NAVAL INSTITUTE.

Vol. XVI., No. 1.   1890.   Whole No. 52.

U. S. NAVAL INSTITUTE, ANNAPOLIS, MD.

NAVAL ADMINISTRATION IN ALASKA.

BY COMMANDER HENRY GLASS, U. S. Navy.

In August, 1880, the writer was ordered by the Navy Department to proceed to Sitka, Alaska, and relieve Commander L. A. Beardslee, U. S. Navy, in command of the sloop-of-war Jamestown, then stationed at that place. The Jamestown had been hurriedly fitted out at Mare Island Navy Yard, in the spring of 1879, no steamer being then available for the duty, and sent to Sitka to preserve order among the Indians, and to prevent threatened conflicts between them and the white residents and traders and miners scattered about that portion of the territory.

From the impossibility of the Jamestown's cruising about the tortuous channels and sounds of Southeast Alaska, she remained quietly moored in the inner harbor of Sitka during the entire period of her service on the station, small detachments of officers and men being moved from point to point, as found necessary, by means of the steam launches with which the ship had been supplied when fitted out for that special service.

On reporting for the duty assigned, it was soon found that much more was involved than the simple command of a vessel of war, the commander of the Jamestown representing in himself the whole machinery of government in the territory, so far at least as affording protection to the lives and property of citizens was concerned—something very foreign to the ordinary routine work of a naval officer in time of peace. The only guides for my action were a copy of the letter of general instructions from the Navy Department to my predecessor, authorizing him to take any steps that he might consider necessary for the preservation of good order in the territory, and his various reports from time to. time of his actions, all of which had received the full approval of the Department.

At this time Alaska was absolutely without any form of local government, and was in a condition almost as free from the operation of civil law as the interior of Africa. Our Congress had, since the cession of the territory by the Russian government in 1867, been content with making the whole of Alaska a single district for the collection of duties on imports, with a Collector of Customs at Sitka and deputies at Fort Wrangell and Kadiack Island, extending over the territory certain sections of the Act of 1835, governing intercourse with Indian tribes; these sections, however, related only to the introduction and sale of spirits and breech-loading fire-arms and ammunition, and authorized a contract with the Alaska Commercial Company of San Francisco by which that company was granted for a term of years the exclusive privilege of taking seals on the Prybilof Islands in Behring's Sea. The terms of the treaty of cession guaranteed to Russian subjects, electing to remain in the territory after the change of nationality, all the rights and privileges of American citizens, and the undisturbed possession of their property, but no steps had been taken to secure to them those rights; no courts had been established, nor had any means been provided for the acquisition or transfer of titles to property.

Alaska was for several years after the cession a district of the Military Division of the Pacific, with a portion of a regiment of artillery garrisoning Sitka and one or two other points; but this form of control was abandoned in 1876, and a period of nearly three years elapsed with no exhibition of force, beyond the occasional visit of a revenue cutter, until the administration of affairs was assumed by the Navy Department in 1879.

As illustrating the situation of affairs, the following case may be cited. In the fall of 1879, Commander Beardslee arrested and sent to Portland, Oregon, for trial before the United States District Court at that place, a miner who had in broad daylight, and in the presence of a dozen witnesses in Sitka, deliberately attempted the murder of a person with whom he had some trifling dispute. The wounded man recovered, and, as actual murder had not been committed, the District Judge directed the discharge of the prisoner. The opinion of the Judge was to the effect that in territory such as Alaska, under the exclusive jurisdiction of the general government, and not organized, only the statutes defining piracy, and such special acts as had been passed by Congress for the government of the territory, were operative, and that, as the offense charged did not come under any of these, the court had no jurisdiction.

' The Collector of Customs at Sitka and his deputies were the only civil officers in the territory, and could exercise no authority beyond their routine duties under the Treasury Department. Hence it will be seen that the entire responsibility for the security of life and the preservation of good order in Alaska at that time devolved upon the commanding naval officer. That the difficulties of the situation were not greater must be ascribed to the orderly character of the white population at Sitka and other settlements, and to the generally tractable disposition of the Indians.

Sitka had in 1880 a white population of between two and three hundred, made up of a few Russians who had elected to become citizens of the United States, in accordance with treaty provisions, Russian half-breeds, traders, and a few miners who had been attracted there in the previous year by reported discoveries of rich gold-bearing quartz ledges on Baranof Island near Sitka. The Indian village, or *rancho*, separated in part by a stockade from the white settlement, contained usually in the winter season nearly a thousand persons, living in a straggling row of houses along the beach, a single house often sheltering twenty or thirty men, women, and children. The Indians had a sort of tribal, or rather family, organization; each family, of which there were four at Sitka, having its distinguishing name and crest, or *totem*, such as the Raven, the Wolf, the Bear, and the Whale, and each being under the nominal control of a head man or chief. The powers exercised by these chiefs were very vague, and even the manner of appointment or succession to leadership in the different families did not seem to be regulated by any fixed principle; but generally the office was hereditary in the richest and most powerful families.

Commander Beardslee, whom I relieved, had made while in command a careful study of the Indian question as it presented itself at Sitka, and displayed great tact and firmness in controlling the Indians and in preserving a good understanding between them and the white settlers. He employed the chiefs as a sort of irregular police force, taking them into the pay of the navy as scouts, thereby greatly increasing their importance and influence among the tribes, the very indefiniteness of their powers assisting him materially. This system, somewhat extended, was continued; the chiefs were put into uniform, assimilated to that of the naval service, their respective districts of the Indian village defined, and they were made, as far as possible, responsible for the condition of affairs in the village and the conduct of the members of their tribes.

It being impracticable to take the Jamestown into the interior channels on which the Indian villages are nearly always located, and the great extent of the territory to the westward making easy and rapid communication impossible, her influence was in a great degree confined to Sitka, and that port became the scene of efforts to improve the condition of the Indians and establish order among them. These efforts were afterwards extended to other points in the territory, and the seat of government was shifted from the quarter-deck of the Jamestown as occasion demanded.

Affairs went on quietly enough for some time after my arrival at Sitka, and my attention was directed chiefly to preparing the ship for the coming winter and to settling occasional disputes on shore that were submitted to me for arbitration. But with the approach of bad weather and the long nights of the dreary winter in that latitude, the Sitka Indians began to return from their summer hunting and fishing camps, and it was soon apparent that some active measures must be taken to suppress the disorder that at times reigned supreme in the village and often turned the night into pandemonium. Fighting was of almost daily occurrence, frequently attended by serious wounds to the parties engaged, and the howls of drunken Indians, rioting and dancing about their camp fires, could be heard on board the ship at almost any hour. To remove, as far as possible, the danger of collisions between the white people and the Indians, the latter were compelled to leave the limits of the white settlement each day at sunset, and no white men were allowed to go into the Indian village at any time except under orders for duty.

Drunkenness was found to be very prevalent, the Indians having been taught by white men to distill a wretched sort of rum, called

Hoo-che-noo, from molasses supplied them for the purpose by the traders. The process was very simple, and required only a crude apparatus that the Indians could readily procure. A mixture of molasses and water, fermented by the use of yeast made from the roots of certain plants, was put into a boiler, usually made from an old coal-oil can, and set over the fire in the middle of the house. The worm was made of a bent tin tube, or in some cases an old gun barrel served this purpose; a stream of water was usually directed on the tube, and a wooden bowl received the spirit as it trickled out. The process could be carried on day and night, and it was subsequently found that in the village of fifty or sixty houses over two hundred of these primitive distilleries were in operation in December, '1880'

This condition of affairs is not so much to be wondered at, taking all the circumstances into consideration. The short winter days of that latitude give the Indians little opportunity for hunting and fishing, and the long nights are without any amusements known to civilization. Ignorant, intensely superstitious, and compelled to endure several months of enforced idleness, with no mental employment, gambling and drunkenness naturally became habits of these Indians, and they eagerly embraced the means of intoxication so readily supplied them.

Becoming convinced that some measures to end the scenes of drunkenness and violence constantly presented in the village were absolutely necessary, it was decided to commence at the bottom and stop the importation and sale of molasses. As no law existed for this, a meeting of all the citizens was called at the Custom House, the situation and its needs fully explained, and all the traders were urged to enter into an agreement not to import molasses into the territory from that time, and not to sell any large quantity of molasses at one time to any Indian. Several of the traders showed a reluctance to sign an agreement that would cut off a considerable portion of their trade, but upon a significant hint from one of the miners present that, if the commander of the Jamestown would send all his officers and men on board ship and promise to know nothing of what was going on ashore for a few hours, it would be put out of the power of any one to sell molasses to Indians in Sitka, all hands came into the agreement, which was thereafter faithfully observed. But as it was known that many of the Indians had bought considerable quantities of molasses and had it stored in their houses, in readiness

for making the usual winter supply of fire-water, this agreement of
the traders was only a beginning.

The chiefs were next called together and told that the manufacture
of Hoo-che-noo must immediately stop, but all declared their
inability to put down what they nevertheless loudly declared to be
the ruin of their tribes. One of the chiefs, Ana-bootz, who was
looked upon as an authority by the others, suggested making a raid
on the village and destroying all the stills and molasses as the only
effectual means of dealing with the evil. All agreed that this would
be the only means of stopping drunkenness among their people, and
they united in a request that it should be adopted. This had been
done by Commander Beardslee in the previous year, but on that
occasion the houses of the chiefs had not been subjected to search
and their stills had not been molested, hence their readiness in advocating the raid.

After waiting for some days for any uneasiness among the Indians
to subside, the chiefs were assembled on board the Jamestown and
told that the raid had been decided upon as they had requested, and
that all preparations had been made to carry it out successfully.
Three of the ship's boats, fully manned and armed, were in readiness, under command of the executive officer, for landing, and the
entire marine guard of the vessel sent on shore. All being ready,
signal was made from the ship, the marines were marched into the
village at the double quick and posted in rear of the line of houses
to prevent any stills being carried out and secreted, and the boats
pulled in for the beach. The blue-jackets were landed in three
parties, in the center and at each end of the village, and being provided with axes and hammers, commenced at once their work in aid
of the cause of temperance. Every house in the village was carefully searched, no exceptions being made. The Indians looked on
in sullen astonishment, but no resistance was attempted by any one,
and finally many of those whose stills had been first destroyed
joined actively in the work, and were of assistance in hunting out
concealed stills and stores of molasses, showing a childish glee at the
discomfiture of those who had supposed themselves secure. In all,
over two hundred of these stills were found and destroyed, with a
great deal of molasses prepared for distillation. In one house five
stills were found ready, or in operation, with six barrels of molasses,
the owner of the house being a distiller of some prominence in the
community, the quality of whose Hoo-che-noo enjoyed a great

reputation. He was allowed to sell the molasses not needed as food for his family.

The night following this raid was the quietest one the village had enjoyed for several months, and from that time no serious affrays ever occurred among the Sitka Indians.

A few of the stills escaped the general destruction, but they were discovered and broken up within a few days, and the manufacture of Hoo-che-noo became a lost art at Sitka.

The principal winter employment and amusement of the Indians having thus been destroyed, it was necessary to give them something else to do and think of, and this was done in establishing a general system of cleaning and draining the village, which was in an indescribably wretched condition.

The village was built on sloping ground in front of some low hills, which, in the almost constant rain of the winter season, sent down small streams of water that in many cases entered the houses in rear, and converted all the spaces between them into mud-holes, while all the filth and refuse from the village were thrown on the beach. The Indians were set to work, under direction of some of the officers and leading men of the ship, digging a deep ditch in rear of the houses, with outlets at intervals running down to the beach; earth was heaped up about the sides and in rear of the houses to prevent the water's running into them, and the beach was cleaned up, smoothed, and neatly gravelled down to low-water mark. The sections into which the village had been divided were in charge of their respective chiefs, and daily inspections were made to ascertain that the orders given were properly carried out.

Of course, there were some cases of holding back, and even of resistance to the new order of things, but a judicious system of confinement in the guard-house for a day or two at a time, with the infliction of small fines of blankets, the usual currency among the Indians, soon convinced the most conservative among them that it was wisest to join in the march of improvement. A very few days sufficed to make a decided change for the better in the appearance of the village; the Indians themselves soon began to appreciate the benefits resulting from their enforced labors, and no trouble was found in continuing the system.

Finding a great deal of sickness existing in the village, a dispensary was established on shore under charge of the senior surgeon of the vessel, who showed great interest in the efforts to improve the condition of the Indians, and who willingly treated all cases reported to him. This, and the improved cleanliness of the village, resulted in a few months in a perceptible decrease in the mortality, especially among the Indian children.

With the change in the condition of the village there was soon noticed a change in the habits and appearance of the Indians themselves. More and better articles of clothing were bought from the traders; about the village there were seen fewer Indians whose clothing consisted of a single blanket; cooking stoves were in some cases bought and used instead of the open fires with kettles hanging over them; lamps appeared in some of the houses, and, above all, soap, to which the Alaska Indian is usually a total stranger, became an article of steady and increasing consumption.

A number of new and smaller houses were commenced, many of the Indians showing a disposition to abandon the usual communal system of living practiced all over Alaska.

Frequently a family consisting of several generations will be found living in common in a single house. The usual Indian house in Southeast Alaska is a large square structure, built of hewn timbers planted upright in the ground, with no partitions. The sleeping quarters are usually narrow, raised platforms running around three sides of the house. The fire for warming the house, cooking, and giving such light as they need at night, is built on the ground in the middle of the house, the smoke escaping through a hole in the roof left for that purpose.

Under such circumstances any improvement in the moral condition of the people is almost impossible. Those who showed a disposition to abandon this mode of living were usually the younger members of a family, who would leave the older ones in possession of the family house and set up for themselves in a smaller one as near at hand as possible. The separation of families was encouraged as far as practicable, and assistance given to those desiring separate homes : their houses were planned for them, and they were often assisted by some of the men from the Jamestown in putting them up.

At the time spoken of, the Alaska Indians were almost entirely without education of any kind, only a few faint attempts having been made in that direction in the territory. The Russian Government had caused a few Indians to be educated, with the purpose of using them as missionaries of the Greek Church among the different tribes; but no attempt had been made to educate and improve the

Indians as a race. The New York Board of Home Missions had sent out two or three missionaries to Alaska, and a teacher sent out by the same body had attempted to establish a school for Indian children at Sitka in 1879. The first teacher had resigned and been replaced by a young lady specially educated for this work, who was making a brave effort for success, but under the most discouraging circumstances, when the improvements in the Indian village were commenced. A visit to the school, established in an old building originally intended by the Russians for school purposes, afterwards used as a hospital for a time, then abandoned and fast crumbling to decay, showed a picture of earnest, faithful work that could apparently result only in failure unless a change could be brought about. The schoolroom was bleak and desolate, the furniture consisting of a table and a few rough benches, and with no means of instruction save a few elementary books not sufficient in number for the few children present. The number of children attending the school varied so greatly from day to day, and their ages were so dissimilar, that no systematic teaching seemed possible.

Interest was excited in the undertaking—the enthusiasm of the young teacher being contagious; all school books and appliances in the trading stores at Sitka were bought,—a meagre assortment at best,—and orders sent to San Francisco for a larger supply, the fund accumulated from fines imposed on the Indians for petty misdemeanors being used for this purpose. Compulsory education was decided upon and put into operation at once. A census of the village was taken and all children between the ages of five and fifteen years were enrolled and required to attend school, the number reaching nearly two hundred.

To make this system effective several practical difficulties had to be met and overcome. From the number of children living frequently in the same house there was, of course, great difficulty in distinguishing them. Even where they had any names beyond those of the families to which they belonged, the teacher was unable to use them in mustering her scholars, and the Indian heads of families would have paid little attention to an order to send their children to school unless some means had been found to enforce obedience. To meet these difficulties each house in the village was numbered and the number of children of each sex noted. Then one of the men of the ship was directed to make a number of circular tin badges equal to the number of children enrolled for school purposes. On each of these badges was stamped the number of a house, with the sex and number of each child belonging to it, commencing with the oldest. These badges were worn suspended about the neck by a cord, and became a ready means of identifying the children. A muster was held every morning and all absentees reported, noting carefully the numbers of the houses and the number of children that were absent from each house. These reports were handed to the officer in charge of the guard-house on shore after school hours, and on the following day the oldest man in each house from which absentees were reported, who was presumed to be the owner and head of the family, was arrested and brought before the commander. If no valid excuse was presented for keeping the children away from school, the head of the family was punished for disobedience of orders. The usual punishment was a fine of a blanket or restraint in the guard-house for a day. The sum realized from the sale of blankets collected as fines was applied to making improvements about the school.

Only a short time was needed to convince the Indians of the value of education, or at least of the advisability of having their children attend school regularly, and it became a common thing to see an Indian taking several children up to the school building for delivery to the teacher. In many cases the parents, after taking their children to school, would remain as interested spectators, the total number present on some days reaching nearly three hundred men, women, and children.

The success of this effort to introduce education among the Indians, and the readiness with which some of the children learned, led to the establishment at Sitka of what was intended to become a manual training school for Indian boys, under the general management of the Board of Home Missions, a portion of the same building being used. This undertaking was encouraged and assistance given by the officers and men of the Jamestown in repairing the building, fitting up a dormitory for the boys, and laying out a garden. When the school was fairly established in July, 1881, it had twenty boys, varying in age from eight or ten to twenty years, who had gladly left the Indian village to obtain the instruction given, while living in a greater degree of comfort than was possible in their former homes. The expense of maintaining this school is borne by the Board of Missions; it has since been enlarged and a department for girls added, and the latest reports are encouraging for the future.

While this school is reported to have been attended with some degree of success, it is to be regretted that the instruction given has not been of a more practical character, as was originally intended. Had it become really a school for teaching useful trades to boys and girls, far more good would have been accomplished in improving the conditions of life among the Indians than by all the teaching of catechisms and church dogmas that could ever be given them. It is to be hoped that the very liberal appropriation made by the last Congress, for the support of education in Alaska, will be utilized in the establishment of training schools for boys and young men where they may learn trades useful to them in the present condition of the territory and in its future development, and that other instruction, except in rare cases, will be confined to the English language alone. Even the ordinary branches of a common school course cannot be advantageously taught, except in very few cases, and an effort to do so will be productive of no lasting good.

Singularly enough, as investigations into the habits and customs of the Alaska Indians were pursued, a well-established system of slavery was found to exist in the territory, although over thirteen years had elapsed since the date of cession to our government. The slaves were held by a title of ownership as absolute as ever existed in any of our Southern States, the owners having even the power of putting slaves to death at their pleasure, and in times quite recent it was not an unusual occurrence for one or more slaves to be sacrificed at the burial, or rather cremation, of the former master; to accompany him in his future life.

This next required attention, and the names and ages of all persons held in slavery, with the names and families of the owners, were ascertained as far as possible. In Sitka alone nearly twenty slaves were found, varying in age from mere children to old men and women. The children were descendants of slaves, as, since the American acquisition of the territory, the Sitkas had been engaged in no active wars in which prisoners had been made, the usual custom having been to kill or make slaves of all captives not ransomed.

After informing the leading men of the village that no form of slavery could exist in any territory of the United States, the entire population of Sitka was assembled on the parade ground, and a formal order read declaring slavery abolished in the territory. It was also declared that any person attempting to exercise any right of ownership over another would be severely punished. Taking advantage of the superstitious reverence in which any written document is held by the natives of Alaska, a paper had been prepared for each of the liberated slaves and delivered in presence of the former owners. These papers gave the names and ages of the holders, and recommended them to the protection of all officers of the government. Attached to each paper was the seal of the Jamestown, with the signature of the commanding officer.

Some of the interested parties were inclined to contest this very summary deprivation of their property, but these cases were few and confined to those who had acquired slaves by purchase. A few decided words caused all to submit to the inevitable, and in no cases were any of the new freedmen molested. Many of the latter soon took advantage of their newly acquired freedom, and returned to their native villages, from which they had been taken as mere children. The same order was published among other tribes, and was, it is believed, generally obeyed, and the slaves in that portion of the territory liberated.

The Indians in all parts of Alaska are grossly superstitious, and stand in abject fear of their Shamans or witch-doctors. These Shamans are said to be selected for their peculiar office at their birth, and the cardinal requirements for fitness for it seem to be never to cut or dress the hair or allow any water to touch the person. Before final admission into the ranks of the regular practitioners, the candidate is subjected to various severe ordeals testing his endurance of hunger and exposure. They are supposed not only to have the power of discovering witches and counteracting their spells, but also of calling up their own familiar demons to annoy or punish any who offend them ; hence the great terror in which they are held by all Indians. An Alaska Indian would prefer at any time to engage a grizzly bear in single-handed fight rather than to touch one of the charms used by a Shaman in his incantations.

This superstition was discouraged as far as possible, and the witch-doctors were warned that they would be severely punished if found at any time practicing their profession. The usual course of procedure was, when an Indian was ill, to call in one of the Shamans, who, after some incantations and beating on tom-toms, would declare the sick person under the spell of some witch, and exact a fee for discovering the witch and dissolving the spell, the fee being in proportion to the wealth and importance of the patient. The fee being agreed upon, the Shaman usually selected as the witch to be denounced some helpless old crone, or member of the community without friends, who would be arrested and put in close confinement, beaten and starved until confession was made, the witch died, or the sick person recovered. In some cases the witches were burned alive.

During a short absence from Sitka, in January, 1882, a case of witch-denouncing occurred there, of which I was promptly informed on my return. Fortunately, the accused witch was rescued by some of the miners and others in Sitka, who heard of the case before he was put to death. An investigation left no doubt of the guilt of the leading Shaman of the village, and he was arrested and confined in the guard-house just as he had about completed all his preparations for leaving Sitka in a very hurried manner. All the Indians were assembled in front of the guard-house, the witch-doctor was brought out, and the case and the absurdity of his pretensions explained through an interpreter. It was announced that the Shaman's hair would be cut off close to his head, that he would be scrubbed thoroughly, to deprive him of the supernatural powers he claimed, and then be kept at work for a month, and afterwards banished from the Sitka settlement. He was first invited, however, to test his powers, in presence of the Indians, in bringing any plagues he chose on the commander and his officers and men. The sentence was carried out, to the great delight of all the Indians present, but banishment was not found necessary, as the Shaman was not proof against the ridicule to which he was subjected, and left the village of his own accord at the expiration of his confinement. This case seemed to have a good effect in breaking down the witchcraft superstition, and no more cases occurred for some time ; none, at least, that were made public.

A peculiar Indian trait is shown in the patience with which tribal feuds are kept alive for many years, with occasional outbreaks of active hostilities. In one case a sort of intermittent war had lasted for nearly eighty years between the Sitkas and the tribe living at Fort Wrangell. The origin of the war, which had caused the loss of a great many lives on both sides, was unknown to the present generation, but it was a point of honor with both tribes to carry on hostilities as occasion served until an equality of losses had been secured. According to the universal custom among the Alaska Indians, the death of any member of a tribe, if occasioned in any manner, however innocent, by another, demands the death of a member of the offending tribe of equivalent rank, or the payment of a number of blankets equal to the computed value of the man to his tribe.

The balance had, in the war mentioned, apparently inclined to one side or the other from time to time by the murder or capture of single individuals or small hunting and fishing parties, an equality never being secured, until neither tribe dared to approach the territory claimed by the other.   As this state of affairs was unfortunate for the Indians, and annoying to the white men coming into the territory as prospectors and to establish fisheries, it was decided to make an effort to put an end to it.

The tribe at Fort Wrangell was directed to send three of their leading chiefs to Sitka for a conference with the Sitka chiefs, a safe-conduct being furnished them for the purpose.   On their arrival at Sitka they were called on board the Jamestown, where they were met by the three oldest Sitka chiefs, and peace negotiations opened, presided over by the commanding officer, assisted by an interpreter. After grave deliberations on the part of the chiefs, lasting several days, during which as accurate a computation as possible of the losses on both sides was made, it was decided that the two tribes were as nearly even on the general result as they could hope to be, and a formal treaty of peace was drawn up and presented to the chiefs for signature.   All gladly acquiesced in the decision and the terms of the proposed treaty, the chiefs signing it with their totems. Copies of the treaty were exchanged and carefully preserved by each tribe, and the terms faithfully observed.   Feasting was indulged in for some days, and the Fort Wrangell Indians returned to their homes, very happy over the result of their mission.   Other tribal feuds were ended in the same manner.

In all dealings with these Indians they were found to be proud, sensitive, and generally truthful, with a high sense of justice.   They have great respect for white men of character and force, and are easily controlled when convinced that any system established for their government is just and equal for all in its operations.   A case in point occurred in the summer of 1881, soon after the military post was established at the mining camp of Juneau City.   A chief of one of the tribes living at some distance in the interior made his appearance at the post, accompanied by a number of his warriors, to ask the assistance of the officer in command in the adjustment of a claim made by some members of his tribe against one of the traders.   The claim, on investigation, was admitted by the trader to be just, and a satisfactory settlement made promptly by him.   The chief was highly delighted with the course taken, and expressed his entire conversion to the new mode of settling difficulties. He explained that it would have been quite easy for him to have obtained satisfaction in the usual way,—by killing the trader on the first opportunity that occurred,—but that having heard of the newly established system of administering justice, he had decided to give it a fair trial first, reserving to himself, however, the right to fall back upon the old custom, had justice or a hearing been denied him.

The Indians show a great desire for improvement in their surroundings and for acquiring property. Indeed, the possession of property is with them one of the chief claims to consideration. With such characteristics, the race, if properly treated, may become of great use in the future in the development of the fisheries and mines of Southeast Alaska, which are already valuable. But if they should be subjected to the course of treatment usually shown to Indians on government reservations, the result could not fail to be disastrous ; it would certainly entail great expense on the government and lead to the extinction, in a short time, of a very interesting people.

Late in the fall of 1880 some discoveries of gold-bearing quartz ledges were made at a point on the mainland, about one hundred and fifty miles from Sitka, but within the territory of the United States. The exaggerated reports of the value of these discoveries, sent to San Francisco and other points on the Pacific Coast, led immediately to a rush of prospectors and miners to the territory ; men who, in the majority of cases, had lived for years the rough life of mining camps, and were accustomed to little restraint except that imposed by the bowie-knife and revolver and the occasional administration of lynch law. A small mining town called Juneau City, after one of the original locators, was soon formed in the vicinity of the quartz ledges. While waiting for the winter snows to melt and expose the ledges sufficiently to allow prospecting, these men, who generally indulged in the most extravagant ideas of the richness of the country, wrought themselves into a high state of excitement over their prospects of wealth. The town was regularly laid out, town lots claimed and held at high figures, and conflicting claims, in some instances, soon led to threats of violence. Some of the miners, knowing the condition in which Alaska had been left so long by our Congress, without civil law or government, openly said " there is *no law* in Alaska," and proposed to exercise the law of individual might in settling cases where their claims were involved. In such a condition of affairs a slight spark might easily have produced an explosion of serious character, and it was manifestly the duty of the commander of the Jamestown to use the force at his disposal in the prevention of disturbances, and not wait for an actual outbreak before taking action.

For this purpose the miners and others were called together early in May, and after the condition of affairs had been explained, a proclamation was read and posted in a conspicuous place, putting all inhabitants, white as well as Indian, under military law for the preservation of order, the security of life and property, and the punishment of acts of violence. This proclamation, by its own terms, avoided interference with any rights to property already acquired, or to be acquired, or with any statutes in force, and was subject to the approval of the President.

Naturally, some opposition was manifested by the most turbulent portion of the small community to an innovation on mining camp procedure of so startling a nature, but the better disposed persons, who were in the majority, supported it, wisely regarding the military administration of justice as preferable to the irresponsible rule of Judge Lynch.

A reservation for government purposes was located in a position commanding the mining camp and the Indian village near at hand, and in a few days the miners saw the entirely new spectacle of a naval officer stationed among them with a force of blue-jackets and marines under his command sufficient to repress any disorder. The executive officer of the Jamestown, Lieutenant-Commander Charles H. Rockwell, was selected for this duty, and a force assigned him consisting of twenty-five officers and picked men, with a boat howitzer and a Gatling gun. Mr. Rockwell, by his cool and firm administration of affairs, under general instructions from the commanding officer, and the just settlement of many disputes referred to his arbitration, soon won the confidence of all, and the miners became rapidly accustomed to naval discipline. Affairs went on with the regularity and order of a well-disciplined man-of-war, and it was soon a subject of remark among the miners of longest experience that Juneau was the most orderly mining camp ever known on the Pacific Coast, while it enjoyed the singular good fortune of being one in which no murder was ever committed by a white man.

Subsequently, when in command of the steam sloop Wachusett, which vessel could easily reach the mining settlement at any time, seeing that the desired effect had been produced, and needing the services of the officers and men on board ship, it was decided to withdraw the force from shore, and the community was so informed. The miners at once, without exception, united in a petition to have the naval force remain, even those who had in the beginning been loudest in their opposition, saying they preferred naval administration of justice to any other.

Although, as has been said, the Jamestown could do no cruising in Southeast Alaska, her influence was not confined entirely to Sitka and the mining camp, but was extended quite widely. By means of the steam launches, with which the ship was supplied when fitted out for this special service, many other points were reached, and in July, 1881, parties of officers and men were scattered about over nearly four hundred miles of coast line from north to south, engaged in keeping order and suppressing disturbances among the Indian tribes. This duty became much easier and was attended with far less exposure to officers and men when the Wachusett replaced the Jamestown on the station in August of that year.

Southeast Alaska consists of a narrow strip of coast, ten marine leagues in width, with innumerable islands, some of large size, separated from the mainland by narrow channels, and extends from the middle of Portland Canal, in latitude 54° 40' North, to Mt. St. Elias, the initial point of the boundary line between the territory and British Columbia, which extends thence due north to the shores of the Arctic Ocean. It possesses a peculiarly equable climate, with an annual mean temperature much higher than found elsewhere in so high a latitude, due to the influence of the warm waters of the Kuro Siwo, or Black Stream of Japan, a portion of which, deflected by the Aleutian Islands, strikes the coast about Sitka. The thermometer rarely falls as low as 0° Fahr. in this portion of Alaska, and the lowest temperature recorded on board the Jamestown at Sitka in the winter of 1880–81, the thermometer being in an exposed situation, was only 19° Fahr., the mean for the three winter months being a little below 34° Fahr.

The natural conditions produce an excessive fall of snow and rain, and this, combined with the mild temperature, gives rise to a dense vegetation reaching everywhere from high-water mark to the snow limit on the mountains. The sides of the mountains, everywhere very steep, are covered with a dense growth of fir, spruce, alder, and cedar, with thick undergrowth that makes the country as impassable almost as a tropical jungle.

The timber is generally low in growth and not of much value, but in some of the valleys, especially on the larger islands, good timber is found in considerable quantities, the fir and a species of yellow cedar being the most valuable.

The fur trade of this portion of the territory is of no great value at present, the most esteemed furs being the sea otter and the silver-gray fox, but only a few skins are taken annually.

The fisheries are at present the most valuable resource of South-east Alaska. Salmon of fine quality is found in the greatest abundance in all the streams, and has already become an important article of export, there being several large establishments where the fish is packed for the San Francisco market and for foreign trade. Halibut, cod, and herring are also very abundant, while all the small streams are filled with delicious trout.

No portion of Alaska will ever become useful for purposes of agriculture, and very little of the surface can be utilized for pasturage. From the shortness of the summer no grain will ripen, and the luxuriant grasses that grow so rapidly in the long mid-summer days cannot be properly cured for hay. A few potatoes of fair quality are produced at Sitka and at places to the southward, and in the gardens about Sitka and the other settlements there may be seen growing fine beets, turnips, cabbages, and cauliflowers, but other vegetables do not mature. No fruit is found growing anywhere in Alaska except a few varieties of wild berries.

Of the mineral wealth of Alaska very little is yet known, no sufficient explorations having been undertaken. Some of the quartz ledges about Juneau have proved to be valuable, and are now being worked successfully, and a considerable amount of gold has been obtained from placer claims. Of most of the discoveries of 1880, upon which such bright hopes were founded, but little more is known than at the time the original locations were made. The failure of Congress to provide any effective form of civil government for the territory, with the consequent doubt concerning the acquisition of valid titles, has deterred capitalists from engaging largely in mining operations, except in one case, where the venture has proved largely remunerative. The infrequence of mail communications, the distance from points of supply of mining machinery and appliances, and the great cost of transportation, have also contributed in a great measure to this result.

The disadvantages under which the territory has labored for so long a time are now being remedied to a great extent, and as it becomes known, and its resources are developed, there is every reason to anticipate that the wisdom shown in the acquisition of Alaska will be fully demonstrated.*

*A more extended account of the climate, resources, and population of Alaska may be found in the valuable report of Mr. Ivan Petrof, published in Report of the 10th Census of the United States.

U. S. NAVAL INSTITUTE, ANNAPOLIS, MD.

POWDER IN GUNS.

[A continuation of " Velocities and Pressures in Guns," Vol. XIV, No. 2.]

By Lieutenant J. H. Glennon, U. S. N.

In a paper entitled " Velocities and Pressures in Guns," Vol. XIV, No. 2, the action of powder in guns was treated mathematically, with the tacit assumption of a certain premise which, though not new, being in fact the very foundation of Messrs. Noble and Abel's treatment of the work done by powder in guns, at first appears doubtful. The question is whether or not a progressive powder, if all burnt in a gun, and the gases produced have gained mechanical and thermal equilibrium by the time the projectile reaches the muzzle, will give as great a muzzle velocity, weight for weight, as a powder all of which is burnt before the projectile begins to move. This point can be elucidated in several ways, assuming, as in that paper, no radiation or conduction of heat to the gun or between the portions of the gas. In other words, the portions of gas are assumed to expand adiabatically.

One comment has been made and may be noticed here, namely : if, immediately around the surface of the burning powder grains, gas at high temperature and pressure exists, and the rest of the gas is, or may be considered, in thermal and mechanical equilibrium at a lower temperature and pressure, then the lower pressure, being that of the gas directly in rear of the projectile, would be the force accelerating it. Looking into the subject, however, we see that if only the low pressure exists at the base of the projectile, the high pressure must be somewhere else, as at the face of the breech-block ; and if the low pressure may be considered by itself as accelerating the projectile in one direction, so may the high as accelerating the gun in the opposite direction. Newton's law on the equality of action and reaction would then not apply, remembering, of course, that the mean forward motion

of the powder gas (a motion partly action and partly reaction) is
limited by the relative motion of the projectile and gun.

To proceed, suppose that we have a quantity of gas occupying a
non-conducting cylinder $OMCD$, Fig. 1, that the pressure of this
gas is represented in Fig. 2 by LB and the volume by OD. Inter-
pose a non-conducting piston AB between the two halves of the
volume of gas.

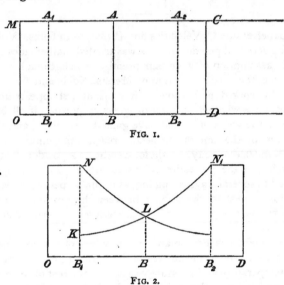

FIG. 1.

FIG. 2.

Now suppose that we push the piston AB to the position A_1B_1
(Fig. 1). The pressure of the gas to the left of the piston will increase
from LB (Fig. 2) to NB_1, following the pressure-volume law $pv^n =$
$p_0v_0^n$, where n equals $\dfrac{c'}{c}$, the ratio of the specific heats of the gas, or
approximately 1.4. The work done in compressing the left half will
then be represented (in proper units) by an area NB_1BL (supposing no friction). Part of this work, LBB_1K, will be done by the
other half of the gas. The external work necessary to push AB to
the position A_1B_1 will therefore be represented by NKL. If now
the piston at A_1B_1 is released it will start towards AB, and when it
reaches that position it will have a *vis viva* represented by NKL,
will pass the position AB, and will then encounter retarding pres-

sures equal to those formerly accelerating it. It will be brought to rest at the position A_2B_2, determined by $BB_1 = B'B_2$, will then return to A_1B_1, then back again, and so on, as long as there is no external work done.

Now connect the piston by means of a piston-rod, or otherwise, with the outside, so that it may do external work. It will come to rest when it has done the external work represented by NKL, and not until then. It will then be in its original position AB, and the pressure on each side of it will be the same as originally, so that if it were removed the gas would be in thermal and mechanical equilibrium.

Evidently, while the piston is in motion from B_1 to B_2, the cylinder head CD will be subject in succession to all the pressures represented by the ordinates of the curve KLN_1, and not alone to the pressure B_1K, with which the pressure to the right of the piston begins.

If now the mass of the piston is made smaller, its velocity at B will be increased, its mechanical energy or *vis viva* still remaining KLN. In the limiting case, where the mass of the piston is infinitesimal, its velocity will be infinite, the mean velocity for the travel of the piston will be infinite, and the distance B_1B_2 will be traversed in an infinitesimal time. Suppose that we take this limiting mass for the piston, and suppose, simultaneously with its release, that we release the cylinder head CD, allowing it free motion along OD, the mass of CD being supposed appreciable. Evidently CD will start under an accelerating force KB_1, and in an infinitesimal time will be subject to all the pressures KB_1 to N_1B_2. The piston AB will make an infinite number of vibrations in a short finite time, and will finally take up a mean position between O and D. The piston AB might then be removed, as, its velocity being finite, it will have no finite *vis viva*, the gas will be in equilibrium, and all the energy represented by NKL will have been transmitted to CD. The piston of infinitesimal mass of course does not exist; more properly a piston should be supposed which by its mass and motion would represent the mass and motion of the gas itself. In this case equilibrium would be gained, though not so quickly as with the infinitesimal piston. A little explanation possibly will be necessary to connect the assumed case with the case of fired gunpowder.

A charge of gunpowder is not all ignited at once. The particles of gas have a certain temperature and definite tension at the instant of combustion. Suppose a small quantity of the powder ignited instantaneously, and suppose for the sake of simplicity, in lieu of other gas, that gas from just the same small weight of powder (expanded against pressure) already fills the powder chamber. The pressure of the last may be represented above by B_1K, and the pressure of the newly burnt powder by B_1N. The piston AB represents the outside surface of the last burnt powder gas, this gas being supposed to expand without loss of heat by radiation or conduction. CD represents the base of the projectile supposed free to move when the powder is burned. When the two parts of the powder gas are in equilibrium, the energy KLN will have been transferred from the last burned gas to the projectile. This will be in excess of the work that would be done by the same expansion in total volume, assuming both portions originally in equilibrium at the pressure LB. It will be the same with each successive portion, so that when all arrive at equilibrium the pressure will be the same as if the combustion of all had been simultaneous.

It is necessary to remember that the pressure of a gas is exerted equally in all directions, and that it does just the same quantity of work in expanding from one volume to another in all directions as in expanding in only one. The newly formed gas at the surface of the grain expands in all directions, and the vibrating piston AB ordinarily corresponds exactly with a vibrating spherical (oval or ring) surface around the grain, the gas inside of which finally gains equilibrium with the remainder of the gas in the gun.

In deducing the general equation (" Velocities and Pressures in Guns ") of motion in the bore of a gun, a mean temperature T is assumed to exist in the gases, such that the pressure may be represented by $P = Rx\frac{T}{V}$. Now, while powder is burning in a gun no such equilibrium does exist, though the pressure P is fixed and definite ; given P, we can very readily find a mean temperature for the gas if we solve this equation.

Given the work already done, we can also find a value for T. The question is, *when no equilibrium really exists*, are both answers the same, or should they be? To answer this let us look at Fig. 2. The pressure KB_1 would furnish the temperature of the expanded gas, and by subtracting this from the temperature of combustion of powder, the work already done, KB_1 would equal $Rx\frac{T}{V}$, where x is the weight of the expanded gas (as in " Velocities and Pressures in Guns "). Is the pressure KB_1 that this T determines the accelerating pressure? No, unless all the powder gas is in equilibrium; as before stated, the accelerating pressure is a compound of all the ordinates from B_2N_1 to B_1K. If, however, no powder is burning, all the portions of gas being supposed at the same temperature and therefore the same pressure, it is. In other words, the general equation is true at any point if w, the weight of powder burned and in equilibrium, becomes constant at that point.

We assume, in mathematically treating this subject, that each elementary portion of the powder gas gains equilibrium in the element of time after burning. This is simply one way of saying that the powder gas gains equilibrium just as fast as it is evolved.

Imagine, to look at the subject in another way, the bore of a gun subdivided into a number of smaller gun-bores of equal size, and suppose each similarly loaded with its proportion of powder, the same for each of the smaller calibers. Now, imagine that these bores are so small that we can consider in each the combustion of the small quantity of powder contained as instantaneous. Further, suppose the bullets are of such different sizes that, though the small charges are ignited at different times, the bullets will all reach the muzzle of their respective barrels at the same time. Ignite the charges in succession, as determined by the size of the bullets. The pressure in each barrel at the muzzle will be independent of the time of arriving there, following the law $pv^n = C$, where C is some constant and will evidently be the same in all the barrels. The work done in each barrel will be precisely the same; and the total work for all the barrels will be precisely the same as if all the powder had burnt instantaneously in the original gun and then expanded to the muzzle. If now the size of the bullets be changed, the same equality will hold, provided that each bullet reaches the muzzle at the same time. There would then be an equilibrium of pressure throughout the gas if at the muzzle all the barrels were suddenly merged in one.

If, however, the bullets had not been properly proportioned there would be no such equilibrium, some of the projectiles would arrive late at the muzzle, and when merged the tensions throughout the gas would be variable. It is thus plain, if equilibrium is gained, that only one fixed quantity of work will be done for a fixed expansion of the same mass of powder gas (of fixed quality). Some light is thrown on the subject by Professor J. Clerk Maxwell in his "Theory of Heat" (edition of 1883), page 189, where he says: "If the system under consideration consists of a number of bodies at different pressures and temperatures, contained within a vessel from which neither matter nor heat can escape, then the amount of energy converted into work will be greatest when the system is reduced to thermal and mechanical equilibrium by the following process.

1st. Let each of the bodies be brought to the same temperature by expansion or compression, without communication of heat.

2d. The bodies being now at the same temperature, let those which exert the greatest pressure be allowed to expand and to compress those which exert less pressure, till the pressures of all the bodies in the vessel are equal, the process being conducted so slowly that the temperatures of all the bodies remain sensibly equal to each other throughout the process.

During the first part of this process, in which there is no communication of heat between the bodies, the entropy of each body remains constant. During the second part the bodies are all at the same temperature, and therefore the communication of heat from one body to another diminishes the entropy of the one body as much as it increases that of the other, so that the sum of the entropy remains constant. Hence the total entropy of the system remains the same from the beginning to the end of the process. The work done against mechanical resistance during the establishment of thermal and mechanical equilibrium is greater when the process is conducted in this way, than when conduction of heat is allowed to take place between bodies at sensibly different temperatures."

The case of powder in guns as we are treating it is precisely similar to the case cited, with the exception that the 2d expansion does not take place, as the 1st will bring the portions of gas (treated as the above-mentioned bodies) to the same temperature and same pressure at the same time. But the entropy of the powder gas will determine its temperature at the muzzle, and therefore the work done; in short, it would determine the isentropic curve, according to which the gases would *finally* expand in the case of either instantaneous or progressive combustion.

M. Sarrau is probably correct in his statement that for any particular gun, charge, projectile, shape of powder grain, and material of powder, there is a size of grain that gives a maximum velocity, or that does in the gun the maximum amount of useful work. It is impossible, however, to discuss this maximum if we regard only primary causes. According to the most favorable treatment, the progressive powder can at best only equal the instantaneous. The maximum powder then cannot be much better than any other powder completely burned and equalized in a gun. If the loss of velocity due to secondary causes is less with the progressive powder than with the instantaneous, we might expect the most progressive powder which can be completely burned in a gun to bring this loss to a minimum; in other words, the powder, all of which is completely burnt just as the projectile leaves the muzzle, cannot be far removed from the maximum powder. If then we make $\frac{l}{l_\tau} = KY_1 = 1$ (page 403, Proceedings U. S. Naval Institute, Vol. XIV, No. 2), and solve for r, we will have very nearly the time of burning of the grain of maximum powder in open air. All grains smaller than this will be of quick powders, all grains larger, slow powders, the other conditions of loading remaining the same.

M. Sarrau's method of finding the maximum powder by means of the first derivative of an empirical value of v (formula 15), which is true for slow powders only, is of course inadmissible. Generally now in guns maximum powders cannot be used, as the pressures would be too high.

Messrs. Noble and Abel really treat only of quick powders in guns. In finding the velocity due to powders of the same material, they use for any gun a certain factor of effect, which they multiply into the theoretical velocity (according to their method) due to the weight of the charge.

They tacitly assume as a fundamental principle, without question or attempt at proof, that the muzzle velocity of a certain projectile will be the same in the same gun if the weight of charge and the material are the same, no matter what the form or size of the grains may be. This of course can be true only when all of the charge is burnt in the gun.

M. Sarrau in his later works really treats, on the other hand, of slow powders only. From certain premises he obtains an equation of motion, equation 1 (p. 318, Proceedings Naval Institute, Vol. XIV, No. 2), and for w substitutes a function of the travel of the shot. When, however, w becomes equal to W, or when all the powder is burnt in the gun, it is plain that w ceases to be a function of the travel of the shot, it being then a constant. M. Sarrau's treatment then will not hold for quick powders or those which are entirely burned in the gun. In " Velocities and Pressures in Guns " it is readily seen that M. Sarrau's velocity formula for slow powders is deduced and certain numerical functions calculated, identical in value with corresponding ones of M. Sarrau, starting with Messrs. Noble and Abel's fundamental principle for velocities with quick powders. The point of difference of these authorities is that each practically omits treatment of the half of the subject treated by the other, and that the value of the exponent n is assumed by M. Sarrau as 1.4, and by Messrs. Noble and Abel as slightly greater than 1.

Similar guns are those of which the bores are, or may be considered, similar right cylinders. They are similarly loaded when the volumes of powder chambers and the weights of the charges and projectiles vary as the cubes of the calibers, and the powder grains are similar in figure, their corresponding linear dimensions being proportional to the calibers. It is readily seen that "*in similar guns similarly loaded, the velocities and pressures corresponding to distances passed over proportional to the caliber are equal.*"

Taking formula (4), (page 400, Vol. XIV, No. 2, Proceedings Naval Institute), we see that the quantity in parenthesis is abstract, and the same for the same travel in two similar guns similarly loaded; it is readily seen that the term outside the parenthesis reduces to the same quantity in the two cases, when w equals W the full weight of charge; that is, when all the powder is burnt, as with quick powders.

For slow powders $w = WaKY_1(1 - \lambda KY_1 + \mu K^2 Y_1^2)$ (page 403, "Velocities and Pressures in Guns"); w for the two guns is *proportional* (in place of equal) to W, and the principle is proved for v as before. The proof for pressures is similar.

In the formula $pv^n = $ a constant, which represents the instantaneous powder process, I will state that I do not consider $n = 1.4$ as final, or necessarily the best value; n is greater than 1 and probably not greater than 1.4. Tables could just as readily be calculated for 1.2 or any other value as for 1.4. This latter is simply true for the perfect gas. Nor is it certain absolutely that the velocity of combustion varies as the square root of the pressure (static). Any other power of the static pressure could be followed out just as readily, however, if further experiment should indicate another value. These values, however, lead directly to the best (probably) existing empirical formula for velocity, and therefore cannot be very far from the truth.

It then appears that the accelerating pressure when powder is burning in a gun is composed of two quantities, one of which we may call the natural pressure, a static pressure corresponding, to take a familiar case, somewhat with the barometric pressure of the atmosphere; the other quantity depends on the rate of evolution of the powder gas, and corresponds almost exactly with the pressure on a surface due to a wind. Suppose a very low barometric pressure forward of a shell, a high barometric pressure in rear with a strong wind blowing directly into the chamber of the gun, and the case is not dissimilar to what really must take place when a progressive powder is burnt in a gun. When the powder is all burnt, to carry on the similarity, the wind will almost instantly die out and the gas will exert a pressure due to its weight, volume and temperature alone, a purely static pressure.

The quantity depending on the rate of evolution of the powder gas plays a still more important part with explosive compounds than with powder. If a finite quantity of explosive could be burned and equalized (as regards temperature and pressure) in an infinitesmal time, the pressure would be infinite. This is not realized, but is approached, when explosives are detonated. It undoubtedly partly, if not entirely, accounts for the great differences observed between the results of primary and secondary explosions of the same explosives.

It must be remembered that there are a number of different ways in which the subject of powder in guns may be treated, and in choosing any particular way the reasons that lead to that choice should be given. No theory or practice can make a progressive powder do more work than an instantaneous powder, though experiment seems to show that the former may do more of *useful* work. The theory chosen makes them equal as to work, and the experimental method of determining coefficients (by means of velocities in two guns) assumes them equal as regards *useful* work. In this latter the method is slightly in error. A factor of effect may be involved in the above-mentioned coefficients (A and B of formula (15) or M and N of (20), "Velocities and Pressures in Guns"). If this factor is found to vary slightly with the size of guns, the coefficients for any powder for a certain gun should be determined from guns of approximately the same size. For fairly approximate values of the coefficients to cover a number of guns of different sizes, guns of nearly the extreme sizes should be chosen. Dissimilar guns or dissimilar loadings should always be chosen, in order to make the ratios of the two terms of the formula as different as possible for the different guns, thus permitting a more accurate solution where the given quantities are at best only approximate.

Many powders now-a-days are not uniformly dense, being most dense on the surface and correspondingly slow burning, and least dense in the interior. If the velocity of combustion of powder varied inversely as the density, the formulas for velocity and pressure as deduced would apply equally well to these powders, provided that layers of uniform density were regularly arranged in the powder grains. This law, however, does not hold, and this fact may be expected to cause some error in the use of the formulas of "Velocities and Pressures in Guns."

It may be remarked that with slow-burning powders, such as is German cocoa powder when used in the 6-inch B. L. R., there is an interior layer in the grain that might as well be non-explosive, so far as giving velocity to the projectile is concerned.

A part of the work done by powder is done on the gun itself. In fact, the gun is only a projectile of certain mass (or virtual mass), and a pressure and velocity curve could readily be constructed for the travel of the gun while the projectile is in the gun, just as for the travel of the projectile during the same time. The work done on the gun in this way is of course small relatively, and taking account of it would generally be a useless refinement.

It may be stated, in closing, that no method dealing with pressures in guns as purely *static* can account for the velocities obtained. The pressures are *partly static*. The resistance to the projectile in air is quite similar to the pressure in the gun, and, as well as the resistance to a ship moving through the water, is very largely *dynamic*. The greater the velocity with which the explosive gas is evolved, the greater the pressure; just as in the air, the greater the velocity of the projectile (or the more quickly the air ahead is compressed and that in rear expanded), the greater is the resistance.

U. S. NAVAL INSTITUTE, NEWPORT BRANCH,

JANUARY, 1890.

NOTES ON THE LITERATURE OF EXPLOSIVES.*

BY CHARLES E. MUNROE.

No. XXII.

United States Letters Patent No. 409943, August 27, 1889, have been granted Stephen H. Emmens for " Gun and Projectile for High Explosives," in which he claims to have invented " a new and useful improvement in apparatus for utilizing high explosives in warfare, (No. 2)." This invention is additional to his improvement in apparatus for utilizing high explosives in warfare,† set forth in his application for U. S. Letters Patent, filed January 27, 1888, serial No. 262172.

The present apparatus is especially designed and adapted for use by infantry and in small boats ; and the invention consists in a novel combination of parts, whereby he claims to be enabled to lighten the apparatus to any required extent, and to support a relatively short powder-tube for the propelling charge within the guide bore of the torpedo by means of a tube in line therewith, that incloses the firing device, and is in turn supported by a simple stock in the form of a stake or the like. A drawing accompanies his specification representing an elevation of a torpedo-gun and a bird-torpedo, illustrating the invention.

The " firing mechanism," as he terms it, especially distinguishing the present weapon, comprises a wooden stock fitting into and sup-

---

*As it is proposed to continue these Notes from time to time, authors, publishers, and manufacturers will do the writer a favor by sending him copies of their papers, publications, or trade circulars. *Address Torpedo Station, Newport, R. I.*

† Proc. Nav. Inst. **15**, 289 ; 1889.

porting a metallic tube, and axially perforated and slotted at its front end to accommodate within said tube a rod, which is bent at right angles at its rear end to form a trigger that projects outward through a bayonet-joint slot in the stock and tube. The front end of this rod is fixed in a piston which carries a firing-pin, and between said piston and the front end of the stock is a spiral spring; hence when the trigger is pulled back and turned into the holding-notch of the bayonet-joint slot, said spring is compressed, and when the trigger is released the spring urges forward the piston, with its firing-pin, to explode the propelling charge. This is contained within a short "gun-cartridge shell" or powder-tube, which fits into the front end of said tube and is coupled thereto by a pair of bayonet-joints. Preferably the "powder-tube," as it is termed, is provided with a primer-recess in its breech end, and with an axial ignition tube extending forward from said primer-recess to the front of the propelling charge, which may be of any suitable explosive. The ignition-tube is filled with gunpowder, and the recess is provided with a suitable percussion primer. When the latter is exploded by the firing-pin a sheet of flame is produced within the tube, which ignites the propelling charge at its front end, so as to insure its perfect combustion and an effective discharge of the weapon.

The bird-torpedo comprises a tube fitting closely over the metallic tube and powder tube above mentioned, and plugged at its forward end by the screw-stem of a conoidal torpedo-head. In an external annular charge-space immediately behind the head and around said tube, cartridges of emmensite, or other high explosive, are arranged, side by side to form the high explosive charge. By using cartridges of different lengths the size of the charge may be varied to any required extent. These cartridges are held in position by a collar and a cylindrical jacket, and they are fired by a time-fuze. The rear end of the torpedo-tube is provided with three equidistant wings. An annular screen is fitted to the metallic tube of the torpedo-gun immediately in front of the trigger to protect the hand of the person discharging the weapon from any escape of heated gases.

The drawing shows the weapon planted in the ground for firing, but any other mode of mounting may be employed, e. g. the stock may be clamped in a holding-tube on an ordinary swivel stand or gunwale attachment.

In conclusion he claims as his invention: In combination with a thimble-shaped projectile having an axial guide tube open at its rear end and surrounded by an annular charge-space for high explosives, a short powder tube and a tube inclosing a firing device, coupled together end to end and fitted to the interior of said guide tube, and a stock fitted at its front end to the rear end of said firing device tube, substantially as described.

The *Ann. des Mines* 5, 197–376; 1888, contains a very valuable " Report of the French Commission for Explosive Substances on the Use of Explosives in Presence of Fire-damp," in which the authors state that having satisfied themselves, both from the results of previous investigations and in some preliminary experiments, that mixtures of coal-dust and air were not as dangerous as mixtures of marsh-gas and air, they have restricted themselves to the latter and more dangerous mixture. The experiments were conducted in a boiler which, by suitable arrangement, could be completely filled with a gaseous mixture of known composition, and the explosive under examination was suspended in the middle of the boiler and fired by the aid of electricity, the effect on the gaseous mixture being ascertained by the alteration in pressure observed on a pressure-gauge.

The marsh gas was prepared from sodium acetate and soda-lime and stored in a gasometer; it contained per cent 10.8 of air, 7.9 absorbed by bromine, no carbonic anhydride, and 81.3 of methane (by difference). The mixture introduced into the boiler contained generally about 10.3 per cent of methane; this approximates to the most explosive mixture, whilst a 6 per cent mixture verges on the limits of non-inflammability.

The first series of experiments were tried with unconfined explosives freely suspended in the gaseous mixture, this being the most dangerous condition; such experiments are useful to ascertain the actual safety of an explosive.

The explosives investigated were: *Ordinary powder; dynamite* containing 75 per cent of nitroglycerol and 25 per cent of silica; *Paulilles ammoniacal dynamite,* a mixture of varying proportions of nitroglycerol, ammonium nitrate, and of a carbonaceous substance destined to utilize the excess of oxygen produced by the detonation of the nitroglycerol and ammonium nitrate; *gun-cottons* of the general formula $C_{24}H_{40-n}N_nO_{20+2n}$, in which the maximum value of n should be 12, but which in practice does not exceed 11. The following table represents the number of cubic centimetres of nitric

oxide evolved from 1 gram of the various gun-cottons when examined
by Schloesing's method ; the value of n indicates the composition :

			$n =$	cc. N_2O_2.
Endecanitric cotton, .	. . . . .		11	214
Decanitric	"	. . . . .	10	203
Enneanitric	"	. . . . .	9	190
Octonitric	"	. . . . .	8	178
Heptanitric	"	. . . . .	7	162
Hexanitric	"	. . . . .	6	146
Pentanitric	"	. . . . .	5	128
Tetranitric	"	. . . . .	4	108

Military gun-cotton (205 cc. N_2O_2) corresponds approximately to
the decanitric; *mining gun-cotton* (193 cc. N_2O_2) to the enneanitric;
and a third gun-cotton (173 cc. N_2O_2) investigated, to the octonitric.
Tests were also made with *blasting gelatine* consisting of gun-cotton
and nitroglycerol, and with *dynamite gelatine* consisting of blasting
gelatine mixed with some such substances as a mixture of potassium
or sodium nitrate with charcoal or sawdust. All these when detonated completely under the above conditions, generally ignited the
gaseous mixture, and must therefore be regarded as dangerous for
use in mines where fire-damp may be expected. The last two compositions did not detonate completely in cold weather. Dynamite
and also gunpowder even ignited the gaseous mixture when detonated
in a Settle's water-cartridge.* *Hellhoffite*, a mixture of dinitrobenzene with 1½ or 2½ times its weight of nitric acid (made on the
spot where employed, but not considered sufficiently convenient
for use in mines); *Favier's explosive*, 9 parts of mononitronaphthalene with 91 of ammonium nitrate; and *Bellite*, about 20 parts of
dinitrobenzene and 80 of ammonium nitrate; either did not explode
at all, as was the case with the first two when freely suspended
in air, or exploded incompletely, as in the case of bellite. Under
exceptional circumstances, finely pulverized Favier's explosive was
made to explode and then ignited the gaseous mixture; but the
detonating mixture used with this explosive is, however, of itself
dangerous. Two powders consisting of gun-cotton and barium
nitrate ignited the gaseous mixture, whilst a powder consisting of 60
parts of gun-cotton (175 cc. to 185 cc. N_2O_2), 30 parts of barium
nitrate, 6 parts of saltpetre, 3 parts of gelose, and 1 part of paraffine,

* Proc. Nav. Inst. 14, 766 ; 1888.

only inflamed the gaseous mixture twice out of 14 times. Various experimental mixtures were now investigated. Dynamite with an equal weight of either crystalline sodium carbonate, sodium sulphate, ammonia alum or ammonium chloride, did not ignite the gaseous mixture, but with larger proportions of dynamite the ignition of the fire-damp generally ensued. Some interesting experiments were made with mixtures of coal-dust and dynamite. Three coals, containing respectively 40, 27.86 and 23.60 per cent of volatile matter, mixed with equal weights of dynamite, did not set fire to the marsh gas when exploded, even a mixture containing 4 of dynamite to 1 of the first coal in one experiment did not ignite it, but a mixture of 2 parts of dynamite with 1 part of a lignite (braunkohle) containing 62.4 per cent of volatile matter, ignited the inflammable gaseous mixture. Mixtures of 20 of dynamite with 80 of ammonium nitrate detonate freely in the air without causing ignition of the marsh gas ; in this case both substances explode, but the temperature of the explosion of ammonium nitrate is lower than that of the dynamite, and therefore moderates the activity of the latter in the mixture. Mixtures of gun-cotton and ammonium chloride require so much of the latter to make them safe that they ultimately cease to explode at all. Mixtures of gun-cotton and ammonium nitrate become safe when the former, giving 173 cc. of N_2O_2, does not constitute more than 20 per cent of the mixture. From these experiments it was concluded that simple explosives susceptible of igniting fire-damp will always fire gaseous inflammable mixtures, but that it is possible to produce binary explosives in which the temperature of detonation is sufficiently reduced to be safe.

It is pointed out that, although the temperature of ignition of marsh gas is 650° C., the ignition is greatly retarded by admixture with another gas, such as air or even oxygen ; the retardation amounting to 12 seconds even if the temperature of the whole volume of gas is at 650°, whilst at lower temperatures it is still longer. It is for this reason that the immense temperature developed by an explosion does not ignite fire-damp, for this temperature is reduced with such great rapidity, some thousandths of a second, that there is not time to ignite mixtures of fire-damp unless the initial temperature is sufficiently great, according to the observations of the present investigators about 2200° or over.

It is interesting to note that explosions not igniting fire-damp will ignite mixtures of coal-gas and air, these having a lower ignition point ; and explosions igniting neither marsh-gas and air, nor coal-gas and air, may ignite a mixture of hydrogen and air. The heat generated by explosives is next studied, and methods and formulae are given for its calculation.* Calorimetric experiments with various explosives freely suspended in the air yielded results, some of which are tabulated below.

	Weight of Cartridge, Grms.	Fulminate in Detonator. Grms.	Increase in Pressure after Explosion, Metres of Water.	Heat in Cals. for 100 grms. of Explosives calculated.	
				From the Increased Pressure.	From Theoretical Data.
Dynamite.................	50	1.5	0.83	97.0	97.2
Military gun-cotton (205 cc. of N₂O₂)	50	"	0.72	88.0	104.0
Mining gun-cotton (193 cc. of N₂O₂)	50	"	1.48	172.0	103.0
Blasting gelatine (Paulilles),	40	"	0.56	80.2	153.5
" "	50	"	0.69	80.6	"
Bellite	50	"	0.28	32.7	100.2†
" ......'.............	50	3.0	0.34	39.8	"
Pyroxylin powder (Moulin-Blanc)................	50	1.5	0.60	70.0	82 dissoc. of BaCO₃.
" "	65	"	0.61	54.7	" "
58 gun-cotton + 42 barium nitrate	50	"	0.70	82.0	81 dissoc. of BaCO₃.
20 gun-cotton (173 cc. of N₂O₂) + 80 ammonium nitrate................	50	"	0.15	17.5	80
25 " + 75 "	50	3.0	0.41	48.0	91
35 " + 65 "	50	1.5	0.43	50.0	102
60 " + 40 "	30	"	0.35	68.1	96
80 " + 20 "	30	"	0.63	123.0	82
90 " + 10 "	30	"	0.94	183.0	76
20 dynamite + 80 ammonium nitrate..........	50	"	0.225	26.4	50
67 dynamite + 33 ammonium alum............	50	"	0.445	52.0	53‡
50 dynamite + 50 ammonium chloride	50	"	0.42	49.2	10‖
60 " + 40 "	50	"	0.44	51.5	35
67 " + 33 "	50	"	0.53	62.0	49

* Proc. Nav. Inst. 15, 306 ; 1889.

† Calculated for the proportions $175C_6H_4(NO_2)_2 + 825NH_4NO_3$.

‡ Allowing for complete combustion of the dynamite and simple dehydration of the alum.

‖ Complete combustion of dynamite, and decomposition of ammonium chloride into $HCl + N + H_3$.

Wherever the heats found coincide with the theoretical, complete detonation has ensued, and therefore these results confirm and explain the behavior of the various explosives already noticed. In explanation of the apparent anomaly of the lower nitrated gun-cottons giving higher temperatures than the more strongly nitrated ones, it is shown that the proportion of combustible gas in the gaseous mixture produced by the detonation increases as the proportion of nitrogen in the cotton diminishes; therefore the lower nitrated cottons should yield a more inflammable mixture which, by its combustion, would produce a great increase in temperature, and might even account for those temperatures which exceed the theoretical in the above table.

A similar behavior is observed with mixtures of gun-cotton and barium nitrate. A gun-cotton indicating by titration 180 cc. of N_2O_2 always ignites marsh gas, because the temperature of detonation, about 2000°, is increased by the combustion of the inflammable gas produced by the detonation; whereas the pyroxylin powder containing 60 per cent of this gun-cotton, 30 per cent of barium nitrate, and 6 per cent of potassium nitrate ignites marsh gas only exceptionally, owing, it is suggested, to the oxygen supplied to the mixture in the form of nitrates being sufficient to burn enough material so as to dilute the gaseous mixture produced by the detonation and render it non-inflammable; but by adding more barium nitrate, so as to provide enough oxygen to consume all the gas and admitting the subsequent dissociation of the carbonate, the temperature of detonation is raised to 2550°, which is sufficiently high to ignite fire-damp. With regard to force it is shown that the same amount of power may be produced at a lower price by the use of larger quantities of dynamite and ammonium nitrate than by employing smaller quantities of the more expensive dynamite.

Irregularities in explosives, such, for instance, as incomplete detonation, which produces indefinite reactions, are a source of danger, and are more common in mixtures than in simple explosives. These reactions vary under different circumstances, such as the amount of fulminate used for detonation, the temperature of the charge at the time of explosion, etc., therefore an explosive should not be considered safe if, amongst its several modes of detonation in the open air, there be one susceptible of igniting marsh gas. The most suitable substance for reducing the temperature of detonation of an explosive is ammonium nitrate, as other substances, such as sodium carbonate, or sulphate, or alum, or ammonium chloride, only decompose partially and their effect is uncertain. Mixtures worthy of further consideration are: the pyroxylin powder; gun-cotton (173 cc. N_2O_2) with 80 per cent of ammonium nitrate, giving a force of 7550 grams and a detonating temperature of 1920°; dynamite, with 80 per cent of ammonium nitrate, $f = 6260$ grams, $t = 1500°$; bellite, $f = 2993$ grams, $t = 2186°$; Favier's explosive, $t = 2120°$. The last two require to be further experimented with. Ammonium nitrate reduces regularly, as the proportions of it in the mixture increases, both the temperature and power of dynamite and highly nitrated gun-cotton; but with octonitric gun-cotton these reductions do not set in until 70 per cent of nitrate is present. Curves illustrating this behavior are given in the report.

Experiments were now made with the more dangerous explosives confined in tubes of lead or tin, closed below and open at the top. The explosive was placed and rammed in the tube, on a plug of marl or sand 5 to 6 centimetres thick. It was then tamped with clay or sand, or, in some cases, coal-dust to the thickness of 10 to 12 centimetres. Tubes of various thicknesses were used and exploded, as in the previous experiments, in the boiler. The portion of the tubes surrounding the explosive was blown to powder, whilst the upper and lower portions were generally found nearly intact at the bottom of the boiler. The following table of results illustrates the effect of the thickness and quality of the tube on the temperature of explosion.

| | Description of Tube. | | | Effect on Mixture of Marsh-Gas and Air. I = Ignition. N = No ignition. |
| | Diameter. | | Material. | |
Explosive.	Internal. mm.	External. mm.		
Dynamite, ...	27	40	Lead	N.
"	30	40	"	I.
"	30	35	"	I.
"	25	40	Tin	N.
Mining gun-cotton, . .	32	42	Lead	N.
" " . .	30	35	"	N.
Blasting gelatine, . . .	25	40	Tin	I.
Dynamite gelatine, . . .	27	40	Lead	N.
Ammoniacal dynamite, .	25	40	Tin	N.

Other conditions being the same, the safety of the explosive increases with the thickness of the tube, owing, it is assumed, to the fact that much of the heat of detonation is converted into mechanical force, and so the temperature is reduced below the dangerous limit. Dynamite loses more than one-third of its initial heat of detonation in shattering the tube and distributing the debris. To ascertain the effect of the density of a charge, the explosive was enclosed in a thin glass tube and fixed by wads between the clay plug and tamping in the metal tube so as to leave an annular space between the metal sides of the tube and the exterior of the glass tube. Under these circumstances it was observed that the work done on the tube was greatly diminished, owing, it is suggested, to the expansion of gases within the tube, and instead of one-third, only one-fifth of the initial heat of the detonation of the dynamite was absorbed. The same result is observed with loosely packed explosives; for example, dynamite simply poured into a 32–42 mm. lead tube and tamped, ignited fire-damp, whereas under the same conditions with the dynamite rammed in, no ignition ensued. This shows the importance of ramming and tamping well when using explosives. Altering the shape of the tube simply varies the work imposed on a unit volume of gas, the larger the proportion of surface the greater the imposed work, but does not affect the density of the charge. Further experiments show that dynamite alone or mixed with 80 per cent of ammonium nitrate, or 33 per cent of ammonia alum, or 40 per cent of ammonium chloride; bellite; pyroxylin powder (Moulin-Blanc); and gun-cotton (173 cc. N_2O_2) mixed with even 40 per cent of ammonium nitrate, explode completely when confined in 25–40 mm. tin tubes, bellite even in a 30–35 mm. lead tube; but blasting gelatine requires a stronger tube to ensure its complete detonation.

With regard to modes of igniting the charge, when a fuze (Bickford) is used, care must be taken not to place it in contact with the explosive, for under such circumstances gas and flame may be projected through the tube into the external air, and so produce ignition of the fire-damp. Other objections to the Bickford fuze are the uncertainty of its composition and the danger attached to lighting the exposed end. When electricity is used, preference should be given to currents of low tension, so as to avoid sparking and possible ignition of fire-damp. Detonating cords made of gun-cotton have been found uncertain. Lauer's friction cap has not yet been investigated by the Commission. Various detonators have been examined, and, with the exception of those of Ruggieri and Scola, and ordinary 1.5-gram fulminate caps, they do not ignite mixtures of marsh-gas and air when exploded by themselves. Strengthened fulminate caps differ from the ordinary ones chiefly in being made of stouter metal and having the fulminate more compressed, and covered by a little metallic plate with a hole in the centre to allow access of fire; the gases from the explosion of these caps have, therefore, more work to do, and lose temperature, and consequently a strengthened 1.5-gram cap, as might be presumed by the above hypothesis, does not ignite a mixture of marsh gas, or even coal-gas and air, but does ignite a mixture of hydrogen and air; a larger charge, however, 5 grams for instance, produces sufficient heat to ignite fire-damp. By imposing more work on the gases of detonation of an ordinary fulminating cap, a similar effect is produced. Thus, by winding fine copper wire closely round the outside of an ordinary 1.5-gram cap, this ceases to ignite fire-damp when detonated.

The conclusions arrived at have already been mentioned above, but special stress is laid on the importance of making the explosive do as much work as possible; and bearing in mind the uncertainty attached to the use of even the safest explosives, and the advisability of avoiding their use in presence of inflammable mixtures of marsh-gas and air, the selection of the safest explosive " ought to be considered as diminishing the danger of explosion considerably, but it ought not to be regarded as suppressing it absolutely."

Binary mixtures of ammonium nitrate with dynamite or gun-cotton or dinitrobenzene are recommended from a point of view of safety, although it is suggested that experiments should be made on the method of manufacture of these mixtures, on their durability, and on their behavior in mines. All the results of the numerous experiments are tabulated, and notes containing interesting theoretical considerations are appended, in which methods of calculating the force and temperature obtainable from explosives are given, and the results of such calculations are also tabulated. An objection to the use of water cartridges is the large bore-holes required.

Subsequent to the drawing up of the above report further experiments were made, using a mixture of coal-gas and air, thus avoiding the expense of the manufacture of marsh gas. The most explosive mixture of coal-gas and air contains 15 to 16 per cent of the former. In the experiments it was more convenient to use a mixture containing 10 per cent; the ignition temperature of this mixture appears to be about 2100°. This was ascertained in the same way as the ignition temperature of marsh-gas and air, namely, by exploding in it various cartridges of which the temperature of detonation, etc., could be calculated, the cartridge giving the lowest temperature and igniting the mixture being the one selected for fixing roughly the ignition temperature of the mixture.   This mixture of gas and air is not much more inflammable than marsh-gas and air; therefore it is considered that it would not be too severe a test to make all explosives destined to be employed in presence of marsh gas satisfy the condition of not igniting such a mixture of coal-gas and air when detonated while freely suspended in it.

In the first report it was suggested that the moisture of the atmosphere aided the explosion of fire-damp, but further experiments have not confirmed this suggestion.

The statements regarding dynamite are not altered in the second report, but as regards gun-cotton it is shown that, although the gases from octonitric cotton burn in the air when the cotton is exploded unconfined, nevertheless when this explosive is confined in a tin tube, 25 mm. internal diameter, 31 mm. external, or thicker, not only does the gas not burn, but also the detonation is incomplete.   By testing mixtures of dynamite and ammonium nitrate containing from 90 to 40 per cent of the latter, it is shown that the detonating power increases as this proportion decreases, and complete detonation of the unconfined mixture does not take place until 50 per cent of dynamite is reached; even increasing the amount of fulminate used with the 20 per cent dynamite mixture scarcely improves the extent of detonation. In a 25 mm. by 31 mm. tin tube the condition of affairs is not much better, but in a 25 mm. by 40 mm. tin tube complete detonation of both dynamite and nitrate ensues.   The gases evolved from this mixture do not react on one another; the same is the case with mixtures of dynamite and alum, ammonium chloride, etc.   It is suggested that in such mixtures the non-explosive constituent, the one absorbing heat on decomposing, is decomposed only slightly, or not at all, during the propagation of the explosive wave, their decomposition only taking place in a closed tube.   Various mixtures giving gases which react on one another were examined, such as mixtures of ammonium nitrate with combustible substances, gun-cotton, dinitrobenzene, ammonium picrate, naphthalene tar, heavy coal-oils, colophony, etc.   In the open air they behave sometimes as if the explosive wave, excited at one point, exhausted itself before traversing the whole mass of the explosive; at other times, as if the explosive wave passed through the whole mass of the explosive, but stopped at the decomposition of the explosive constituent or constituents, the gases evolved being dispersed before reacting mutually; therefore the detonation in the open air is incomplete. When, however, the gases are confined, as, for instance, when the explosive is put in a sufficiently strong metal case, then the reaction between the gases takes place, and the effect of a complete detonation is obtained. There are numerous results, tabulated, illustrating these points. Mixtures of about 80 per cent of ammonium nitrate with gun-cotton, or 83 per cent with dinitrobenzene, do not detonate completely either in the air or in thin (25x31 mm.) tubes, but do in 25x40 mm. tubes. Mixtures of ammonium nitrate and picrate in the open air give an effect equal to the decomposition of the nitrate, but in the tube the result is equal to the decomposition of both. With tars, etc., the result is never much above that attributable to the ammonium nitrate.

Experiments were now made with increasing quantities of safety mixtures to test their capability of igniting the inflammable gaseous mixture; the various results are tabulated and discussed. In the following table is given various data relating to the mixtures which proved safe; other mixtures already referred to are either too expensive or inconvenient, or improper as explosives, as far as the present results indicate.

	Force in kilos.	Heat from 1 kilo in cals.	Grms. of Fulminate required for Detonation.
Dynamite (for comparison),	8.490	1.109	...
30 of dynamite, 70 of ammonium nitrate,	6.750	.600	0.50
20 " 80 " "	6.260	.530	0.50
20 of gun-cotton,* 80 of " "	7.610	.810	1.50
15 " " 85 " "	7.000	.700	1.50
12½ of dinitrobenzene, 87.5 " "	8.000	1.006	1.50
10 " " 90 " "	6.700	.750	1.50

Other mixtures worthy of consideration are those containing mononitronaphthalene, or naphthalene, and ammonium nitrate, but their behavior is at present uncertain. Finally, attention is called to mixtures of ammonium copper nitrate and ammonium nitrate. The heat of formation of this salt $N_2O_6CuO, 4NH_3$ is for $(N_6 + H_{12} + O_5 + CuO)$ 164.1. The products of decomposition have not been studied, but the mode of disengaging most heat is as follows:

$$N_2O_6CuO, 4NH_3 = 3N_2 + H_2 + 5H_2O + CuO,$$

* About octonitric.

disengaging 516 cals. per kilo. When exploded in the air, white fumes of ammonium nitrate and of oxides of nitrogen are evolved; therefore the products are uncertain. Its complete combustion by ammonium nitrate is represented by the following equation:

$$Cu(NO_3)_2, 4NH_3 + NH_4NO_3 = 4N_2 + 8H_2O + CuO,$$

or proportions of 76 to 24 of nitrate, heat produced per kilo = 655.5 cals., whilst the force = 6.090 kilos, and temperature of detonation 1750°. This was therefore considered as an interesting explosive in connection with safety in fire-damp. In the open air or in tubes the detonation of the salt does not appear complete according to the above equations, either alone or when mixed with ammonium nitrate; in fact, is almost arrested in the open air when the nitrate present is double the copper salt. It is believed it would explode completely in a mine bore-hole; with regard to its safety the results are very satisfactory, even with mixtures containing only 25 per cent of nitrate, but the mixture containing 80 per cent of nitrate is safe in coal-gas, detonating at a temperature of 1300°, with a force of 5500 grams.

In conclusion they point out that safety mixtures of dynamite and ammonium nitrate should not contain less than 80 per cent of the latter. The other mixtures must fulfill the conditions of having a temperature of detonation lower than 2200°, of not yielding combustible gas after complete detonation, of having sufficient force to avoid use of larger quantities of the explosive, of being sufficiently explosive to minimize missfires, of being kept sufficiently long without undergoing change, and they must not be too expensive. Mixtures fulfilling these conditions more or less are those with ammonium nitrate, and (1) a maximum of 75 per cent of ammonium copper nitrate; (2) 6 per cent of naphthalene; (3) 15 per cent of octonitric gun-cotton; (4) 10 per cent of mononitronaphthalene; (5) 10 per cent of dinitrobenzene. They recommend trials of these in mines.

To avoid danger as much as possible, the mixtures should be well mixed, and the smallest proportion of the dangerous substance, consistent with efficiency, should be used.

As safety depends on the instantaneous admixture of the gas of the detonation with the surrounding atmosphere, it is therefore dangerous to blast with a large quantity of explosive in a confined space.—*J. Soc. Chem. Ind.* **8**, 415–419; 1889.

H. Lohmann, in *Zeits. f. Berg.*, *Hütten- und Salinwesen* 83; 1889, gives a "Report on Further Experiments with Explosives at the Experimental Level of the König Colliery, near Neunkirchen (Saarbrücken), in Respect to their Behavior with Fire-damp and Coal-Dust," in which he states that *an explosive* consisting of a mixture of dynamite with 40 per cent of soda crystals gave very good results, both as regards safety from colliery explosions and the production of large coal. Samples of "carbonite" gave better results than before, and the cartridges were better made. Water cartridges— the explosives surrounded with water—gave good results as to safety.

"Ammonia dynamite" consists of 40 per cent of ammonium carbonate, 10 per cent of nitre, and 50 per cent of a mixture of nitroglycerol and kieselguhr, containing sufficient of the latter to make the mass plastic but not too soft. The introduction of ammonium carbonate is less advantageous than the oxalate. A good mixture is 45 per cent of ammonium oxalate, 15 per cent of nitre, and 40 per cent of the kieselguhr dynamite. It seems possible that by some alteration in the composition of ordinary black powder it may be possible to use it in collieries with safety, ammonium oxalate being used in place of carbon.

"Securite" gave bad results as regards safety from explosion. The Rottweil powder works has consequently much altered and greatly improved this explosive.

"Roburite" was used with success in workings which were free from fire-damp. Analysis failed to show the presence of chlorine in the explosive obtained from Witten, although it was supposed that some of this element was present.

As regards risk from explosion, "ammonia dynamite" proved to be very safe; "kieselguhr dynamite" and "gelatine dynamite" were bad and about equally so; "roburite" is better than "gelatine dynamite," but not so good as "ammonia dynamite," although it is possible that this explosive may have been recently still more improved. "Carbonite," the improved "securite," and "soda dynamite" afford great safety. "Gelatine dynamite No. 2" is as good as ' No. 1," and Favier's explosive, one sample of which consisted of dinitrobenzene, mononitronaphthalene, ammonium nitrate and nitre, also gave very good results.—*Loc. cit.* 419.

English Patent No. 9164, June 23, 1888, has been granted T. G. Hart for " Improvements in Explosives for Use in Firearms," which claims a form of gunpowder consisting of compressed potassium chlorate, granulated, and permeated " with a saccharine solution or other suitable hydrocarbon liquid."

Silesit is the name given to an explosive for which Austro-Hungarian Letters Patent No. 2219 were issued November 12, 1887, to Drs. Pietrowicz and Siegert. It consists of potassium chlorate 60 parts, antimony sulphide 10 parts, and sugar 30 parts.

English Patent 13789, Sept. 24, 1888, has been granted T. M. Justice for " An Improved Explosive Compound."
' This invention relates to an explosive composed of " a suitable nitrate " coated with a hydrocarbon, such as paraffine or naphthalene, and rendered sensitive to detonation at the desired time by admixture with potassium chlorate.—*Jour. Soc. Chem. Ind.* **8**, 818; 1889.

English Patent 11102, July 10, 1889, has been granted W. H. A. Kitchen and J. G. A. Kitchen for " A New Explosive."
This invention relates to a mixture of chlorate of potassium, coal dust, and resin or sulphur.

We have previously noted that when picrate of potash is used in rockets and bombs a curious whistling sound is produced.* U. S. Letters Patent No. 411714, September 24, 1889, have now been granted Antonio del Grande for " Preparing Pyrotechnic Compounds," in which he claims the method of preparing potassium picrate by reacting on a solution of picric acid with magnesium carbonate, and mixing with this an aqueous solution of potassium nitrate, by which potassium picrate is precipitated.

English Patent 13360, Sept. 15, 1888, has been granted T. Chandelon for " Improvements relating to the Manufacture of Explosives," the chief feature of the invention being the use of organic picrates, together with nitrates, or chlorides or both, to form an explosive compound.
Mixtures are described of picrate of benzene, picrate of nitrobenzene, picrate of dinitrobenzene, and picrate of nitronaphthalene with nitrate of ammonia or other nitrates; also of mixtures of the above picrates in which the nitrate has been replaced wholly or partly by a chlorate.

* Proc. Nav. Inst. **15**, 85, 86; 1889.

H. von Asboth proved the "Presence of Pyridine in Amyl Alcohol," *Chem. Zeit.* **13**, 871–872, by dissolving picric acid in the alcohol, when crystals of what proved to be pyridine picrate crystallized out. These crystals melt at 144.5°, and are decomposed by strong acids or bases with the separation of pyridine. Heated gently, they are partially decomposed; strongly heated, they decompose. Their formula is $C_5H_5NC_6H_2(NO_2)_2OH.$—*J. Soc. Chem. Ind.* **8**, 134; 1889.

English Patent 370, January 10, 1888, has been granted S. H. Emmens for "The Preparation of Certain New Compounds from Picric Acid, and the Manufacture therefrom of Explosives, partly applicable also to Colored Fires."

The crystals deposited, after treatment of picric acid with red fuming nitric acid of sp. gr. 1.52, are called by the inventor "Emmens" acid, and are an ingredient of his explosives.

The liquor remaining after the deposition and removal of these crystals will yield a second crop of similar crystals, and will deposit simultaneously a quantity of lustrous flakes. These flakes, heated in water, separate into two new bodies, one of which enters into solution and forms crystals unlike the above, while the other body fuses and remains below the water undissolved.

These acid crystals and residuum are also used as ingredients of the explosives and a mixture of " Emmens " acid, and these two new bodies are called " Emmens " mixture. Explosives are formed by fusing " Emmens ' acid to a nitrate together, or " Emmens " mixture and a nitrate, and such explosives are called " Emmensite." The mixtures have to be exploded by detonation.

English Patent 10722, July 24, 1888, has been granted A. Nobel for " Improvements in Explosive Compounds." This invention relates to the manufacture of an explosive consisting of nitrate of ammonia and picrate of ammonia in the proportion of three parts of the former to one part of the latter.

Gum arabic, dextrin, or an analogous substance is added in the proportion of about ½ per cent of the material, with a view of hardening it.

English Patent 6560, May 2, 1888, has been granted A. Nobel for ' Improvements in Explosive Compounds." This invention relates to the treatment of nitrated starch and nitrated dextrine for the production of explosive compounds to be used in place of gunpowder.

For this object nitrated starch or nitrated dextrine is incorporated with nitrocellulose, and both dissolved in a suitable solvent, such as acetone, etc.   By means of such solvents, which are afterwards distilled off and recovered, a very perfect incorporation can be brought about.

Other explosive substances or oxidizing agents, such as picrates, chlorates, nitrates, etc., may be incorporated with the aforesaid materials, the quantity of the solvent being so far increased as to allow of a complete and uniform distribution of the substance to be incorporated.

U. S. Letters Patent No. 411127, dated Sept. 17, 1889, have been granted Hudson Maxim, of Pittsfield, Mass., for a " Method of Producing High Explosives."

The object of this invention, as stated, is to prepare a high explosive in a form so that, while retaining all the explosive power of its constituents, it will be in such a compact, solid, and therefore stable condition as will permit it to be readily and safely handled and will adapt it to be projected from ordnance with gunpowder; that is to say, it is the purpose to prepare the explosive so that it may be packed in shells and other projectiles and safely fired from guns as shells charged with ordinary gunpowder are commonly fired.

To this end the invention consists in dissolving gun-cotton or nitrocellulose in a proper solvent which is capable of being evaporated, adding to the dissolved nitrocellulose nitroglycerine, and then evaporating the volatile solvent from the mixture.

In carrying out his invention he first thoroughly dissolves gun-cotton in the solvent, which usually will be a liquid, such as acetone or ethylic acetate, or a mixture of sulphuric ether and alcohol.   A sufficient quantity of the solvent will be employed to thoroughly reduce or dissolve all, or practically all, of the gun-cotton.   The nitroglycerine will then be added and thoroughly incorporated with the dissolved nitrocellulose.   The product thus produced is now to be spread out and the volatile solvent evaporated therefrom, the residue being a comparatively hard, dense, and rigid mass, the components of which are maintained in a stable and safe condition.   He claims that this explosive has all the high power of gun-cotton or nitrocellulose and of nitroglycerine, cannot readily be exploded by concussion, and may be worked and shaped for the purpose of charging shells or projectiles.   The presence, however, of the nitroglycerine in the explosive renders it capable of being exploded by ignition or by a detonating charge of fulminate of mercury, or by electricity. The quantity of nitroglycerine, however, is not sufficient to make it susceptible of exploding by any ordinary form of concussion, such as it would be subjected to in handling or in being discharged from a gun with gunpowder.

The quantity of nitroglycerine in the explosive will, of course, depend upon the desired sensitiveness of the compound, and, accordingly, as the explosive contains more or less of nitroglycerine will it be necessary to employ a detonator of less or greater energy. In practice the object will be to add only such quantity of nitroglycerine as will insure the proper instantaneity of explosion.

He states that other solvents may be employed, such as vapors of various liquid solvents or a partially dry solvent; or even a dry solvent may be used, such as camphor; and he therefore does not limit himself to any special form or kind of solvent. So, also, the nitroglycerine may be mixed with a solvent capable of dissolving the nitrocellulose, and the latter be dissolved in conjunction with the nitroglycerine.

Aware that gun-cotton has heretofore been reduced to a very tough, dense, and solid mass by solvents, he does not claim such process, as the explosive thus produced has not the capacity of exploding with the certainty of his explosive, as he states, and that it has not the proper degree of sensitiveness to the detonating charge, and the instantaneity of its explosion depends largely upon the conditions of its confinement. He is also aware that gun-cotton has been dissolved in nitroglycerine, thus producing what is known as "nitrogelatine"; but this is not his process, for the resulting explosive is an unstable jelly-like mass, very susceptible of explosion, while his explosive has all the advantages of those last named products with none of their disadvantages.

He does not claim therein the compound produced by the process described, the same being previously claimed in his application, Serial No. 318670, filed July 25, 1889, but desires to reserve his right to claim said compound.

What he claims as new is:

1. The herein described process of producing high explosives, consisting in dissolving gun-cotton or nitrocellulose in a suitable solvent, adding to the dissolved nitrocellulose nitroglycerine, and then evaporating from the mixture the volatile solvent.

2. The process herein described for manufacturing explosives, which consists in dissolving nitrocellulose in a volatile solvent, combining therewith nitroglycerine, and evaporating the volatile solvent therefrom after the admixture of the nitroglycerine.

Austro-Hungarian Patent No. 2387, November 25, 1887, has been granted D. Johnson for "Gunpowder from Nitrocellulose." For sporting powder the mixture consists of dinitrocellulose 68 parts, barium nitrate 25 parts, potassium nitrate 6 parts, and ultramarine 1 part. These are mixed with water, corned and dried, and they are then treated with a mixture of camphor 10 parts and benzene 50 parts, when they are warmed to about 100° C., to convert the whole. For military powder the proportions are dinitrocellulose 35 parts, barium nitrate 60 parts, and charcoal or lampblack 5 parts.

English Patent No. 14027, September 29, 1888, has been granted C. O. Sundholm for "Improvements in the Manufacture of Dynamite." The object of this invention is to render dynamite not liable to injury or deterioration by water or moisture, and even capable of acting effectively whilst immersed in water. The invention consists in employing as the absorbent of the nitroglycerine the substance obtained by carbonizing without access of air, a siliceous earth naturally containing organic or carbonaceous matter.

According to the *Journal Chemical Industry* 11, 350–352, Romite, which is the invention of a Swede (Sjöberg), is prepared, according to one method, by treating 10 parts of ammonium nitrate and 7 parts of potassium chlorate with a solution of 1 part naphthalene in 2 parts of rectified paraffine oil. Among the advantages claimed for romite is included safety in manufacture, transport and use, but the writer of this article calls attention to the widely known fact that chlorate compounds constitute unsafe and unstable explosives, and refers to the spontaneous ignition of some romite whilst being conveyed from Sweden to Germany for experimental purposes, under the superintendence of the inventor himself.

English Patent 7497, May 22, 1888, has been granted H. J. Allison for "Improvement in Explosive Compounds," the object of which is the production of a blasting powder, to consist of a mixture of ordinary granulated black blasting powder with nitroglycerol, which shall preserve the granular form and structure of the former and not be reduced to a pasty, sticky or deliquated mass during mixing. This is effected by mixing starch paste with the materials of the black blasting powder (charcoal and a nitrate), when on drying and granulating this powder possesses a cellular or honeycombed structure, and readily absorbs nitroglycerol while retaining its full granular form.

A Swedish engineer, J. W. Skoglund, has invented a new explosive which he calls Grakrut. According to the official reports this gray powder has been used with 25 millimeter as well as with Nordenfelt's machine guns. The former, with 70 per cent of the new powder against 100 per cent (or the usual charge) of ordinary powder, gives 33 per cent greater initial velocity, without the pressure in the gun being increased more than 5 per cent. With 62 per cent (ordinary charge weight) of gray powder, the initial velocity was increased 24 per cent without any perceptible increase in pressure. With a charge of 74 per cent (ordinary charge weight) the initial velocity was increased 40 per cent without the gun being subject to any undue pressure. With regard to the important question of smokelessness, the report states that while with Nordenfelt's machine guns smoke of ordinary powder remains for twenty-five seconds, the gray powder only leaves a transparent steam, which is only visible for five seconds.—*Sci. Am.* **312**; 1889.

English Patent 14623, November 11, 1886, amended October 5, 1888, has been granted C. E. Bichel for "Improvements in the Manufacture of Explosive Compounds."

The inventor prepares a "cheap and harmless explosive" by first sulphurizing resin oil or other hydrocarbons, such as wood or coaltar or their distillates boiling between 120° C. and 200° C.; secondly, mixing the above with nitrates or chlorates. He also employs the above sulphurized hydrocarbons mixed with nitrobenzene, nitrophenol, nitrotoluene, etc., and with explosive nitro-compounds such as nitroglycerol and nitraniline, in order to render these latter more stable.

The sulphurizing is carried out by distilling the hydrocarbons in combination with sulphur, and it is claimed that the compounds resulting from the above treatment "exhibit the peculiarity that they mix readily with nitro-compounds, whilst in the case of the products of distillation not sulphuretted, even when they appear to mix, a subsequent separation takes place which renders the production of a stable product impossible."

English Patent 11751, August 15, 1888, has been granted C. Lamm, the object of the invention being to protect and preserve explosives which contain hygroscopic salts.

In carrying this out, carnauba or other palm wax is employed mixed with Japan wax, naphthalene, paraffine, etc., as a waterproof coating for the explosive when made up into blocks or cartridges.

English Patent 12424, September 13, 1887, has been granted E. D. Müller for "Improvements in Explosive Compounds," in which the patentee proposes, in order to cool the explosive gas mixtures, to mix various explosives, such as nitroglycerine, with 15–65 per cent of salts containing more than 5 molecules of water of crystallization, preferably soda crystals, the alums, the sulphates, etc. of sodium. These mixtures are formed into cartridges, and at the explosion, the water being converted into steam, cools the gases down to such an extent that an absolute security against the ignition of the explosive gases is obtained. For warm climates magnesium sulphate is preferable to soda crystals or Glauber salt, and different mixtures and salts are selected according to the advantages they offer in special cases.

English Patent 16920, December 8, 1887, has been granted A. Nobel for "An Improved Explosive Compound." The inventor finds that ammonio-nitrate of copper ($4NH_3Cu(NO_3)_2$) possesses explosive properties, and although it cannot be fired by a spark or Bickford fuze, it develops great energy when detonated.

The ammonio-nitrate of copper is not sensitive to concussion. It is claimed that a very short flame of comparatively low temperature is emitted on detonation, not capable of igniting fire-damp under conditions where gunpowder or dynamite would cause it to explode.

It can be used for propelling projectiles after it has been granulated in the same manner as gunpowder is granulated, but has to be fired by a special primer of gunpowder. In filling cartridges for firearms with granulated ammonio-nitrate of copper, the gunpowder primer is so arranged as to be fired directly by the cap, and then the charging is completed as with ordinary gunpowder.

For use in mines it is simply filled into ordinary paper cartridges and fired by a detonator.

English Patent 5061, April 5, 1888, has been granted H. Müller for "Safety Fuze and Igniting Apparatus for Setting Fire to Explo-

sive Charges without Danger from the Presence of Fire-damp,"
which is effected by means of an envelope of wire, gauze, or similar
material fixed on to the tube in a special way.

English Patent 5624, April 16, 1888, has been granted C. A.
McEvoy for "Improvements in Fuzes," the object being to form a
fuze that will be ignited by immersion in water. This is effected by
means of sodium or potassium, or a mixture of both, suitably attached
to the fuze.

U. S. Letters Patent No. 414662, November 5, 1889, have been
granted Franz L. and Hans Firmann for a "Percussion-Fuze for
Blasting." This invention relates to fuzes for blasting and like purposes (such as the firing of cannon) wherein the primer or explosive
is exploded by being struck by a hammer.

It provides a percussion fuze which can be safely transported,
handled and used. The hammer is normally partly retracted, being
so held against the tension of its spring by a safety stop which prevents it moving forward far enough to reach the primer. When the
pull-wire is pulled the hammer is drawn fully back, the spring fully
compressed, and on increasing the pull the wire is disconnected, so
that the hammer is released and flies forward against the primer.

U. S. Patent No. 408096, July 30, 1889, has been granted W. H.
Frazer for a "Cap for Fuzes,' the cap being made of rubber partially
vulcanized, but yet sufficiently flexible to be slipped over the end of
a fuze. The inventor seeks by this means to avoid the danger which
attends the fixing of the ordinary metallic blasting cap on the leading
fuze by the usual method of crimping.

In the "Report of Committee of the Society for the Protection of
the Interests of Chemical Industry in Germany," it was proposed, on
behalf of the percussion-cap manufacturers, that *denaturized spirit*
be mixed with 0.025 per cent bone-oil or 0.5 per cent oil of turpentine,
and, in any case, with 0.025 per cent pyridine bases, for the production
of fulminating mercury, to the exclusion of wood spirit, which lowers
the yield and deteriorates the quality of the product. According to
the proposal, the most suitable substance to use in the case of the
fulminating mercury manufacture would be the ethereal distillate,
which forms a by-product in the manufacture of fulminating silver,
as the presence of bone-oil and oil of turpentine gives rise to an
undesirable coloration in the manufactured article.—*Chem. Zeit.* 12,
249.

In a " Note on a Plan for a Nitrogen Iodide Photometer," M. G. Lion proposes, in the *Compt. rend.* 109, 653–654; 1889, to follow the suggestion of M. Guyard,* and use the property of nitrogen iodide of being decomposed by a ray of light, as means of measuring the photometric powers of different sources. The novelty of his plan consists in using two burettes, each of which contains the reagent, and which are connected together by a capillary tube containing an index of mercury. Each separate beam of light falls on one of the burettes, and any difference in intensity causes a difference in rate of evolution of the nitrogen, which therefore causes a displacement of the mercurial index.

The *Am. Chem. Jour.* 10, 1888, contains a note by Dr. J. W. Mallet upon the " Influence of Light upon the Explosion of Nitrogen Iodide,"† stating that the instances of explosion of nitrogen iodide, when *under water*, which he has previously described‡ (*Am. Chem. Jour.* 1, 4–9; 1879), took place when the direct rays of the sun fell upon the explosive, though the explosion was precipitated by friction or concussion. There were but two cases of such explosions, and, as the result of his experience, Mallet finds that under ordinary circumstances nitrogen iodide, while wet, exhibits no extraordinary sensitiveness, and may be safely worked with, only becoming highly dangerous on drying.

C. F. Cross and E. J. Bevan have examined the " Acetyl Derivatives of Cellulose " produced by the method of Franchimont, *Compt. rend.* 92, 1053, and from the experimental data which they cite, the soluble product of the original reaction appears to be a pentacetyl cellulose. The formation of such a derivative would, of course, have a very important bearing on the constitution of cellulose, and hence the writers reserve any more positive statement regarding the composition of the product until their investigation has been more extended.

Erwig and Koenigs (*Ber. Berl.* 1464 & 2207; 1889), have, however, obtained a pentacetyl dextrose by using this same method.— *Chem. News* 60, 163; 1889.

C. E. Guignet states, in the *Compt. rend.* 108, 1258–1259; 1889, that "Colloidal Cellulose" may be produced by immersing in

* Proc. Nav. Inst. 14, 441; 1888. †Ibid. 14, 441; 1888. ‡13, 424; 1887.

sulphuric acid of 50° B. filter paper which has been previously treated with HCl and HF, or carded cotton of the finest quality which has been thoroughly dried, care being taken to avoid a rise in temperature while the cellulose is immersed in the acid bath. The cellulose forms a transparent, gelatinous mass, which is not affected by contact with a large excess of acid, but is rapidly converted into dextrine at 100°. When the acid has been completely removed by washing (which is most thoroughly done by finishing with alcohol and drying at the lowest possible temperature), the colloidal cellulose dissolves in pure water, forming a slightly milky solution which is readily filtered, deposits no precipitate even after several hours, and is not affected by boiling. If immersed in sulphuric acid of 60° for a short time, or in acid of 55° for a longer time, it becomes insoluble in water. When treated with nitric acid it forms nitrocellulose in the same way as ordinary cellulose, and becomes slightly less transparent. The author regards parchment paper as a cellular tissue, the pores of which are filled with colloidal cellulose.

English Letters Patent No. 16330, November 28, 1887, have been granted C. Stockes for " Improvement in the Manufacture of Non-inflammable Celluloid Matters," which consist in incorporating a quantity of a metallic salt, such as protochloride of tin dissolved in alcohol, ether, or other suitable solvent, with the celluloid or similar substance made from nitrocellulose.

The appearance of a paper by Mr. V. H. Veley upon one aspect of the subject affords C. F. Cross and E. J. Bevan an opportunity for placing on record some experimental results which have accumulated regarding the "Conditions of Action of Nitric Acid" (*Chem. News* 60, 13–14; 1889). They have already cited in *Watt's Dict., New Ed.*, Art. Cellulose, their observation that the presence of urea arrests the specific action of nitric acid upon the liquefied cellulose, changing it into one of simple hydrolysis; and they give here the experimental results upon which this observation was founded, together with subsequent ones, which go to show that with the graduated increase of urea there is a gradual departure from the characteristic action of the pure acid, and the results are so very evident to the senses as to make a very striking lecture demonstration.

These results led them to extend the investigation of the influence of urea—*i. e.* of the elimination of nitrous acid—to other actions of nitric acid, taking first metallic copper, when the solvent action of the nitric acid was almost completely arrested, and these with similar experiments afforded sufficient evidence that the interaction of nitric acid and the metals is another instance of dependence upon the *tertium quid*, which in this case could scarcely be other than nitrous acid.

As the extension of this investigation to the action of mixtures of $HNO_3 + H_2SO_4$ promised to throw further light on the method of formation and constitution of the explosive derivatives of the celluloses and compound celluloses, a series of comparative nitrations of both cotton and jute fiber were made with and without the addition of urea to the acid mixture, and the evidence as to its influence on the reaction was uniformly negative, a result which strengthens the view that the derivatives in question are alcoholic nitrates. When, on the other hand, the formation of a nitro-derivative is involved, the presence of urea may have a very important influence, as shown by Dr. Perkins, Jr., in a recent note on the "Nitration of Anthracene in Presence of Urea" (*Chem. Soc.*), where this modification of conditions proved sufficient to determine a simple substitution in the carbon nucleus.

Frederick J. Smith having to deal with electrolytic gas in a research on explosions in tubes, and believing that some error would be introduced by the use of rubber tubes as usually employed, has devised a flexible tube which seems almost wholly free from the disadvantages of the diffusion of some gases through rubber.

A length of small rubber tube is drawn through another tube of equal length but of greater diameter ; each end is furnished with a small length of glass tube fitting the smaller tube; one end is bound off with cord, then the space between the inner and outer tube is filled with water or any liquid required ; the other end is then bound off. By this means a "Water-Jacketed Flexible Tube" is easily constructed.—*Chem. News* 60, 187 ; 1889.

W. Michelson contributes to the *Ann. Phys. Chem.* 37, 1-24 ; 1889, a paper on the "Normal Velocity of Inflammation in Explosive Gaseous Mixtures," in which, after reviewing the work of Davy, Bunsen, Schlösing and Mondésir, Berthelot and Vieille, and of Mallard and Le Chatelier, many of which have been resumed in these Notes, he gives the data obtained from a large number of experiments with mixtures of illuminating gas, hydrogen, carbon monoxide and methane with air, and with oxygen, and the mathematical and graphical discussion of the same. The paper is accompanied by numerous tables, illustrations of apparatus used, and photographs of the flames produced.

M. Neyreneuf contributes to *Ann. Chem. Phys.* **17**, [6], 351–377; 1889, a paper entitled "Researches on the Chemical Harmonica," or singing flame, that contains some fine illustrations of the beautiful figures developed by explosions of gaseous mixtures when burnt in this way.

Berthelot and Petit give in the *Ann. Chim. Phys.* **17** [6], 80–106; 1889, the results of their determinations of the "Heats of Combustion of Carbon in its Different States: Diamond, Graphite and Amorphous," from which we extract the following tabular summary, in which the results obtained by Berthelot and Petit are compared with the famous ones of Favre and Silbermann:

Author.	State of Carbon Ash-free.	Gram-units of Heat per gram.	Kilogram-units of Heat per molugrams.
B. & P.	Amorphous (wood charcoal), .	8137.4 cal.	+ 97.65 Cal.
F. & S.	" "	. 8080. "	+ 96.96 "
B. & P.	Graphite, ...	7901.2 "	+ 94.81 "
F. & S.	"	{ 7796. "	+ 93.55 "
		{ 7762. "	+ 93.14 "
B. & P.	Diamond (Cape), ...	7859. "	+ 94.31 "
B. & P.	" (Bort), ...	7860.9 "	+ 94.34 "
F. & S.	::	{ 7770.* "	+ 93.24 "
		{ 7878. "	

In the same volume of the same journal the same authors give on pages 107–140 the results of their determinations of the "Heats of Combustion and Formation of the Nitriles," from which we extract the following tabular summary:

Heats of Combustion.

Name of Compound.	Formula.	State.	Kilogram-units of Heat per molugrams.
Formonitrile, . .	H.CN	liquid.	+ 152.3 Cal.
Acetonitrile, . . .	CH₃.CN	"	+ 291.6 "
Propionitrile, . .	C₂H₆.CN	"	+ 446.7 "
Benzonitrile, . .	C₆H₆.CN		+ 865.9 "
Benzyl cyanide, .	C₆H₆.CH₂.CN	"	+1023.8 "
Oxalic nitrile, . .	Cn.CN	gaseous.	+ 262.5 "
Malonic " . .	CH₂.Cn.CN	solid.	+ 395.1 "
Succinic " . .	C₂H₄.Cn.CN	"	+ 546.1 "
Glutaric " . .	C₃H₆.Cn.CN	liquid.	+ 699.8 "

* Number preferred by Favre and Silbermann.

*Heat of Formation from their Elements.**

Name of Compound.	Kilogram-units of Heat per molugrams.
Formonitrile, ...	+ 23.5 Cal.
Acetonitrile, ...	+ 0.5 "
Propionitrile, ...	+ 8.7 "
Benzonitrile, ...	− 33.1 "
Orthotolunitrile,†	− 34.8 "
Benzyl cyanide, ...	− 27.9 "
Oxalic nitrile, ...	− 73.9 "
Malonic nitrile, ...	− 43.2 "
Succinic nitrile, ...	− 32.0 "
Glutaric nitrile, ...	− 22.8 "

The signs and numbers in the above table differ somewhat from those in Berthelot and Petit's final table, but agree with those given under each experiment.

F. Stohmann, C. Kleber, and H. Langbein, *J. pr. Chem.* [2], **40**, .77–95, have determined the " Heat of Combustion " of :

	Heat of Combustion. Calories.	Heat of Formation. Calories.
Durene, $C_{10}H_{14}$, ...	1393.9	+ 29.1
Pentamethylbenzene, $C_{11}H_{16}$, . . .	1554.1	+ 31.9
Hexamethylbenzene, $C_{15}H_{18}$, . . .	1712.2	+ 36.8
Diphenyl, $C_{12}H_{10}$, ...	1494.3	− 19.8
Naphthalene, $C_{10}H_8$, ...	1233.6	− 17.6
Anthracene, $C_{14}H_{10}$, ...	1694.3	− 33.3
Phenanthrene, $C_{14}H_{10}$, ...	1693.5	− 32.5

The following conclusions may be drawn from these numbers:

(1). The heat of combustion of the solid homologue of benzene increases, on an average, 155.1 cals. for the addition of each methyl-group.

(2). The substitution of the phenyl-group for hydrogen in benzene, to form diphenyl, increases the heat of combustion by 717.0 cals.

(3). The displacement of a hydrogen atom in the hydrogen molecule by the phenyl-group makes the heat of combustion of the benzene molecule thus formed 718.5 cals. higher than that of the hydrogen molecule.

(4). The formation of diphenyl from benzene, with separation of

---

* Carbon supposed in state of diamond $= + 94.3$ cal. † $CH_3.C_6H_4.CN$.

two atoms of hydrogen, takes place without any apparent thermal effect.

(5). The thermal values of anthracene and phenanthrene are practically identical.

(6). The conversion of benzene into naphthalene and naphthalene into anthracene raises the heat of combustion some 458 cals. in each case.—*J. Chem. Soc.* **56**, 1042; 1889.

Berthelot and P. Petit have examined the "Thermo-chemical Relations of the Isomeric Nitro-camphors," *Compt. rend.* **109**, 92–95; 1889, using for this purpose specimens obtained by MM. Heller and Cazeneuve in their study of these bodies. The previous researches of Berthelot have shown that camphor belongs to a special class of aldehydes characterized by the property of combining either with the elements of water to form acids, or with a quantity of oxygen three times as great as for ordinary aldehydes which give rise to bibasic acids with peculiar properties. These derivatives may be represented by diamylenic acid which belongs to the fat acid series, on the one hand, and cymenic acid, which belongs to the aromatic series on the other. The camphenic series thus forms the connecting link by which we pass from the fat acid to the aromatic series. It is quite to be expected then that there should be two nitro-derivatives one of which should be allied to nitro-ethane, and which Berthelot styles *a-camphor nitrate*, and a second which should possess a phenylic function and which he styles *phenol nitro-camphor*.

For the first of these, *a*-camphor nitrate, which has the formula $C_{10}H_{15}NO_3$, these chemists have found the heat of formation from its elements to be + 89.1 cal., and from camphor (solid) and nitric acid (liquid) + 7.3 cal. This last number is of precisely the same order as the heats of formation of the nitric esters, that of ethyl nitrate being + 6.2 cal. and of nitroglycerine + 4.7 cal. × 3. Hence it is not surprising that these chemists should have found that this camphor nitrate detonated immediately when projected in fine drops into a red-hot tube, and that the superheated vapor detonated also.

The phenol nitro-camphor forms a hydrate of the formula $C_{10}H_{15}NO_3.H_2O$, and this body gave for the heat of formation of the anhydride from its elements + 125.2 cal., and from camphor (solid) and nitric acid (liquid) + 43.4 cal. The last number is very much larger than that for its isomer, and it is related to that of the nitro-derivatives of the aromatic series, which is about 36 to 38 cal. It is quite to be expected that this nitro-camphor would be less explosive than the former, and the experiments verified this expectation, for when projected·in drops into a red-hot tube it was destroyed with carbonization, but without detonation, and the same proved true of the superheated vapor.

In discussing the state of "Unstable Equilibrium of the Atoms," *Zeit. physikal. Chem.* **3**, 145-158, E. Pringsheim says that in substances which readily undergo sudden decomposition or isomeric change the atoms are supposed to be in unstable equilibrium, and he points out that the heat developed in such cases is usually very large, and much greater than the energy required to effect the decomposition; hence, if partial dissociation, in the sense of the Clausius hypothesis, is brought about by sudden raising of the temperature or some other disturbance, the heat generated serves to continue and complete the decomposition.—*Jour. Chem. Soc.* (Abstr.), **672**; 1889.

P. Cazeneuve, *Compt. rend.* **108**, 857–859; 1889, has effected the "Reduction of Nitro-camphor to Nitroso-camphor" by boiling 300 grams of chloronitro-camphor and 1500 grams of 93° alcohol for about one-half hour with a copper-zinc couple made the action of a solution of 100 grams of copper sulphate on 600 grams of granulated zinc. The filtrate was then distilled to dryness, the cupric oxychloride removed by treatment with warm dilute chlorhydric acid, and the residue recrystallized from alcohol, when the nitroso-camphor was obtained as a white, crystalline solid, insoluble in water, slightly soluble in cold alcohol, but more soluble in boiling alcohol or benzene. It alters when exposed to light, becomes greenish, and gives off nitrogen oxides. It is neutral to litmus and orange III, but acid to phenolphthalein. It does not melt before decomposing, but at 180° it suddenly becomes green, entumesces, and evolves nitrogen oxides. If thrown on red-hot platinum it detonates.

A. Behal, in a paper entitled the "Conversion of Methylbenzylidene Chloride into Triphenylbenzene," *Bull. Soc. Chim.* **50**, 635-638, states that the liquid product of the action of phosphorus pentachloride on acetophenone, after standing for eight months, exploded spontaneously, and crystals of triphenylbenzene were formed at the same time. On investigating this phenomenon, the author found that water determined the decomposition of the original substance into hydrogen chloride, monochlorcinnamene, and acetophone, the latter compound having resulted from the action of water on the previously formed monochlorcinnamene, and that if the latter body were saturated with dry hydrogen chloride and heated in a sealed vessel at 40°, it exploded, with formation of triphenylbenzene crystals.— *J. Chem. Soc.* **56**, 998 ; 1889.

In the study of the " Pyrimidines " (Metadiazines), *Ber.* **22**, 1612– 1635 ; 1889, A. Perines has obtained the dichlornitroethylpyrimidines, $C_7H_7N_2Cl_2NO_2$, in the form of a yellow powder which explodes when heated.

Among the bodies obtained by D. S. Hector in his study of the " Action of Hydrogen Peroxide on Phenylthiocarbonide," *Ber. Berl.* **22**, 1176–1180 ; 1889, is the silver nitrate derivative of dianilidoorthodiazothiole, $C_{14}H_{12}N_4S.AgNO_3 + H_2O$, which decomposes on exposure to light and explodes when heated ; and the nitroso-derivative, $C_{14}H_{11}N_4S.NO$, which turns yellow when dried and explodes when heated.

Among the derivatives of " Orthonitrosulphanilic Acid " obtained by J. Z. Lerch, *Chem. Centr.* **286** ; 1889 ; *Listy Chem.* **13**, 85–89, is the orthonitroazobenzeneparasulphonic acid, which crystallizes in long yellow pyramids that explode when heated.—*J. Chem. Soc.* 880 ; 1889.

In describing the "Action of Hydriodic Acid on Allyl Iodide," *Bull. Soc. Chim.* **50**, 449–451, H. Malbot says when hydrogen iodide is passed into allyl iodide some propylene is liberated and great heat is evolved, eventually causing the propylene iodide which has accumulated to decompose with explosion. The explosive substance may be destroyed almost as soon as formed by alternately passing the current of acid and then warming the liquid ; an explosion is thus prevented, but at the same time the conversion of propylene iodide into isopropyl iodide is hindered, and nothing is obtained but propylene gas and a residue of iodine, with a little carbonaceous matter. If the allyl iodide is mixed with isopropyl iodide, the reaction is much less violent, but still little if any allyl iodide is converted into isopropyl iodide. If, however, the mixture is cooled with ice and salt, the conversion is nearly complete.—*Jour. Chem. Soc.* **56**, 766 ; 1889.

K. Buchka states (*Berl. Ber.* **22**, 829–833 ; 1889) that the "Preparation of Metanitrotoluene" can best be effected by dissolving pure metanitroparatoluidene, prepared by Gattermann's method, in alcohol (3 parts) and concentrated sulphuric acid (3 parts), and adding drop by drop to the cold solution a saturated aqueous solution of sodium nitrite which contains rather more than the calculated quantity of the nitrite. As soon as all of this is added the solution is kept for some time, then carefully warmed until the evolution of nitrogen has ceased and the liquid has assumed a dark brown color. The alcohol is evaporated, the product distilled with steam as long as oil passes over, and the metanitrotoluene extracted with ether. The product thus obtained distills entirely between 228° and 231°, and solidifies when cooled, melting again at 16°. The yield is from 66–84 per cent of the theory, though one experiment gave 90 per cent.

As the result of the "Action of Methyl Diazoacetate on Ethereal Salts of Unsaturated Acids," E. Buchner has obtained the methyl acetylenedicarboxylodiazoacetate, $C_3HN_2(COOMe)_3$, and he has prepared from this ester the acetylenedicarboxylodiazoacetic acid, $C_6H_4N_2O_6$. The silver salt of this acid, he finds, explodes slightly when heated, yielding an oil which is probably either $C_3H_4N_2$ or $C_6H_8N_4$.—*Berl. Ber.* **22**, 842–847 ; 1889.

"Diamide Hydrate and Other Salts" have been obtained by Curtius and Jay, who have continued their researches upon the compound diamide N_2H_4, which was isolated by the former chemist in 1887. They have obtained the hydrate by distilling diamide hydrochloride with caustic lime in a silver retort, and they find it to be a fuming, corrosive liquid, with a strongly alkaline reaction and a taste like ammonia. It appears to be the strongest reducing agent known, and when dropped on mercuric oxide it explodes violently. The study of the salts shows that when any salt of diamide is mixed with a nitrite, free nitrogen is evolved almost explosively.—*Berl. Ber.* **22**, 134 (*Ref.*); 1889; *J. prk. Chem.* **39**, 27, 107.

From a brief but interesting notice of the life of Dr. J. Peter Griess, *J. Soc. Chem. Ind.* **7**, 612–613 ; 1888, we learn that he discovered the first one of the diazo-compounds (that curious group of explosive bodies) in 1858. His work on the azo and diazo bodies was continued from that time up nearly to the day of his death, and may be regarded as classic.

In his study of the " Nitroso-compounds of Ruthenium,"* *Compt. rend.* 108, 854–857; 1889, A. Joly finds that when brown ruthenium sesquichloride is treated with large excess of nitric acid it yields a red nitrate, and when this compound is treated with successive quantities of hydrochloric acid and boiled for a long time, it yields a pale crimson solution of a chloride which, when evaporated at 120°, leaves a brick-red crystalline mass of the composition $RuCl_3NO + H_2O$. When heated at 440° in carbon dioxide or in a vacuum, this salt decomposes rapidly with evolution of oxides of nitrogen, and leaving a residue of ruthenium sesquichloride and dioxide.

Solutions of the nitroso-sesquichloride or of the double alkaline salts are not precipitated by alkalies in the cold, but if the solutions are mixed with sufficient alkaline hydroxide or carbonate to combine with three atoms of chlorine, and are then boiled, a pale brown gelatinous precipitate is formed and the supernatant liquid is neutral. After drying at 150° it forms a black mass with a vitreous lustre, which has the composition $Ru_2O_3(NO)_2 + 2H_2O$. This body is not decomposed by heat at temperatures below 300°, but at 360° it slowly decomposes in a current of carbon dioxide, yielding black graphitoidal oxide, Ru_4O_9, and above 440° it decomposes explosively with incandescence and evolution of oxides of nitrogen.

H. N. Warren, in treating of " Magnesium as a Reagent," states that, as a reagent in the dry way, magnesium reduces all metals with the exception of the alkalies and alkaline earths, and may possibly aid in the decomposition of the latter, provided a suitable combination of the same could be obtained.

An exception to reduction in this line, however, may be mentioned in the powerful manner in which it attracts molybdic anhydride when fused with the same, the combination being so intense as to be accompanied with loud detonations.—*Chem. News* 60, 187–188; 1889.

In discussing the " Formation of Deposits of Nitrates," A. Müntz and A. Marcano (*Compt. rend.* 108, 900–902; 1889) have previously attributed the formation of the deposits of nitrates in South America to the enormous deposits of the excrements of birds, bats, etc., which occur in the immense caverns in the Cordilleras (Abstr. 1885, 1042). They now state that the formation of deposits of nitrates can, in fact, actually be watched.

* Proc. Nav. Inst. 15, 92; 1889.

In many large caverns there are no remains of birds, but the soil is charged with nitrates and is found to contain enormous quantities of the bones of mammals.   These bones are very friable, and consist of calcium phosphate, with very small quantities of organic matter. Calcium carbonate is absent, and has, indeed, been converted into calcium nitrate, which is found in the earth by which the bones are surrounded.   Bone caverns are numerous in Venezuela, not only in the littoral mountains, but also in the flanks of the Cordillera of the Andes.   In some cases the deposit exceeds 10 metres in thickness, and the earth is highly charged with the nitric ferment.   The quantity of calcium nitrate present varies from 4 to 30 per cent, and the quantity of calcium phosphate from 5 to 60 per cent.   These observations confirm their previous conclusions as to the origin of the nitrates.

According to the *Sci. Am.*, July 20, 1889, a family in Haverhill, Mass., were recently seated about an open grate, in the parlor, in which a fire was burning, when an explosion took place which broke a window screen and a pane of glass in a window some feet away and filled the room with smoke, while the report was so loud as to be heard at quite a distance.   Examination showed that an ornamental tube, about one inch in diameter, which arched the fire-place was broken, and that the explosion had been caused by the action of the heat upon resin with which the tube had been filled.   It is the custom of the trade to fill tubes which are to be bent with resin, so as to prevent them from collapsing during the operation, but it is also the custom to melt the resin and pour it out after the bending is completed.   Evidently this had been neglected in this case.

The Boston *Herald*, May 26, 1889, says that those who use the flash lamp in instantaneous photography should be sure and buy the pure magnesium flash powder, as the article commonly known as "flash powder" is a composition containing picric acid, a highly explosive substance.   This mixture was harmless before the flash lamp was invented, but in that species of lamp where an air-bulb is used, there is apt to be a suction after the flash which will draw back the flame into the reservoir containing the powder.   Several bad accidents have happened to amateurs through neglect to use the pure magnesium powder.

A flash powder which gives a yellow light for use in orthochromatic photography is made from one part of metallic magnesium and five to seven parts of sodium nitrate.

The following books are announced:

"L'artillérie actuelle en France et à l'étranger. Canons, fusils, poudres et projectiles." Paris, 1889.

"Hand-book of Gunpowder and Gun-cotton." Wardell, London, 1888.

"Vorschrift über die Versendung von Sprengstoffen und Munitions-gegenständen der Militär- und Marineverwaltung auf Landwegen und auf Schiffen (Sprengstoff-Versendungsvorschrift) nebst militärischen Ausführungsbestimmungen." Berlin, 1889.

"Anleitung zum Zünden von Bohrlochladungen durch Friction u. s. w." Von Johann Lauer. Vienna, 1887.

"Anleitung zur Bestimmung der Bohrloch-Ladungen für Sprengungen in Schlagwetter führenden Gruben." Von Johann Lauer, Vienna, 1887.

"Lauer's Vorschläge zur Verhinderung von Explosionen u. s. w." Von Ed. F. Csánk, Vienna, 1887.

"J. Lauer's Frictionszündmethode." Von J. Mayer. Reprinted from Oester. Zeit. Berg- und Hüttenwesen, 1887.

The *Mittheilungen des Artillerie und Genie-Wesens*, Parts 8 and 9, 1889, contains a Bibliography of the Periodical Military and Technical Literature for the first half of the year 1889, in which we find thirty-two titles of articles on explosives, besides a considerable number on the use of high explosive charges in shells.

"Mining Accidents and their Prevention," by Sir Frederick Abel, with a discussion by some thirty-six experts, to which is appended the laws governing coal-mining in each of the United States and in Great Britain and Germany (431 pp., 1889), is published by the Scientific Publishing Company of New York.

U. S. NAVAL INSTITUTE, ANNAPOLIS, MD.

FLEET TACTICS.

By Lieutenant Richard Wainwright, U. S. Navy.

The first question that arises in entering upon the subject is whether or no modern arms and modern inventions have so changed all methods of naval warfare as to make the lessons drawn from history useless, and also if it is impossible to lay down any rules for the guidance of naval commanders besides the few points of minor details that illustrate the handling of a few vessels or single vessels —in fact, minor tactics, with the gun, ram, and torpedo as weapons. If it is true that the march of progress has so altered the conditions of warfare as to make the study of naval history useless, then we have no guide except what can be found in the study of modern naval peace manœuvres and a few recent naval battles, of which that of Lissa is the most important; but even from these slight data it is possible to deduce certain rules, and it will be found that these rules agree with those deduced from historical examples, and that it is only minor tactics which must be radically changed with the improvement of the weapon.

There has always been an outcry raised against laying down rules for the conduct of warfare. Military rules have been opposed because of the numerous elements of uncertainty that enter into every problem, and in the case of naval rules the opposition has been still stronger, because of the additional uncertainty introduced by the necessity of contending with the elements. The great master of the art of war, Jomini himself, says that "war, far from being an exact science, is a terrible and impassioned drama, regulated, it is true, by three or four general principles, but also dependent for its results upon a number of moral and physical complications." In another place he says, "war in its *ensemble* is not a science but an art. Strategy, particularly, may indeed be regulated by fixed laws resembling those of the positive sciences, but this is not true of war viewed as a whole. Among other things, combats may be mentioned as often being quite independent of scientific combinations, and they may become essentially dramatic, personal qualities and inspirations and a thousand other things frequently being the controlling elements. Shall I be understood as saying that there are no such things as tactical rules, and that no theory of tactics can be useful? What military man of intelligence would be guilty of such an absurdity?" Baron de Jomini lays down general rules for the guidance of the military student, and, by analyzing numerous campaigns and battles, shows how he has derived the rules and illustrates the battles. The march of improvement on land has been great, and yet Jomini is still an authority of the highest value. Because the improvement in the weapons at sea has been great, and because all rules are difficult of application to warfare on the sea, the elements of uncertainty being numerous, shall we therefore say that naval warfare not only is not a science, but also is not even an art? that it is a mere matter of brute force? that no rules can be used as a guide, and that the naval commander must trust to the hard fighting of his individual vessels, and that the best work an admiral can do for his fleet is to command the largest vessel and make it conspicuous in the fight? One might answer in the language of Jomini, What naval (military) man would be guilty of this absurdity? were it not for the high naval authority cast upon the side of no rules—not openly and plainly, yet with language that will bear no other construction.

Many officers beyond doubt think that there is no better method to conduct a fleet than to carry it on the shortest road to the enemy, trusting to hard fighting to win the day. They say that on the sea all operations and manœuvres are plainly visible to the enemy, whereas on land there may be doubts in the mind of either side as to the position and strength of the troops opposed, and a portion of the line may be strengthened or weakened without the knowledge of the opposing force. Strategy, with its political, geographical, and military or regular strategic points, remains now, as throughout the history of war, the same, but tactics are useless. The advocates of brute force are not confined to those who pin their faith on any one weapon; there are adherents of the gun, the ram, and the torpedo, who point to the sentence extracted from Nelson's famous Trafalgar order, " No captain can do wrong if he places his ship alongside that

of an enemy," as the best rule of the greatest of naval commanders,
and yet every line of that order, of which the above is a short
extract, shows Nelson to be a believer in tactics and a great tactician, and throughout it he is striving to line out the course to be
pursued by his fleet, in accordance with the various positions of the
enemy, so that he may combine his strength upon a portion of the
enemy's, the aim of all tacticians. The tactics to be pursued in any
given case must somewhat depend upon strategical considerations.
It may frequently happen that it is not to the best interests of the
country to risk a decisive battle, for the resources of the enemy may
be so superior that the loss incurred in defeating him will be ruinous
to us while not felt seriously by him. If our enemy be decidedly
inferior to us, it may be wise to endeavor to overwhelm him at all
points at once; but where the two fleets are nearly equal, we can
hardly expect that our fighting qualities will be so far superior to his
as to give us a decided victory. If both sides rush together and
engage in a *mêlée*, chance will rule; but the result must be disastrous to both fleets, and the victor of such a combat can have but
little fighting value left. In most of the great battles won by the
English fleets they gained some decided tactical advantage in the
outset through superior handling of their fleets, and in those days
their superior seamanship and gunnery, and higher state of discipline
made an English vessel far superior to a French, Spanish, or Dutch
vessel of the same size. We know that many of their commanders,
notably Collingwood, paid great attention to rapidity of fire. The
fire was generally reserved until close quarters, and then poured
in by broadsides; the opposing crews, less skillful and fatigued by
firing at long range, could not compete with the English in rapidity
of fire, and their guns were silenced. The slaughter was usually
far greater on the enemy's vessels than on the English, and, although
courageously fought, they were obliged to surrender. This great
slaughter was largely due, undoubtedly, to the rapid firing, and some
may have been due to superior ballistics. Because of this superiority, they adopted the tactics, when to windward, of cutting through
the enemy's line and engaging him to leeward, thus preventing him
from slipping away to leeward when disabled. Can we expect to
have this superiority to all enemies?

In 1812 our officers and vessels were superior to the English,
and our crews at least their equals, but the fact remains that in spite
of our victories in the naval duels, if the war had continued longer, we

.would have been without a vessel able to keep the sea. It would be egregious vanity for us to assume any individual superiority over European vessels or fleets as now situated, and if our only dependence must be on hard fighting,—unless we contemplate building a naval force equal or superior to our possible foes,—we might as well cease our endeavors to build up a navy. It certainly seems improbable that for many years to come we can expect to have a large fleet. We may hope to have a small fleet of battle-ships, and if this must rush at the enemy whenever he is sighted, we may expect to beat him once, we may hope to defeat his second fleet, but for a third encounter we would certainly have no fleet. There may arise occasions when, because of the political situation, it may be necessary to risk our small fleet; that is, an immediate victory may serve to prevent another nation from joining the enemy, or may serve to induce an ally to join us, but certainly under ordinary occasions it would be foolhardy to attack the enemy at once and insure our final destruction, while on these extraordinary occasions it certainly would be wise to endeavor by manœuvres to gain some advantage over the enemy, and, while destroying him, preserve as many of our vessels as possible.

To begin with, let us see if we cannot find some rules by which we may be guided, and from which, if we follow them, we may hope for advantage, and which, if we violate them, will plainly give the enemy the advantage if he adopts the correct course.

There will be three cases to consider in the conduct of a fleet ; when meeting one of equal strength, when meeting one of superior strength, and when meeting one of inferior strength.

When meeting an equal force, the first thing to be considered is whether an immediate and decisive action is necessary to the interests of the country, or whether it is advisable merely to delay the enemy until we can receive reinforcements ; or, is our fleet so important to us that it must not be risked until we have reasonable certainty of a complete victory? Undoubtedly the most satisfactory course for any commander to pursue is to press on to an immediate engagement; his responsibilities are lighter, and his anxiety is shortened. He has only to have confidence in his own courage and that of his officers. Even if he should be defeated he can fight hard, injure his enemy, and leave his courage unimpugned. On the other hand, if he wait, he must be vigilant day and night ; he must restrain those who, desirous of ending their anxiety, will try to force a fight. He knows that if by an unfortunate accident he should lose touch of the enemy he will be subjected to adverse criticism. There will always be those under him whose only idea of fighting is to have it out at once, win or lose, and until he has achieved success they will believe he is wanting in courage. The adventurous side will always be popular. Still, the commander of a fleet of the United States will in many cases find it his duty to risk little in order to preserve his fleet in fighting condition, and win the game by strategy and tactics in place of hard fighting. Suppose we had but a fleet of eight battle-ships—and it must be some time to come before we shall have that force: the enemy, say England, having many vessels, sends a fleet of eight battle-ships to attack our coast. If we were to attack that fleet, rush out and meet it half way, we might be victorious; in fact, I think we would be victorious, but at the end of the battle I should expect to find our fleet utterly unfit for further contests. The next fleet that came over would have free play on our coast as far as any opposition from our fleet was concerned. Now, history shows us how a small fleet may be handled to advantage in the defense of a coast where the principal points are fortified; and I think common-sense reasons may be shown by which a waiting game may be made to pay, and how a flanking fleet may be a powerful defensive weapon in the hands of an able commander.

The first and easiest case to handle is where the two opposing fleets are equal and both desire to bring on an engagement. Still the question is sufficiently complicated. What shall be the order of battle, and will the fleets stand off and use the guns, or will they close for a ramming encounter? Again, we have those officers who care little about the formation but who would attempt to ram their enemy at all hazards, and there are some good arguments on their side. They have only to concentrate their mind on one idea; straight for the enemy is all they need think about. Their mind is made up, and immediately the enemy is sighted there is no hesitation on their part. If the enemy hesitate or flinch, they will undoubtedly be caught at a disadvantage and probably rammed and defeated. This would undoubtedly be the better course to pursue if you were certain of the pluck of your officers, but doubtful of their skill and your judgment. We certainly can always depend upon the pluck of our officers, and by proper training we should be able to depend upon their skill.

Now for the method of attack. Jomini says that no rational man would, under ordinary circumstances, attempt to attack along the whole extent of an enemy's line; that there are always three points for him to choose from, the right, center, or left, two of which may be equally good and the third can be rejected. On land it is usual to endeavor to extend the front equal to that covered by the enemy, to prevent outflanking, because troops struck in the flank are crushed in detail and rolled up unless strongly supported; but the movement of vessels is more rapid and free, and there is not the same danger from having the flank overlapped by the enemy. Upon sighting the enemy the commander would ordinarily be able to determine immediately the point upon which he intended to press his attack. Before indicating the circumstances that will guide him, it will be first necessary to open the vexed question of the relative importance of the modern weapons, the gun, the ram, and the torpedo. It is here necessary to state I am a firm believer in the gun as the first and foremost weapon. And it seems to me that it is the only position to be maintained logically. I believe that within certain limits you can say what the gun will do, what the ram may do, and what you hope the torpedo will do. And in starting from the other end you can draw the limits of the power of the weapons very closely. For the torpedo the degree of accuracy is poor, the range is short, the number of shots few, and the damage is not greatly in excess of shell. With the ram, while the range is still shorter and the probable damage is greater, the delivery of the blows must be few and far between. The number of blows from the gun are practically unlimited; for the smaller calibers the rapidity with which they can be delivered is very great, and the destructive effect of the larger calibers almost as great as that of the other weapons. I believe each has an appropriate time for its use and its distinct tactical value; besides this, there is a strategic value that may overbalance the tactical one. It must be remembered that the subject of fleet actions is now being discussed, as in individual duels there is room for a different discussion of these weapons. In the case of the gun, the commander of a fleet who has had an opportunity properly to exercise his fleet, and who finds it manned with well trained gunners, should be able to calculate very closely the damage he may reasonably expect to inflict with his guns, according to the composition of the enemy's fleet and the state of sea and weather. He should be able to select a range and concentration of fire that would insure his damaging the enemy. In this selection, and in so handling his fleet, his personal skill would be clearly manifested. With highly trained gunners, good guns, and well protected ships, he would leave less to chance by using the gun than if he selected the ram. If he should select the ram, he can make the best arrangement to rush to the attack, but after that, he can have but little management of his fleet, and his success or failure must largely depend upon individual skill and the accidents of a close encounter, and it would be out of his power to remedy any defects in the execution of his programme after it was once initiated. Certainly a commander who is confident of his own skill and of his power to wield his fleet as a single machine, would prefer that weapon upon which he can base his calculations and throw out the element of chance as far as possible. He would first use the gun until, his calculations proving correct, he had disabled a portion of his enemy's fleet; then the great value of the ram would be seen; it certainly becomes a highly powerful weapon when wielded against a disabled antagonist. At this time its value far exceeds that of the gun, and a good commander would push his advantage home with the ram. Much has been claimed for the torpedo, but whenever it has been practically tested its performance has fallen far below the expectations of its advocates. As at present developed, it has little or no tactical value in fleet actions. Not that I would advocate doing away with the torpedo in battle-ships; only I would be willing to sacrifice very little to the weapon, and it would not alter in the slightest degree the tactical disposition of the fleet or the manœuvres in battle. It may help to disable an enemy, and in the case of a vessel already disabled it may enable her to injure the vessel that is about to ram; but it is hardly powerful enough to prevent ramming, and is too uncertain in its action and effect to change the manœuvres of the vessels engaged. Many will say that in selecting a range at which to engage, concentrating the fire and finally charging the disabled vessels with the ram, the enemy can take the same advantages; this is true of all strategy and all tactics, and at sea, where all motions are plainly visible to the enemy, he would be aware of each manœuvre and could make the proper reply. The only thing is to be prepared and make the best moves according to sound tactics, and if the best reply was made, the terms would still be equal; but if a weak move should be made, and a strong reply by the enemy, the latter would then have an advantage. In studying the tactics of all naval battles, the moves made by the successful commander are readily comprehended, and the reply is plain to the student; he wonders why the defeated commander failed to make it, until he realizes that what is so plain on paper becomes more difficult when undertaken amid the excitement and confusion of a battle when resting under the responsibility of conducting a fleet. He realizes that the commander who despises tactics and considers it too simple a matter to be worthy of study, meets his antagonist without preparation, and for want of a proper reply is defeated, while the successful commander, having the few simple rules well in mind, and having considered their application to all the possible positions of the enemy, is ready to make the strong move and take every advantage of his enemy, and defeats him. The old method of doubling up on the enemy's line was always easily answered by starting to double back on his line when he made the move, but this reply was generally neglected. And in all history there is very rarely seen a case at sea or on land where more than two or three strong moves are followed by the same number of sound replying moves; although plain and simple to the student, to the commander in action they become difficult, and sooner or later the more skillful gains a decided advantage. Perhaps the finest campaign in all history to study and see the best move followed by the strongest reply is the advance of Sherman upon Atlanta, and Johnson's retreat upon that city. Johnson had the weaker force and must retreat if properly pushed, and yet he was too strong to be attacked rashly. Sherman would extend one flank or the other, and by threatening Johnson's flank, or threatening to cut him off from his base, would force the Southern army to retreat. Johnson would retreat at the right moment and fall back to the next strategic point with great skill. Had Johnson made a mistake, he must have suffered a crushing defeat; had Sherman made a false move, if not defeated, his advance would have been checked and he must have awaited the arrival of reinforcements.

The range at which it is best to engage now becomes a question for consideration, and this can be readily fixed for our present guns by reference to the "Text-Book of Ordnance and Gunnery," by Lieutenants Meigs and Ingersoll, and the "Tactics of the Gun," by Lieutenant Meigs. With our present guns and with the present type of battle-ship, 2000 yards should be the opening range under ordinary circumstances, as then the probable target would catch one fourth of the shots fired. The effective range being about 1100 yards, the inside range, or where the gun is no longer supreme, is affected by three things: the range of the torpedo, from 300 to 400 yards; the magazine rifles and machine guns, their fire becoming dangerous at less than 500 yards; and the danger of being rammed, which depends upon the tactical diameter of vessels engaged. This latter is the most important limitation to the range to be selected, for if within the space where the enemy may ram, it becomes necessary to turn your bows to his vessels, unless you have superior speed and intend to run away. Leaving this out of the question, if you come within this distance, a ramming conflict becomes inevitable, and the gun is subordinated to other weapons. Therefore the range will ordinarily be in the neighborhood of 1000 yards, not over 2000 and not under 600 yards. Should the enemy's vessels be slightly protected in comparison to ours, or should his guns be of less power, it may be advisable to engage at greater ranges so as to take full advantage of our superiority. The angle of presentment is the next point to be considered, and this is regulated by the distribution of the guns and armor. The modern tendency is to armor the entire exposed surface and give strong bow fire; then, approaching end on, or nearly end on, will allow the use of full strength of fire and expose the least surface with the armor at the best angle; but when intending to carry on a gunnery combat there will be very little time occupied in passing from the outer to the inner limit of the gunnery range, so that the angle of presentment must be regulated so as to keep the fleet between the limits and yet as much less than 90° as possible.

The next question is the order of formation. The fleet should be so formed as to give full opportunity for engaging with the battery, and yet it must be a flexible formation, readily maintained and easily reformed when once broken. The first thing to be observed by a fleet is close order, and this should be very close, so as to give good mutual support in case of ramming. Then if your vessels are nearer together than the enemy's, you have a greater force than his on some part of the line, and you move your vessels in less space. A formation in column is generally bad, for many reasons, and certainly in very close order it becomes dangerous if steaming at high speeds. Should one vessel become disabled, the line might be thrown into disorder and collisions might take place. A formation in line is bad, as it does not permit a full development of the fire of the battery, and should the enemy pass through, your shot would be almost as dangerous to friend as to foe, and any vessel disabled and dropping astern would be without protection. I believe that a bow and quarter line, or *échelon*, is the formation that most nearly answers

the requirements, and this in spite of the adverse criticisms in
"Modern Naval Tactics." Admiral Randolph makes the following
remarks on bow and quarter line: "It is hard to preserve; the
opportunities, however, for using the battery are very good. It is not
the most efficient for mutual protection from ramming, as the sternmost of two vessels receives none from her bow neighbor, and the
leader of two, not the utmost possible from her quarter neighbor.
Furthermore, it is not a good formation for ordinary steaming, as it
is hard to maintain your station, and is difficult to change direction
in, although it permits great liberty of movement of the vessels themselves."*

In three years' experience in the North Atlantic Squadron I have
seen the vessels frequently cruising together in most of the prescribed
formations, and do not believe that it is hard to preserve this formation or difficult for vessels to maintain their stations. Commander
Colby M. Chester used on the Galena, attached to the squadron, a
small instrument that greatly facilitated keeping in position in any
formation except column of vessels. It consisted of an alidade on
deck with a rigid connection to the engine room below, where a
pointer was attached that moved with the alidade. When in position,
the pointer was set at the zero of the scale with the alidade on the
mainmast of the guide. A man was stationed to keep the alidade
on the mainmast of the guide, so that the position of the pointer
showed the engineer on watch how the ship was maintaining her
bearing with the guide, and when it was necessary to diminish or
increase the number of revolutions of the propeller. All that is
necessary beyond this is for each vessel to have the error of the
compass accurately known and to steer a fairly straight course. To
change the direction of the line of bearing, bring the fleet into column
of vessels in natural or reverse order by a simultaneous change of
course to the direction of the line of bearing ⅛ or ⅜ of a circle. Then
give the leading vessel, the guide, the direction of the new line of
bearing as a course; this will be taken by each vessel in succession.
When all are on this line, give the new course, making an angle of
45 degrees with the new line of bearing; this will be assumed by the
vessels simultaneously and they will be in bow and quarter line.
The direction can be changed by wheeling while in this formation as
readily as when in line abreast, but wheeling with an extended front
of vessels is an awkward manœuvre. Even if the bow and quarter

---

* "Modern Naval Tactics," Commander Bainbridge-Hoff, U. S. N.

line is a difficult formation to maintain, we must increase our skill in
holding positions. As we cannot afford to reject fine instruments of
precision because sailors have been accustomed for a long time to
the handling of coarse ones, so we cannot afford to reject formations
or evolutions because we lack experience. The officers and men
must be educated up to the required point. With every improvement introduced in the Navy have come objections from some that
the instrument was too intricate to be placed in the hands of sailors;
still, so far, neither the sailor nor the officer has failed when called
upon for additional skill, knowledge, or intelligence; all that is needed
is time to adapt themselves to the instrument or method that they
are required to use. Commander Bainbridge-Hoff says: "Formations *en échelon* present to the enemy a line of rams and a line of fire.
Yet it is a dangerously bad formation if it is attacked in its line of
bearing, as it is not flexible." I think he has gained the impression
of the want of flexibility in the formation from the old method of
handling a fleet *en échelon*, viz., forming it only from line abreast and
requiring a return to line abreast to change the formation; this is all
changed in the new evolutions with which he has had so much to do,
and certainly there is no want of flexibility if properly handled. As
to its weakness, in all formations there are points of weakness and
points of strength against an attack. In line abreast, which is strongly
advocated by some, an attack upon either flank would find the weak
point of the formation. To meet this by a change of direction, there
being sufficient time to make the change, would take no longer *en
échelon* than in line abreast. And to meet it by change to column
would require the vessels *en échelon* to alter their course forty-five
degrees against ninety degrees for those vessels in line abreast. The
fact that this formation presents to the enemy both rams and a line
of fire, and permits great liberty of movement to the vessels themselves, more than balances any disadvantages there may be in this
formation. The introduction of the ram and torpedo caused many
to remodel their ideas of naval tactics, and a rather sudden departure
was made from old formations and manœuvres. It may be that this
was too sudden, and that where so much must be speculation
and derived from paper arguments, we may still be able to take
some lessons from history. Sir Howard Douglas carefully derives
his steam tactics from experience with sailing vessels and from the
history of naval warfare when sail was the motive power. All this
was thrown aside when the ram was re-introduced, but more modern

experience has shown, I think, that his arguments were well founded and need only slight changes to adapt them to the present time. Thus it is no longer necessary to bring the broadside to bear upon the enemy in order to deliver the full strength of the battery. By consulting the valuable diagrams of Lieutenant Chambers in General Information Series Nos. 5 and 6, it will be seen that the average vessel can deliver as effective a fire at an angle of fifty degrees from the keel as at ninety. Because of the ram, it is advisable to meet the enemy nearly head on, and it may be advisable to alter the angle of the keel with the line of bearing from forty-five to something nearer ninety, and this may be done without sacrificing any strength of fire. The group system has been strongly advocated, and many ingenious evolutions have been laid down, but the system is cumbersome and sacrifices a portion of the offensive strength of the fleet to increase the power of defense; its advocates have been misled by the undue prominence given the ram. In groups of three, a portion of the fire strength is sacrificed, and there will be a strong tendency to split into groups after the first rush of the attack. There are times when the group system may be advisable; such as when our fleet is somewhat weaker than that of the enemy and it is necessary to make a strong showing and to run unusual risks in our attempt to defeat the enemy; then the unarmored vessels may be brought into the line of battle, and their inherent weakness somewhat remedied by the formation of groups, preferably groups of twos.

The ideal fleet is one composed of eight line-of-battle ships in four divisions of two each, right, center, left, and reserve divisions. This forms groups of two, but very pliable groups; the divisions keeping together in fleet formation as long as possible and then separating into pairs; each division to have its fast cruiser and sea-going torpedo-boat, and the fleet to have in addition such specially devised boats, rams, and dynamite cruisers as may be deemed advisable; six vessels to cruise together in *en échelon*, cruisers sent out as scouts, and the reserve with torpedo-boats, etc., in the rear.

There is good authority for placing unarmored vessels in the line of battle. Still, believing as I do that the gun is the principal weapon, and that it would be allowable to place them in the line only in some such exceptional cases as mentioned above, the argument of Admiral Sir Geoffrey Hornby, K. C. B., seems unanswerable. The Admiral says: " You may put an ironclad in any position you please, and if she is attacked by three of these unarmored vessels she will make a run at one vessel or the other, attack her with her guns, and threaten her with her ram, and it must be death to that vessel if she touches her in any way. The two other vessels, you say, will follow and disturb the action of the ironclad; but it must be recollected that the ironclad will always be firing at them, one gun at least, and with shells, the bursting power of which must necessarily, as far as we know, destroy the fighting power of the cruiser. If any one contemplates what the bursting of a shell filled with 37 pounds of powder in the center of any unarmored vessel would be, I think he would agree there would be no more fight left in the ship. It is no answer to say that the ship is armed with a gun as heavy as the ironclad has; the *morale* of the people will be destroyed, and they will be unable to continue the action against the concentrated force to which they are opposed. For that reason I hold that the unarmored ship cannot contend against the armored."*

The improvement in rapid-fire guns, and the adaptation of the system to larger calibers makes it a necessity to protect the people working the batteries against shell from these guns. Therefore the unarmored vessels will cruise in advance and on the flanks, acting as scouts and giving warning of the approach of the enemy, and then taking up positions well away on the flanks. The fleet will be so manœuvred as to bring the line of bearing in such a direction as to make an angle of 45° with the enemy's course, and the course of the vessels so varied as to draw the line towards the enemy and also to give full development to the fire. If he is in line and shows signs of intending to charge through, fix the course so as to bring the bows of the vessels to the bows of the enemy. The fleet will then be *en échelon.*

With the fleet right in front, under ordinary circumstances, the reserve division will support the left flank, as it is most in danger of being doubled on; and should it be evident that the enemy have no intention of adopting ramming tactics, they can be used to lengthen the firing line. In all cases such torpedo-boats and rams as may be attached to the fleet will shelter themselves behind the battle-ships until the confusion of the enemy or the smoke affords good cover for an attack. When the left is clear of the enemy, all go about with the same helm as pre-arranged or signaled, and stand down for the enemy again. If the enemy decide to pass our line firing, the fleet will be so handled as to keep about one thousand yards from the enemy's vessels, and

* "Modern Naval Tactics," Commander Bainbridge-Hoff.

at the same time to give ample opportunity to the batteries, the effort being made to draw the fleet past one flank of the enemy without incurring the dangers of an oblique attack. In the days of sailing vessels and smooth-bore guns the oblique attack nearly always proved ruinous to the attacking party, for his leading vessels were exposed to the fire of the entire line of the enemy, and were badly cut up before the guns of the rear vessels could be brought to bear. At the present time there is less danger, as the vessels are a much shorter time under fire, owing to their great speed, and nearly the whole strength of fire can be brought to bear as the guns are now distributed. Still, an oblique attack tends to induce the enemy to concentrate his fire upon the leading vessels. The fire of the heavy guns of the fleet will be concentrated upon one or two vessels of the enemy, according to the distance, state of sea, and weight of fire of the fleet, taking into account the probable number of hits and the disposition of the armor of the opposing vessels, the idea being to insure as nearly as possible the disabling of at least one of the enemy's vessels, rather than distributing the damage over several vessels. Should some of the enemy be more lightly armored than others, they should be selected as the target, if so placed in the line as to be readily aimed at. One vessel disabled will tend more to throw the enemy's line into confusion than if the hits were distributed; besides, with distributed hits the damage may be repaired without reducing the fighting efficiency of the vessel. The fire of the lighter calibers, particularly the quick-firing guns, should be left to the commander of each vessel, he pouring it into his nearest available enemy, so as to keep down the fire of the heavy guns, damage torpedo tubes, and riddle unarmored ends.

If the enemy decides to try ramming tactics and attacks formed in column, he will charge through some portion of the line; the reserve and torpedoes will form in rear and strengthen that portion. The concentrated fire of the fleet should be able to damage the leading vessels of the column. The reserve must be ready to fill any gap in the line that may be caused by vessels being disabled, the most likely point being where the enemy charges through the line, for the vessels there will be liable to be rammed, and will receive the fire of the enemy as they pass through. The torpedo-boats will be ready to attack any disabled vessel of the enemy, and with the reserve must make a strong attack on the rear vessels, should the others succeed in getting through without being disabled. If the enemy attempts to ram with his vessels in line, the ordinary position for the reserve would be in rear of the center division.

The question of weather or lee gage has given rise to some controversy. If it were possible to hold the weather gage during an action without sacrificing other points, it would be well to hold it; for to windward you can see your own ships and gain occasional glimpses of the enemy, while to leeward you can see but little of either. But with steam fleets in action, the interchange of positions is likely to be rapid and frequent; it would then be ·preferable to start from the leeward position, as then, the formation being well preserved, the position of your own vessels and that of the enemy would be known, and after charging through you would have the weather position clear of the smoke in which to reform the fleet, and the enemy would have the drifting smoke to add to his confusion. At least this is the conclusion I have reached after witnessing many target practices and some sham attacks where blank charges were used. "Parker's Steam Tactics" is the only authority I can quote to support my idea as to the position of the admiral in command of a fleet in action. All former experience seems to be against me, and yet I find myself unable to relinquish the opinion that the line of battle is as improper a position for the commander of a modern fleet as it is for the general of an army. The history of former naval battles shows us that the admiral arranged his plans before the action, informed his captains, and after hoisting the signal for going into action, closed his signal-book and turned his attention to fighting his flagship until the close of the action, when he resumed his office of admiral. The following appears to be the idea most generally held: "The admiral's flag belongs on the largest and most powerful vessel, which should outshine all others as a brilliant example in the heat of battle. History shows that the admiral must himself be in the midst of the fight. If he wishes to be the vital element in his ship and to make the best use of all the components of his command, then he must put himself in a position to see everything and to be seen of all."*

How an admiral can see everything and make the best use of all the components of his command, and at the same time outshine all others as a brilliant example in the heat of battle, is hard to conceive. It sounds well, but it is not practicable. There is now greater opportunity for the tactician to handle a modern fleet, and an admiral can

* "Modern Naval Tactics," Commander Bainbridge-Hoff.

be better employed than he would be if setting an example by the way he fights his flagship. When we have to fight modern vessels we shall have no lack of captains to fight them boldly, bravely, and skillfully ; but if the combat is to be with the fleet combined, and not a confused *mêlée* of separate conflicts between two or three vessels, the fleet will require the undivided attention of the admiral, and he should be so placed as to be able to concentrate himself upon the entire action, and not have his attention especially drawn to the conduct of one vessel. Being in the most powerful ship, he would be most likely to disable an enemy, and in the charge through might be detained to give the *coup de grâce*, while his fleet, having passed through, would be without his direction when reforming. Again, to him is confided the policy of his government and the condition of the entire force at its disposal, and he will have carefully estimated what risks it is wise to run with his fleet ; and the exigencies of the case may require him to withdraw his fleet from action, which would be a difficult task if he were in the thick of the fight, even if he were able from this position to judge of the condition of affairs. I firmly believe that the admiral should withdraw as much as practicable from the heat of the battle by taking his station on a strongly protected despatch-boat, in designing which everything must be sacrificed to speed and buoyancy. During the artillery portion of the combat he can shelter his vessel under the battle-ships, and in the charge through escape at the flanks. He may run as little risk as is consistent with keeping his fleet well in sight, unless in cases of desperate emergency, when, as a general may sometimes choose to lead a last charge, he may go on board one of his vessels of the line and lead the attack upon the enemy. Undoubtedly, it will require some moral courage for an admiral to take his station in battle on a despatch-boat, at least until the idea receives more general approval. He will think of the long line of illustrious men who have always been in the thickest of the fight. Some will point to the time at Mobile when Farragut, rushing ahead with his flagship, saved the fleet and won the battle ; but, on the other hand, if Admiral Byng's attention had not been distracted by a collision, he might have signaled to his rear squadron to engage, and perhaps have gained a victory off Minorca ; at least he would have saved his reputation and his life. Still, if the admiral has had practice with his fleet, and finds his captains skillful in manœuvring, and the fleet under his control acting as a machine, he will not only take pride in handling it, but will have confidence that in its combined action lies the secret of success. And he will feel that his station on a despatch-vessel, while it does not withdraw him from danger, gives him a position from which he can see and manœuvre his fleet, so that it becomes a mighty engine of destruction in his hands, and not a mere conglomeration of vessels. He will prearrange his plans as far as possible, and fully inform the captains of his intentions and desires. Many points may admit of preconcerted action, particularly which helm is to be used in turning after the charge through; but, for the skillful manœuvring of the fleet, signals will be wanted, and we need further practice with battle signals. Flag signals must be frequently obscured during an action, and some other method of communicating with the fleet must be adopted. I think the so-called Japanese day signals would be found to work admirably. These signals are light wooden bombs, projected into the air from mortars. When they explode they throw out shapes, which can be greatly varied and are readily distinguished at a considerable distance. The smoke lies most thickly near the water, and these signals, thrown well up into the air, would be usually visible from deck, and always from an elevated point aloft. The numbers and combinations necessary for battle signals would be few. Large distinguishing flags or shapes should be carried at the highest points by all the vessels of the fleet, so that the admiral may readily judge of their positions. These would be generally visible from aloft.

An attack having been decided upon, and the force of the enemy being equal to ours, what point of his line shall we attack? There are many different situations which may arise that will require each a different answer. If the enemy should be formed in column, particularly if he shows signs of intending to ram, the head of the column would seem the best point upon which to concentrate the fire of the vessels, for, by disabling the leader or leaders, the column would be thrown into disorder. As in all cases when it has been decided to bring the enemy to close quarters, if he is determined to charge through and try the ram, we must present our bows, and cannot prevent his using the ram, so in this case we must endeavor so to set the course of our fleet as to cause him to charge through at or near our center. This will give our vessels all along the line the best opportunity to fire at the leading vessels of his column, and should he be able to pierce our line near either flank he might be able to detach and destroy the flanking vessels. He should never be allowed when in column to pass along either flank of our fleet, for then he would have the advantage of concentration upon the flank vessel. If he attempt this by withdrawing or refusing the squadron on that flank until his center has passed, we can head for his line and charge through. But the best point for us to make would be to oppose our center to his leading vessels, when we could combine the fire of six of our vessels on his one or two during the advance, the two in the reserve division being ready to meet with their fire the vessels as they pass through. Here the enemy would be able to pour the fire of their eight vessels into our center vessels; still, at this point, each vessel of the enemy will meet with the fire of four of ours, the two between which they are passing and the two of the reserve squadron.

It is true that the enemy having reserved their fire from necessity, being in column, will be able to use their heavy calibers, but at this range the smaller calibers, the quick-firing and machine guns and magazine rifles will have the advantage; the smoke will be great and the heavy guns must be fired unaimed and amid a shower of missiles from our four vessels. I think the chances will favor our damaging his leading vessel and escaping with our center vessels uninjured. If the leading vessel is injured, some of our torpedo-boats or rams should finish its career. The enemy may charge through our fleet without ramming; it would seem as if it would be natural for our vessels to crowd imperceptibly a little towards the flanks, leaving him a free passage, and it would certainly seem a most dangerous attempt for any of his vessels to sheer out of the column and attempt to ram; if he should charge through, our two center vessels and reserve squadron, with torpedo-boats and rams, if we have any, should block the way for the rear of the column and attempt their destruction. But should the shock come, and his leaders meet our vessels with the ram, their speed must be suddenly checked; the other vessels must sheer to continue their course, his formation must be broken, he must meet our rams at a disadvantage, and we should succeed in doing far the greatest injury. I firmly believe that the attempt to charge in column or columns will always prove fatal to the fleet making the attempt; that their complete success at the point of shock by ramming their opponents would be their own undoing, as in arresting the advance of the leaders the mass would be thrown into confusion, and would in turn become the prey of a well-formed line. The impetus of the body is not increased by its mass; it must fly apart at the first shock unless it would be its own instrument of destruction; it cannot submit to further compression with safety. Then the leaders should be damaged by our concentrated fire and would have therefore less than an even chance at the point of shock.

If the enemy be formed in line and he decide to ram, our line should be so manœuvred as to bring our center opposite to one of his flanks—to his right flank if our bow and quarter line is formed right in front, and to his left flank if our fleet is left in front. This would enable the fire of our fleet to be concentrated on his flank vessels, and if he in turn concentrate on our flank vessels they will be farther away, being refused, than if we attacked on the other flank. The reserve will strengthen the flank that must charge through the enemy. Such of the fleet as clear the enemy's flank should change direction simultaneously 12 points towards the enemy—that is, to the left if right in front, to the right if left in front—and follow up his engaged wing. If this wing has continued its course, our vessels will again change course simultaneously 4 points, when in their rear, and follow them up. If the enemy attempt to turn immediately after passing through our line, our fresh wing would strike them in the flank when they were disordered and when they were probably weakened by the attack of the first wing. The first wing should go about together as soon as the last vessel was clear of the enemy's line, and be ready with the reserve to meet the enemy's fresh wing. The torpedo-boats and rams will keep in rear of the wing that clears the enemy, and change course and move down upon the enemy with that wing.

If the enemy intend to engage in an artillery conflict he will be in some extended formation, as line or *échelon*. Then selecting the same flank as before, that is, his left, if we are right in front and *vice versa*, we try to draw our line along this flank; if he make the correct reply to this move the fleets will pass on parallel lines and all will depend upon accuracy and concentration of fire. Then on most occasions the fleet would go about, change direction 16 points, and pass again, watching for the opportunity to strike with the ram when vessels were disabled or the fleet disorganized. In all cases the gun would be our first and principal weapon, and whether he charged our line or passed firing we should attempt by concentrating the fire of our vessels to injure seriously one or more of his. If forced to ram by him, our concentration of fire should give us the advantage before the shock; if not forced, we should gain a decided advantage before using the ram. Never waste our ammunition at great ranges and pour in the lighter missiles at close quarters.

The formations and evolutions necessary for such tactics are very simple; column, line, and *échelon* are the only formations necessary, and the change from one to the other by simultaneous or successive changes of course. But the formation in column should never be a complete bow and stern line; each vessel should hang a little on the quarter of its leader, thus enabling a very close order without risk of accident; the proper signal to be made for each to take the starboard or port quarter of her leader, or to alternate, as the case might require. The formation for attack being as near a bow and quarter line as the circumstances permit, will be probably something between the line and the column, never reaching either limit, but adopting the course between the two that will tend to draw the fleet along one flank of the enemy; the line of bearing of the fleet being at an angle of about 45° with that of the enemy, rarely being parallel to his, and never at right angles; the course of the vessels always being brought nearly or directly opposite to that of the enemy when within ramming distance, that is, presenting ram to ram.

Strategy may require us to attempt the complete destruction of the enemy and thus somewhat modify our tactics. Our fleet may be masking the enemy's while we have another engaged in blockading, bombarding a fortified port, or forcing a landing; in such cases we could not afford to allow a large proportion of his fleet to escape, even if we destroyed the remainder. We must bring his entire fleet into a decisive engagement. Now, while preferring to start with an artillery combat if we are certain of preventing an escape, should there be any uncertainty, we must bring the ram into play; but even then we should be guided as in the first two propositions, and if he be in column, charge so as to bring the head of it at our center; and if he be in an extended formation, charge with our center abreast one portion of the line. In both cases we shall concentrate our fire upon the selected point, the only difference being that we must make greater efforts to ram in the charge. If we have superior speed, or have a sufficient force of swift torpedo-boats and rams to force a fleeing enemy to turn, then we may endeavor to injure his flank vessels with artillery before charging.

Again, strategy may require us to avoid for the time a decisive battle. Our fleet may be opposed by the masking fleet of the enemy, while he has another fleet engaged in blockading or attacking one of our fortified ports or endeavoring to force a landing of troops.   If we meet the combined fleets it must be our endeavor to prevent him from separating and engaging us with one fleet while the other fleet proceeds about its undertaking.   If he succeeds in opposing us with his masking fleet alone, we must endeavor to escape from this fleet without too great a sacrifice, and then proceed to interrupt the operations of the other fleet.   In neither case can we afford to risk a decisive engagement at once.   In the first case we should run the risk of being overwhelmed by a superior force, and in the second, even if we should defeat the masking fleet, it would be at the risk of so crippling our own vessels as to leave his remaining fleet free to carry out his hostile intentions.   Strategy here requires us to play a waiting game.   In the case of the combined fleet we must proceed against them with our cruisers well in advance, our battle-ships next, with rams and torpedo-boats in the rear.   Push forward the protected cruisers to resist the advance, as a game of long bowls with its chances of accident, we being near our repair shops, is in our favor.   Let them fall back only when too closely pressed by a superior force, being careful to keep beyond the range of rapid-fire and machine guns.   When they fall back they will disclose our line of battle, which will retire when the advance of the battle-ships of the enemy threatens to bring on a general engagement.   Encourage the enemy by every means in our power to expend ammunition and fuel.   At night bring forward the rams and torpedo-boats.   By risking some of these and sustaining the attack with one or two battle-ships or armored cruisers, fast ones, the enemy may be forced into expending considerable ammunition.   These attacks need not be pushed home and can be undertaken advantageously with but little idea of damaging the enemy's vessels, but undertaken mainly with the expectation of forcing him to reduce his supply of ammunition, a most serious thing for any fleet undertaking a distant expedition, for not only must reserve supplies of ammunition and fuel be carried with the fleet, but also the vessels must have time and opportunity to refill.   The enemy would be able probably to defeat a real attack from torpedo-boats and rams, with the aid of electric lights and rapid-fire guns, but he will be more than human if he does not fire away when they are first sighted and before they incur imminent danger, especially if he be stirred up by a few large torpedoes from a dynamite cruiser.   Here will be one use of a torpedo-boat designed to carry pneumatic guns, for at four thousand yards distance one could throw in among the fleet enormous charges of dynamite, and even if they failed to damage the vessels their moral effect must be great. It will certainly require skillfully executed manœuvres to harass the enemy and yet prevent a general engagement; but after all it is only a question of ratios. At the speed of modern vessels, the enemy pressing steadily onwards, the speed of our retreat must be great and yet it must not become a flight. The ratio of the speeds must be carefully regulated so as to check the enemy and then avoid him. If he makes hot pursuit he may be made to cover much ground, with the consequent expenditure of fuel, by frequent changes of direction. The enemy's fleet would be likely to have with it a number of auxiliaries to carry coal, ammunition, and possibly troops, and the expedition might be seriously crippled by the destruction of these vessels. Many attempts might be made for this purpose both during the day and at night. Cruisers might be sent around the flanks if the unarmored vessels of the enemy kept well in the rear; torpedo-boats and rams might get around the flanks or through the fleet at night, or our fleet might charge through the fleet of the enemy and seriously damage the auxiliaries before they could escape. After a false attack had been made on the fleet with torpedo-boats, a second one might be made and pressed home, being closely followed by our battle-ships, which, charging through the enemy's line, could attack the transports, etc., with every hope of destroying them before their own battle-ships could come to the rescue.*

If our fleet is masked by the enemy either by being caught or pressed into port, then it must make every endeavor to escape the masking fleet so as to attack the active fleet of the enemy. It then becomes a matter of escaping blockade, the only question being whether it would be advisable to escape singly or in fleet. Single vessels might escape and again combine at some selected rendezvous, but such a separation of the fleet would be dangerous ordinarily and might result in disaster. The fleet should attempt to force their way out together. By repeated attacks with torpedo-boats and false alarms created by apparent attempts to escape, the enemy can be continually harassed; he will be forced eventually to send some of his fleet away for coal, and possibly for ammunition. Then let the fleet seize the proper moment, possibly a dark night or during bad weather, and, attacking first with torpedo-boats, push through with the fleet.* -The harbor may be such that it will be necessary to have

---

* This is one of the few occasions when torpedo-boats would be used first in an attack; they being sent ahead to soften the enemy's line and make the charge through practicable.

range lights showing to guide the fleet. Should this be the case, it would be well to show lights in this vicinity for several nights before the attempt to escape is made, and pretend to signal with them to boats or vessels in the harbor, so that when really used the enemy would not know their character. Otherwise a vigilant enemy would readily pick up ranges leading out of the harbor and know that a real attempt was to be made. Perhaps the most difficult case to encounter would be where the masking fleet could seize a harbor near the fleet that they were masking, and, without attempting to blockade our fleet, merely keep up a close observation with cruisers and torpedo-boats, being ready to slip and pursue when our vessels attempted to escape. But even here we could make a vigorous attack upon his observation vessels, and frequently make a pretence of escaping, forcing him to pursue, until the proper moment came for an actual attempt. Certainly there would be ample opportunities for the exercise of the talents of skillful seamen and tacticians on both sides. If the enemy equaled us in these points, we should escape unless his fleet was much the stronger. Having escaped, we must attack his other fleet and prevent the accomplishment of his designs. Our fleet would probably be in telegraphic communiction with the point attacked, and we might be able so to time our escape as to reach the threatened point at a time when his fleet was engaged with the fortifications and when we could strike him at a disadvantage. Having made our escape, we should have but little time to wait before attacking, for, as soon as the masking fleet lost touch of ours it would naturally look for us near the point attacked. Now comes the moment in which the bold commander will always delight; there is no longer room for hesitation, the points for and against an attack can no longer be weighed, we must make a vigorous attack, and, damaging the enemy, either drive him off or cut through to the harbor defenses. If we fail to drive him off or so injure his fleet as to prevent the further prosecution of his design, we are defeated, and it will be no advantage to the country to preserve the vessels from harm if their fighting strength is nullified. Heretofore we have refused a decisive engagement to preserve our vessels for this crucial moment; now we must do the greatest possible damage to the enemy even if we sink our entire fleet. But we should command success, for if our attack be well timed he must not only engage the forts but must also be caught between the fire of our harbor defenses and our fleet. At the moment when the enemy has pushed close in to attack the forts or is trying to pass into the harbor, then if our fleet strike him in the rear we shall have done the best that can be expected of strategy and tactics, and our fate will depend upon our seamanship and strength.

We may have more than eight battle-ships cruising together; if they are over ten, I believe they can be more successfully fought if they are divided into two fleets. When the number of auxiliaries accompanying a fleet is taken into account, and the extended line of battle, even when in very close order, is considered, it would seem that, if too close a combination were used, tactics would be almost impossible, and the difficulty of seeing and obeying signals would lead to confusion. If sailing together, they should be divided into separate fleets and manœuvred separately, but having the same object in view. When together, the ordinary sailing order should be in parallel lines of bearing, the right vessel of the rear fleet being directly astern of the right vessel of the leading fleet, the distance between lines being at least one-half a mile. This will make an excellent formation for battle, unless the enemy advance with extended front, and from want of speed or because of our position we cannot charge through his line. If this be the case the fleets must be so manœuvred as to extend our front to give a strong line of fire. If we can charge through an extended front of the enemy while we are in parallel lines, we should be able to destroy or disable some of his vessels. This formation combines the advantages of the column and the line, as we have depth with a sufficiently extended front to deliver a strong fire. The distance between lines should be enough to allow each preceding line to be well clear of the enemy before the following line charges through. The leading vessels should not attempt to ram unless forced by circumstances, or unexceptionable opportunities offer, so as to leave a clear space for the last line, which, after charging through, should turn and stick to the enemy and ram and use torpedo-boats freely, the leading line or lines turning when well clear of the enemy and moving in such a direction as to attack the remaining vessels of the enemy not heretofore closely engaged. If the enemy form his fleet in compact order, then we must endeavor to draw our line across his front and strike him in the flank if possible. In handling a large number of vessels the reserve fleet will be a most important point, and the commander who has his reserve well in hand and can throw it upon the enemy when his formation is confused by a close attack should win the day.

In this discussion of naval tactics there is no attempt to provide for many of the cases that may arise in battle; it is only intended to show that there still is room in naval warfare for tactics, and the chances will favor the tactician, other things being equal; that the gun first, and afterwards the ram, are the principal weapons, and that the fundamental principle of naval as of military tactics is to combine our force against a portion of the enemy. The evolutions and manœuvres must be simple, the tactics must be plain; the enemy may readily answer every move, but should he fail to be a tactician or make a wrong move we can seize the advantage and should defeat him. On the other hand, if we fail to adopt any tactics and rely on the prowess of individual vessels, our fleet will be a confused mass, and a skillful enemy will surely hold us at a great disadvantage.

'There have been great strides made in naval weapons, but in looking over the field with an eye to tactics, it appears to me that the love of novelty and invention has carried us too far in both guns and torpedoes. Our high-power guns are magnificent weapons, but are they entirely designed to meet the work they must accomplish? Their great range is of no advantage in an ordinary naval combat, as the necessarily limited supply of ammunition will prevent us from firing at an enemy until the probable chances of a hit are within reasonable limits. If, as seems highly reasonable, the opening range hereafter will be about 2000 yards, cannot we save in weight of gun without greatly lessening the penetration or increasing the height of the trajectory, diminishing the efficacy and the accuracy of fire? As all know, this weight might be used to great advantage in many ways—in more guns, more ammunition, more armor, or more coal. The supposed terrible effect of torpedoes has led to their adoption on board any and every vessel. It seems to me that torpedoes on the ordinary type of vessels are likely to prove more harmful to friend than to foe; that their main use should be on specially designed vessels; for in armored vessels and above-water tubes, and in any kind of tube in unarmored vessels, the rain of missiles from rapid-fire and machine guns that will be poured in before the range of the torpedoes is reached will make it probable that they will explode on board, while only possible that they will explode under the enemy. The chances should be at least the other way before a weapon is generally adopted.

Our great need is to study strategy closely in time of peace, and to adopt simple formations and evolutions, in order to ascertain carefully the limits of our weapons. Much must be left to conjecture, and can be demonstrated only in battle; and even then tactics can only be outlined, a few broad general principles laid down, and the rest left to the admiral of the fleet. It would be impossible to provide for every case, and he should not be hampered with binding instructions. With the general plan of attack in his mind, the tactician will be ready to take advantage of every mistake of his opponent, and will hold victory in his grasp.

ARMOR FOR SHIPS: ITS USES AND ITS NATURE.

BY SIR NATHANIEL BARNABY, K. C. B.

[Reprinted from the Proceedings of the Institution of Civil Engineers.]

At the commencement of the Russian war in 1853-4, several points in relation to the defense of ships against projectiles had been established, by experiment, to the full satisfaction of the naval experts. It had been settled that war-ships must be built of wood; iron was an altogether unsuitable material, as the captain of H. M. S. Excellent reported, 1850, that "whether iron vessels are of a slight or substantial construction, iron is not a material calculated for ships of war." Seventeen iron ships, which had been in process of construction for fighting purposes, were thereupon condemned as useless for war service. It had been decided that protection by armor-plates was of no value. Iron plates 6 inches thick, placed in front of the wood, had failed to protect it. Admiral Sir Thomas Hastings had reported that wrought-iron plates (6 inches thick) riveted together and fixed over the planking of a ship's side, would give no protection at 400 yards against shot fired with 10-pound charges from 8-inch guns and from heavy 32-pounders. The favorite guns were therefore shell-guns.

The armament of a three-decker of 121 guns consisted of one-half of 65-cwt. shell-guns and one-half of 32-pounders, with one 68-pounder pivot-gun. The shell-guns were placed on the lower and middle decks. In a two-decked ship the lower deck was armed throughout with shell-guns, and the other decks with 32-pounders of from 42 to 58 cwt. In some cases the main deck was partially armed with shell-guns. The frigates were armed with 8-inch 65-cwt. shell-guns on the main deck and with 32-pounders on the upper deck. All shot were of cast-iron. No one knew what the result of an action with these shell-guns would be, but all agreed that the struggle would be short.

In the autumn of 1853 a Turkish fleet, of seven frigates and some smaller vessels, was attacked at Sinope by a Russian fleet of six line-of-battle ships (three of them first-rates), two frigates, and two or three smaller vessels. The Turkish ships had no shell-guns. They were armed with 24-pounder shot-guns. In less than five minutes a Russian ship of the line, Grand Duke Constantine, destroyed a battery and a Turkish frigate anchored near it, chiefly by shell from the lower deck. A short time afterwards the Ville-de-Paris, another Russian ship of the line, blew up, by shells, another Turkish frigate. Of the crews in the Turkish fleet, above three thousand perished. Of the ships, only one small steamer escaped.

In the allied attack upon the forts at Sebastopol in October, 1854, the Agamemnon, London, Sans Pareil, and Albion took up positions within 1000 yards of Fort Constantine, which was armed with one hundred and four of the heaviest guns and was supported by three other smaller heavily armed batteries. In about an hour the Albion, London, and Sans Pareil were so injured that they had to haul off to a greater distance. The Sans Pareil and London soon afterwards resumed their stations to support the Agamemnon; the Rodney, Queen, Bellerophon, and Arethusa also came to her support. The Queen was almost immediately set on fire by a shell, and had to be towed out of action. The Agamemnon and the remaining vessels kept up the cannonade, which had commenced at two o'clock P. M., until it was dark, when they drew off. They had in all three hundred and ten men killed and wounded. The Albion and Arethusa were so damaged that they had to be sent to Malta for repairs. The Albion, Retribution, London, and Queen had all been on fire, chiefly from the effects of shells with time-fuzes.

The destructive effect of the shells and other incendiary projectiles used by the Russians in these engagements led to the immediate use of armor, notwithstanding the previous adverse decisions of English gunnery officers. The adoption of armor led at once to the use of iron in the hull by English architects, in spite of the series of experiments at Portsmouth which have been referred to, and which had brought about the condemnation of the fleet of seventeen iron ships.

The first move in the direction of armored plating was made by the French, in the month succeeding the unsuccessful attack of the allied fleets upon Sebastopol. In an imperial order to the French Minister of Marine, dated from St. Cloud, 16th November, 1854, the Emperor Napoleon III. pointed out that in warfare there must be, in addition to courage and ability, even chances. Actions on land are avoided so long as there is no chance of success. So in naval warfare, if the fleet were risked, it would be in the hope of destroying that of the enemy. An enormous capital is hazarded in order to destroy that which has cost the enemy as much. But, he went on to say, so soon as the fleet is employed in the attack of a fortification the proportions are entirely altered, for not only will a ship be found inferior to a land battery, because a ship offers a large object to strike while the land battery occupies but little space and is protected by parapets, but also the stake is materially different. So it happened that in the Black Sea twenty-five thousand sailors and three thousand guns could not seriously injure the Russian fortifications, and that indecisive attacks were made at other places, entailing serious damage to the ships without doing any material harm to the enemy. Under the existing conditions risks were incurred for nothing. If the ships threw their projectiles at 2000 metres they consumed their ammunition at a dead loss, and gave a false idea of the power of the fleet. If they approached nearer, they exposed the state to sacrifices too considerable in proportion to the object in view, for it would generally be perfectly senseless to risk the loss of a fleet for the destruction of a few forts. To remove this difficulty he proposed to create " *une flotte de siège,*" capable of producing decisive effects, and at the same time of lessening to the state the chances of loss of men and money. The course taken to meet the Emperor's views was to build floating batteries with numerous light broadside guns, and covered from end to end with iron armor-plates.

Table I. shows what changes in distribution armor has undergone since 1854.

There are two periods. In the first period the armor was employed, as has been seen, to make the chances of loss of capital more nearly even in a contest between ships and forts. The ships or floating batteries were to be small, numerous, and armored. In the second period the armor was to protect sea-going ships in contests with the sea-going ships of an enemy. That period is still current, and has reached a fifth phase, a phase which is in fact only a repetition of the first. The illustrations have been confined to the ships of the French navy, as the English are in no way responsible for them and can perhaps regard them dispassionately.

TABLE I.—PHASES OF ARMOR.

First Period, Floating Batteries, 1854–58.

Second Period, 1858–1888.

GLOIRE.

1856, 1st Phase.   Armor (side), 4¾ inches.
5500 tons.
Speed, 12¾ knots.
Percentage of weight devoted to
armor, 15.

MARENGO.

1869, 2d Phase.   Armor (side), 7 inches.
7750 tons.
Speed, 13½ knots.
Percentage of weight devoted to
armor, 17.

BAUDIN.

1880, 3d Phase.   Armor (side, etc.), 22 inches.
11,200 tons.
Speed, 15 knots.
Percentage of weight devoted to
armor, 33.

TAGE.

1886, 4th Phase.   Armor (deck), 3 inches.
7000 tons.
Speed, 19 knots.
Percentage of weight devoted to
armor, 14.

DUPUY DE LÔME.

1888, Return   Armor (side), 4 inches.
to 1st phase.   6300 tons.
Speed, 20 knots.
Percentage of weight devoted to
armor, 15.

The first phase of the second period was entered upon in 1860, when a sea-going ship, the Gloire, was launched, clothed from end to end in exposed parts with iron armor. Some of this armor, 4¾ inches thick, was impenetrable to the heaviest guns then carried at sea. The second phase soon followed. It was marked by growth

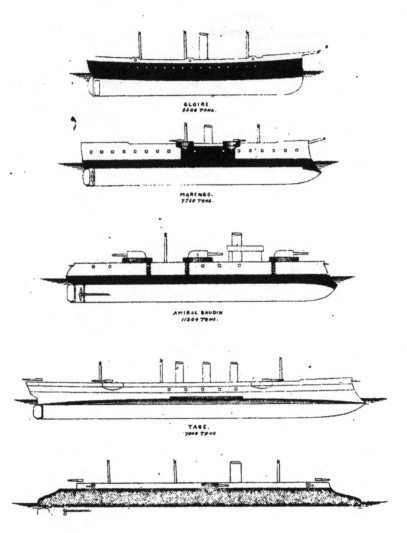

GLOIRE
5600 Tons.

MARENGO.
7750 Tons.

AMIRAL BAUDIN
11204 Tons.

TAGE.
7000 Tons

Speed, 12½ knots.
Percentage of weight devoted to
    armor, 15.

MARENGO.

3d Phase   **Armor (side), 7 inches.**
7750 tons.
Speed, 13½ knots.
Percentage of weight devoted to

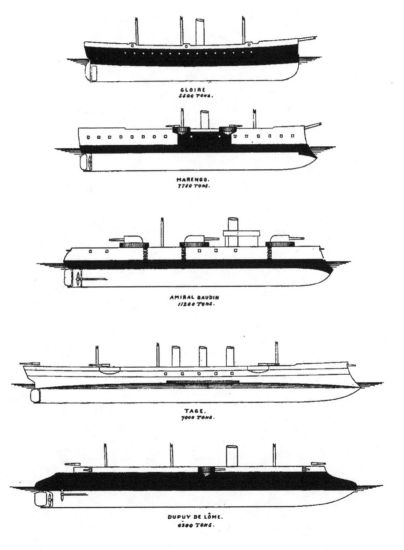

GLOIRE
5500 TONS.

MARENGO.
7760 TONS.

AMIRAL BAUDIN
11200 TONS.

TAGE.
7000 TONS.

DUPUY DE LÔME.
6300 TONS.

in the weight of the gun, by improvements in the powder, and by alterations in the form and character of the projectile to enable it to perforate armor. It was characterized, on the other hand, by increases in the thickness of the armor and by improvements in its quality. But the ship did not increase in size. The British sea-going ironclads of the first phase, commenced in 1861, are larger than those in the Admiral and Victoria classes built twenty-five years later.

To resist growth in the size of the ships for twenty-five years, while the gun and the armor were making strides in advance, was no easy matter. And, while this growth in the gun and the armor went on, torpedoes were introduced, and torpedo-boats. Higher speeds became thereupon necessary and had to be given to the ships.

In the second phase of this second period the French ships underwent a remarkable change, rendered necessary by the foregoing considerations. The French divided their battery, as the English had already done, into two parts. They continued to protect one part with armor, and the other they left unprotected. The relative extent of these two batteries was subjected to continual modifications. The number of guns protected became less and less in succeeding ships, while the number of those left unprotected increased.

At last, taking the Amiral Baudin for example, launched in 1883, and the Formidable, launched in 1885, there are from eight to sixteen machine-guns and twelve 5½-ton guns fought without armor-cover from any direction. The only guns having any armor protection are three in number, each of 75 tons, and these are protected only as to the carriage of the gun; its breech, its charge, and its loaders are without any armor-cover. The signal stations and the officers in command are also unprotected by armor. Admiral Touchard, the French Controller of the Navy in 1873, gave the reasons for this reduction in armor. Referring to the Gloire, Marengo, Hercules, and Koenig Wilhelm, he says: "These, then, are the instruments for fleet-fighting of to-day. Is it to be believed that they will continue to occupy the first position, that they will be the *ne plus ultra?* No, for the gun goes on always increasing in power; and in France, as in England, Russia, and Prussia, we are no longer contented with guns of 10 and 11 inches—we go up to 12 and 13 inches, and there are laid down in the dockyards ships for fleet-fighting plated with armor from 9½ to 12 inches thick, and from 9000 to 10,000 tons displacement. Where shall we stop in this strife? Where will it end? I see the end," he goes on to say, "in the abandonment of armor for the guns. Sooner or later this will be the end for the armored cruising ship. Let us admit, if you will," he says, "that the battery should for the future be plated with armor of 10 inches, as it soon will be, the question of abandoning it will nevertheless be resumed upon these two considerations: 1. Is battery armor of 10 inches penetrable? 2. If it is penetrable, does it not become rather a danger than a protection?" And then he goes on to say: "And now let us see what will take place in a battle between fleets. The squadrons being in order of battle, if both freely accept the contest, there will be collision by ramming, either with or without preliminaries. In any case it will suffice that one of the two should be resolved for it to make ramming the first act; let us mark the opening phase of the fight. The combatants advancing towards each other would be ranged in one line, or in several lines, or in squadrons; but, whatever the order of battle, there will be formidable collisions and grinding of the sides; they will cross each other's courses, and then, putting the tiller hard over, will return to the encounter. Again will they strike and again pass on. Woe to the vessel, which, disabled by the first attack, or turning less quickly than her adversary, exposes her flank! Like the Re d'Italia at Lissa, she will be sunk by a single blow.

"It is here—in this initial phase—the ships perpetually fouling each other, that the gun appears on the scene in all its might. The manœuvre is foreseen. A single order, clear, concise, energetic, follows in three words: 'Prepare for ramming.' At this command the guns are pointed low; the captain of each, his eye fixed on the sights, with bent body and extended arm, waits! · He watches the moment when his adversary will pass before the muzzle of his gun— the moment, fleeting as the light, which he must seize for firing. Silent and motionless, each man is at his post, but lying full length upon the deck. The captain alone remains erect; suddenly the guns and musketry burst forth, and the vessels crash into each other with their sharp-pointed prows, and gliding by each other prolong the contact along their sides until they reach their sterns, which they avoid in order to protect their rudders and screws. After this first encounter, the combat will always be carried on at short distance, if not broadside to broadside, the gun and rapidity of evolution playing the principal *rôle*."

At the same date Admiral Sir John Dalrymple Hay, who had been President of the the Armor-plate Committee, said:

"I come to the conclusion at which Admiral Touchard has arrived, that it is impossible to conceive of any ship but a ship of the very first class, ships of which I do not think many are likely to be constructed, which will be entirely and completely covered with impenetrable armor. If the armor is not impenetrable, then it is worse than useless."

Of the several parts demanding protection he gave the following, with his opinion of the order in which they stood as to importance: 1. The magazine; 2. The propelling machinery; 3. The mounting for the heavy guns; 4. The water-line; 5. The men.

The commander-in-chief at Portsmouth in 1876, Admiral Sir George Elliot, said, in an address circulated among Members of Parliament, that vast sums of money had been uselessly expended in clothing ships with outside armor, because this feature of protection against the gun attack had not been submitted from time to time to the independent judgment of competent critics. He considered that "the evidence of the superiority of the gun, and the development of the efficacy of the ram and the torpedo, had deprived us of sufficient excuse of late years to continue to fight the losing game of armor against guns." He complained that the Admiralty had treated the ram and torpedo attack as secondary, and that its efforts should have been directed toward the protection of ships under the water-line rather than above it.

These French ships, Amiral Baudin and Formidable, mark the close of the third phase of the second period.

The fourth phase in the French navy, marking the total abandonment of side-armor in fighting ships of considerable size, was not reached until ten years after Admiral Sir George Elliot had declared it to be necessary to abandon side-armor, even in the largest ships, in the terms quoted above. The building of the Tage marks this phase. This vessel, called a protected cruiser, is far larger than the fully armored line-of-battle ships of 1858, and is armed with sixteen guns, capable of perforating 8 inches to 10 inches of iron armor. Although much larger than the Conqueror or Hero, the only armor is the bomb-proof deck over the machinery and magazines.

Writing four years ago an article on shipbuilding for the "Encyclopædia Britannica," the author said:

"The use of armor has arrested the development of the shell. But it is not inconceivable that its abandonment in front of the long batteries of guns in the French and Italian ships will invite shell attack, and make existence in such batteries, if they are at all crowded, once more intolerable. It remains to be seen whether in that case exposure will be accepted, or a new demand made for armor, at least against the magazine gun and the quick-firing gun. If exposure is accepted, it will be on the ground that the number of men at the guns is now very few, that the gun positions are numerous and the fire rapid, and that, if the guns had once more to be fought through ports in armor, the number of gun positions would be reduced, and the fragments of their own walls, when struck by heavy projectiles, would be more damaging than the projectiles of the enemy."

What is here called the fifth phase is the fulfillment in 1888 of this anticipation of 1885. There has been a new demand made for armor against the magazine gun, the quick-firing gun, and the shells with high explosives, and the French are once more clothing their ships with thin armor.

Having thus rapidly traced the history of the application of armor to ships-of-war during the last forty years, the author will now examine the weapons.

There can be no doubt that the immediate cause of the change in French policy is the assumption that the shells charged with gun-cotton, melinite, bellite, lyddite, or some other numerous high explosives, will be fired from ordinary guns.

Shells with bursting charges of gunpowder are ineffective against thick armor. To perforate such armor, the walls of the shells must be made so strong that the contained gunpowder charge would fail to burst them.

More than five years ago the Italian government experimented with gun-cotton as a bursting charge for shell. At about the same date English artillerists were endeavoring to perfect a delayed action fuze, so as to delay the bursting of the shell containing high explosives until after the armor had been perforated. Since then every government appears to have had experience, at the proof butts, of the difficulty of making a trustworthy controllable fuze for this purpose. When this fuze has been established, shells charged with high explosives will come under control, because bellite and lyddite appear to be in themselves perfectly manageable. The French authorities have satisfied themselves that they have brought melinite under control. The introduction of 4-inch side-armor over the batteries proves this. The quick-firing gun will not account for it, because 4-inch armor is perforable at 1000 yards by the 14-pounder quick-firing guns; and here quick-firing guns, 45-pounders, 55-pounders, and 100-pounders, are now being prepared. Moreover, this 4-inch armor will make it even easier for common shells of large caliber, with bursting charges of gunpowder, to be exploded among the guns' crews. The fragments of the armored walls will be in themselves very destructive.

But it has been ascertained that a delayed-action fuze is not at present procurable, and that the shock caused by 4-inch armor will explode shells with high explosives, fired from moderate-sized guns, on striking. This is why 4-inch armor has been adopted.

The effect of a burst between decks of these high explosives is a matter of experience to some. The French have pursued the subject diligently, and they apparently consider it better to have the armored walls broken into fragments and scattered about the decks than to endure the explosion of melinite between decks, breaking the beams and frames, upsetting the guns, and paralyzing the crews. But if this is so, and there be any likelihood of having to attack ships like the large French battle-ships Amiral Baudin, Formidable, Amiral Duperré, and others, it is of the utmost importance to be prepared to make such an attack.

The author knows, from an inspection of the drawings of targets which have been experimented upon with a single charge of gun-cotton, that a number of heavy 16-inch deck beams, covering a length of deck of 12 feet, have been broken completely across and lifted 12 feet above the original level. They were broken across immediately over the charge, and from 10 to 14 feet away from this point (at the knees) they were broken again, so that the decks and beams were piled up in a huge wreck at least 12 feet high.

When the large undefended structures in the French ships are considered, the seriousness of this attack becomes evident. And it is the artillerist who must be first impressed. It is of more consequence to be able to inflict damage in war than to be able to resist it. The best defense is to be found in a vigorous attack.

It must be understood that the powerful ships in modern navies are not protected, so far as their batteries are concerned, by the armor which the French think necessary. They are unarmored from their lower decks upwards.

And now to examine the question as to what is likely to be the outcome of this re-introduction of side armor. Is there any chance that there will be an approximation to finality in the use of such armor?

Look first at the quick-firing gun and see what it will be capable of doing against 4-inch armor. Prior to the introduction of the machine gun and the quick-firing gun, the most rapid fire of guns, capable of perforating an unarmored side, did not reach two rounds per minute. The introduction of the Nordenfelt 1-inch machine gun with four barrels raised the rapidity of aimed fire with steel bullets to 120 rounds per minute. These bullets perforate a plate ¾ inch thick at 200 yards. The single-barrelled quick-firing gun of 1.85 inch caliber, firing 3-pound shot or shell, can get off fifteen aimed rounds per minute. The shot from this gun perforates a steel plate 1.8 inch thick at 1000 yards. The 2.2 inches single-barrelled quick-firing gun throws 6-pound steel shells at the rate of fifteen aimed rounds per minute, perforating more than 2½ inches of steel at 1000 yards, and 4 inches at close range. Armor-piercing projectiles from quick-firing guns will soon follow. The 6-inch 100-pounder is capable of perforating nearly 10½ inches of iron armor at 1000 yards range, and with steel shell can just get through 9 inches of compound armor at 200 yards. From five to six aimed rounds can be got off from this gun per minute. A further development of the quick-firing system is the automatic gun.

The 3-pounder, 6-pounder, and 50-pounder automatic guns can be fired, with rough aiming at short ranges, at the rate of fifty, thirty-six, and fifteen rounds per minute. With deliberate aim they can be fired with nearly half that rapidity.

It is certain that, apart altogether from these quick-firing guns, it will only be necessary to put up targets of this thin armor and expose them to the fire of heavy projectiles (shot and shell) in order to show the frightful wreck behind the target which occurred years ago, and which led to thicker and ever thicker armor. But there will be this difference in favor of the gun, that the projectiles are heavier and stronger, the velocities are higher, and the explosives are more powerful.

Experiments have hitherto indicated the certainty that armor will be blown in by charges exploded against it, and that bursting charges with high explosives will be carried through armor probably 12 or 14 inches thick. Now come the questions for the naval administration:

1. Is full advantage to be taken of these high explosives in any war which may break out within the next two or three years? To this question the author can give no answer.

2. How is the new attack to be met in ships? The answer of the French has been seen. They would, if they could, cover their batteries with 4-inch armor.

Another defense is to take care to have a superiority of fire. Any armor carried by an adversary reduces the number of effective shots which can be delivered, because heavier guns must be employed and the fire must be slower. But 4-inch armor is so penetrable that there will be no difficulty in delivering effective blows. Moreover, the fact that the enemy has armor reduces his armament very seriously. If he should have to fire the guns through port-holes cut in the armor, he will hardly be able to secure his battery from the intrusion of the projectiles against which he has armored himself. In the French ship Dupuy de Lôme no guns are fired through ports. The guns are mounted on an unprotected upper deck. The armor protects them only by keeping out the high explosive shell which might burst under them.

A third way of meeting this new attack is to multiply the ships with the same expenditure of capital, and thus to limit the loss of capital in a disaster arising from the use of these powerful agents of destruction. If this third way is admissible, it is most clearly to the advantage of England to pursue it. This country, above all nations, needs numbers of ships.

To other powers the significance of individual ships is of most concern. To the English it is important above all things to be present in many places with ships capable of attacking any enemy. For every ship-of-war having 14 knots speed now in the British navy there are one hundred and thirty-three ships to protect. (Table II.)

TABLE II.

	England.	France.	Italy.	Russia.	Germany.	United States.
Ships-of-war capable of steaming upwards of 14 knots an hour......................	92	68	43	22	34	12
Number of merchant ships of 100 tons and upwards to each of above vessels	133	22	40	56	61	303
Merchant steamers of 14 knots speed and upwards, and of 1500 (gross) tons measurement and upwards..........	87	22	4	..	12	..

If the British government determines to spend money upon invulnerable ships, the difficulty in increasing our relative numbers will be perpetually growing. There is no difficulty, except in finding the money, in making an invulnerable ship. Ships can be built and navigated, which no torpedo, nor ram, nor gun which can be worked from any ship now in existence can fatally wound. In such a ship every man may be absolutely protected, high explosives notwithstanding. But there will be so few of them that commerce and the colonies may be lost for want of ships, and there will only be the satisfaction that the sailors have been protected in such ships as exist.

In order to put forward in a concrete form the author's view as to the type of fighting ship most suitable for the present needs of the British navy, he has brought forward a design. It is for a ship of 3200 tons displacement, costing one-fourth of the so-called first-class battle-ship of to-day. A sufficient number of ships of this type could probably be built and armed in two years, the guns being included in the contract. It may be said that nothing can be done upon such dimensions and such cost which can entitle the ship to be called a battle-ship. It may be well remembered that the seventy-four-gun line-of-battle ship of fifty years ago had a total displacement of only 3000 tons, and that the eighty-gun ship of the same period had only 3500 tons. This design comes between the two, and will cost as much as three seventy-four-gun ships, although, as has been already said, it will only cost one-fourth of the modern first-class battle-ship.

It is difficult to get the sailor to appreciate a money basis. He has to fight in the ships. He sees that his personal chances of life in the smaller ship are less than they would be in the ship costing four times the money, and he naturally values his life and his chances of personal victory more than he does gold. He cannot be expected to see that if four British ships, costing for the four a million of money, destroy or capture one ship of the enemy costing the same amount, at the sacrifice by foundering of the ship which he might command, that that would be an altogether good result.

It is necessary to bring to the front the enormous increase in the value of the material of war which is entrusted to each captain and to each man in a modern ship. In the line-of-battle ship of fifty years ago, the value of the material for a single command would have been about £100,000, and the value per man in the crews of such ships not more than £150. In the ironclad of 12,000 tons of to-day, the value of ten of the former line-of-battle ships is entrusted to each captain, and not less than £2000 to each man in the crew. In the ship of 3200 tons, which is furnished in illustration of the principle of giving preference to numbers of ships rather than to individual costliness, each captain would still control an expenditure twice that required for the line-of-battle ship of fifty years ago, and each man in the crew would represent from £1000 to £1500.

The description of the ship may be very brief.

The speed suggested is 17 knots an hour on the measured mile as a minimum. This is two knots higher than the speed of the largest armored ships of France.

There is an under-water deck. It is kept wholly under water, and may be, upon the dimensions shown (but with some corresponding variations in displacement), of thicknesses of from 1 inch to 3 inches. This deck is covered from end to end by a solid raft of packing material manufactured by the Woodite Company, which rises above water several feet at the sides and six inches in the middle. This packing material is practically non-absorbent of water. It will exclude for twenty-four hours 96 per cent of the water which would enter any empty compartment. It requires an appropriation of weight of from 4 to 5 per cent of the displacement. By this means the floating power of the ship is preserved against sudden destruction under artillery fire. The raft-body can only be slowly destroyed.

For the defense of the fighting position, steel-faced armor 9 inches thick is employed. Within this there is further armor protection, 6 inches thick, over the directing station for the officers.

The order in which protection is given is as follows:

1. Propelling and steering machinery and magazines.
2. Floating power.
3. Officers at their fighting stations.
4. Guns and gunners.

In close action the whole crew may be under armored cover. The armament is central, with an unrestricted all-around fire from every gun. The guns may either be a pair of 22-ton armor-piercing guns, as shown, or a greater number of lighter guns for firing high explosive shells. In either case the mounting is revolved, and the guns are trained and fired by the officers in the central position. The gunners have only to load the gun in their own separate compartment of the revolving mounting and to connect the firing wires.

The design and the estimated cost of the ship have been prepared by the Naval Construction and Armaments Company, assisted by Sir Joseph Whitworth & Co.

What the author desires to say is that such a ship may attack fortresses with as good a chance of immunity from destruction from torpedoes and rams as the ship of 12,000 tons ; and for the same money there will be four ships to one. The same is true of the defense of ports. As to fighting at sea with other ships, each of these four ships will find perforable to its guns 99 per cent of the area of the sides of those armored ships which have side armor. The ships of 12,000 tons, armed with 67 and 69-ton guns, will find only one-half per cent more perforable to their guns. There are four batteries and eight heavy armor-piercing guns in the four ships, and only two batteries and four heavy armor-piercing guns in the 12,000-ton ship, which has cost as much. There are four captains for independent as well as for united action, four securities against breakdown or damage by the enemy, and the power of occupying four positions on the field against one in the 12,000-ton ship. The principle holds for ships of larger size, say one-third, or of one-half the cost of the largest ship of the enemy.

But the peculiar feature in this design, that is, a single central battery, becomes less advantageous as the ship is increased in size. It is, perhaps, only applicable with undoubted advantage where at least three ships and three batteries can be brought against the very large ship of the enemy, who may have as many as three armored positions in one ship.

If it should be said that this dispersion of force entails a risk of destruction in detail by encounters with units of greater force in the hands of the enemy, that argument simply goes to show that organizing skill will be required to ensure the presence of the united forces where they are needed.

The position is, that to satisfy the demand for superiority in individual ships, an expenditure of one million of money has been reached for a single-armed ship.

On examining the ship as the French ships have been examined, it will be discovered that they are most seriously exposed to the attack of the powerful weapons which are now in rapid course of development. The naval authorities have to decide for themselves whether they will concur in still further enlargement in individual ships, or will endeavor rather to meet them by combining the forces of smaller ships. It appears to the author that there is no difficulty in taking this latter course, and that it has many advantages.

The principle of subdivision is consistent with perfect seaworthiness, with a speed as high as that of the largest ships, with the control of weapons which can be used with fatal effect upon the most powerful ships of the enemy, and with such powers of endurance as will enable the smaller vessels to receive injuries from the largest ships without necessarily fatal results.

In this exposition of the uses of armor, it has become apparent that fighting ships must continue to use it. When armor is employed in the form of comparatively thin horizontal plating, experiment seems to have shown that steel, low in carbon, is the best material. When it is employed in the form of a wall, either upright or inclined, and comparatively thick, the value of a hard face becomes very marked.

There are three distinct modes of manufacturing thick armor. They are described by Mr. Alexander Wilson, the original inventor of compound armor, and the chairman of the firm of Charles Cammell & Co., as follows:

A large wrought-iron plate, built up of many thicknesses, is passed through the rolls, and is then, whilst red-hot, pushed horizontally into a huge iron casting-mould, revolving on trunnions. When the plate is securely fixed, the mould containing it is turned up into a vertical position and liquid steel is poured from a ladle into a trough, which distributes it in small streams into the cavity between one side of the wrought-iron plate and the side of the mould, precautions being taken to prevent it from flowing elsewhere. When the steel has become solid, the whole mass of iron and steel, which is fused together, is taken out of the mould, and after re-heating, again passed through the rolls until it is reduced to the thickness required. It is subsequently bent at the hydraulic press to the curvature of the ship, and is then placed on the planing or slotting-machines to be cut to the finished sizes, the drilling of the bolt-holes being the final operation. This is the plan pursued by Messrs. Charles Cammell & Co.

The method adopted by Sir J. Brown & Co. is somewhat different. It consists in taking a wrought-iron plate, after allowing it to cool, and placing a thin steel plate a few inches from its surface by means of wedge-plates round the three sides, small steel studs being placed at several points to prevent the two plates from coming too close to each other in the furnace. The whole mass is then heated, taken out of the furnace, and lifted by a crane vertically into the pit, where the molten steel is poured between the two plates from a ladle and trough. The plate is afterwards rolled, or otherwise compressed, bent, and machined.

Mr. Wilson describes the successive phases of the manufacture of the forged steel plates at the Creusot works as follows :

I. The running off, by means of the Siemens furnace, of an ingot proportionate to the weight required for the plate when finished ; the latter representing only about 55 to 60 per cent of the steel cast.

2. The heating of the mass in gas-furnaces, and the forging of the plate under the eighty-ton hammer.

3. The shaping, either by hammering or with the hydraulic press, followed by annealing in a coal furnace.

4. The trimming, either with a planer or a saw, when the metal is cold.

5. A first vertical oil-tempering at a high temperature, followed by another tempering at a lower temperature, and often by annealing.

6. The finishing of the plate by the planing off of the ends.

The most successful plate ever fired at in England or elsewhere, so far as the author can learn, was tested at Portsmouth on the 24th of March, 1888.

It is a steel-faced plate 10½ inches thick, manufactured according to their process by Charles Cammell & Co., and attacked by 6-inch forged steel shot of 100 pounds weight, and having a muzzle velocity of 1976 feet per second, and a striking energy of 2723 foot-tons.

The same firm manufactured, at the same time, a steel plate. Mr. Wilson says it is the " first steel plate which has ever withstood such a severe test without breaking up," and that certainly it has proved much superior to any plate manufactured at Creusot which has been subjected to artillery fire. It was tested, under similar conditions to the preceding one, at Portsmouth on the 19th of May, 1888.

Three shots were Holtzer forged steel shell, of 100 pounds weight, 17⅝ inches long ; two were Palliser chilled cast-iron, of 98 pounds weight.

The Schneider plates have, he says, invariably cracked through at the second shot, and in the generality of cases at the first, and this with chilled cast-iron projectiles ; but the plate under notice, which was subjected to the same attack as the compound plate, not only withstood the two Palliser chilled-iron shells, simply pulverizing them, in fact, but also stopped and kept out the three Holtzer forged steel projectiles, and this without showing a crack. The plate is the first on record which has succeeded in stopping these forged steel shells; and it was sufficiently tough to do this without being itself fractured. Still, when the results are compared with those obtained against the compound plate tested on the 24th of March, 1888, it will be seen how the hard face of the latter gives it a marked advantage in resisting the perforation of projectiles.

These results are against direct firing; but against oblique firing the advantages of the compound steel-faced system are far more prominently demonstrated.

The author has treated the subject broadly, and, with the kind assistance of the armor-plate makers, has enunciated the whole aspect of the armor question. He has considered not only what armor is, but how it is employed. It is not a naval question in the first place; it is one of engineering science. It must be remembered that civil and mechanical engineers have originated all improvements in the types of ships, including the introduction of steam, of the screw, of iron and steel in their construction, and of the use of armor. The great improvements in ordnance and torpedo attack are also due to them. The attitude of the engineer towards questions of war material differs from that of a member of a political administration, and from that of naval and military officers. The engineer is mainly interested in the question of the development of the powers of the weapons of war, and he is constantly thinking how they may be extended, and how the maximum of power may be got out of a given expenditure. The tendency with the administration is to estimate the comparative value of available resources of rival governments, and not to look beyond the immediate future. It may be seen from the speech of Sir John Pakington, when introducing navy estimates on the 25th of February, 1859, and when the French ironclad sea-going ships were being built, how easy it is to undervalue new forces. This administration undertook the building and conversion at that time of sixty-seven wooden line-of-battle ships and frigates; and they were all in the hands of the dockyard authorities, in various stages of production, a few months afterwards. But not one of them was suitable for the circumstances of the time. Many of them were never finished.

The author would agree cheerfully to any proposals by the Government, as to the size of the ships and to the use of armor for them, if they were laid for approval before some competent technical committee for a month.

The times are critical, and, if the author were responsible for the steps which must be taken by the administration, he would insist upon having the soundest independent judgment which could be obtained.

The author's successor at the Admiralty pleaded for inquiry into the subject in *The Times* newspaper in 1884 and 1885. He has himself pleaded privately, but through the proper channels, with each administration in turn from that time to this. The difficulties surrounding the question are not lessening; they are sensibly growing.

The author feels satisfied that if such a body as the Council of this Institution had been consulted in 1868, 1871, and 1875, dates when naval ordnance was in a crisis and under serious debate, it would not have happened that the question of breechloading would have been kept closed from 1866 until 1879. He believes that if such a body had been consulted in 1859, it would have appreciated better the significance of the four ironclad ships then building in France, and the Government proposals would have been wisely modified.

So far as the author knows, the proposals of the Admiralty (which he has not seen) as to large ships and the armor for them may be sound and wise. If they are, it would be very unlikely that such reference as has been suggested could cause inconvenience or delay.

The author is not of those who consider it possible to protect sailing ships and slow steamships by means of cruisers in a severe war. Indeed, nothing has been said in this paper about regular fast war cruisers. Long before he left the Admiralty, and before the design of Sir William Pearce for a fast cruiser was sent in, a cruiser of 22 knots speed, and to hold 2000 tons of fuel, had been designed at the Admiralty; but it was decided not to build it. The views which prevailed were very well stated publicly by Mr. George Rendel, who was then, happily, a member of the Board of Admiralty. He said :

"As far as I understand the question, an alternative is to add to the navy a number of very swift partially armored vessels of war, especially built to act as the guardians of our commercial ships. Now, there are one or two considerations which seem to make that course a very difficult one. In the first place, the number of vessels required would be something enormous. If we make the supreme effort required to provide them, we have to remember that not only would their maintenance be extremely costly, but they would rapidly depreciate in value, from the progress in science. As an example, if we assume that that course had been followed some fifteen years ago, we should certainly have had vessels engined in a manner which would have rendered necessary a complete refitting of engines and boilers a few years later, because it would be absolutely essential to maintain such vessels at the maximum point of efficiency, and for this purpose compound engines must have been substituted for the engines previously used. Again, another period of a few years, and we should have found ourselves in the presence of the fact that such vessels must be of steel to cope with their rivals of the day. Again a reconstruction of the fleet. On the other hand, if it were possible to make the commercial navy self-protecting, or to a large extent self-protecting, we should be sure always to have, in vessels selected for that purpose, the latest improvements, and to have them at a very trifling cost, because naturally ships would drop out of the list as others of greater power came in. Then, instead of viewing with as much alarm as pride the growth of our great mercantile marine, we should view it with unmixed satisfaction. We should remove the perhaps greatest pre-occupation of those who have to direct our naval policy and naval construction; in short, the object to be obtained is so important that we cannot but think that a very careful examination of the question ought to be made."

But this view does not prevail now. Cruisers of 9000 tons are being built of the same type as the French Tage. And yet those who look ahead can see at least two most important elements of change coming into view.

Look, moreover, at the composition of the squadrons in the manœuvres in 1888. While England had nearly as many 14-knot ships in her mercantile marine as in the Royal Navy, and almost all of them on the Admiralty list as auxiliaries, yet not one of them was engaged.

The Admirals reporting on the operations draw attention to the proved necessity of having an ample supply of swift cruisers attached to a fleet to keep touch with the enemy, and as this had been abundantly proved in 1885, it is difficult to understand why no use was made of them.

The author feels that the difficulty in providing a suitable armament for ships will, for a good many years to come, be even greater than that of procuring ships, and that the merchant auxiliaries will be the last for which provision will be made.

To many minds it seems that the hope of the future, for peaceful sea-traders, lies rather in lessening than increasing the individual superiority of the special ship of war. They would spare no efforts to raise the character and strength of the fast mercantile ships. But it must be admitted that there is no prospect of a diminution in the use of armor in regular fighting ships. The evident tendency is towards its introduction into every fighting ship.

Recognizing the soundness of judgment of both parties—of those who would improve the regular ship-of-war, and of those who would develop the latent powers of the auxiliary naval forces—there is another question, and to this the author has endeavored chiefly to confine himself.

That question is, Ought England at the present moment to move still farther onward in increasing the size and cost of heavily armored ships requiring four or five years to complete, or ought this country rather to endeavor to increase rapidly the number of protected ships, capable by reason of their speed and armament of taking part in any engagement with an enemy, however powerful? And then there is the further question, Is advantage taken of all the national resources in the manufacture of guns and modern projectiles and explosives, which will certainly be needed much more than ships are needed?

DISCUSSION.

Sir George B. Bruce, president, said he had much pleasure in moving a vote of thanks to Sir Nathaniel Barnaby for his paper on a question so full of interest and importance to the country. However great might be the divergences of opinion upon the question, there could be no difference of opinion as to Sir Nathaniel Barnaby's right to speak on the matter, and there could be no doubt that everything he said was worthy of the most careful and attentive consideration, not only in the Institution, but throughout the country.

Sir Edward James Reed hoped the Institution would not think him obtrusive, as a member of the Council, in taking an early part in what must, he imagined, prove a very elaborate and interesting discussion. He was specially charged to discuss another paper on the following day, and the day after Sir Nathaniel Barnaby would read a very interesting and important paper at another institution. He was afraid therefore that if he did not avail himself of the present opportunity he might not be able to speak to the Institution on the subject. The President had used words with reference to the paper which had been upon his own tongue, namely, that it was a paper of the greatest interest and importance, and, as he had already intimated, he thought it would furnish abundant material for debate, for it not only dealt with the main question in a very broad manner, but with a number of minor questions which were sure to excite great interest. He thought it was due to the author and to the Institution to discuss a question like that with perfect frankness and dispassionateness. He therefore hoped that what he said would be carefully watched for the purpose of ascertaining if he in any way passed beyond fair, legitimate, and real criticism of disputed matter. He . wished to guard himself on that point, because he came from another place in which a good deal of heat was generated in connection with the subject. The author had given a series of what were described as "phases" of armor. They consisted of five circles, the first and last wholly shaded. If it were not out of order, he should like to ask the author if he would be kind enough to explain what it was that regulated the proportion of shaded to unshaded area in those circles. He had studied them as well as he could, but he had not been able to see on what principle these diagrams were constructed. They were shaded and partially shaded circles for the purpose, he supposed, of indicating to the eye in a simple manner the relation between the armored and the unarmored portions of a ship. He supposed, therefore, that the shaded part always represented armor, and he might be right in assuming that it represented armor above water ; if so, his difficulty was to ascertain what the unshaded portions represented. Were they the areas of the unarmored sides of the ship?

Sir Nathaniel Barnaby explained that his intention was to show the portion of the ship above water which in commencement was completely clothed with armor as represented by the shaded circle. In the second phase some of the armor came off, a part of the fighting batteries of the ship being left uncovered. In the third phase there was still more uncovered ; in the fourth phase still more, and then the cycle of changes was suddenly ended by a complete covering again of the side with armor. He thought he had made that point clear.

Sir Edward Reed said that his difficulty was with the diagram of the ship. The area of the unarmored visible sides seemed much larger in proportion to the armored than the unshaded part of the circle indicated. However, the author's explanation would serve all the purposes he had in view. He desired to make a remark about that method of illustrating the state of the armor question. Although he had not the smallest doubt that the author in compiling the paper had travelled along some clear line of thought, he was entirely at a loss to understand the grouping of the vessels. The circular disks were given to illustrate what might be called the rise and fall and re-rise of the armor. But then this extraordinary thing happened. The author had introduced into the line of progress a couple of cruisers, and they followed apparently in the line with the line-of-battle ships that had gone before them. The author, from considerations which would be present to most minds, and no doubt from a desire to avoid as much as possible controversial points, had selected his illustrations from the French navy, and that had resulted in his introducing into the series of vessels the Dupuy de Lôme. On the previous evening, in the House of Commons, Sir Edward Reed had mentioned that ship; thereupon the First Lord of the Admiralty sprang up and said that if the Dupuy de Lôme system was considered to be of any importance, he was prepared to combat that idea. The vessel, therefore, was knocked out of his mind on that occasion by a very high authority on those subjects; yet he found it in the paper occupying a position illustrative of the progress of the armor question. But if the author had taken illustrations from the British navy instead from the French, it would have been impossible for him to have taken the series of vessels he had chosen, or anything like them, because while it was true that the English were building huge vessels of 9000 tons, resembling the Tage more than any others adduced, it was impossible to put them into one series of ships illustrating the growth or otherwise of armor, because simultaneously with the building of those unarmored ships the Admiralty was about to build eight ships of 15,000 tons, and two others of very considerable size, for which was claimed more perfect protection by armor than had ever before existed. So that, if that series had been taken from the British navy and from the succession of battle-ships as illustrating the progress of the armor question, there would have been, instead of the Tage and the Dupuy de Lôme (vessels covered with 4-inch armor only), the first-class battle-ships now proposed to be built by the Government, and occupying a position of an entirely different character from that which the other vessels could claim. He consequently feared that the outside world (which was a pretty good-sized section) might be seriously misled by the course which the author had taken. He should have expected to find not a mere solitary cruiser of the French navy, and a cruiser so bad that the First Lord of the British Admiralty did not hesitate to denounce it the moment it was mentioned in Parliament, but such first-class battle-ships as were now proposed. In that case the meeting would have been led from the Warrior down to the first-class line-of-battle ships about to be constructed, instead of from the Gloire to the Tage and the Dupuy de Lôme.

Sir Nathaniel Barnaby said that he had explained that when he wrote the paper he did not know what the Admiralty's propositions were. He knew nothing of the ships in question.

Sir Edward Reed did not think that affected his position, because when the paper was presented the author must have known that the Admiralty intended to continue building armored ships, and ships of a large class. All that he wished to guard against was the supposition that the course of the armor-plate question in this country was in any way represented by the series given in the paper. It was nothing of the kind. If, instead of the 4-inch armor-plate vessels, the ships about to be constructed with very thick armor were taken, the cruisers would be relegated to a very different category and fall under a very different line of thought and discussion. There was nothing in the paper to show that the Tage only came into the series as a cruiser, and that the Dupuy de Lôme followed in that capacity, and that these vessels did not in any way represent the progress and growth of the armor question, as far as he knew, in France, and most assuredly not in England. The paper cleared up what had been mysterious to him for a long time, and it raised a question of the most serious and vital character touching the navy. He had never been able to understand how it was that men of the highest skill and competence in their profession should have produced ships with the extremely limited amount of armor which characterized a great many vessels in the navy. But the explanation was furnished by the author, who stated that, in his judgment, it was of more consequence to be able to inflict damage in war than to be able to resist it. He could quite understand that any one with that idea dominant would indeed sacrifice the material which would permit of resistance, and would devote himself to the development of the attacking force of the ship, which was the chief characteristic of many modern vessels.

That was a principle right enough, he admitted, if a man were fighting a man, or even an army fighting an army, although even then he thought it would be rather bad strategy never to be seeking to use defensive means that were offered, and always trusting to a

superior attack. But in the case of a ship of war the principle seemed
to him to be as false as anything could possibly be, and for this
reason, that in approaching an enemy on the sea, the latter had with
him an ally of the most persistent, ever present, determined character that could possibly be manifested, namely, the sea. If the defense
were slighted and the enemy allowed to perforate the ship in such a
manner that the sea could enter and dispose of it, the enemy would
have the advantage not only of a powerful but of an omnipotent and
over-mastering ally, who scoffed at all superiority of attack, giving
no recognition to the detailed advantages possessed by a ship, but
sending everything belonging to it to the bottom of the sea. That,
in his judgment, was the natural consequence of the theory pro-
pounded. The paper, as it appeared to him, denied to the country
what it had a right to demand. The author had very properly said
that the question was no other than an engineering one. He had
reproached the sailors with having too little regard for money and
consideration of cost. Sir Edward Reed reproached the author with
having had a great deal too much consideration for the cost. He
thought the country had a right to ask its engineers, in designing a
battle-ship for the defense of this island (which was in itself little more
than a big ship, although happily not so sinkable as some ships had
been made): "Will you send its protectors out into the Channel on
a platform which the enemy shall not be able to send to the bottom
readily, which he shall not be able to send to the bottom at all by
the attack of light and inexpensive guns, and which, when the worst
comes to the worst, he shall only be able to send to the bottom by
bringing out enormous and costly guns which shall tax his skill and
strength to produce?" There was a striking acknowledgment in
the paper that in saying that he was not asking the engineer anything that he could not do, because the author not only laid it down
that men could be sent into the Channel upon an invulnerable platform, but asserted that everything could be made invulnerable.
"There is no difficulty," he said, "except in finding the money, in
making an invulnerable ship. Ships can be built and navigated
which no torpedo nor ram nor gun which can be worked from any
ship now in existence can fatally wound." (He should like to go a
little further and say, which no number of torpedoes or rams or
guns can fatally wound.) "In such a ship every man may be absolutely protected, high explosives notwithstanding." When a railway
engineer was asked to build a bridge across a river he said to his

company, "If you want to bridge the river I will make the bridge, and it will cost you so much"; in like manner why should not the naval engineer say, "If you want a safe ship I can build it, and it will cost you so much." He did not go so far as the above words of the author, to which he knew he did not commit himself; he did not want so much protection as there described, but he wanted the platform protected; he had invariably demanded it, and if a contrary resolution were proposed in the Institution and voted for by every man present, he would nevertheless go on demanding first-class line-of-battle ships so protected that the enemy could not send them to the bottom unless he brought out guns which would severely tax his pocket and his strength to produce. He wanted to know what, as a naval constructor, he had to do with cost. People seemed to think that there was a body which restrained and limited the expenditure upon the navy. That was a delusion. On the contrary, there sat within a few yards of the Institution a body which would sanction any expenditure that a Minister could be got to ask for; and if the House of Commons were told that the Government was going to send men and officers out into the Channel for the defense of the country on invulnerable and unsinkable platforms, there never would be anything more than that which happened now; somebody who thought that the country could do better without any ships at all, or at any rate that it could do with a much fewer number, would make a motion on the subject, but very few indeed would vote for it, and the great majority of the House of Commons—of every House of Commons that he had known, and he had sat in Parliament for fifteen years continuously—would cheerfully grant the money. He saw present a distinguished officer who had been quoted in the paper, Admiral Sir John Hay, Bart., and one of the finest records of his Parliamentary career was his maintenance, from a naval officer's point of view, of the very proposition he was now laying down, that it was not his business to restrain and hamper his profession with questions of money. Nor was it so with the naval architect, and for this good reason, that the thing to be protected, namely, the existence of the independence of the country, was worth protecting at a very considerable cost. Then there was another proposition growing out of that part of the subject which he hoped the meeting would allow him to state. He must express some surprise at it, because he always felt that behind any opinion of the author's there must lie a strong thought of some kind; but he was at a loss to understand what the author could have behind his words when he pointed out that it was for other countries, rather than for this, to care about a strong individual ship; that this country had not the same reason for desiring to have strong individual ships as other countries. He differed entirely from that proposition. He did not deny that the other part of the author's contention was thoroughly sound and good, namely, that plenty of ships were wanted, and that the power of England should be maintained in a great many parts of the globe. That, he admitted, was a necessity, but it was not the first necessity. The first necessity was the inviolability of England's shores. What was the state of the case as between this and other countries? A sea-port town or two might be destroyed in Russia and tribute levied upon them; what did it matter? The effect was hardly felt throughout that empire. Several towns might even be destroyed in France, yet such was the wealth and recuperative power of that country that no mortal blow would be inflicted by anything that could be done to its ports. But in this country nearly all the great towns and cities were ports, or nearly ports, and therefore the first essential to its existence, in a maritime war, was to be able to keep the mastery of the Channel by means of ships individually strong, and as compared with other powers, so to say, omnipotent. And simultaneously with that, a multitude of ships should be secured for the multitudinous services which the navy was called upon to perform all over the kingdom. But, notwithstanding what he had said, he was extremely disposed to go a long way with what seemed to be the main contention of the author, namely, the multiplication of small ships, individually far from weak, for attack or for defense, and of moderate size and cost. The reason was that he could not get the invulnerable ship built; he could not get the safe platform constructed. It seemed to be the determination of every constructor who had power and responsi-bility, to make it his care not to give to the ship an invulnerable platform which the enemy could not easily dispose of. At the present moment this country was in a position of being about to vote eight millions of money for eight ships which did not to his satisfaction fulfil that primary condition, but were distinctly made in such a way that they could be destroyed without the attacks of the very power-ful guns to which he had alluded. Since, therefore, the proper ships were not to be had; since ships costing a million of money were exposed to going down to the bottom of the sea because they had been denied the necessary armor of defense—the necessary height and depth—he felt persuaded by the paper into the course of prefer-ring the multiplication of ships of less size and cost and taking the chance with them. The vessel proposed by the author was, to his mind, one of an extremely interesting character. Regarding the sea as the enemy's most destructive ally, when once allowed to enter in sufficient quantity, and seeing that the invulnerable platform could not be obtained, the question arose whether, as the ship must be destroyed, it was not better to adopt the author's design. An idea in a vague way had sometimes passed through his own mind whether a vessel should not be produced defended in regard to the engines and boilers by an under-water deck, then filling it up from that deck to a substantial height above water, so that the enemy could not sink it by fire except by the comparatively slow process of blowing out that solid but light material. It was a conception which seemed to him to be extremely full of interest and thoroughly worthy of study and consideration. The thing about it which he did not like (it was a mere matter of detail) was that the author had not in his judgment carried that solid substance up high enough above the load-line. He could not reconcile his mind to vessels which sank when a little water was let into them. He thought they ought at any rate to be able to take some water on board and still float. If there was only a margin of 6 inches, as allowed in the paper, for the solid material above the load water-line, he thought that was not sufficient. This was very much the same question as that of the belt. Why he detested the belt for a line-of-battle ship when narrowed down, as it always was, was this: If the belt was 6 or 8 feet high, or even 4 or 5 feet, a cer-tain quantity of sea could be admitted, or an extra quantity of stores taken on board, and still the vessel would remain a safe, satisfactory, fighting vessel; but if the belt was designed to be 2 feet 6 inches above water, and if it came out, as it often did, only 1 foot 3 inches above water in the finished work, the margin was gone. A desperate effort was going to be made against that in the new ships. He believed the Board of Admiralty had determined to exercise a little direct control over the designing, and the intelligence and knowledge displayed in Parliament encouraged him to believe that many things would be done if it had full scope. One of the instructions to the constructor was, " For the future you must allow a margin of displace-ment," and he believed that 4 per cent was what had been imposed upon him. That was a very wise thing; but then the constructor, although he had given 4 per cent margin of displacement, had pro-

vided for a very small coal supply; and before the ship was finished
he was told to double it, and, when doubled, the margin was all gone.
He had no doubt that that was what had been done and was how it
happened that a first-class British ironclad, costing a million of
money, to defend the interests of the country in the Mediterranean,
with 13,000 indicated H. P., was to be allowed 900 tons of coal.
There was an odd remark in the paper which he hoped he might be
excused for noticing. The author had quoted from Mr. George
Rendel, and he had introduced the quotation by stating that Mr.
Rendel was then "happily a member of the Board of Admiralty,"
and, as an illustration of the happy effect which he had on the Board,
it was stated that when the author designed a cruiser of 22 knots
speed, and to hold 2000 tons of fuel, it was decided not to build it.
He did not wish to say anything reflecting upon Mr. Rendel, but if
his influence on the Board of Admiralty was to prevent fast cruisers
from being built for the British navy, the inference he should draw
would be not that it was a happy circumstance that he was on the
Board of Admiralty, but that it would have been a much happier one
if he had been far away from it. The fact was now acknowledged
that the author had been for several years urging the Admiralty to
constitute a body to inquire into that great engineering question upon
which millions of money were being lavished from the accidental
causes of the moment, and the author had mentioned what was
perfectly true—that Mr. White, his successor, like himself, had
advised independent inquiry. What was the result of it? An expenditure was about to be incurred, not quite so much as some people
supposed, but of about sixteen millions, upon ships from the design
of one gentleman. He was a thoroughly competent man, and he
would always bear the warmest possible testimony to the ability of
the present Director of Naval Construction. He was no doubt one
of the ablest designers; indeed, the only doubt he had about him
was whether he was not a great deal too able. He had lived long
enough in the world to be more afraid of very clever people than he
was of moderately clever ones. But the situation of things at present
was that the late Chief Constructor of the Navy was inviting inquiry,
that Sir Edward Reed, a predecessor of his, was urging inquiry, that
the present Director of Naval Construction was urging inquiry—
that all inquiry was denied, and that the millions were about to be
expended upon the designs of a gentleman who was an apprentice
when Sir Edward Reed was Chief Constructor, and who, with all his

ability, might possibly be very beneficially restrained by the bringing of other minds independently and freely to bear upon the subject. Millions were to be lavished upon these hole-and-corner designs. He joined with the author most cordially, first, in believing that if the Council of the Institution had been invited to deal with the question from time to time, great public evils and public waste would have been averted, and that if such a body were now called in to advise the government of the country before launching out into such an immense expenditure on designs which he condemned, and which the late Chief Constructor most delicately, but nevertheless most powerfully, called in question, an inquiry might be raised which would save all that expenditure, and which would lead the country, as it ought to be led, by sound, good, broad, professional advice, along that path which ended either in the vindication of its greatness and the preservation of its power, or in overthrow and destruction, under conditions which would terminate its career in an unjustifiable neglect of the known resources of civilization.

Sir Nathaniel Barnaby asked permission to make an explanation concerning the material proposed to be employed as solid raft packing for the defense and stability in small fighting ships. Since the paper had been read, his attention had been called to the fact that specimens of the material had been sent out by the Woodite Company, which, in the conditions in which they were sent, were very far from being non-absorbent of water. He ought to have observed that himself, but he had not done so. The material, as he accepted it, was non-absorbent of water, and he submitted the specimens, which, without any envelop or external waterproofing, were non-absorbent. In order to reduce the price and the weight of the material, the manufacturers had produced what they thought to be a better result. They produced blocks not weighing more than 16 pounds per cubic foot, and they subsequently "waterproofed" those blocks, as shown by the specimens. That latter material when waterproofed weighed, as he had said, about 16 pounds to the cubic foot, and cost £45 per ton. It was impervious to water so long as its jacket remained unbroken. If its jacket was broken, the internal packing admitted water. It did not absorb water into its own structure, and it did not appear to suffer disintegration. The material thus presented two aspects: (1.) A material impervious to water, and weighing less than he gave as a weight which could be accepted. (2.) A material very much lighter when unjacketed, which was pervious to water, but which was intended to be jacketed and was then water-tight.

Vice-Admiral E. H. Howard said it had been stated by the author that he could build four of the vessels described in the paper for one of the monster ironclads designed by the Admiralty, and that he thought was a very important point. Superiority in numbers was a great advantage, giving, as it did, different points of attack. Besides, the four vessels would have four independent heads. The present practice was to make everything dependent on the captain of the ship and to put everything under his control. The responsibility, therefore, of the captain of one of the large ships would be very great indeed. In the case of the four vessels that responsibility would be divided, and for his own part he would much rather have command of the four ships than of the one ship.

Captain C. P. Fitz-Gerald, R. N., said he had the misfortune to be opposed to the author on more than one occasion, but he knew that the author would be the last man to deprecate fair and honest criticism, and he would be able to reply to any observations that were made. He had also had the misfortune to differ from Vice-Admiral Howard with regard to his statement that four weak ships were better than one good one. How he was to command four ships in action he did not know ; any captain would probably find it sufficient to command one. If the four ships could be worked under the direction of one brain, there might be some soundness in such strategy, but he held that to be impossible. Signals in the heat of action would be of very little use. The four ships would be governed by different brains, and it would not be possible to bring them all to bear upon the decisive point at the decisive moment. It had been laid down as one of the first principles of war, either on shore or at sea, that there should be the greatest possible force at a given point at a given time. The principle had been a favorite maxim of Napoleon's, and had been followed by all the great captains of all ages. With the suggested division of the responsibility in the four ships, he did not see how that happy conclusion could be arrived at. He was a believer in what had been called the moral effect in war. No doubt the words had been much misused ; but still the fact that an army or a fleet thought itself superior either in weapons or in any other way to the enemy it had to fight, was admitted on all hands to be a very important factor in the case. Whether it was a magazine rifle or a breech-loader, or a more powerful ship, it gave confidence to the men who were going to fight. He knew that it was the fashion to say that the whole business of naval war was being reduced to a matter of mechanical engineering. He ought to speak with caution on the subject before civil engineers, because naturally their minds would be concentrated upon the question of engineering. Naval battles were, in their view, to be fought by machinery, by the turning of a crank, the pulling of a handle, the discharge of a torpedo, or the firing of a gun. He did not believe in that, and he maintained that the personal element—the fighting men themselves—would enter as largely as ever into the question. To go on any other principle would be making a very great mistake. The effort to put the eggs in different baskets was, he believed, a departure from sound principle. The question had been argued from many points of view, and the Spanish Armada had been cited as an instance. The English were said to have had then a larger number of quick-working small ships than the Spaniards, and could therefore buzz around them and attack them as they liked. Professor Laughton, however, had exploded that theory. In a recent lecture he had shown that not only had the English more guns, but heavier ones, than the Spanish, and that they destroyed the enemy by superior seamanship and superior artillery fire, and in other ways into which he need not enter; the English certainly had superior speed and mobility, which was another point of great importance. The author had put forward a ship of 3200 tons, going at a speed of 17 knots an hour. Such a ship might possibly attain that speed, but under what circumstances? In dead smooth water. It was known that a small ship could not keep up its speed like a big ship in rough water. The speed fell off at once, and it was impossible to compare a ship of that kind with one of those magnificent designs of Mr. White's recently brought forward at the Institution of Naval Architects. If a seaman were told that the small ship could keep its speed under the conditions that it was likely to meet with in the ocean, he simply would not believe it. As to the point of tactics, it was all very well to say, "If you bring out your one monster I will attack you with five ships; I have so many more guns and so many more chances than you." It should be remembered that these five ships would get in each other's way and would very likely run into each other. Besides, a battle-ship did not go out to fight alone; there was usually a fleet, say of half-a-dozen ships, and such a fleet would require thirty small ships to attack it. Those thirty small ships could not all be on the same piece of water at the same time, and could not fulfill the first principle of strategy, that of bringing to bear the greatest force at the decisive moment. The author had quoted a passage from Admiral Touchard, and had thus given it the imprimatur of his authority, expressing sentiments with which Captain Fitz-Gerald could not agree. He also said, " It is difficult to get the sailor to appreciate a money basis. He has to fight in the ships." No doubt it was difficult, and he hoped it would remain so. Naval men did not care about the money; the taxpayer had to provide that. Sailors wanted the best of everything to do the best work; and he believed that the policy of building cheap ships, knowingly inferior to those of foreign nations, was a totally wrong policy. The author said: " He sees that his personal chances of life in the smaller ship are less than they would be in a ship costing four times the money, and he naturally values his life and his chance of personal victory more than he does gold." He did not suppose that the sailor would think much about · his life when once he was engaged in fighting, but he did value his chances of victory. It was all very well to put him in an inferior ship and say, " It is true you will be sunk, but it will not matter, because there are others to take your place." That was a false position to place him in, and it was depriving the service of one of the first principles of warfare, namely, that of giving confidence to the men in the ships in which they fought. To endeavor to impress the men with the money value of the ships was a great mistake. The captain, whether of a torpedo-boat or of a line-of-battle ship, carried into battle with him his life, his honor, the honor of his country, of his flag, and of his profession ; and he thought that was quite enough, without impressing upon him the fact that his ship cost a million of money. He therefore came to the conclusion that the only policy for England, as a rich nation, was to have the best of everything, and to pay for it. He could quite understand a poor nation building a large number of cheaper ships and trying to multiply its chances in that way; but England had to protect the commerce of an enormous empire, the richest that the world had ever seen, and the envy of other countries, and it was absurd for her to haggle over the price of defending it. Whatever might be the cost, it could bear no reasonable proportion to the riches to be defended, and the only possible way of defending them was by absolute supremacy on the sea, both as to number and as to the individual fighting ships.

Admiral H. Boys said the paper began with a matter of almost ancient history which he well recollected. The tenor of the author's remarks appeared to be that naval officers in those early days, instead of endeavoring to advance the science of shipbuilding, were rather disposed to check it. It was stated that "the captain of H. M. S. Excellent reported in 1850 that whether iron vessels are of a slight or substantial construction, iron is not a material calculated for ships of war." He well recollected those experiments. They were made by firing at one of the first iron-built ships, at a distance of about 400 yards, with 32-pounders. The material of the ship was called iron, but from the fractures and the débris thrown about the ship, the results were somewhat similar to breaking a pane of glass with stones. He thought the officers who made the trials were quite right in saying that that was not a material that should be used in the construction of war ships, and in doing their best to prevent its use. With regard to the statement that "seventeen iron ships, which had been in process of construction for fighting purposes, were thereupon condemned as useless for war service," he could only say that that was the first time he had heard of it. As to the question of vulnerability of the water-line of ships against artillery : when ships were perfectly steady and in smooth water, perforation under water was almost impracticable, because immediately the shot touched the surface of the water it ricocheted and skimmed above it. In smooth water there was no reason why the water-line should be protected with armor more than any other parts of the ship. With regard to rolling motion, he believed that the rolling "period" of vessels of that description was about ten or twelve rolls per minute ; or say one roll in six seconds. Considering that a ship would roll both ways, its dangerous zone would be out of water only half the time, and that would make it three seconds. Then it should be remembered that the attacking ship would be rolling, so that the three seconds would be reduced to one and a half second, which would be all the time that the vulnerable part of the water-line would be exposed to fire. It should also be borne in mind that ships would not be rolling the same way at the same time ; if one rolled one way while the other rolled the opposite way, the guns being fixed to the deck could not be pointed to the water-line at all. Then there were two other points to be considered. There was the probability of the swell or wave intervening, as occurred with a ship to which he once happened to belong. A slaver was taken by a ship of which he was gunnery-lieutenant. It had to be destroyed, as it could not be sent to port. It was a nice "innings" for him to have a day's practice to sink the ship. The slaver was anchored 200 yards from the steam-frigate, both ships rolling from 5° to 10°. The vessel was fired at with 68-pounders and 10-inch guns. He anticipated that it would be possible to sink it at once, but the frigate expended the entire allowance of ammunition without sinking it, after all; it was not possible to get a shot in between wind and water, and it was found necessary to take gunpowder on board and blow the vessel's side out. In olden days, when ships fought under sail, there was the question of weather-gauge. Ships were then always under pressure of sail, and they heeled over; the lee ship had the weatherbends thoroughly exposed to fire from the windward of the numerous guns carried by ships of those days. The guns perhaps could not be aimed as accurately as those of the present day; but considering that there might be fifty or sixty guns peppering away at a ship's weather-side, the chances were that some of the shots would enter below the water-line, and even then the vessel need not necessarily sink immediately. It seemed to be implied that those on board ship had no means of pumping the water out, or taking steps to keep the ship afloat, supposing she should be dangerously wounded. The other point was the personal error of the firer in taking a snap-shot from one rolling ship at another. Altogether, he considered the risk to a ship from the penetration of projectiles at the water-line had been very greatly exaggerated. There was a point referred to at p. 107 of the paper in which he could not agree with the author. "It is not a naval question, in the first place; it is one of engineering science. It must be remembered that civil and mechanical engineers have originated all improvements in the types of ships, including the introduction of steam, of the screw, of iron and steel in their construction, and the use of armor." He would quote only a few lines that appeared a week or two ago in a journal which took another view of the subject. "Science, with her usual pedantic imbecility, has been building ships for some time under the inspiration of algebra and with utter disregard for certain simple practical considerations. The sea, after all, is a very rough road at times, and knocks about whatever goes upon it without mercy." He could not agree that the question was not a naval one. The naval officer was as necessary to the engineer as the engineer was to the naval officer, and the two could not be dissociated. The screw was the invention of a naval officer, Captain Smith; at any rate it was introduced into practical use by him. It was at the suggestion of naval officers that many of the improvements introduced had been made, and he believed it would continue to be so. The author had stated that he felt satisfied "that if such a body as the Council of this Institution had been consulted in 1868, 1871, and 1875, dates when naval ordnance was in a crisis and under serious debate, it would not have happened that the question of breech-loading would have been kept closed from 1866 until 1879." He would appeal to any naval officer if, during that time, the muzzle-loaded rifled guns were not perfectly efficient. The greatest artillerist of the day, who was one of the brightest ornaments of the Institution, and who did more to reform gunnery than any living man, Lord Armstrong, though he had first instituted the breech-loading gun, reverted to the muzzle-loading gun and made it a serviceable weapon. He did not say that he was at present an advocate for returning to muzzle-loading guns, but at the time he referred to they were certainly the best obtainable, and it was the introduction of slow-burning powder requiring long barrels which could not be loaded inside the ship that had led to the present breech-loading system. The muzzle-loading guns were, on account of strength and simplicity, the most suitable for any naval operation; they were accurate enough, penetrating enough, and rapid enough in firing, and there was at that time very good reason for adhering to them. He differed from Captain Fitz-Gerald as to the tactics of smaller vessels in competing with large ones. The ram of a vessel of 3000 tons, going at 8, 10, or 12 knots an hour, was quite sufficient to destroy the biggest ship that was ever contemplated. As to the class of ship proposed by the author: in action there must be four or five of those ships to compete with one of the larger new ships. With respect to the ram, one was as good as the other. He would, for the sake of illustration, call the large one Goliath, while the four smaller ships might be called Monday, Tuesday, Wednesday, Thursday. By a preconcerted arrangement, and as a point of honor, it should be agreed between the four ships that on whatever day of the week an action might be brought on—if, for instance, on a Monday—the ship called Monday should run any risk in order to hamper the movements of the Goliath, if necessary to get across her bows to be rammed. He maintained that even a 14,000-ton ship, with one of 3000 tons hanging to her stem, must be stopped; then the Goliath would be at the mercy of the other vessels, Tuesday, Wednesday, and Thursday, to ram her on either side. These were the main arguments in favor of numbers, as proposed by the author, against size.

Mr. W. H. White observed that when Sir Nathaniel Barnaby spoke on such a subject as that under discussion, everything he said must be listened to with respect. When a man had spent forty years of his life almost entirely in endeavoring to strengthen the navy by the addition of efficient ships, and when he had often to keep his mouth shut when anything but fair attacks had been made upon his professional work, clearly when he came forward in public and gave his views upon the past, present, and future, all were bound to attend to him. He was the more careful to say that at first, because nearly all that he was about to say would be in the sense of differing from the author. When the author came forward, it was to be assumed that he came for the purpose of provoking discussion, and of getting as far as possible the opposite points of view clearly stated. First he wished to take strong exceptions to the historical " phases " represented in the diagrams. And he did so for this reason: On reaching the year 1880 there was a change in the type of ship. The Gloire, to begin with, was distinctly a battle-ship; the Marengo in 1869 was still a battle-ship, and so was the Baudin in 1880. After that time the battle-ships were conspicuous by their absence, and they began again. It was not in the same sequence at all. The series began with a cruiser, the Tage, built for a distinctly different service, and the Tage developed into the Dupuy de Lôme, again a cruiser. Of course it might be said that that was merely playing with names, and when ships of 7000 tons were built, to talk of them as not being battle-ships was incorrect. He did not wish to bandy words about it, but he would put the case as it presented itself to him. There was now building in France a battle-ship, the Brennus, and if that vessel had taken the place of the Tage, at the date of 1887–88 there would have been altogether a different condition of things. He of course knew more than he could say about the Brennus, but taking what had been published about the ship, it had a thick-armored belt and strong-armored gun stations, and the question of protection of the side above the thick-armored belt was one which he believed had not yet been definitely decided. That led him to a point which he desired to make most clearly, namely, that in discussing the matter a distinction must be kept between battle-ships and cruisers. He had no doubt that if Captain Fitz-Gerald and other gallant officers found themselves in vessels called cruisers and came across a vessel

called a battle-ship, which, because of size or some difference of construction, they thought they could safely engage, they would do so.
But that was not the point. The point was that there were afloat
in other navies certain great ships possessing features which could
only be attained in ships of large size ; other nations had been and
were still producing battle-ships, and the question was how could
those vessels fairly and with the greatest certainty of success be met
on equal terms ? He did not say that numbers should not be developed and many cruisers built, but the point he was trying to make
was that it was unwise on any mere *à priori* argument, obtained even
from the highest source, to be content to rest the defense of all that
was most valuable to the empire upon ships which were individually
and distinctly inferior to those possessed by other countries. If that
principle were admitted as the principle on which the Admiralty
policy should proceed, it would be seen that there was room enough
in the navy for battle-ships and cruisers, and he wished to point out
that in the Admiralty programme which was now before the country,
and which included seventy ships, there were only ten that came
into the category of battle-ships, eight of them being very large
ships, of which so much had been said, and two, vessels of more moderate size. Sixty out of the seventy vessels belonged to the class of
cruisers. Nine of them would be large cruisers and approximately
the same size as the Tage, but swifter, more heavily armed, and
better protected. Thirty-three would probably be vessels of about
the same displacement as the design exhibited by the author, and
the remainder would consist of smaller swift torpedo gunboats. In
view of the almost absolute want of experience in naval warfare under
modern conditions, was it not unreasonable that this country should
proceed on any merely doctrinaire course, rather than that in the
creation of its new modern fleet it should have all classes of
ships represented? He would pass by for the present any consideration as to whether the designs were what they should be, the
best that could be produced; on that point he would say a word
further on. The question was whether as a matter of policy, so
much being unknown, it was not the wisest and the best thing to
provide against all risks. At the risk of repeating what he had
recently said, and in order to complete his remarks, he desired to
state in the briefest way what were the principal advantages
belonging to increase in size. Captain Fitz-Gerald had referred to
one, namely, the maintenance of speed, and of that there could be no

doubt. Then there was a further and very important thing—steadiness of gun platform, of which also there could be no doubt. Then there was the enormous increase in the power of the concentrated attack, which came from having a large armament carried in a single ship under a single direction. Further—and that was the point often overlooked—there was the fact that a large ship was much less likely to be destroyed or put out of action by a single heavy attack, whatever form it might take, than a small ship. It was conceivable that a single heavy shell bursting in the vessel proposed by the author might make it unworkable. Notwithstanding the packing material and the under-water deck, a single heavy shell exploding therein with such explosives as were now known might for all fighting purposes destroy the ship, but with a ship that was very much longer and very much larger, and which had its armament very much more widely distributed, clearly the risk of destruction by a single heavy attack was greatly reduced.

He believed that in fortification the doctrine that he had tried to illustrate took the form of saying that a large place was a good place to defend. He thought that those points for the large ships were clearly points that could not be ignored. Of course they must be paid for. Suppose that four or five ships could be built for the cost of one of the big ones; if what he had put forward was correct, it did not matter whether or not four or five could be built for the price of a large one. If the large one was necessary it must be constructed. What he wanted to present to the meeting was the idea that for the English there was no necessary choice between the two alternatives. It was sometimes put to them as if their finances had run so low that they must multiply small and could not afford big ones. He did not consider that that was the real condition of things. There was room in the navy, with its great range of duties, for all classes of ships, and there might even be room for the class suggested by the author. In criticising the design he was quite sure that the author would take in good part what he was about to say. To his mind there could be no defense whatever for putting a protective deck in a ship of war wholly under water, because, for all practical purposes, the bringing of the deck above water did not sensibly increase the risk of its being directly attacked; but if it was put entirely under water, clearly the space contained below the protective deck in no sense assisted the maintenance of buoyancy and stability, when the upper works were damaged in action; the maintenance of these qualities depended chiefly upon wood or woodite packing above the deck. Further, he conceived it to be a doubtful principle, notwithstanding the high authority which recommended it, to put relatively large weights into the form of packing material which added little or nothing to the defense. The woodite, according to the author, weighed from 4 to 5 per cent of the displacement; that must mean about 140 or 150 tons weight put into the woodite for the single purpose of protecting the buoyancy and the stability within moderate angles of inclination. Supposing the protective deck to be elevated, as was now almost universally done, so that it came above the water-line for a considerable portion of the length of the ship, the need for the packing material was reduced without prejudice to the maintenance of buoyancy and stability. Supposing further that the same weight which was put in the packing material was put into coal, there was a defense which, as was well known from experiment, was of enormous value against many forms of attack; the coal was a usable material in the ship, and it might happen that the possession of that additional coal in such a ship would mean the fulfillment of a duty that would otherwise be impossible. His opinion, therefore, taken for whatever it might be worth, was entirely adverse to the adoption of a type of ship which depended for flotation upon packing material which was not in a useable form like coal, and which added but little or nothing to the defense. And it was to be noted that in some foreign navies, where the opposite rule had been adopted a few years ago, a change had been made in the later ships in the direction he had indicated. The author in his paper had not made any allusion to the Admiralty designs, but of course it was impossible to discuss the question without referring to the designs which were now before the public. Sir Edward Reed, in opening the discussion, distinctly did not avoid reference to those designs, and he therefore hoped that he should be in order in making a few remarks on this subject. The objection made to the designs for the large battle-ships had taken this form, that the ships were too large and too costly and had too much armor, according to the author's view, and that according to Sir Edward Reed's view they had not armor enough. Might it be permitted to those who had to do with the designing of these battleships to hope that between two such authorities they had struck the golden mean. He spoke simply as representing the Admiralty, which, as was well known, had not proceeded lightly on the course

that had been followed. Never in the history of the navy, as far as
he had known it, had there been more careful or even as careful
deliberation given to any question. The magnitude of what was
contemplated had been fully understood and appreciated, and the
country was in a position of knowing what had been done by the
papers that had been presented to Parliament. Sir Edward Reed
had said, and Mr. White distinctly challenged the assertion, "all
inquiry was denied" about the new designs, "and that millions were
about to be expended upon the designs of a gentleman who was an
apprentice when Sir Edward Reed was Chief Constructor, and who,
with all his ability, might possibly be very beneficially restrained
from the bringing of other minds independently and freely to bear
upon the subject. Millions were to be lavished upon these hole-
and-corner designs." " Hole-and-corner designs ! " Designs prepared after months of consideration by the Board of Admiralty !
Designs which were submitted to and selected by such a meeting of
experienced naval officers as had never been before summoned.
Designs which he had recently described in public in the presence
of Sir Edward Reed, challenging him to justify any of his criticisms
on their character. He would leave it to the Institution to judge if
such a description as " hole-and-corner designs " applied to a course
of proceeding in which the Admiralty had taken the only means of
publicity in its power, and had brought to its assistance the best
naval advice it could obtain. Perhaps he might be permitted to add
one word on the personal reference without offense. He had always
thought that it was the privilege of members of the Institution, among
whom he had the honor to number himself, that biographical notices .
should be reserved till they became obituary. Sir Edward Reed
had done him the honor to bring to the notice of the Institution the
fact that he was once an apprentice. It was perfectly true, and he
was delighted to think that it was true. He thought that all who
went in for applied science were the better for knowing the use of
tools. But he might also add that he was not so young as he once
was, although he should never, during Sir Edward Reed's lifetime,
be as old as Sir Edward was. With all his rapidity, he could not
overtake the start which Sir Edward Reed had attained by the accident of birth. But for twenty-two years he had been continuously
engaged in that not very pleasant occupation, designing and building ships, and he thought he might say without egotism that he had
had special opportunities of acquiring information, and he hoped in

the public interest that he might prove capable of usefully applying information so gained.

With that remark he would pass away from the personal reference, and simply say that at the Institution of Naval Architects he had dealt with every point of a technical nature which had been raised by Sir Edward Reed; that he had dealt, he thought, to the satisfaction of the members present, with all those points, not as matters of opinion, but as matters of fact. With reference to what Sir Edward Reed had said in regard to coal supply, that given to the new designs was determined by the Board of Admiralty, and was in excess of that of nearly all the battle-ships afloat in the English navy, and greatly in excess of that of nearly all battle-ships of foreign navies. He wished to say one word as to the history of a design in the Admiralty, because he should extremely regret that a matter of public importance should degenerate into anything of a personal nature. The naval members of the board, having before them existing types of ships, desired to embody in some new ship certain features of armor, armament, speed, protection, and coal supply. Those features having been decided upon, it became his business, as the responsible naval architect of the Admiralty, to embody them as best he could in a ship that should be seaworthy, habitable, and capable of attaining the speed and traversing the distances which the Board of Admiralty decided upon as desirable. To speak, therefore, of millions of money being lavished upon his designs, under these circumstances was, he thought, scarcely a complete description of what happened. The designs which had been adopted, it could not be too often repeated, had been most carefully selected by the most experienced officers of the Royal Navy, both at the Admiralty and elsewhere; and, whatever might be their merits or demerits, it was not fair or correct to describe them as emanating from his brain, or to attribute their existence to any initiative of his own. He need say no more about those great ships, except to add that he was certain that even the author, with his recommendation of smaller ships, less costly and less armored, would admit that the addition of those large ships to the navy must greatly increase its force. With regard to cruisers he must again differ from the author, with great respect. When he spoke of the defense of the mercantile marine and appeared to recommend, on the authority of Mr. George Rendel, allowing the mercantile marine to take care of itself, he overlooked the fact that in foreign navies there were great numbers of specially designed

men-of-war cruisers which had to be met by something. He agreed
with everything the author had said about the desirability of arming
and using the splendid merchant steamers which the country possessed, whenever circumstances of war should arise. It was a perfectly distinct question what these vessels should be used for; how
best they could be employed was a matter that would be decided
when the time came; but he was convinced that there would be
work enough found for the greater number of them without ex-
pecting them to take the part of improvised cruisers belonging to
the Royal Navy. But if they should be employed, the question to
be asked was, were these vessels expected to meet the men-of-war
cruisers of foreign navies? Was it to be expected that the very best
of those vessels would be capable of staying to fight, on anything
like an equal chance, a vessel built specially for the purpose of
fighting, a vessel such as the Tage, or the Dupuy de Lôme, or the
Cecile? Why had those vessels been developed in the French navy?
why had the armored-clad construction of the French navy been
almost standing at pause, while money had been spent and every
effort urged on to complete such vessels as those? Simply because,
as was perfectly well known, it was no breach of official confidence,
for it could be read in the French newspapers; the idea which the
French authorities had in their mind was to use those vessels, in
time of war, to destroy English commerce and the mercantile
marine. How were they to be met, if in the English navy there were
not to be found vessels equally fit to fight and certain to overpower
the French vessels, if they met them? How was that to be done
unless the English also constructed large and swift cruisers powerfully armed? The author had referred to the fact that cruisers of
9000 tons were being built for the Royal Navy. He thought the
country had good reason to rejoice that that was the fact. The
vessels now being built, the Blake and the Blenheim, if they fulfilled
the intentions of the design, would be superior in coal endurance
and in fighting qualities to any vessel existing in the French navy,
and surely such a fact as that was one which justified their existence.
Then, as to their being so large, he could not help saying that he
thought the author knew why they had become so large, why the
Blake was a vessel of 9000 tons, because it steamed 22 knots an hour
and had a weight exceeding 1800 tons assigned to coal and dispos-
able weight, independently of armament and equipment. The load
of protection, armament, equipment, and coal closely approached

4000 tons. If the Admiralty were content to give up speed, if it were content to let vessels that were being built abroad run the chance of being met by a vessel that could not catch them, then it could reduce size and save money; not a saving to the country on the whole, perhaps, but a saving in one direction. Passing from those somewhat controversial matters, he would refer to a subject which was strictly germane to the paper, and in which he happened to have a special interest, and certainly some special information; he referred to the subject of the manufacture of armor and to what had been lately done in this country in the development both of steel-faced armor and steel armor. Nearly three years ago he was informed in the public press that he was an obstructive. That was not the first time, but it then happened that he was again declared to be an obstructive. The subject was armor, and he was obstructive because he was said to be still recommending to the Admiralty the continued use of steel-faced armor, and was blindly shutting his eyes to the magnificent results which were being produced at Creusot. Safety was only to be had by giving orders for armor to the establishment at Creusot. English steel-makers had been left behind, and the French were head and shoulders above them. But he believed that if the English steel-makers were given a chance they would prove that their hands had not lost their cunning, and that in England armor could be produced, whatever it might be made of, that should be as good as any made abroad. He was glad to say that the view he then put forward was approved by the Admiralty, with the result that a series of experiments had been undertaken that had been in progress for more than two years, and had proved that England could produce steel armor as good as any produced abroad—he believed better—and steel-faced armor, which was one of the grandest defenses a ship could have. The author had alluded to those experiments, and particularly to the experiments made against the armor-plates supplied by firms at Sheffield whose reputation had been made in that branch of manufacture, firms which had done good work for the navy, and had given as honest and true a defense as was possessed by any war-ships afloat. The Admiralty had not only obtained from Messrs. Cammell and Messrs. Brown splendid specimens of steel-faced armor to be used as standards, but those very firms had made steel armor of great excellence; and firms like Messrs. Vickers had produced steel armor in many respects of a marvellous quality. This showed clearly that if the need arose for re-enforcing the possibilities of the recognized armor-plate Sheffield firms, the necessary resources were to be found in this country without going abroad. The question was not any longer one as between steel-faced and steel armor. Steel-faced armor in the trials had behaved magnificently. The trials had justified their practice up to date, and it was now known that steel armor as well as steel-faced armor could be used and the most splendid results obtained. Those plates had been fired at with projectiles made in France, and he did not know why, if the defense could be so satisfactorily accomplished, English firms should not also make the means of attack.

Sir Frederick Abel said it might seem to be the climax of presumption for a chemist to take part in such a discussion, but perhaps there might be some little appropriateness in the remarks of the Chief Constructor of Her Majesty's ships being followed by some observations from one who had taken some considerable part in developing agents for the destruction of ships, and it was from that point of view only that he desired to say a few words, some of which would be words of friendly criticism. With reference to the point in which he took the most interest, namely, the effect of shells against armor-clad ships, he confessed he came to the conclusion that a naval man, in reading the paper, must be somewhat puzzled as to what he was to expect from the effects of the so-called high explosives against those ships. The author had stated (p. 9) that after the Italians had made some experiments with gun-cotton shells, "English artillerists were endeavoring to perfect a delayed-action fuze so as to delay the bursting of the shell containing high explosives until after the armor had been perforated. Since then every government appears to have had experience, at the proof butts, of the difficulty of making a trustworthy controllable fuze for this purpose." He then went on to say, "When this fuze has been established, shells charged with high explosives will come under control, because bellite and lyddite appear to be in themselves manageable." He further stated that "the French authorities have satisfied themselves that they have brought melinite under control. The introduction of 4-inch side armor over the batteries proves this." In looking for the proof, he found it stated further on that "it has been ascertained that a delayed-action fuze is not at present procurable, and that the shock caused by 4-inch armor will explode shells with high explosives, fired from moderate-sized guns, on striking." Then it was stated (p. 11) that

"Experiments have hitherto indicated the certainty that armor will be blown in by charges exploded against it, and that bursting charges with high explosives will be carried through armor probably 12 or 14 inches thick." Those statements appear to me to be hardly reconcileable, and he should be glad to hear what the author had to bring forward in the way of facts as to how they could be reconciled. He might state it as a fact, however, that delayed-action fuzes were no longer a matter of theory—they had been actually developed. More than one thoroughly reliable delayed-action fuze was now attainable; and if the effective application of high explosives depended only upon the provision of a thoroughly reliable fuze of that nature, the naval officer might depend upon having, in the next war, high explosives brought with perfect efficiency to bear even against thick armor. But it was known, on the other hand, that the provision of such a fuze was not the only element in the successful application of high explosives to the destruction or penetration of armor-clad ships. The author had pointed out that 4-inch plates would cause the explosion of high explosives when thrown against a ship in the form of armor-piercing projectiles. It was true that he stated in another place that they would be carried through 13 or 14 inches of armor; but, as a matter of fact, it had been shown that 4 inches of armor could cause the explosion of armor-piercing shells containing high explosives. He would venture, however, to caution constructors as well as naval officers against drawing too rash a conclusion from that fact. It had been already proved by experiments that 4 inches of armor would not always keep out high explosives; that such armor could be penetrated by armor-piercing projectiles which would carry the explosive charge through. Not only that, but still further progress had been made, and similar results were now obtained against 6-inch armor. Chemists were only on the threshold of the application of high explosives, and he ventured to think, therefore, that it would be necessary to exercise very considerable caution before laying down the rule that armor even exceeding 6 inches in thickness, whether steel-faced or steel, would of a certainty keep out high explosive shells.

Captain Orde Browne said, a fact on which it had been very difficult to get information had been now stated, namely, that shells charged with high explosives might penetrate 6 inches of iron, and that it was desirable that vessels should be divided into two classes, battle-ships and cruisers. Did not the question consequently arise whether armor ought not to be treated in two different ways? A cruiser carried armor that might allow a high explosive to pass into the interior much more easily than a battle-ship, for it might probably be a long time before a high explosive could be got which would pass through 18 inches of armor such as the battle-ship carried. The cruiser, as he understood, was intended to carry on a fight only for a short time. He did not suppose it would be expected to engage a land fort, for example, under ordinary circumstances. There were cases in which a line-of-battle ship might have to do that. He believed the author considered this to be a total misuse of ships, but such things would sometimes happen. If distinction were made between the two classes, did it not follow that a harder class of armor should be provided for the weaker ship in order to make sure that it should keep out or break up shell with a high explosive under any circumstances? The first object with the cruiser was to be sure to preserve it from a shell with a high explosive entering inside and then bursting; and he believed that was to be done by hard armor, at the expense, no doubt, of what had never before been allowed in this country, namely, the plate breaking and being held up to a certain extent by bolts behind. He had heard Mr. Vickers speaking of the two plates he had submitted for trial, one harder and the other softer, and saying that he did not know which was best. The question was, what for? If the plate had such a thickness that without doubt it would stop perforation, and that impact of shells was expected to happen again and again, then, of course, the softer the better. If a weak ship was to be exposed a short time, the great object was to make certain that no high explosive should get into the interior. Was it not worth while to allow armor to break to pieces if it were thereby made certain that it would keep high explosives out? In this country it had been continually insisted that the back of the armor should be soft, because unless it was soft, whatever theoretical power of extension as well as tenacity it had, it would not extend but would break. The Admiralty had insisted upon its extending as much as possible and upon its breaking as little as· possible at the back, and the result had been that the plates had necessarily been soft at the back. He was therefore anxious that experiments should be made with a harder class of armor, with a view to those thinner clad vessels which were now found to be open to the entrance of shells charged with high explosives.

Mr. B. Martell said he merely rose to ask a question. The author had made a statement (p. 19) which he thought, unless some further explanation were given, might be misleading. "The most successful plate," he said, "ever fired at in England or elsewhere, so far as the author can learn, is shown in Figs. 5 and 6. It is a steel-faced plate, 10½ inches thick, manufactured according to their process by Charles Cammell & Co., and attacked by 6-inch forged steel shot of 100 pounds weight, and having a muzzle velocity of 1976 feet per second and a striking energy of 2723 foot-tons. Mr. Wilson says "it is the first steel plate which has ever withstood such a severe test without breaking up, and that certainly it has proved much superior to any plate manufactured at Creusot which has been subjected to artillery fire." Seeing that the subject was of national importance, and that manufacturers had been devoting their enterprise and investing an enormous amount of money in perfecting the best steel armor-plates that could be produced, it was greatly to be regretted that the author had not mentioned all the experiments, referred to by Mr. White, that had been made by the Admiralty with plates manufactured by the various firms throughout the country. The plates manufactured by one steel manufacturer alone had been expatiated upon, while those produced by other manufacturers for the same purposes were not even referred to. Considering the high position occupied by the author, it might almost be considered an act of injustice to those who had competed, to find that no allusion had been made to their plates. He had seen official reports of the experiments, and he should like to ask whether the author had seen any official report bearing on the subject, so as to be sure of the accuracy of his statements. In one of the reports allusion was made to a steel plate that had been experimented upon at the same time. Five shots had been fired at a 10½-inch plate by the French shells to which Mr. White had referred. He shared Mr. White's surprise that the French should have been called upon to make shells to fire at English armor-plates. Two of the five shells were cast-iron, and they crumbled into pieces on striking the armor-plate; the other three, which were steel, made an indentation to a certain extent, but, to show the tenacity and the resistance of the armor-plate, they rebounded so that they struck the shot-guard some 20 feet distance, and rebounded back again to the steel armor-plate. He wanted to know why it was that no information had been given by the author on that subject.

Mr. W. H. White said that the question was not one which the author could answer. The facts were simply these: The Admiralty, in entering upon the experiments, gave to each firm that submitted a plate the guarantee that it should be fired at under the same conditions, and that the firm should have the report on its own plate. Although Mr. Martell was no doubt accurate in his statement that he had seen a copy of one report, that would be owing to the action of the particular firm in question.

Mr. B. Martell hoped the author would use his influence to obtain a copy of the official report for the good of the country, in order that a proper comparison might be made of the experiments. When the manufacturers of the country were incurring such an enormous expense and striving so much to arrive at the perfection of armor-plating, everything should be done to encourage them, and at least put them on a fair footing with each other. When such statements were made as those put forward by the late Director of Naval Construction, they would have an enormous effect throughout the country, and would not, it was thought, appear fair to those who had entered into the competition.

Mr. E. Robbins said he had come to the conclusion that in order to meet all requirements a new departure was essential. Instead of steel plates, he had formed the idea of having interwoven bands held on not so much by rivets as by an elastic binding concrete. The bands would be combined with wood and metal, and would form an elastic body which would throw off shot.

Admiral J. H. Selwyn remarked that the paper appeared to touch almost every question known to naval architects and engineers who occupied themselves with ships or guns. He would not refer to all the points, but would only mention six in which he thought they were still behind in scientific work. There had been an enormous amount of rule-of-thumb work, which consisted in a system of trial and error on the large scale, and sufficient attention had not been paid to the scientific basis of action. It was now generally agreed that long-continued speed was an absolute necessity for all ships, and in order to get that (even more valuable for cruisers than for battle-ships), strong boilers were wanted. But were the boilers strong? An advance had been made during the last five or six years in compound engines, though, unfortunately, ship-owners had been persuaded that the essence of the whole was in the engine and not in the pressure on the boiler. He had known a high-pressure boiler, one of the much-abused boilers of the Wanderer, on Mr. Perkins' principle, which had since been at work for eight years, day and night, at 350 pounds pressure, which was tested at 2000 pounds on the square inch when made, and which was to-day capable of taking 500 pounds pressure without injury. He considered that no proper boilers would be constructed as long as that part of the subject was not closely investigated. Without those high-pressure boilers, neither a high economy nor a high speed could be attained. That high speed, to be of value, should not be produced by a puff of forced draught; it should be produced over a long time. To get that, the use of concentrated or fluid fuel was wanted. He had been long at work on that subject. He was delighted to see some progress in armor, but it had only proceeded from solid steel and solid iron to sandwiched steel and sandwiched iron; it had not proceeded on a well-defined scientific basis. He was glad the author saw his way, if only money enough was given, to produce vessels which could not be sunk by shot, torpedo, or ram. Sir Edward Reed asked the question whether it could be done with many torpedoes, many guns, and many rams; and he did not think that that question had yet been answered. The nearest approach to it would be when the whole power of the engine could be devoted as well to pumping out the ship as to propulsion. Strong and far-reaching guns were also wanted. He had made it his business to look into the subject of wire guns, and a question had been asked in Parliament on the subject. Engineers, he thought, would be surprised to hear that so long ago as 1884 the Ordnance Select Committee recommended a riband-gun as having demonstrated completely the durability and the reliability of the wire system for very high pressures; the guns having been made, not by Mr. Longridge, but by other persons, after they had found out that the rule-of-thumb was of no good, and that his formulas were valuable. A wire gun at Woolwich fired a shot $12\frac{1}{2}$ miles without any injury, at an angle of 40°, and with a charge which was nearly double that which the 9-inch gun of the same caliber, a service gun, could possibly have stood. Why that gun had not been brought forward he did not know. He believed that the system of wire-gun making would be brought forward some day as a commercial manufacture, on account of its extraordinary cheapness and durability. No proper result would be obtained from strong guns without strong powder. So long as cubes of powder were used, many of which would not burn even after they got to the muzzle, and at night were seen burning afterwards, which also tended more to erode the gun

than any gas, so long scientific principles would not be brought to
bear on the question.   With regard to the six points to which he had
referred, everything that was desired could be done in respect to
number one by Perkins' boilers; to number two, by concentrated
fuel; to number three, by a scientific construction of armor; to number four, by hydraulic propulsion; to number five, by wire-guns, and
to number six, by perfect combustion, as Colonel Brackenbury had
called it, of " the spirit of artillery, or good, strong powder."

With reference to the subject of armor-plating, the author had
stated that Mr. Alexander Wilson was the inventor of compound
armor.   In 1870, Admiral Selwyn first drew the attention of naval
architects to the construction of armor on the analogy of the anvil.
He did not know whether sandwiched armor might or might not
have been invented before, but sandwiched armor never could do
what properly constructed armor would do.   He was not ambitious
of claiming priority of invention in compound armor as now made,
but he maintained that a most important investigation was now possible, based on experiments which had given a cone of dispersion
that had sufficed to spread the impact of perhaps 9 or 10 inches
diameter over a diameter of 48 inches, making a bulge at the back of
the plate of 2 inches deep, which had crushed up the oak behind it,
but so spread the number of tons force received on the point of
impact over a larger area of 48 inches in diameter, instead of 9 inches,
or as 63.6 to 1809.6 square inches, as to bring it down to not much
more than 7 tons on the square inch on the oak backing of the
armor.   That was a magnificent result ; but the plate cracked in
consequence of its not being properly proportioned, and what he
urged on the engineers was the question whether the most important
molecular movement of the steel particles was not more valuable
than any great thickness of steel face or wrought-iron backing.
The plates now made by Sir John Brown were, according to the
Ordnance Select Committee, a little different from what the author
had described, being made by wedging, 4½ inches from the back of
a 16-inch plate, a steel face supported partly round three sides, and
pouring the cast metal in between the two from above.   That cast
metal had been brought to such a heat that it united both with the
steel face and with the wrought-iron back, and the result on its being
fired at with a Holtzer or Firminy armor-piercing shot was such a
cone of dispersion as he had mentioned ; and which ought to show one
of two things—either the steel particles had separated from each

other entirely in their descent through an average of 2 inches over 48 inches of diameter, or they were in perfect cohesion.   In one case, the steel could be no longer steel, but powder; and in the other it would be evident that the great molecular movement could take place amongst the steel particles without disruption, and a great absorption of force, or "shot work," in that manner be obtained, making it possible to have much stronger armor with much less weight, and probably at considerably less expense.

Admiral Sir George Elliot was a supporter of the views of the author to a large extent. He did not propose to dilate on the subject of the papers that had been written, but would content himself with offering a few remarks, from a tactical point of view, with regard to the character of this new design.   He considered it a most valuable novelty in the designs of ships of war, and one upon which the efficiency of the future fleet of the country, and indeed the safety of the empire, depended.   He looked upon the question of a battleship solely from the point of view of a great fleet action.   If any attempts were made to depart from that view and it was endeavored to give a battle-ship other qualities, its fighting efficiency would be to a certain extent lost, and the battle would be therefore hazarded when it arrived.   The object should be to gain victory.   The fate of the empire would probably depend upon the result of some great fleet action, and not upon fights between squadrons.   Decisive battles would be fought, not between two different powers, one of which was striving to run away, but between two powers that had come out on purpose to fight.   That had always been the case, and it ever would be.   So far, therefore, he differed from the author, although he quite understood that he had been pressed into giving his ship the speed he had given.   A battle-ship did not require a 17-knot speed, a 14-knot speed was ample; it would give a certainty of a 10-knot speed in battle.   That was all that was required, and all speed given beyond that was so much loss in fighting power, and so much risk of losing the battle when it took place.   The new 14,000-ton ships were designed for 17- or 18-knot speed, but if they were built for 14-knot speed they would have a far greater armor protection, and far greater safety and efficiency in battle.   It was said that great speed was required in case the enemy, when sighted, should try to run away.   In olden times, when an enemy was sighted, the frigates were let loose like greyhounds, to attack the rearmost ships of the fleet and bring on an action, and owing to many causes there would always be sluggish ships in every steam fleet.

Nowadays cruisers should be made as swift and powerful as possible, and he, for one, should not object to see the first-class cruiser with 14,000 tons displacement, because battles between single ships or small squadrons would always be fights in which the gun would predominate, and in which the ship could escape from the ram and torpedo. The author had spoken of unarmored ships being combined with a battle-fleet, and therefore having to run the same risks as the battle-ships. He denied that altogether. The cruisers that would be attached to the fleet would form a separate squadron, like frigates of old, and would be kept out of action until the opportunity occurred for their coming and rendering assistance probably to the crippled ships of their own country; but they would not be placed in the line of battle. It was impossible to go into this question, which was a very large one, as fully as it ought to be dealt with without trespassing too far on the patience of the meeting, and he would confine his remarks to those which he could, perhaps, best offer as a tactician. The whole question was whether the gun power was to be concentrated in one vessel or divided between a number of vessels, thereby multiplying the units of a fleet and reducing the risk of great disaster. He was not prepared to say that he would now hold to the size of the ship that had been brought forward; but, taking the ship altogether, he thought the author had done everything in his power for a ship of that size. For himself he should be inclined, for reasons that he would state, to advocate a vessel of twice the displacement as the limit of the size of the battle-ship of the future. In that way he would have two ships of 7000 or 9000 tons for one that the Admiralty was going to build of 14,000 tons. That would enable him to make the protective deck sufficiently thick to ensure safety to the magazines and boilers. The vessel should be of deep draught in order that the armored deck should be as horizontal as possible. If the vessel was of shallow draught, height being required for the engines, a turtle-deck became necessary, thus making it more easily penetrable by shell. He should like to see a 6-inch deck horizontal, which would prevent the penetration of any shot from any gun that was carried. The great fault that he had to find with the Admiralty design was the want of recognition of the destructive effects in a fleet action of the ram and torpedo. He believed that in the first great battle at sea more ships would be destroyed by the ram and the torpedo than by the gun. It had been so customary to look to the gun as the ruling power that that idea was not readily given up; but he thought that when officers were called upon to report upon the nature of the battle likely to be fought, they should have before them a plan of campaign, and also come to some decision as to the character of the fighting of the future. On these points he did not think that any settled plan had been determined. The whole thing depended upon the future mode of fighting. Was it to be as formerly in line of battle, in close order in one, two, or three lines? If that was the case, he could only say that the prospects of destruction from the ram and the torpedo were far greater than could be expected from the gun. Taking into consideration the number of ships that would be brought together in close contact, it appeared to him to be impossible that they could escape the torpedo. It was said that the torpedo was erratic and did not go straight. Torpedoes, however, would be cruising about in the middle of the fleet in one direction or another. It was said that they might hit the friend as well as the foe, but still the danger existed. There would, he imagined, be two torpedo gunboats attached to every battle-ship that went into action, besides the torpedoes which the ship carried, and he did not see how it would be possible to escape them. The ship might not be sunk, but one blow would open a compartment and let in sufficient quantity of water to put it out of trim, and then, being disabled, it would be at the mercy of the nearest opponent. The Admiralty had not recognized the destructive effect of the ram. The best proof of that was that the bows of all the ships were unarmored. The bow of every French ship was armored. If two fleets met, and one had stronger bows than the other, the ships with the strongest bows would have the advantage. They might not strike end on, but they would hit the bow somewhere, and the strongest would come off best. The shorter and the handier ship would have the opportunity of planting its spur on the bow of the other. If there was the smallest oblique angle of the ram on the bow, there was at once an advantage; the handier ship would have that advantage, and that point had not been sufficiently recognized. It was stated that the Russians were going to build enormous armor-clads; he only hoped they would do so. He did not know what their size was to be, but his belief was that other nations would try and multiply their torpedo-boats and their smaller vessels, and trust more in the future to the ram and torpedo than to the gun. In his opinion the days were gone by for big ships, big guns, and side armor. He held that opinion twenty years ago and had not altered it. It was said that officers in the future would not have the pluck to ram. History did not record that French officers were wanting in pluck, and he believed what an old and distinguished officer had said to him, that the great work in future in naval battles would be ramming. If so, the proposed monster ships would be a decided mistake; the bigger the ship, the more easily it would be rammed, and the same blow would disable a big ship as would disable a smaller one. When it was said that the smaller vessels would be destroyed in detail, it was forgotten that those smaller vessels would all the time be endeavoring to undermine those enormous citadels with whom they were opposed. With regard to the raft body, he believed, as the side armor was penetrable, that this mode of protection would be far the safest, as a ship would have to be very much riddled before water enough would be let in to disable it; but let one shell or shot go through the armor into a large compartment, or reach the magazine or boilers, and the vessel would be at once totally disabled. He would only add that every ship intended to ram should be armor-clad round the bow, with struts across so as to support the armor. In that case the vessel, if it rammed another, would not, as had hitherto been the case, do itself as much injury as the vessel it struck.

Mr. T. Nordenfelt said he thought the author was quite right in laying stress upon the cost of ships of war, because if the nation could afford to build very large and powerful ships very heavily armed with heavy guns, and could send them out to fight the sea-battles referred to by Admiral Sir George Elliot, it would be able to hold its own on the seas, and would have a distinct and decided advantage over powers who could not afford it. He thought a fleet of eight 14,000-ton ships was a grand and admirable thing. It might be expensive, but it might carry the day in battle. Some of them might be lost, but the others might gain the victory. Mr. White had very carefully and cautiously said that such vessels as had been recommended by the author might even find room in the British navy. Those words "might even," however, did not apply to foreign nations. When they had adopted it, perhaps the British might follow suit. With many foreign nations it was a necessity that the ship should be able to carry heavy guns. Some of them could not afford to build heavy armed ships. England, besides the grand nucleus of 14,000-ton ships, required also a large number of ships which could carry heavy guns for the purpose of the very

many points of defense that were necessary. Even England, however, could not afford to have 14,000-ton ships at all the out-stations,
coal-stations, and depots; but needed guns powerful enough to penetrate other ships that might be met with at those stations. About
ten years ago he remembered that a smile spread over the faces of
some of the authorities at the Admiralty when the Italians commenced their eight large ships, beginning with the Duilio and ending
with the Umberto; but even to-day those ships would give them the
chance of doing enormous damage at sea, and they might live somewhat longer than their opponents. The word "battle-ship" was,
he thought, somewhat limited in the meaning set upon it by naval
men. It was nowadays difficult to know what a battle-ship meant.
The new cruisers of the Mersey class, carrying 9.2-inch guns, 22
tons, had to do battle for their country like every other ship, and
they would no doubt join in a big battle; indeed they could not be
kept out of it. Most of the smaller cruisers had no protection except
the steel deck against vertical fire or against the explosion of shells.
The great object of the paper, as he understood it, was to advocate
a ship something between the cruiser, so called, with a steel deck,
and an armored ship. He had made a ship which protected the
guns, the gunners and officers, the machinery and all the vital parts.
Mr. Nordenfelt's view of it was that it was a gun-carriage—a means
of supporting and carrying the guns swiftly enough to overtake large
ironclads, even if not swiftly enough to escape cruisers. But it was
not wanted to escape cruisers; they had big guns and armored
turrets, and the object was to catch them; and he thought that a
speed of 17 knots an hour was perfectly sufficient for what he might
call a mobile gun-carriage. The author had said that such ships
could meet an ironclad if they were four to one. If there were four
ships carrying 9-inch guns or other guns, and if they could, from
being handy, choose their position of attack against the big ship,
even if one or two of the smaller ships were lost, the big ship would
probably be disabled before the other smaller ships were lost. It
had lately been proposed by a foreign nation to have ships of the
type recommended by the author with 6-inch quick-firing guns, two
on each ship in action at the same time, and all capable of an all-
around fire. With four ships capable of more speed than the big
ironclad to be attacked, each ship having always two quick-firing
guns in action, that meant a firing power of from fifty to seventy
shots per minute, each shot being a 100-pound shell containing a

high explosive, thus giving from 2 to 3 tons of shell metal fired
during each minute against the ironclad. What would an ironclad
with a very large percentage of its surface unprotected look like in a
few minutes after such an attack? It could only fire with its big
guns one shot every two or three minutes in addition to the fire from
the auxiliary armament. He would rather fight on board one of the
smaller vessels than on the big one. Time was a very important
element in work of that kind. He had had the honor of being the
first man to introduce into this country officially the system of quick-
firing guns, and had in fact given them the name "quick-firing
guns," and he believed that such guns had to a great extent revolu-
tionized naval construction. But it was easy to go too far and to put
too much weight in armor. Sir Frederick Abel had spoken with
reserve about high explosives, but Mr. Nordenfelt thought that Sir
Frederick could have told the meeting that it was a fallacy to suppose that any plates of 4 inches must necessarily cause on the outside
an explosion of high explosives. If a high-explosive shell could not
to-day penetrate 4-inch or 5-inch plates, it would do so to-morrow,
and long before the ships were ready. Field artillery, cavalry, and
infantry attacked the enemy without any protection whatever, because
if they had protection they would lose their mobility. And it was
the same at sea. Fortifications, temporary and fixed, were wanted
ashore. Big ironclads were needed, no doubt, in the same way for
resisting the fire of the enemy. But there was room also for other
ships, and he was perfectly certain that gallant sailors would prefer
to feel that they were on board a floating gun-carriage which gave
them an enormous power of attack, rather than be in another kind of
carriage in which the defense was more studied than the attack, and
in which they had to be protected by thick armor for a time. No
protection was perfect. If a shell hit a big ironclad when it happened to heel over so as to admit the shell below the armor, it would
be most seriously injured, and a ram or a torpedo might sink it.
The protection could not be perfect, and he therefore preferred the
means of powerful attack to the means of protection. He wished to
say one word on the subject of armor plates. Captain Orde Browne
had said that it was understood to be necessary for the armor plate
that it should have a part of it, the hinder part, which would hold
together the pieces in front when the front pieces resisted the attack
of the shells. No doubt that was so at the time. Mr. Alexander
Wilson did a great deal of good by his patent for compound armor

in days when there was a want of experience in steel-making. The solid steel plates made at the Creusot Works by Mr. Martell and by others were very brittle, flew to pieces, and laid bare the side of the ship, and Messrs. Cammell and Messrs. Brown had benefited the various navies during the last decennium in enabling them to use armor of which the front part resisted shells, the hind part welding the pieces together. If steel-makers had succeeded in making solid steel plates of such a quality that they not only resisted the shells sufficiently but were able to hang together, surely that was better and safer than the older plates. He admitted that compound plates could be made harder at the surface, and consequently tend to break up the shells, but shells were now made that did not break up even when fired at a sharp angle. What was wanted was the thickness of solid steel with sufficient percentage of carbon to resist penetration. Some remarkable results were exhibited in the photographs of plates made by Mr. Vickers. If such solid steel plates could be made in large quantities, they would be more useful than the plates which had not so complete a resistance. He liked formerly a hard steel face even if it were shattered in breaking up the shell, because it was not likely that two shots would hit in the same place. But still better than that was a plate which did not let the shot pass through and which did not fall off when struck. Admiral Howard had spoken of the advantage of small ships in comparison with large ironclads, and he had mentioned one point that could not be too much dwelt upon, namely, that there were four brains and four captains to take charge of the same power or the same money that was placed under the charge of one man in a larger ship. Another officer had spoken about splitting the brains into four, but that was not the point. The point was the question whether the power of the single brain could deal with the responsibility and the many complicated questions that continually arose when commanding a ship of war, or whether the intelligence and experience of four captains of ships would not be preferable.

Mr. T. E. Vickers wished to refer to the question of the material of which armor plates should be made. He submitted for inspection photographs of two solid steel plates which had been tried in the recent series of experiments made by the Admiralty and alluded to in the paper. The plates were all 8 feet by 6 feet by 10½ inches thick, and were tested on board H. M. S. Nettle, at Portsmouth. The gun used was a 6-inch gun charged with 48 pounds of powder and 100-

pound projectile. The plates were supported in a framework against wood backing, which represented the side of a ship. The gun was placed at a distance of 30 feet from the plate. Photographs were taken before being fired at and after each shot was fired, and any fine cracks were marked with chalk so as to be seen plainly on the photographs. Afterwards, the back of the plate and the wood backing were photographed in order to show to what extent the plate had answered its purpose in protecting the ship. The steel projectiles were also photographed. Five rounds were fired at each plate; the projectiles used for the first, second, and fifth rounds were the well-known Holtzer steel armor-piercing shells; at the third and fourth rounds ordinary Palliser chilled-iron shots were used. The first of the two plates was delivered to the Government before the steel plate of Messrs. Cammell mentioned in the paper, but was tested after it. The three Holtzer shots, after penetrating the plate to a considerable depth, rebounded against the bulkhead through which the muzzle of the gun protruded, and the Palliser shots were broken into small pieces. The throwing back of the steel shots was unprecedented, as far as English experiments were concerned. The second plate was made at a later date in order to see if the good results at the first trial could be repeated. Advantage was taken of this second trial to try if a somewhat harder plate would give a better result. This plate was penetrated to less depth than the preceding one, and the Holtzer shots were thrown back with greater force and were also broken; the plate itself was rather more cracked than the first one, but not seriously so. The whole of the plate remained in position, although one corner was found separated when the plate was removed. The second plate was fired at on a frosty day, and it was possible, if not probable, that the increased cracking was due to the lower temperature. The conclusion he derived from these experiments was that the hard nature of the steel which could be used in the front portion of the compound plate gave a better resistance to the projectiles than the first portion of the steel one did, but that this advantage was lost by the comparatively feeble resistance of the iron portion behind it. The steel plate continued its resistance to the last, as shown by the fact that the steel projectiles rebounded with considerable energy. He believed that in thicker plates the advantages of steel would be still more marked.

Mr. Sydney W. Barnaby said he proposed to address his remarks not to armor, but to that part of the paper which was to him of greater interest, namely, the new material introduced by the author as a substitute for armor at the region of the water-line. It was astonishing to learn that with an appropriation of 5 per cent of the displacement of a vessel, a solid belt of that material could be placed at the water-line, which would exclude 96 per cent of the water that would otherwise enter through perforations at that region. The author had shown how he would use that material for the protection of an armored ship-of-war. He had listened with much interest to the criticisms which had been passed upon that design, and he thought they might be summed up in this, that the ship was small. Every naval architect would agree that there was no fault in a design which was more easy to remedy than smallness. If the only fault that could be found with a man was that he was too young for his post, the obvious answer was that he would grow out of it. In like manner, if the severest stricture upon a design was that it was too small, the obvious answer was that it was easy to enlarge it. If the author were told that more coal was wanted in the ship, he would say, " It will take so many more tons displacement ; I can do it." If he were told that it was necessary to increase the auxiliary armament, he would again reply, " It will take so many more tons, but it can be done." But in the case of a ship of 14,000 tons displacement, when the critic began by saying that the ship was much too large, then that there was not half enough coal, and then that certain portions of the ship were coated with 5 inches of armor where there ought to be 15 inches; in order to satisfy him, it would be necessary to take a clean sheet of paper and begin the design over again. The author had alluded to the Admiralty policy of taking up merchant cruisers and using them as fighting ships. He had stated elsewhere that it was quite possible to build such ships with their machinery below a bomb-proof deck, in which case they would be as easily defensible as a regular ship-of-war. It appeared to him that if such ships were built like the present Atlantic liners, but with their machinery covered by a $1\frac{1}{2}$-inch deck, and with the new material stowed between the protective deck and another thin deck 2 feet above it, as in the author's design of a war vessel, such a ship would be as well defended as a cruiser of the Tage type ; and he would ask the author whether he did not think that it might be wise for the Admiralty to build such ships protected in that way, and that instead of hiring inefficient war cruisers from the merchant navy, it might let out these vessels to the ship-owners for trading purposes. Of course the packing material would not be put in until the ships were required to fight, and the space occupied by it would be available for cargo. All that the owner would have to put up with would be a certain permanent load imposed by the protective deck and a large engine-room staff, because two or even three screws would probably be required in order to keep the engines below deck. The gain to the country would be twofold: first, the present fast ships which, it would be remembered, were the only vessels which had a chance of running the gauntlet of hostile cruisers and bringing in the food supply, would not be withdrawn from the trade routes at the very time when they would be most useful; secondly, very efficient war cruisers might be made self-supporting, or even a source of profit to the nation instead of a constant drain upon its resources.

Dr. W. Pole, Honorary Secretary, while fully admitting the great public importance of the design of armored ships generally, remarked that the branch of the subject which most nearly concerned engineers was the construction of the armor, and its mechanical efficiency in affording protection. The author of the paper, though he had given some particulars, had said but little on this head, and it was somewhat surprising to find that, although for thirty years the subject had been so prominent as a national question, it had come so little before the Institution. The nature of the material to be used, its mode of manufacture, the design, form, and dimensions of the armor, the manner and amount of its resistance to shot and shell, the connections and fastenings, and many other points, were all questions purely belonging to the civil engineer; but, strangely enough, there was scarcely a reference to them in the whole of the minutes of proceedings. The Government had, however, thought it expedient from time to time to call in the services of engineers to their aid, and as Dr. Pole happened, at an important time, to have taken an active part in this way, he would ask leave to put on record some account of what had then been done. The author had alluded to the early opinions authoritatively given, about 1850, by British naval authorities, against the use of iron for naval war purposes. But the French were not so hasty, for a year or two afterwards, at the outset of the Crimean war, the French Emperor suggested the use of thick plates of hammered iron as a protection to floating batteries, which it was proposed to use in attacking Cronstadt, and an experiment was made in 1854 at Portsmouth by firing at a 4½-inch plate with wood backing. The result of this was that three such batteries were ordered by the

French Government and an equal number by the English. When the time came for their use, the British batteries were not ready, but the French ones were, and they were employed in the attack on Kinburn on the 17th of October, 1855. They were exposed to a heavy fire, at a distance of 700 yards, for about three hours, unsupported by the fleet, and at the end they came out practically uninjured. Such a test as this was conclusive, and from that date the utility of iron armor in protecting ships-of-war became an accomplished fact. The French lost no time in carrying the idea further, for they at once began building a large sea-going ship-of-war, of 5500 tons, protected on this principle. This vessel, now become classical under the name La Gloire, was launched in 1860, and was clothed from end to end in exposed parts with iron armor, some of it nearly 5 inches thick. Meanwhile the Kinburn success had gone far to dispose of the prejudices of the British dockyards, and the question of the claims of iron was re-opened by several experiments made at Portsmouth and at Woolwich, in 1856-1857 and 1858, with iron structures of various kinds. And, moreover, as the Government could not allow the French to steal a march upon it by the new-fashioned armor-clad man-of-war, it was decided to keep pace with them by constructing a similar ship, protected in the same way. This was the Warrior, a fine steamer of 9200 tons, which was launched in 1861. But, when the application of iron armor began to be seriously entertained, the Government found that it involved a multitude of questions, chiefly of an engineering character, which required the most careful and patient investigation, and it wisely decided that the whole subject should at once be submitted to a scientific committee, on which the engineering profession should be represented. This body, which afterwards became well known as the "Iron Armor-Plate Committee," was appointed by Mr. Secretary (afterwards Lord) Herbert, in January, 1861, and it consisted of six members; namely, Captain Dalrymple Hay, R. N., chairman, Major Jervois, R. E., Brevet Colonel W. Henderson, R. A., Dr. Percy, W. Fairbairn, and Dr. Pole. The members of the committee were in active work till the middle of 1864; they reported frequently and fully their proceedings to the Government, and their reports were printed in four folio volumes, with full illustrative plates. These documents were withheld from publication at the time for reasons of public policy; but after this interval there could be no impropriety in giving to the Institution a brief account of their proceedings, and of the results that followed therefrom.

The object intrusted to them was to investigate generally the application of iron to defensive purposes in war. Their first duty was to make themselves acquainted with the knowledge already existing on the subject, and this they did in two ways. In the first place they collected and reported on all the records they could find of previous experiments or investigations. These led to a general belief that iron was capable of forming a good protection, but still very little useful practical knowledge had been acquired, either as to the quality of material most efficient for the purpose, or the most advantageous mode in which it should be applied. The next step was to learn the views held on the subject by those persons best acquainted with it, and the committee accordingly invited the evidence of a great number of scientific and mechanical men who were likely to possess knowledge bearing on the subject. Among the witnesses were Sir William Armstrong, Sir Joseph Whitworth, General Lefroy, Captain Andrew Noble, Mr. C. W. Lancaster, and Mr. John Anderson, as eminent authorities on modern artillery; Mr. Scott Russell, Mr. Laird, Mr. Samuda, Mr. John Grantham, and Captain Hewlett, on naval construction; Professor Wheatstone, Mr. Mallet, Mr. Nasmyth, Captain Inglis, on the general scientific and mechanical treatment of the question, and all the most eminent iron manufacturers in the kingdom as to the properties of the material. The information given by these witnesses was extremely valuable, but their opinions were very conflicting on the novel points before the committee. Having thus obtained all the information possible, the committee began its own more special labors, and these consisted of original investigations, chiefly of a practical and experimental character, which were directed to two classes of objects: viz., first, the determination of a great number of points affecting the general application of iron for defensive purposes; and, secondly, to the direct trial, by ordnance, of special constructions, in the shape of targets designed and prepared for the purpose. The first and most important point was the nature of the material best adapted to resist the impact of shot. On this there was little positive knowledge, and the witnesses held great diversity of opinion. Some recommended soft and tough iron, others, steel of different kinds, and others, combinations of steel and iron. The committee tried an extensive series of experiments, which led to the conclusion that the most suitable material was simple malleable iron, the best kind being that which combined in the greatest degree the qualities of softness and toughness. It appeared that the use of defensive armor-plates against shot was to resist the blow in such a way as would do the least mischief to the structure, and that the best mode of accomplishing this would be to make the plates of soft, tough material, which would allow the " work " residing in the shot to be absorbed in indenting and battering them without producing fracture. The committee directed much attention to the mode of manufacture of armor-plates. This, at the commencement of its labors, was quite a new thing, as no masses of malleable iron of such magnitude had previously been made. The first plates were made by hammering, but powerful machinery was soon erected for rolling them, and the rolling proved preferable for obtaining the quality required. The committee gave every encouragement and stimulus to the makers to improve their manufacture as much as possible. With this view the plan was adopted of inviting them to see the experiments and to study carefully the effect of shot on plates, giving them at the same time the benefit of all available information which could throw light on the subject. These opportunities were of the greatest use, for by no other means whatever could any such complete insight have been obtained into the nature of the problems to be solved and the conditions necessary for their solution, and further, this system of openness gave universal satisfaction to the trade. The iron-makers availed themselves willingly of the facilities afforded them, and the result was that, whereas at the appointment of the committee there were very few iron-masters who could make thick armor-plates, and those of very uncertain quality, there were at the end of the committee's labors many excellent makers, the dimensions had much increased, and the quality had immensely improved. At first the committee could scarcely rely on the quality of any plates above 3 inches thick ; afterwards, 5 and 5½ inches became common thicknesses of approved quality, and some of the best samples were thicker still. The chemical composition of the various irons was accurately determined, and various metallurgical data were obtained. Careful comparisons were made with foreign irons, and it was fortunately found that armor-plates could be made with iron obtained from British ores, by the ordinary processes of smelting, and manufactured with mineral fuel, and not only equal but superior to foreign plates of much more costly production.

Another very important point of the committee's inquiry was the form in which iron could most favorably and economically be used for defensive purposes. Various opinions were held on the subject :

some authorities recommended a peculiar arrangement of bars, which previous experiments had favored; certain eminent engineers and ship-builders suggested a number of thin plates, well fastened together, a form which was thought to present great advantages for naval purposes in affording support to the structure, and at the same time offering great facility and economy of construction; while a great number of other schemes for bent plates, corrugations, ribs, bosses, and many other ingenious contrivances had their supposed advantages strongly urged. More than four hundred plans were submitted to the committee; these were all examined and reported on, and some of them were tried; but through all these complications the simple result was arrived at, that the best application was a single plate of uniform thickness, with the surface perfectly plain. It was found that a large size of plate was advantageous as giving fewer joints, although small plates were more trustworthy in quality, as well as much cheaper. At first it was supposed to be an advantage to connect the edges of adjoining plates by grooving and tonguing, and this was done in the Warrior, but it was shown to be a mistake, doing mischief instead of good, and was therefore abandoned. Another point of great uncertainty was as to the best mode of fastening armor-plates to the ship's side. Numberless schemes were devised for doing away with bolts passing through the plate, but experience showed that, with certain precautions, bolts and nuts were perfectly efficient. One of the latest experiments, however, gave remarkably good results with the French plan of large wood-screws holding into the backing behind. Great interest attached to the question of the backing most suitable for ship armor-plates. When these were first used, they were fixed directly upon the hull of timber ships, and when they were first applied to iron ships (in the case of the Warrior), it was thought expedient to imitate the former conditions by placing a thick layer of wood between the armor-plate and the skin of the vessel. Many objections were raised to this: the wood backing was said not only to be unnecessary but to be absolutely prejudicial, as liable to decay, and to destruction by fire and by shell. It was necessary that the committee should thoroughly test this, and elaborate experiments were designed and carried out for this purpose; but the committee was unable to recommend that this wood backing should be dispensed with, as it appeared to perform important functions for which no thoroughly efficient substitute could be found. The nature of the projectiles was a subject intimately connected with the resistance of the plates, and a great number of experiments were made in conjunction with the Ordnance Select Committee, to determine the influence of variations in the weight, velocity, form, and material of the projectiles as regarded their effect on the armor. Discussions of the results obtained were given in two memoranda by Dr. Pole in January, 1862, and January, 1863.

In addition, however, to the determination of general principles, it was the duty of the committee to make trials of actual structures on the largest scale, for it was impossible to tell by any à priori reasoning or by any deductions from small trials how the armor of ships would behave under fire. The committee remarked:

"One of the most striking deductions from our experiments is the difficulty even for those most conversant with the subject to foretell what are likely to be results of untried schemes, however promising they may appear. The effect upon iron structures of heavy shot from modern ordnance is totally unlike the result of any other kind of mechanical action, and has conditions essentially peculiar to itself; experiments, therefore, on a great scale are absolutely essential to determine the merits of many plans still awaiting trial."

It was therefore determined to make targets on the actual scale, representing portions of the sides of ships armored in particular ways, and to ascertain their value in the most positive and unequivocal mode by firing at them with heavy artillery to the point of complete destruction. One of the earliest of these was a target representing a portion of the Warrior frigate, which, as already stated, had been built somewhat hastily in emulation of the French ship La Gloire. The target was 20 feet long by 10 feet high, and was composed of 4½-inch wrought-iron plates bolted to a wood backing 18 inches thick, behind which was the iron skin supported by wrought-iron ribs exactly as in the ship itself. The target was fired at in October, 1861, with many rounds of shot and shell from the heaviest guns then available, at 200 yards range, 3229 pounds of projectiles striking the structure; and although naturally much damage was done, the resistance offered was considered very satisfactory. This same target was again tried with heavier artillery than before, and the report was that no other target experimented on had offered so good a resistance. Many authorities considered, however, that it would be desirable to try other systems of protection, and accordingly several targets were constructed embodying different ideas. Two of these were on plans by Sir John Hawkshaw, one by Sir Wil-

liam Fairbairn, one by Mr. Scott Russell, and one by Mr. Samuda,
all which, under fire, gave valuable information of some kind or
another. And as the building of armored ships went on with modifications in their mode of protection, targets representing them were
put to the proof, the Minotaur, the Lord Warden, the Bellerophon,
and others being proved in this way. These and other ships were
built expressly to carry out the designs drawn from the information
obtained by the committee, and they represented the best knowledge
and the most efficient protection attainable at that time. In 1864,
at the instance of Dr. Pole, a target was made representing a system
of armor which he had seen applied to a French ship building at
Cherbourg; it involved some novelties, and the trial was particularly
instructive. The attention of the committee, however, was not confined to ships only, but was largely devoted to the application of
iron to permanent land fortifications. There were absolutely no
facts on the subject of any value, and data could only be obtained by
trials on a comprehensive scale. Accordingly, several structures
were made of different plans that appeared feasible, and the result of
the trials led to the development of satisfactory practical designs.

The members of the committee endeavored, throughout their
work, not only to ascertain facts, but to reason upon them, and to
make them useful for practical guidance. And when it is considered
how little knowledge there was when they began their work, and
that when they finished it they left the system of armor both for sea
and land defenses fully and successfully established, it must be
admitted that their four years' labors were not in vain. The great
feature especially of the manufacture of the plates may be said to have
been created under their auspices. In the twenty-five years that
have elapsed since their labors ceased, no doubt great progress had
been made, not only in the magnitude of the ships, but in the force of
the artillery used against them, and the consequently larger scale of
the armor; and it must, of course, be added that the new form of
malleable material produced by fusion, whatever it may be called, had
introduced great changes into the modes of preparation. It would be
the province of others, who had the knowledge and the will, to give
to the Institution information of these more modern improvements.

Sir Frederick Bramwell had nothing to remark on the engineering
part of the subject, but he would like to refer to a question of political
economy. Mr. Sydney Barnaby had proposed, in the interest of the
ratepayers of the country, that the Government should make a certain

class of ships, capable of receiving "woodite" protection, which, when not wanted for the purposes of war, should be leased out to the carrying trade as merchant steamers. If that could be done, he said, such ships would be earning money instead of being a cost to tax-payers in the time of peace; also, that, when war came and ships were wanted to bring food to the country, they would not be withdrawn from the merchant fleet in the way they otherwise would be. Sir Frederick Bramwell desired on this suggestion to remark that, for Mr. Barnaby's proposal to succeed pecuniarily, it would involve that, in times of peace, merchants should hire Government ships, and should not build ships of their own; but as in time of war the Government ships must be withdrawn, and as by the suggestion the merchants would not have built ships, the vessels employed in the food trade would, when war came, be as much withdrawn as would be the ships owned by merchants, and there would still be the withdrawal of food-bringing ships. That would be one result, and the other would be that they would be allowing the Government to embark upon trading, which it was always too ready to do, competing with private capi-talists, investing capital in the supply of ships and interfering with private commerce. He hoped that the Institution would always listen to anybody of the name of Barnaby, and certainly to Mr. Sydney Barnaby, with every respect to matters of engineering, but would be slow in adopting his views as to who ought to supply the merchant navy of the country.

Sir Nathaniel Barnaby thanked the members for the manner in which they had received the paper, with its many faults of omission, some of which had been happily corrected by Dr. Pole, and some faults of commission with which he should have himself to deal. With reference to the remarks of Admiral Sir George Elliot, he thought it was a great mistake to suppose that the rams of ships which had not armor carried out to the stem were therefore weak. The case had been quoted of the German ship the König Wilhelm, which ran into and sank a sister ship, and it was said that the ram was seen to be twisted out of its place, and that was urged against the use of rams as at present constructed. The present form of ram had been very largely influenced by the failure of the German ram that had armor carried out to the end, which armor completely failed to support it. The method was changed, and he contended, having had a good deal of experience in the matter, that it was impossible to have a stronger form of ram than that now used in the British ships-of-war. If any one desired to convince him to the contrary, he would be happy to discuss the matter with him, but he believed it was impossible to have a stronger ram than that now in use, and he was certain that Mr. White would support that view. He regretted that he had not known of the great success which Messrs. Vickers had met with in their attempts to enter the field of armor-plate makers. That, however, was not his fault. The Admiralty had preserved its secrets well, and had resolved to keep undivulged all information as to who could make armor-plates, and, until the photographs were produced, he did not know how great a success Messrs. Vickers had achieved. He had singled out the firms of Messrs. Charles Cammell & Co. for special notice because they were the only ones in England who had made armor for modern ships-of-war, and he might be well excused for singling out those two great firms who had done such splendid service to the country. Captain Orde Browne had asked whether it was not possible to have hard armor for the cruisers to keep out the high explosives, and softer armor for the ships which were more heavily plated. To that, a naval officer near him at once objected very naturally that it was uncertain what explosives the enemy might discharge. That objection to the use of different kinds of armor in cruisers and battle-ships was good also against the use of different principles in applying armor in cruisers and battle-ships, and was the justification for his treatment of the French ships in the diagram. His contention was that the principles of armor defense should be the same in all classes of fighting ships. Look at the table of phases of armor. The Gloire, a French ship of 5500 tons, was designed to receive the attack of the fastest ships of that period, and to resist their shell guns. The Tage, of 7000 tons, was designed to do precisely the same thing. So also was the Dupuy de Lôme. Why, then, should he not compare them in the matter of armor defense? Sir Edward Reed and Mr. White had stated that he must not do so because they were called by different names. The Gloire was called a battle-ship; the Marengo and the Amiral Baudin were known as battle-ships, and the Tage and the Dupuy de Lôme were not. But in what respects would the enemy's guns treat them differently? In the last naval war the principle of defense was the same for all fighting ships. In the very largest ships there was the wing passage completely round the ship in the neighborhood of the water-line inside; and the business of the carpenter and his crew in an action was to watch the whole of that wing passage, and to have plugs with which to stop up the holes that the enemy's shot made. It was inconceivable to him that any arguments could convince an assembly of engineers, called upon to hear and judge, that there was any justification for protecting the crew of one fighting ship with 10 tons of armor per man, and leaving the crew of other fighting ships without any armor, vertical or horizontal.

Sir Frederick Abel had drawn attention to conflicting statements in the paper as to high explosives. Yes, they were certainly conflicting. One set had been put forward by official people who were defending the use of 4-inch and 5-inch armor in new ships as a perfect defense against high-explosive shells. The other set supported his own views, that there could be no approximation to finality in the use of such armor. He had shown how the quick-firing gun and the automatic gun would be effective against it, and that the invention of armor-piercing projectiles loaded with those nitro-glycerine compounds, capable of piercing through even 12 and 14 inches of armor, must be expected. In a paper prepared by him last year for the Naval Construction Company he had said, " 4-inch armor for this purpose is already obsolete before the ships are finished." Sir Frederick Abel had confirmed that. He had stated that experiments had proved that 4-inch armor would not always keep out high explosives; that it would be penetrated by armor-piercing projectiles which would carry the explosive charge through the armor. He had said that if the efficient application of high explosives depended only upon the provision of thoroughly reliable delayed-action fuze, then the naval officer might depend on having in the next war high explosives brought with perfect efficiency to bear against thick armor. In his paper he had asked whether it was only the enemy who was to take advantage of those powerful projectiles. He had also deprecated the use of such armor. He had wondered whether it had been put into the recent large Admiralty designs in deference to a school represented by the gallant officer, Lord Charles Beresford. That officer had stated that the Admiralty had started first of all with the theory that an armored battle-ship should not be sunk or put out of action unless its armor was pierced. He had himself always thought that where armor was used it should be endeavored to make it impenetrable. It was said that such impenetrable armor was useless if the ship could be sunk or put out of action while it remained unpenetrated. But, according to the new school, there need only be put on some thin armor which would be speedily perforated, in order to set everything right. In order to put that armor on, the size of the ship must be increased at the expense of the number of ships; but no matter; the money expenditure per ship might be great, or the speed might be sacrificed, or thickness of armor or armament, in order to get some of that easily penetrable armor which would save the honor of the navy and justify the new school of naval architecture. It was said that the ship of 3000 tons was very small. His object in advising those who prepared the designs was to keep to a ship of minimum size for doing the work. That a ship could be made larger with advantage was no doubt true; but he thought the ship question fairly balanced all the qualities that could be got in a ship of the smallest size capable of fighting an action at sea. The ship was practically as large as the Asia and the Indus, the flag-ships of the Admiral Superintendents at Portsmouth and Devonport. He agreed with those naval officers who thought that any weather which would sensibly affect the comparative behavior of the ship of 3200 tons and that of 14,000 tons, would be found to be not fighting weather for either ship. Mr. White preferred an armored deck rising above the water-line in the middle, and sloping under the water to the sides, to the proposed raft body. He thought he was the first to introduce a deck under the water, with cellular divisions over it, in classes of ships which hitherto had been unprotected. That was in the C class of corvettes in 1878. He thought he was also the first to introduce the mode of defense preferred by Mr. White. That was in the Mersey class in 1883. If he should be the first to be instrumental in the use of the solid raft body in preference to either of those plans, he should think he had made a distinct and most important advance in the art of war-ship building. In bringing his paper before the Institution he had no desire to criticise any Admiralty work. He wished to direct attention to the new forces coming into the field, and to advise that outside engineering opinion should be taken before a large expenditure of money was finally decided upon. Since the Admiralty proposals had been made, they had been considerably discussed. They were said to have met with the acceptance of the navy. He had been listening that afternoon to an excellent lecture given by a naval officer, Captain Reade. The lecture was illustrated by beautiful drawings of ships, and it was a captivating lecture. The lecturer, however, had said some things about existing ships which he did not in the least agree with.

He stated that the new Admiralty designs were quite wrong, and he contended that the ships should have armor all over their sides, 6 inches thick. His view appeared to be that the design which the French had produced, of the Dupuy de Lôme type, was the one most needed for the largest ships. He had read with great interest a paper in the *Nineteenth Century* for May, by Lord Armstrong, and he should be glad to quote a few words from it :

"It will be noted," said Lord Armstrong, "that Mr. White is careful to avoid all questions as to the policy of building armored battle-ships in preference to ships of the cruiser class." He went on to say, "Although I am ready to commend the designs for these new ships, my distrust of the efficacy of all vessels of this armored class in relation to their cost remains unchanged. All the advantage they possess in point of defense is a partial and imperfec protection against artillery fire. As regards rams and torpedoes, they are as vulnerable as ships without armor at all, and they are as liable to perish by the perils of the sea as any other kind of war-ship, while their cost is so great that the loss of any one of them from any cause amounts to a national calamity. Mr. White, in his paper on these new designs, refers to what he calls the 'too many eggs in one basket' argument, but he wisely adds that he leaves that argument to be dealt with by the Board of Admiralty, who are responsible on matters of policy, he being only their technical adviser. The argument referred to is, however, undoubtedly possessed of great cogency, and to my mind the only justification for persevering in the building of such ships is that foreign nations are still doing so. Nevertheless, I maintain that we shall realize a greater amount by chiefly devoting our resources to the multiplication of vessels of the cruiser class in preference to those of the armored class."

Those were weighty words, and expressed almost what he had had in his mind in preparing the paper. He would only add another word. England possessed two sets of very able sons—sailors and civil engineers. He had the highest possible admiration for sailors, whom he knew well, and who were very splendid men ; but when it was decided to enter upon a national policy in such an important stage of affairs—such crisis as the present—he thought that to shut out the great body of engineers, to whom questions of principle might be wisely referred, was a mistake. He thought every one would feel that if it had been possible for the Admiralty to lay before a body of civil engineers the outlines which were put into Mr. White's hands (he was not now criticising the designs), it would have been much better than to enter upon a grave step in policy without such consideration as it ought to have received.

CORRESPONDENCE.

Admiral the Right Hon. Sir J. C. Dalrymple Hay observed that when Sir Henry Chads reported, in 1850, that "whether iron vessels are of a slight or substantial character, iron is not a material calculated for ships of war," he was justified by the character of the ship-plates of which some of Her Majesty's ships were being constructed. Substantial was a relative term, and the most substantial plate they applied to the sides of Her Majesty's ships did not exceed in any case $\frac{7}{8}$ inch. These, under the fire of 32-pounder guns with cast-iron spherical shot and 8 or 10 pounds of powder, were driven in such masses of langrage that it would have been impossible for men to live behind them. It was not till 1854, after the failure of the Anglo-French fleet to effect on the 17th of October any material damage on the sea defenses of Sebastopol, that, on the initiative of Napoleon III., wrought-iron plates were applied to the sides of wooden ships to protect them against shell-fire. The first specimens tried in this country, at Portsmouth towards the end of 1854, were made under the tilt-hammer, and were $4\frac{1}{2}$ inches thick. The spherical shells, at short range from 8-inch guns, broke up on striking, the plates received little or no damage, and the wooden structure was saved. He witnessed these experiments. The next opportunity of seeing an experiment was at Kinburn. The French had three floating batteries there, exposed at short range to smoothbore guns. Some men were killed by shot which entered the large broadside ports; but though the battery which he had an opportunity of visiting on that occasion had been struck sixty-three times by solid shot, no plate had been penetrated or broken, and no bolt had been broken. In 1860 he was named chairman of the Iron-Plate Committee, and was associated with the late Sir William Fairbairn, Dr. Percy, General Sir W. D. Jervois, and Dr. Pole. It might truly be said that nothing was known of the manufacture of iron plates for protection against shot when that committee assembled; when it was dissolved in 1864, it left a record of experiments and opinions which even now, in the advanced stage of science, might safely be consulted. The Warrior, Black Prince, and Achilles were in process of construction, but the Minotaur, Agincourt, and Northumberland were constructed in consequence of the advice of the Iron-Plate Committee; and the rolls by which the plates were constructed were put up by Sir J. Brown in consequence of evidence taken by the Iron-Plate Committee of the superiority of plates so constructed to those made under the tilt-hammer. The turret of Captain Coles led to the "raséeing" of the Royal Sovereign, and afterwards to the experimental fire to which she was subjected from guns of the Bellerophon. During the decade 1860 to 1870, several incidents occurred to make naval men skeptical as to low freeboard, breech-loading guns of large calibre, and the reliance upon a smaller number of large guns in preference to the broadside. Vessels of the Agincourt class had proved exceptionally fine sea-boats. They had assisted in laying an Atlantic cable; they had towed a dock successfully to Bermuda. The Royal Sovereign had been damaged in a contest with the Bellerophon, and the Captain, through being over-canvassed, had perished. One of the most formidable turret-ships ever built was thus lost, owing to the wish of the designer to make it also a sailing ship, against the advice of Sir Edward Reed, then Chief Constructor. The 40-pounder Armstrong breech-loader, with 8-pound charges, was ineffective against the $4\frac{1}{2}$-inch plates, and a 150-pounder breech-loader gun jammed in the breech and was unable to be reloaded, in experiments at Shoeburyness. Thus it came to pass that want of confidence in a breech-loading gun capable of standing heavy charges, and the reliance upon many guns rather than on few, induced many officers of ability, such as the late Sir Cooper Key, to rely upon muzzle-loading ordnance and broadside ships. The resistance to an increase in the size of ships was intelligible. There was no need to build ships much larger than possible opponents.

Since the making of the Suez Canal, the Indian empire was deprived of the support of the Mediterranean fleet if its ships drew more than 24 feet and 8 inches. There was no advantage in increasing the risk on each individual ship, or increasing the cost, unless concentration of power was advisable for other reasons. The author had quoted accurately some remarks of his giving a general assent to Admiral Touchard's expressed opinion. He saw no reason to change the opinion then expressed. In specifying in their order the parts requiring protection, he saw no necessity to alter the respective values of the arrangement quoted. But, in the speech quoted, he previously stated that seaworthiness and speed were above all things necessary. He still adhered to that statement.

It must be remembered that in 1873 the torpedo was hardly developed, and the Polyphemus had not been tried. The torpedo might be guarded against by crinoline, but the ram could only be used or avoided by speed and handiness. This brought him to the special proposal of the paper. If the raft-bottom proposed could be made so as to receive the damage inflicted by the torpedo in such a manner as to lessen the risk of sinking, it would have most valuable results; but this material was, so far as he was aware, not yet even in the experimental stage, and he should be impertinent to offer any opinion on its merits. Small ships of this character would be invaluable, and have many of the advantages claimed by the author; but they did not, in his opinion, do away with the necessity for building large first-class battle-ships. They probably were an improvement on the Riachuelo, though four of them did not seem to be a substitute for the Victoria. How rapidly the march of science caused opinions to develop, and how difficult it must be for an outsider like himself to express a confident opinion, was illustrated by the fact that Sir Edward Reed and the author, two of the most eminent naval architects in the world, and who had each worn the blue riband of their profession as Chief Constructors of the British Navy, were not absolutely in accord on this subject, and neither of them, when in office even, so far as was known, submitted such a design to the public or the authorities. This country had been at war at the same time with France, Holland, Italy, Turkey, Russia, Denmark, Sweden, and the United States. The British navy then consisted of one hundred and forty-six line-of-battle ships in commission, and in 1807 Great Britain seized the Danish fleet at Copenhagen, and in 1808, the Russian squadron in the Tagus, in order that it might have as many battle-ships as all the world besides. This policy was carried out by Chatham and Lord Anson, by Mr. Pitt, Lord Howe, and Lord Barham, and was the only policy which gave this country absolute safety. This superiority was needed not at the end but at the outbreak of hostilities; wars were short, sharp, and decisive, and those best prepared, like Germany on a recent occasion, were successful. Other classes of ships might be extemporized, but a first-class battle-ship took four years to complete. It was better to spend money in building such ships, in case they might be wanted, rather than in paying a ransom after defeat. There were now one hundred and four first-class battle-ships in the world. Great Britain had thirty-four. It was proposed to build ten new ones. To that must be added twenty-six with a speed and gun-power superior to the Italia, with a protection over the vital parts superior to the Italia's guns. As the offensive power of the Italia could be developed by direct fire in a line with keel of all its four guns, and as the bow in ramming was most exposed, so it seemed to him it was essentially needful that a narrow belt of protection at the water-line should be given, as the punching fire of quick-firing guns might possibly materially injure that portion of the ship.

Before closing, he desired to emphasize an opinion he had always ventured to express since the abolition of the Ordnance Office in 1855. The Ordnance Office was essentially a manufacturing depart-ment, though it was overlaid with detail of the Corps of Royal Engi-neers and Artillery. Those corps were justly attached to the War Office, but the construction of guns and the provision of munitions of war should be entrusted to a department separate from the two great spending departments. The nation might then hope to have guns ready in good time for its ships, and identity in make and calibre in the armament of its ships and fortresses. Until this was done he feared the difficulty in providing a suitable armament for ships would continue, and he could conceive nothing more fatal to the naval supremacy of this country than that a ship or a fleet resorting to Gibraltar or Malta should find guns and ammunition of a different character to that which they might require. England must be guided by the course of ship-building in nations which might be pos-sible opponents, and for every battle-ship laid down by foreign coun-tries a faster and more powerful ship should be commenced and rapidly completed here.

Major D. O'Callaghan, R. A., remarked that a tolerably long experience of armor-plate experiments at Shoeburyness led him to hail with satisfaction the proposal to abandon the type of unarmored fast-steaming cruisers, in favor of a vessel, which, while possessing the undoubted advantages of high speed and manoeuvring power, would afford a measure of protection to guns and guns' crews from the dis-astrous effect of an enemy's shell-fire. With respect to ships which depended for their existence on speed alone, Englishmen were not fond of rapid strategic movements to the rear. Such evolutions were at times open to misconstruction, and captains whose orders were to hover on the outskirts of an engagement, and whose *rôle* was to convoy merchantmen or to bombard mercantile ports, would inev-itably turn their blind eyes towards signals and would try conclusions

with ironclads in the thick of a fight, with the certain fate of destruction. The "plucky little Condor" incident would be repeated over
and over again with less happy results. He well remembered the
terrible effect of common and shrapnel shell fired some years ago at
a section of the Shannon at Shoeburyness, and he made up his mind
then that it was almost useless to protect the so-called vulnerable
portions of such a ship—the engines, steering-gear, and magazines—
when the fighting portions, the guns and their crews, were open to
complete and total destruction by the feeblest projectiles. A delayed-
action fuze for high explosives was not, as the author remarked, an
accomplished fact, and shells filled with high explosives were powerless
when fired against armor, unless the latter was very much below the
perforating power of the gun opposed to it. There was little doubt,
for instance, that 4 to 6 inches of hard armor, easily perforable by a
6-inch breech-loading gun, would be quite proof against a shell of
that calibre filled with high explosive which would detonate on
impact, causing the shell to break as harmlessly as an egg on the
face of the armor. On the other hand a similar shell from a 9.2-inch
or 10-inch breech-loading gun would probably overcome such a thin
plate, detonate in passing through it, and though, as far as disruptive
power was concerned, its explosions would probably be more local
and less damaging to structure than that of a powder-filled shell, the
number of man-killing fragments would be greater, and the execution
in this respect proportionately to be dreaded. This effect, too, would,
he took it, be quite independent of any fuze, delayed or direct in its
action, but would be dependent solely on the disparity between the
armor protection and the power of the gun. Steel-faced armor, 9
inches thick, such as the author proposed to use in the new type
cruisers, was just perforable by a 6-inch armor-piercing steel shell at
short range, but it would, he thought, keep out a shell filled with high
explosive fired from a 9.2-inch breech-loading gun striking normally
to its surface, and afford an efficient protection from even heavier
shells striking at an angle. The effect of these projectiles, owing to
the disruptive force being confined to a comparatively restricted area,
might be discounted and localized to a great extent by the employment of traverses, and this, he would suggest, should be an important factor in future construction. On the general question of building four comparatively small ships instead of one large one, it would
be presumptuous for him to express an opinion; but the policy must
surely be sound, since the four ships were practically as invulnerable

as the one bulky ironclad, were more easy to manœuvre, and were susceptible of both independent and combined action. All the eggs were not in one basket, and the loss of one of the baskets would not necessarily involve the destruction of the other three.

Rear-Admiral R. A. E. Scott observed that the author's proposal was to build five small partially-armored ships of 3200 tons each, in lieu of every intended battle-ship of 14,150 tons; or else four ships of rather larger size and higher speed. The former were to steam 17 knots, and the latter, 19 knots an hour, as against the 17½ knots of the new battle-ships. The vessels proposed by the author were to be armed with two 22-ton guns each, and armored so as to protect the crews of these guns, which were to be mounted one on each side of the funnel, and to sweep round the horizon. The lower part of the funnel, together with the uptakes from the boilers, and the vital parts of these ships, were to be protected. The quadrupling of partially armored vessels, having their two guns, their crews, and funnels protected, and capable of steaming 19 knots an hour, would be of great advantage, could sufficient coal-endurance be given them; but unfortunately they could not, upon such small tonnage, carry coal enough to keep at sea for the lengthened periods requisite to defend the more distant commercial arteries, and were not, therefore, the ships which were so greatly needed for the protection of commerce. The utilizing 9-inch armor, so as to afford such a considerable amount of protection, appeared to entail the disadvantage of overheating the gun-chambers by radiation from the funnel; but, on the other hand, the author's calculations as to perforation by the 9.2-inch 22-ton guns of no less than 99 per cent of the area of the sides of foreign ironclads, was the most valuable contribution at the present time. He further stated that the 69-ton guns carried by the present battle-ships would only pierce ½ per cent more of this area, which must at once suggest the query whether the 69-ton guns were valuable in proportion to their weight of metal and of armor needed to protect them and their easily injured machinery. The author had pointed out Sir Thomas Hastings' mistake in condemning iron as the material ˉ for the skin of warships, but the same conclusion had been arrived at by all who witnessed the large pieces of iron which were torn off when the target was struck by projectiles. Latterly, great improvement has been made in the toughness both of iron and steel, but, there was very little experience as to the damage likely to result from shell-fire against the unarmored sides of vessels. Shots were

unsuitable as naval missiles. Properly made steel shells would perforate any thickness that shots would penetrate, and then explode
within the ship, as had been long since proved by Sir Joseph Whitworth. For steel shells, there was a safer and quite as powerful a
material as any which had yet been manufactured abroad, and the
mechanical ability to apply every kind of explosive in the best manner was likewise to be found in the United Kingdom. He believed
that the proposed large battle-ships of 14,150 tons, or a few hundred
tons more for larger coal supply, were the very ships now most
needed for the navy, and the least likely to become obsolete. The
author had directed attention to the imperfect gun-supply, and had
shown the practical value of guns of 22 tons weight; but trustworthy
10-inch guns of about 30 tons would be still more suitable for ships,
as their diameter would admit of the more powerfully explosive
shells which would more quickly complete the destruction of the
sides of an enemy's vessels.

PROFESSIONAL NOTES.

THE CORROSION AND FOULING OF SHIPS.—ANTI-FOULING COMPOSITIONS.

By M. HOLZAPFEL, *Newcastle-on-Tyne.*

[Reprinted.]

The invention of a means for protecting the bottoms or immersed parts of ships against fouling, and the worm in the case of wooden ships, and subsequently against fouling and rust in the case of iron ships, has long occupied human ingenuity. The Phœnicians over 2000 years ago already employed a composition consisting chiefly of asphaltum for this purpose, and relics of the ancient Romans prove that they employed lead and copper sheathing on the immersed parts of their ships. Before iron shipbuilding was introduced, as early as the middle of the 16th century, an anti-fouling composition was patented in this country; but after the invention of yellow metal, compositions in the form of paint were seldom used, till iron shipbuilding necessitated the use of a paint or composition to protect the iron against rust and fouling. Hardly a month now passes in which one or more patents are not applied for under the name of " anti-fouling compositions." This is a proof that the desire to invent a means for preventing the fouling of iron and steel ships occupies many minds. It is therefore surprising that there should be an almost total absence of literature on this subject, and that a perusal of mostly all the specifications which are filed from month to month should prove that even many of those who may be supposed to have given a great deal of time to the study of this subject should, to judge from their specifications, appear to be quite unaware even of the most elementary principles on which alone a successful anti-fouling composition can be based. In 1867 this subject was dealt with at considerable length by Mr. Charles F. T. Young, C. E., who gives the results of a number of important experiments by the British and French Admiralties and others, but who deduces from them the most conflicting and illogical theories.

At this year's meeting of the Institution of Naval Architects, Professor Lewes read a short paper, the outcome of a very exhaustive and painstaking research, but even he seems hardly to grasp the subject and to misinterpret some of the phenomena.

With our daily extending fleet of iron and steel ships, this question continues to grow in importance, and I will therefore endeavor to lay down the broad principles on which a successful anti-fouling composition should be manufactured.

The way to obtain this object is to imitate as nearly as possible the action of copper and yellow metal. When iron ships were first introduced, great difficulty was experienced to apply any substance which would even in a measure preserve the iron and prevent fouling for a reasonable period. Dry dock accommodation at foreign ports then was exceedingly limited, and a vessel could only be dry-docked after the return from her voyage. A man-of-war on commission in foreign or colonial waters, for an extended period would, before she came home and before she could get cleaned, become so foul as to be almost unmanageable and unseaworthy. What the government and private shipowners hoped to attain at that time was a composition or material which could be applied to iron or steel ships and which would last for two, three or four years without being renewed. Many of the various patent compositions, whose manufacturers professed that they would achieve such a result, proved themselves entirely inadequate and barely lasted six or eight months. Others, besides fouling very rapidly, also corroded the iron to an alarming extent. In consequence of this, scientific and practical men endeavored to introduce a method by which copper, yellow metal or zinc sheathing could be applied to iron ships. Such methods did exist and are occasionally even now resorted to, but they could only be carried out at very great expense and at comparative loss in speed. The method adopted is to cover the iron under the water part of the ship with wood sheathing, 1 to 3 inches thick, and to nail the copper or yellow metal over this. There is however this drawback, that water is very liable to penetrate between the wood and iron skin of the vessel, and thus to cause corrosion, which remains undetected till the wood covering is removed. Moreover, the wood covering has to be fastened to the ship by iron or copper bolts, which in either case serve as a contact between the iron hull of the vessel and the copper sheathing, thereby setting up galvanic action, which is highly destructive to the iron hull of the ship.

Mr. Young, in the work above referred to, recommends, as the outcome of his experience and studies, to fasten zinc sheathing to the ship without an intermediate sheathing of wood. The fact that this method is out of use at present, although it was exhaustively tried, completely disposes of it.

The building of dry-docks in all important harbors of the world and the rapidity with which vessels are now cleaned and painted have made it unnecessary to have a composition which will stand as long as yellow metal. In any case it has been found desirable to dock ships at least once in twelve months to sight bottom, and therefore a composition which will keep a ship's bottom clean for twelve months in ordinary trades is at present likely to meet with the best success. It must moreover comply with the following conditions:

Firstly.—It must absolutely protect the ship against rust.

Secondly.—It must have a very smooth surface, so as to reduce surface friction to a minimum.

Thirdly.—It must be quick-drying, so that if necessary a steamer may receive two coats in one day.

Even if in a composition drying more slowly, better anti-fouling properties could be introduced, it is very doubtful whether such a composition would prove a commercial success, because the frequent chafing of our huge iron ships against piers, small craft, and anchor, chains, etc., removes considerable quantities of composition from the ship's bottom and thus exposes the iron to the action of the salt water, and it has been found absolutely necessary to recoat once in twelve months for this reason alone.

It is not difficult to make a composition which will in ordinary trades preserve the vessel against rust where she is not chafed and against fouling for such a period, unless she is exposed to exceptional delay in waters which are very productive of fouling matter.

Now, as to the anti-fouling properties, there are two methods by which they are supposed to be obtained:

Firstly, *exfoliation, i. e.* the separating of small particles of the composition from the main body, by which any animal or vegetable growth which may have attached itself is caused to drop off the bottom of the vessel; and

Secondly, the *poisoning*, by which the fouling matter is supposed to be killed either before attaching itself or after. Some scientific and practical men attribute the anti-fouling properties of copper, yellow metal and zinc sheathing solely to exfoliation, others to the poisoning principle only.

Mr. Young belongs to the former category, and Professor Lewes, without distinctly saying so in his paper, strongly leans in that direction. Both and all are in my opinion wrong, for it is only from the fact that copper and yellow metal *only poison when they exfoliate* that they become anti-fouling, and it is only metals and compositions which in exfoliating produce poisons that are effective anti-foulers.   Mr. Young, as well as Professor Lewes, give instances in their papers of cases where the galvanic action of the salt water on the copper has been neutralized by iron and zinc protecting bars, and where consequently a great deal of fouling was found on the copper.   They both say that when the copper ceased to exfoliate it ceased to be anti-fouling, and they try to deduce from this that it is only the exfoliation which is wanted, not the poisoning.   My argument is this, that when the copper exfoliates it produces a poison (oxychloride of copper) which is highly destructive to the lower animal and vegetable life, and that this poison is the *active anti-fouler* which kills the fouling matter before it can attach itself.   Now, if the exfoliation of copper alone prevents fouling, why does not the exfoliation of metallic iron also prevent fouling?   I may say that it does so to a very limited extent, but not because it exfoliates, but because by coming into contact with salt water the surface of the iron plate is transformed into rust, which also is to some extent poisonous, but not sufficiently to prevent fouling in any but the coldest climates.   Now I think everybody present will bear me out that an iron plate when exposed to the action of salt water for four years will lose at least five times as much of its weight, *i. e.* will suffer five times as much exfoliation as a copper or yellow metal plate : if, therefore, exfoliation were the real and true factor which prevents fouling, iron itself should be a much more effective anti-fouler than copper.

Some people may argue that the success often obtained from the use of zinc white and tallow, which cannot be considered in themselves highly poisonous substances, is a proof that after all exfoliation, if not the only, is still an important factor as an anti-fouler.   But zinc white and tallow when in contact with salt water become highly poisonous, and their chief merit lies in the poisoning principle.   If a mixture of pure vaseline, mixed with chalk, were applied, the exfoliation would be equally good, but I need hardly say that the coating would quickly be covered with fouling matter to such a thickness as to stop the progress of exfoliation altogether.   Exfoliation can only act as an anti-fouler in regard to substances which adhere lightly to a ship's bottom, not in regard to shell which seem to eat into the protective coating till they finally find a firm hold on the bare iron.   I may, therefore, say that exfoliation is not an *active factor* which prevents the adhesion of animal and vegetable life, but a passive agent, which may under circumstances cause the formations which have already adhered to again detach themselves from the ship.   For in dealing with the first developments of animal and vegetable life which constitute fouling, we have not a body heavier than water, which, if it does not find a sufficiently hard hold to keep on the ship would fall off by itself, but we have a hungry and most insidious animal or plant, which will live on anything that is not an active poison.

The poisoning principle is therefore this, that the anti-fouling substance must surround the ship with a thin layer of poisoned water which destroys all animal life that enters it ; and you cannot poison an animal unless it absorbs some poison.   So this poisoned zone gets gradually *absorbed* and must be replaced. This replacement is effected in copper sheathing, as well as in anti-fouling compositions, by a gradual dissolution of the main substance, which, in being dissolved, becomes highly poisonous.   Mercury, copper, arsenic and zinc, in certain forms, are the substances mostly used for compositions, and they must be used in a varnish which will dissolve with sufficient rapidity to admit of these substances continually combining with salt water, and at the same time with sufficient slowness to retain part of them for the length of time for which the vessel will be exposed to the influence of fouling.   We may now take it as granted that a successful anti-fouling paint must contain a considerable amount of poison which will destroy animal and vegetable life in its lowest developments ; and that this paint must slowly dissolve, or corrode, or, if you like, exfoliate.   Whether these poisons are poisons in the ordinary acceptation of the word, or not, does not matter, they must merely be poisons to the class of animals and vegetables which try to attach themselves to a ship's bottom, and they must carry the greatest possible efficacy in the smallest possible volume, and as already stated these animals and vegetables must be poisoned in their first stages of development, in which only they try to attach themselves to a ship or other solid substance, for when they have once attached themselves they grow rapidly, and can stand a great amount of various sorts of poisons, which is evidenced by the fact that a mussel can thrive on a quantity of *verdigris* which would poison a healthy man, also by the fact that large sized shells several inches long are not infrequently seen on copper sheathing where they could not fail to absorb a large amount of poison from surrounding parts of copper.   It will be seen therefore that even *copper* is not a perfect anti-fouler when exposed to trying circumstances, when, for instance, animal and vegetable life in the surrounding water is so strong that all the small particles of poison get absorbed, and before new formations of poison can take place the animals attach themselves to the ship.

Now when they have once attached themselves the poisoning factor generally becomes useless, the animals grow and the ship comes home foul, often to the astonishment of the owner, captain and composition manufacturer, and this not only occasionally takes place with the best compositions, but also with *copper* and *yellow metal*.

I referred above to preparations of mercury, copper, arsenic and zinc.   All of these are supposed to cause corrosion of the iron, particularly copper, which consequently should not be used at all, or only to a very limited extent.   But in each case the anti-fouling coating should be separated from the iron by a coating of anti-corrosive paint.   Many years ago red lead was taken for this purpose, but it has now been almost completely displaced by quicker-drying and more protective varnish paints, which should be so constituted that even if they are exposed to the action of the salt water, their dissolution would be so slow and gradual as to be almost imperceptible.   Many vessels of the mercantile marine have such a solid and hard body of these protectives on their bottom that absolutely no rust can be seen on them except on places where the paint is chafed off.   As to the anti-fouling, or second coating, I have already stated that the varnish conveying the anti-fouling ingredients should be so constituted as to allow of a gradual but very slow dissolution in salt water, so as to set the anti-fouling or poisonous matter free.   Now in varying waters varnishes of various hardness may be used, in the tropics a soft and rather quickly dissolving varnish, in northern waters a hard and slowly dissolving varnish.

In an experience of over 10 years I have absolutely satisfied myself that most mercurial varnish paints on competitive trials will invariably show a better result than any other compound, while the preservation of the iron and the smoothness of the surface are unequalled.   These paints, moreover, are so cheap, and dry so rapidly, that the total expenditure of a steamer for docking, cleaning and painting during four years is considerably less than a single outlay for docking and sheathing with yellow metal, which under favorable circumstances will only last the same period, *i. e.* four years.   It seems to me very doubtful whether a composition lasting longer than 12 months, provided it were dearer than those now in use, would meet with favor, because most shipowners are already under an obligation to their underwriters or the Board of Trade to dock their vessels at least once in 12 months (which, moreover, is desirable in order that the sea-cocks, propeller, rudder, etc., may be inspected), and the cost of anti-fouling paint, if intelligently bought, is a very small one indeed.   The tendency, on the contrary, seems to be to economize further in the cost of the paint, and to dock the ships more frequently, for the prices for dock dues and labor for painting, and the time required for this purpose, are in most parts only one-third or one-fourth of what they were 15 years ago, while the additional speed of a newly cleaned and painted bottom is a great desideratum.   The only improvement, therefore, which may be looked forward to might be the substitution of mercury by some cheaper substance which would be at least as destructive to fouling matter as mercury. I foresee no chance of a compound being invented which would be hard enough to withstand chafing and which could at the same time be used to convey sufficient anti-fouling materials in a suitable and effective manner. All we can therefore expect from the immediate future is comparative perfection within the radius in which we now move, *i. e.* a perfect adjustment of the various gums used in preparing varnishes and of the various anti-fouling materials now known, so that the composition may be quite reliable even in cases where the various paints now used sometimes fail to fully answer their purpose.

DEPARTMENT OF DISCIPLINE.

The following order is published, as it is believed that all interested in the Naval Academy will be glad to see it. Many have recognized the necessity of giving as much weight as possible to officer-like bearing and aptitude for the service, so as to bring forward those who are particularly well adapted to naval duties, standing well in their studies, but not the best students, while possibly the best naval officers. Gradual progress has been made in this direction by giving weight to marks in certain practical exercises, then to the marks received on the cruise for proficiency in the various duties of officers on board ship, and now by the establishment of the Department of Discipline. By weekly changes in the section leaders and daily changes in the officers of the day and cadets in charge of floors, with the frequent practical exercises, each cadet will come under the observation of a number of officers, will receive a number of marks to determine his monthly and term standing; and the result should approach as near to the true value of the cadets as in other studies, and should be as accurate as is possible under any system for ascertaining relative class standings. The coefficient allotted to the new department is sufficiently large to prevent any cadet from taking a high standing in his class unless he stands well in the department.

U. S. NAVAL ACADEMY,
ANNAPOLIS, MD., *January* 25, 1890.

ORDER.

With a view to taking cognizance of officer-like conduct and efficiency in the performance of duty on the part of the cadets, and as being calculated to raise the military tone of the Academy and aid materially in the selection of the most efficient officers for future service in the navy, the Department of Discipline is hereby established, with the approval of the Navy Department, on the following basis:

First. Said department shall be under the direct management of the commandant of cadets, as head of the department, who shall submit to the superintendent monthly, semi-annual, and annual reports of the standing of all cadets in their respective classes in "efficiency and attention to duty," and in "conduct." Said reports shall be made in accordance with the system of marks now established for other departments.

Second. In making up the standing of cadets in their respective classes, the commandant of cadets shall require and make use of daily reports of the efficiency and attention to duty of cadets from the officers on executive duty, and of weekly reports of the same, based on daily marks, from the instructors in all departments. Cases of misconduct and of inattention to duty on the part of any cadet shall be reported, as now required by regulations, and demerits shall be assigned, after investigation, by the commandant of cadets. The column headed remarks in each report must contain a concise statement by officers in charge and by instructors in different departments, of the reasons for assigning marks for "efficiency and attention to duty."

Third. In order to give to officers in charge and to instructors the fullest opportunity for observation, in addition to the daily detail of cadets for special duty, the commandant shall, as soon as the monthly section arrangements are published, make out a detail of cadets as section leaders, to serve as such for one week in turn, changing each Monday.

Fourth. In making up the monthly standing for each class in the department of discipline, the following system shall be observed: The average of the marks for "efficiency and attention to duty" shall be combined with the mark assigned to "conduct" from demerits received, giving an equal weight to each branch. In assigning marks for conduct, which shall be done in accordance with the demerits received during the month, the following system shall be observed: The cadets in each class shall be considered as satisfactory in conduct when the number of demerits received during a month does not exceed—

for the first   class.. 13
  "   second "   18
  "   third   "   22
  "   fourth   "   .... 37 .

For each demerit received during the month there shall be subtracted from 4—

for the first   class............................ 1154
  "   second "   .... .0833
  "   third   "   .... .0682
  "   fourth   "   .... .0405

A negative average *may* thus be obtained.

Fifth. In case any cadet shall have marks for "efficiency and attention to duty" alone, or for "conduct" alone, in any month, he shall not be given class standing in the report for that month.

Sixth. The standing for any term in "efficiency and attention to duty" shall be the mean of the monthly averages in that branch, and the standing in "conduct" shall be the mean of the monthly averages in that branch, and the mean of these two shall be the term average in the department of discipline.

Seventh. The time from the annual examination to the 1st of October in each year shall be considered as a third term for this department, and the final mark for the year shall be the mean of the three term marks.

Eighth. Marks for performance of duty and aptitude for the service on board the practice ship shall be considered as belonging to the department of discipline.

Ninth. The yearly coefficient of the department for each class shall be as follows :

first   class........................ 16
second "   12
third   "   8
fourth   "   4

the coefficient for studies in other departments being re-arranged so that the maximum for the whole course at the Academy, as now established, shall not be changed.

Tenth. Weekly reports shall be made and published of all cadets who are unsatisfactory in "efficiency and attention to duty" or in "conduct."

Eleventh. The department of discipline shall be organized, and the above changes shall take effect immediately on the conclusion of the semi-annual examination for the present year, February 1, 1890.

                                        W. T. SAMPSON,
                           Captain, U. S. N., Superintendent.

THE COMPARISON OF THE INDICATIONS OF TESTING MACHINES WITH THOSE OF THE STANDARD.

Steel and other metals contracted for by the government and by private parties are constantly being tested in many different localities and on many different machines, and a ready means of comparing the indications of the different machines with the standard is desirable.

A very satisfactory comparison was recently made between the Rodman and the Riehlé hydraulic machines of the Department of Ordnance and Gunnery, at the Naval Academy, by means of the spiral pressure gauge used in guns. Five disks were subjected to a pressure of about 20,000 pounds in each machine, and the mean pressures and turns of the spiral compared. The results were as follows:

	RIEHLÉ MACHINE.			RODMAN MACHINE.	
No. of disk.	Pressure.	Turns of spiral.	No. of disk.	Pressure.	Turns of spiral.
1	20,400 lbs.	10.9	6	20,100 lbs.	10.75
2	20,000 "	10.7	7	20,900 "	10.9
3	21,400 "	11.1	8	20,000 "	10.7
4	20,100 "	10.6	9	20,000 "	10.4
5	20,000 "	10.7	10	20,000 "	10.4
Mean,	20,380 lbs.	10.8	Mean,	20,200 lbs.	10.63

The spiral gauge is described on page 168, Proceedings U. S. Naval Institute, XIII, No. 2. Such a gauge, if constructed for use only with a testing machine, would be somewhat simpler and less expensive in construction.

The inspectors should be required to subject the standard disks sent to certain specified pressures in the gauges. The disks should then be numbered and returned to the central office with the record. The whole matter would then be under the control of the chief inspector, who could compare the indications of all the machines with those of the standard.

Such a method would seem to be more accurate and thorough than the present practice of testing different strips from the same plate in the machines to be compared, as well as less troublesome and expensive.

C. S. S.

BALLISTIC-PHOTOGRAPHIC EXPERIMENTS IN POLA AND MEPPEN BY PROFS. E. MACH AND P. SALCHER.

[Translated by LIEUT.-COM. E. H. C. LEUTZÉ, U. S. N., from *Mittheilungen aus dem Gebiete des Seewesens*, Pola.]

Owing to the great interest which the account of former experiments with rifle projectiles excited, we were enabled to make further ones on a larger scale with projectiles fired from cannon. Some of these experiments were made in Pola by Mr. Salcher, under the patronage of the naval authorities of Austria, while others were made in Meppen by Mr. Mach and his son, by invitation of the firm Frederic Krupp. These experiments are in a measure supplementary to each other, as those in Pola were made with larger caliber (9 cm.) and less velocity (448 m. sec.) than those at Meppen (caliber 4 cm., velocity 670 m. sec.), where different shaped projectiles were also introduced. The arrangement of the physical apparatus differed in the two cases and will therefore be described separately, beginning with that of Professor Salcher.

1. Professor Salcher used as head of his "Schlieren apparatus" * a telescope

* See Proceedings, XIV, 1, 1888.

objective belonging to the Prague Physical Institute, which had an aperture of 21 cm. and a focal length of 3 m. The photographic lens was a "rapid rectilinear" made by Dallmeyer. Tin tubes were placed between the illuminating apparatus and the optical parts, to dim the light and keep off the direct rays of the sun. After the apparatus had been placed in optical adjustment, the jar charged and the cannon loaded, an electro-magnetic current was established by a key; this current opened the camera and at the same moment closed another current which fired the gun. The first current was instantaneously interrupted and the camera thereby closed.

The caliber of the steel-bronze piece used was, as before mentioned, 9 cm., its length 23 cm., the initial velocity 448 m. sec., and the distance from the muzzle of the gun to the head of the "Schlieren apparatus" 18 m.

The apparatus fulfilled all expectations, although the experiments in open air take place under much less favorable circumstances than those in a laboratory.

The head-wave appears as a broad hyperbolic-shaped stripe, which is a little further in advance of the projectile than that of the shot of the Werndl rifle. It has, however, the same angle of inclination to the path of the projectile, which is undoubtedly due to the almost equal velocities in both cases. This wave reaches to the cylindrical part of the shot and is therefore limited by v, v, v', v', fig. 1. Inside of this space a peculiar wave, which surrounds the head of the projectile and points to some peculiar phenomenon, will be noticed.

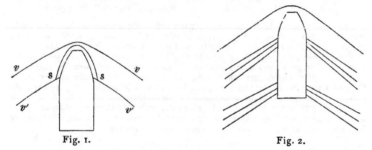

Fig. 1.   Fig. 2.

The photographs show, besides this head-wave, a series of other waves, which emanate from the body of the projectile. As the points of origin of these waves seem to coincide with the position of the expansion rings, it seemed at the first glance that these rings were the cause of the disturbances ; but as these waves also appear with great regularity and equal distances, though in different groups, with smooth projectiles, they must be due to the shape, size, and velocity of the shot. To the little irregularities on the surface of the shot can therefore be assigned only a secondary modifying influence.

These waves seem to be caused by friction, analogous to the waves produced on water by the wind.

The wave lines caused by the 9 cm. projectile are typical. Three diverging stripes emanate from the forward pair of expansion rings and the same from the after ones, but these latter make much larger angles with the axis of the shot (see fig. 2).

At the base of the projectile will be seen the disturbance caused by the air entering into its wake. Behind the Werndl projectile (cal.= 11 mm., v. = 438 m. sec.) this disturbance looks like a glittering stream, similar to rising, heated air. The 9 cm. projectile, with its insignificant difference of velocity, presents a similar aspect. The inrushing air shows a streakiness which makes the picture look as if the projectile was tailed with a bunch of feathers. This similarity results from whirls which really begin close behind the forward expansion ring and surround the projectile in a widening envelop to the base, where they enter into its wake.

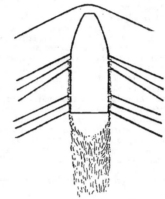

Fig. 3.

2. A special apparatus was constructed for the Meppen experiments. The principle originator of it was Mr. E. Mach, while his son, Mr. Ludwig Mach, worked out the details of construction. Special care was taken in all the arrangements, as the apparatus had to work even during unfavorable weather.

In order to avoid any unnecessary loss of charge, the illuminating jar, which was charged by an induction machine, could not be taken into the circuit until a proper potential had been attained. Immediately after that had to follow the opening of the camera, the firing, and the closing of the camera. The electrical parts of the camera are shown in fig. 4.

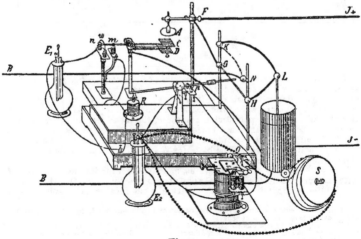

Fig. 4.

From the positive pole J + of the self-exciting induction machine the circuit goes over F, G, H, to the inner coating of the jar L. The potential can be regulated by the position of the plate A. The negative pole J — is connected with the balance beam C of the other part of the potential regulator. As soon as the potential is high enough, C, which is negatively excited, will be drawn to the positively excited plate A, which causes the other end of the beam, which has a wire loop, to dip into the quicksilver reservoirs N, and is then held in that position by the electro-magnet M. By this a connection is made with a battery which is in the drawing indicated by E. The current from E draws an iron core into the coil R, thereby lifting with the other end of the lever (to which the core is attached) the rod GH to the position KN. This interrupts the current between the induction machine and the jar, but shunts it into the circuit which leads to the spark-producing terminals BB. At the same time a current leading to another battery indicated by E_2 is interrupted by lifting the electro-magnetic current interrupter out of the reservoir p. The interruption of this current, which, in connection with M, kept the camera closed, allows a spring to open it, and at the same time strikes a gong at the firing place. The firing and closing of the camera (by hand*) follow immediately. In this manner a proper charge of the jar was assured—an important matter, for had it been too small, the shot might not have caused discharge, and had it been too great, a premature discharge would have taken place.

Everything worked with great accuracy, and the whole transaction occupied less than a second.† The apparatus never failed.

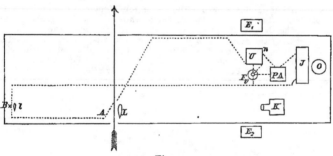

Fig. 5.

The entire arrangement of the installment can be seen in fig. 5. The whole apparatus was enclosed and the light deadened by a wooden hut (14 m. long, 2.5 m. wide), with two openings in the line of flight of the projectile (indicated by the arrow). From the induction machine J near the store O the circuit leads over the potential regulator P, the jar E and the current interrupter U to A, where the shot closes the circuit, thence to the place of the spark B, and further to the outer coating of the jar, and back to the other pole of the induction machine. The batteries E and E, of 8 Bunsen elements each, were outside of the hut.

The optical parts of the apparatus, namely, the illuminating apparatus B, with the achromatic lens l (5 cm. aperture, 30 cm. focal length), the head of the Schlieren apparatus L (21 cm. aperture, 3 cm. focal length), and the cam-

---

* This closing could of course have been done automatically, and would be recommended were the apparatus to be a permanent one.
† The man firing the gun was so expert that it was hardly possible to notice a time interval between the sounding of the gong. We considered that we had to abstain from electric firing on grounds of personal safety, and would advise against it in similar cases.

era K are of course in a straight line. The distances are $Bl = 25$ cm., $lL = 4.5$ m., $LK = 6$ m. The light of the spark B, passing through l, fills the aperture L completely, so that it, L, looks brightly illuminated as seen from K.

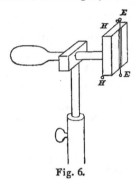

Fig. 6.

The illuminating apparatus is represented in fig. 6. It consists of a parallelopiped of hard rubber, which can be moved horizontally, vertically, and also around the line of sight as axis. A small shallow groove contains the electrodes of magnesium wire, and is covered with a glass plate after the hard rubber has been painted with petroleum. This arrangement obliges the spark to extend in a straight line, and can therefore be made parallel to the edge of the opaque screen * which is placed in front of K. This causes great increase of light, and makes it possible to use a very small spark of very short duration. The capacity of our jar was barely 1000 cm.

The adjustment of the optical parts is made as follows: The different parts are placed approximately at the above mentioned distances, and the center of B, l and L and the objective of K are placed approximately at the same height. The lenses l, L are then removed and the frame of L so placed that its image will be in the center of the ground glass plate of the camera K; the frames of l and B are then so placed that their images will be concentric with that of L. If then the lens l is replaced and a candle-flame is held in front of K, its reflection on l will show if the axis of the lens coincides with lK. The axis of the lens L is adjusted in the same manner.

In order to place B in the proper position, so that the image of the spark will be projected on the screen in front of K, and preferably so that it will be bisected by the screen in the direction of its length, a candle-flame, or, better still, a platinum wire made incandescent in a Bunsen flame is used. This flame or wire is placed immediately in front of the screen and opposite to the center of K, and at the same parallel to the edge of the screen at the place where the image is to appear. If the spark-score B is then placed so as to appear in the well defined part of image of the candle-flame or wire, then reciprocally the image of the spark B will coincide with the candle-flame or wire.† A small after-adjustment of the screen or small movement of K completes the installation. During our experiments the screen was placed vertically and generally to the left, looking in the direction of L.

The muzzle of the gun was distant 12 m. from the place where it caused discharge, i. e. A. The projectile here formed the connecting bridge between two parallel, vertical wires of 0.5 mm. thickness; these were covered with rub-

---

* See XIV, 1, 1888.

† By application of the optical law of reciprocity, the problem is, as in case of a geometrical construction, considered solved; but it becomes necessary to seek for the conditions of this solution.

ber and were about 3 cm. apart. The projectiles used were blunt, caliber 4 cm., length 6, and others pointed at both ends so as to leave at each end a circle of 16 mm. diameter, caliber 4 cm., length 9.8 cm. The velocity in front of the lens L was about 670 sec. m.

The fear that the pressure of air would cause disturbances was found, if not without foundation, at least greatly exaggerated. The shock of the discharge had no appreciable effect on the independently placed parts of small diameter of the apparatus. The illuminating apparatus only, which besides being placed in the sand had to be fastened to the wall of the hut, had to be readjusted. The shock, however, produced no movement in the wall of the hut until too late to disturb the experiments.

The pictures which were taken were on the whole similar to those obtained by firing rifle projectiles, and proved the correctness of the assumption formerly expressed, that the experiments with small models were sufficient provided the velocity was the same. There were, however, some new phenomena; the wave in front of the blunt projectile showed double; the part closest to its head seemed to be apparently an isolated, flat disk of air which seemed to disperse sideways.

The double-pointed projectile showed, besides a single head-wave, three isolated waves which emanated from the sides of the projectile at different places.

We will not give an exacter description or analysis of the pictures, as we have since, during many varied laboratory experiments, obtained still better pictures, of which we will treat later.*

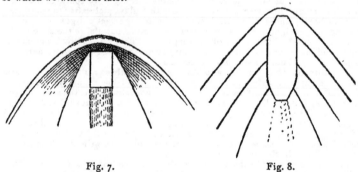

Fig. 7.   Fig. 8.

We have no doubt that even with smaller and less expensive installation, valuable experiments can be made at artillery proving grounds. We also hope that this will be done, and we will be very happy to give any further and more exact information. With this view we have arranged the apparatus so, and given such a detailed description, that even a person with little practice should obtain good result.

We give special weight to the experiments at Pola and Meppen because they proved in presence of several experts that our method can be employed in the study of common projectiles.

* During later experiments it became possible to make quantitative determination.

BOOK NOTICES.

PRACTICAL MARINE SURVEYING. By Ensign Henry Phelps, U. S. N.

‘ While this book was written to fill the need of a suitable text-book on the subject at the Naval Academy, it thoroughly bears out its title and is eminently practical, and will be found useful by all who wish to acquire a knowledge of marine surveying. There have been a number of books written on this subject, but this is the first one that is really practical and by means of which any one with sufficient education to handle the instruments and make the calculations can learn to make a good marine survey. It contains many valuable "wrinkles," some of which have been passed along for years in the Coast Survey, and others that are the result of experience in the survey of the Western Coast under the Hydrographic Office, Navy Department.

AMERICAN RAILROAD BRIDGES. By Theodore Cooper, M. Am. Soc. C. E.

In this work Mr. Cooper gives a history of American bridge-building, together with a description of the existing and accepted types and the methods in use to-day, and it cannot fail to be of interest to all who design, manufacture or use railroad bridges. In the historical portion, under the head of wooden bridges he commences with the "Great Bridge" built across the Charles river, between Old Cambridge and Brighton, in 1660, and carries it along to 1844, when the Pratt truss was patented, in which the tension members were of iron. Under the head of iron bridges he commences with Thomas Paine's letter, written in 1803, and ends with the iron lattice bridge built 1865–66 over the Connecticut river. He then takes up the history of long-span bridges, commencing with the bridge built 1863–64 over the Ohio river at Steubenville, and continuing the history up to the present day. He then gives the theory and practice of designing and proportion, a description of the manufacture of bridges, of the typical American railroad bridges with a discussion of their relative merits, and concludes with a discussion of the failure of bridges. There are a number of plates and some tables giving the result of physical tests of full-size bridge members.   R. W.

BIBLIOGRAPHIC NOTES.

ANNALEN DER HYDROGRAPHIE UND MARITIMEN METEORO-
LOGIE.

17TH ANNUAL SERIES, VOLUME IX. Proof of Poisson's theory of the deviation of ships' compasses by means of observations. Sailing directions for the China Sea. Length of voyages of iron and wooden sailing vessels from Europe to the Bay of Bengal and return. Currents in New York bay and harbor. Pampero in the SE. part of the mouth of the La Plata. Meteorological observations on Butaritari Island in the Gilbert Archipelago. Quarterly weather review. Minor notices: Ice in the South Atlantic; Chemical analysis of the water of the North Sea; Current measurements in the North Atlantic; Currents in the Straits of Belle-Isle, also on the west coast of Newfoundland; Currents near Cape Bon, Tunis, Guantanamo, Cuba, Vera Cruz, Mexico; Currents in the Mozambique Channel; Voyage from Manilla to Zebu; Reports on the harbors of Wellington and Napier in New Zealand and Iquique; Description of a water-spout.

VOLUME X. Remarks on strength and velocity of the wind at Norwegian lighthouses. Remarks on several points on the SW. coast of Africa. The island Santa Maria. Port Adelaide and Port Germain, Australia. Sailing directions for Bahia Blancá, Argentine Republic. Description of and sailing directions for the harbor of Castries. Current and water temperature observations, west coast of Africa. Quarterly weather review, fall of 1885. Minor notices: Sailing directions for the rivers Casamanza and Nuñez; West coast of Africa; Remarks on Bramble Cay and Eastern Fields Reef, eastern entrance to Torres Straits; Remarks on Bulhar, east coast of Africa, also Berbereh; Sailing directions for the river Salum, west coast of Africa; Extension of the Boca Channel, near Buenos Ayres; Remarks on the Paracel Island, China Sea; Remarks on Runs Channel, southern entrance to Moreton Bay, Queensland, Australia; Time signal at Capetown; Anchorage near Hodeidah, Red Sea; Bottle post from different vessels.

VOLUME XI. Current measurements off the new entrance to Wilhelmshaven. Voyage of H. M. S. Olga from Sydney through Torres Straits to Aden. Extracts from the log of Captain Reniers, of the barque Aeolus. The coast of Abacama between Autofagasta and the river Loa, Chili. Dust-storms in the North Atlantic. Quarterly weather review, fall of 1885. Minor notices: Remarks on

Caen, on the French coast, English Channel; Caleta Buena; Extraordinary weather phenomena off Akyab; Running down the trades to the leeward of San Antonio, Cape de Verdes; Sickness resulting from the eating of fish; Bottle post from several vessels.

VOLUME XII. Compensation of compasses for deviation caused by the electric lighting installation. Remarks on harbors on the west coast of Africa. Remarks on Fusan, Corea. Extracts from the log of H. M. S. Carola, voyage from Zanzibar to Aden and return. Passage of Atlas Strait. Sailing directions for the harbor of San Blas, Argentine Republic, also for Khor Ambada, Bay of Tejureh, Gulf of Aden. Deep-sea sounding and exploration of banks in the Pacific. Butterflies in the South Atlantic, long distances from the land. Minor notices: An interesting weather phenomenon in Jubal Straits; Remarks on the Boma mouth of the Malimba (Edea) river; Sailing directions for Gibraltar; also for the river Ponga, west coast of Africa; Depth and bottom off Montevideo; Remarks from the meteorological journal of the German bark Inca; Extracts from the log of the German bark Olga; Voyages of the German ship Ferdinand Fischer, west coast of America.

E. H. C. L.

BOLETIM DO CLUB NAVAL.

Nos. 10 and 11, JULY AND AUGUST, 1889. Hints on the study of elementary naval tactics. Theory of the rudder (continued). On the history of the Brazilian navy. Movements of the squadron.

J. B. B.

BOLETIN DEL CENTRO NAVAL.

JUNE AND JULY. The hydrography of the Parana and Uruguay. The fisheries of Patagonia.

AUGUST. Submarine navigation. A. C. B.

DEUTSCHE HEERES-ZEITUNG.

OCTOBER, 1889, No. 79. Experiments with homing pigeons for military purposes in Italy.

These interesting experiments, which took place between the military lofts of Rome and Civitavechia, a distance of 65 km., have demonstrated the possibility of using the *same* pigeons to perform a *double* service, namely, of carrying messages *from* Rome *to* Civitavechia and *vice versa*. These remarkable results were obtained by feeding certain pigeons in one loft only (that of Civitavechia), and keeping their young ones in another at Rome, thus compelling them to make regular trips from one place to the other to get their food. These experiments are of especial value, as they insure uninterrupted communications with an invested place without fear of exhausting the supply of the carriers.

Naval matters: building of new ships for the German navy; Raising the blockade of the East African Coast at Zanzibar; Explosion of a gun on board the practice vessel La Couronne near Hyères, France.

No. 80. Attack and defense of submarine mines for coast defense. Experiments with homing pigeons (concluded).

No. 81. The annual military manœuvres in Germany, near Hannover, under direct command of Emperor Wilhelm.

The novel features of these manœuvres were, 1st, the use on the field of small movable protected towers provided with rapid-firing guns; 2d, the use of smokeless powder on a large scale both for artillery and infantry.

Description of the torpedo-vessel Vesuvius, of the U. S. Navy.

No. 84. Review by F. v. S. of the prize essay for 1888, on Torpedoes, by Lieutenant-Commander W. W. Reisinger, U. S. N., published in the Proceedings of the United States Naval Institute, Vol. XIV., No. 3.

No. 88. The ironclad Friedrich Karl. Construction of new cruisers for the German navy at Kiel.

, No. 90. Estimates for the German navy for 1890–91=88,781,165 marks.

No. 91. Review of an article by Capt. Zalinski in the *Electrical Review*, N. Y., on the use of electricity to ascertain the position of the enemy in coast-defense operations. Launching of the protected cruiser Jean Bart, at Rochefort, France. Fortifications on the Thames, England.

No. 92. Use of magnesium flash-lights in the navy.

No. 93. Description of the electric submarine boat Goubet, France.

No. 94. England: building and launching of new vessels for the English navy.

No. 95. Krupp's gun trials. New type of the Whitehead torpedo from the Royal laboratorium at Woolwich. Building and launching of new cruisers for the English navy.

DECEMBER, 1889. No. 98. Notes on the defects of the ordnance in the English navy. United States: different opinions by American naval officers on the pneumatic dynamite guns of the Vesuvius (from the *Admiralty and Horse Guards Gazette*).

No. 100. Publication of the official register of the German navy for 1890. England: launching of a new cruiser for the Australian squadron, and other vessels. New English torpedo-stations.

No. 101. Return of the aviso Pfeil from the blockade of the East African coast. New apparatus for launching torpedoes, invented by R. Soutwarth Lawrence, London (Middlesex). United States: further notes and opinions on the Vesuvius.

No. 102. England: ammunition for heavy ordnance. Defects in the construction of the cruisers of the Bird type. H. M.

THE ENGINEER.

OCTOBER 12, 1889. An alleged fraudulent stamping of boiler plates. Manganese bronze. Building vessels in the navy-yards.

OCTOBER 26. Coal armored cruisers. The performance of the Baltimore. Horse-power standards for marine engines.

NOVEMBER 12. Multi-cylinder engines. The strength of alloys at different temperatures.

NOVEMBER 23. Foreign opinions of the Chicago. Horizontal air pumps. Heating by exhaust steam.

DECEMBER 7. Miscalculated ships. Guns and armor. Double screw ferry-boats. Rapid-fire guns for the Lakes.

DECEMBER 21. Steel stay-bolts. Design and construction of chimney-shafts. Forced draft.

JANUARY 4, 1890. Corrosion and scaling of marine boilers. Forced draft. Spontaneous combustion in coal. J. K. B.

JOURNAL OF THE AMERICAN SOCIETY OF NAVAL ENGINEERS.

NOVEMBER, 1889. Graphic determination of crank effort. Evaporators. Tests of Corliss boilers. Contract trial of the Charleston. A compressed air system of forced draft.

A description of a method of using compressed air in a practically open ash-pan applied to a small steam launch at the Mare Island Navy Yard. The arrangement consisted in having a small air compressor of the same diameter as the high pressure cylinder, made out of a 4-inch boiler tube, with composition valve-seats screwed on each end, and the piston driven from the crosshead of the high pressure engine. The compressed air was conducted to the front of the ash-pan, where it was delivered into a nozzle forming a part of the ash-pan door. Several trials showed that with a pressure of $1\frac{1}{2}$ to 2 inches of water, the combustion coal of Wellington coal of good quality was about 30 pounds per square foot of grate. The draft by this system of delivering air into a practically open ash-pan is largely an induced draft. In the particular case mentioned, the air supplied by the pump was not more than ten per cent of that theoretically required for the combustion, the remaining ninety per cent being supplied by induction.

The advantages claimed for this system are that the mechanism is inexpensive and with but little liability to derangement. It is economical and noiseless. Freedom of access to the ash-pan. As compared with the use of a blower driven by an auxiliary engine, there are points in favor of this system. In the first place the blower must supply the entire volume of air needed for the combustion of the fuel, the ash-pan being closed in the usual way in which a blower is used in small launches. The result of the high velocities is rapid wear which requires no little attention. The first cost of the machinery is considerable.

The contract trial of the Baltimore. Measured mile trials at Newport of the Chicago, Atlanta, Boston, and Yorktown. Casualties and repairs. J. K. B.

JOURNAL OF THE ASSOCIATION OF ENGINEERING SOCIETIES.

NOVEMBER, 1889, to JANUARY, 1890. Mills and mill engineering. Electric motors. Notes on a new compound steam turbine. Recent developments in iron and steel manufacture. J. K. B.

MECHANICS.

OCTOBER, 1889. Steam engineering at the Paris Exposition. Compound Corliss engine. Boxes and journals. The Teal portable hoist. Units of power. A new thermometric scale.

NOVEMBER. The Girard-Barre hydraulic railway. The strength of alloys at different temperatures.

In studying some experiments made before the Admiralty, Professor Huwin's attention was directed to the remarkable decrease in the tenacity of gun metal when the temperature was raised. At 350° there was a sudden decrease of tenacity, generally about 50 per cent, and at a temperature of 500° the tenacity had become nil. In the present experiments rolled bars of yellow brass, Muntz metal, Delta metal, gun metal and cast brass, were tried. The results were all plotted in a diagram, and show that in all cases the decrease in strength follows a regular law without any such sudden loss of strength as was shown in the Admiralty experiments. Even at temperatures of 600° to 650° all the bars had still a not inconsiderable tenacity. The ultimate elongation of the bars was also measured. There is a peculiarity in the influence of temperature on the ductility of the bars. In most cases the ultimate elongation diminishes with increase of temperature. With Muntz metal the decrease is regular, with considerable elongation before fracture at a temperature of 650°. With cast and rolled brass the decrease is more rapid, and there is very little elongation before fracture above 500°, while the elongations of gun metal were very irregular.

The indicator and its use. Compound locomotives. The economical production of motive power. The Hertz experiments. The trial of the dynamite gun. A new Swedish explosive.

In Sweden has recently been invented a new explosive which so far has given great satisfaction. It is called "gray powder" (Swedish grakrut), and has, during the summer, been tested at Rosersberg Gunnery School, and also has been accepted for trials in the fleets. According to official reports, the gray powder has been used with 25-millimeter as well as with Nordenfelt's machine guns. The former has, with 70 per cent of the new powder against 100 per cent (or the usual charge) of the ordinary powder, given a 33 per cent greater initial velocity without the pressure in the gun being increased more than 5 per cent. With regard to the important question of smokelessness, the report states that while with Nordenfelt's machine guns, smoke of ordinary powder remains for twenty-five seconds, the gray powder only leaves a transparent steam which is visible for but five seconds.

DECEMBER. Report of the annual meeting of the Society of Mechanical Engineers.

JANUARY, 1890. Steam boiler design. An improved connecting rod. The indicator and its errors. Machine design. J. K. B.

MITTHEILUNGEN AUS DEM GEBIETE DES SEEWESENS.

VOLUME XVII, No. 10. Treatise on the three arms of modern naval battles and the tactics determined by them. Remarks on steel and compound armor plates. Launching of the Russian armor-clad Imperator Nikolaj I. Estimates for the Imperial Austrian navy for 1889–90. Laying down of the Russian armor-clads Gangut and Grosjasci. Armor-clad vessel Siegfried of the Imperial German navy. Attacks of torpedo-boats during the manœuvres of the French navy. Gaubet's submarine boats. Torpedo-boats using petroleum for fuel. Floating docks for torpedo-boats. Repairs of submarine torpedo-launching tubes. Torpedo-boats for the Imperial German navy. Torpedo gunboats for English navy. New air compressor at the Paris Exposition. French torpedo-boats at the Paris Exposition. Apparatus for cutting torpedo-nets. Organization of torpedo-boat divisions in Russia. Sinking of a Danish torpedo-boat. New line of steamers. Exposition at New Zealand. Arctic exposition of Fridtjof Nansen. Literary notices: Torpedoes and torpedo warfare, by C. Sleeman, late Lieut. Royal Navy; Seaman's Handbook, by A. Muchall Viebrook; Treatise on Maritime Sports.

VOLUME XVII, No. 11. Protection and preservation of the bottoms of iron and steel vessels, by Seaton Schroeder, Lieut. U. S. navy. Geometrical solution of problems of electrical discharge. Jury rudders. Trials of the Beauchamp-Towers constant platforms. Improvement on water-tight doors. The Taryan condensing apparatus. Experiments with 36-pound rapid-firing cannon in England. Trials of the pneumatic guns of the Vesuvius. Types of torpedo-boats for the English navy. Experiments with torpedoes at Constantinople. Halpin-Savage torpedo armored gunboat La Grenade. English gunboat Sparrow. English cruiser of the third class Barham. The English battle-ships Royal Sovereign, Repulse, and Renown. Speed of some English war vessels. New English cruiser of the first class. Putting in commission of the Turkish vessel Ertogrul. Turkish torpedo-boat Schanawer. Building of a new Turkish cruiser. Steel-casting in Constantinople. Rebuilding and altering of Turkish armored vessels. Manœuvres of Turkish torpedo-boats. Pump vessel for the Turkish navy. Société centrale de sauvetage des naufrages. Movements of vessels in Algerian harbors. Russian harbor improvement in the Black Sea. The lighting of the coast of the Argentine Republic. New steamer lines. Literary notices: Building of river vessels, iron, wood and composition; Principal harbors in the world; Current observations in the Indian Ocean. New edition of Captain Marryat's novels; German harbors; Explanation of abbreviations, etc., used on home and foreign charts; Naval warfare, a strategetical study of the English fleet manœuvres of 1888; Political and military significance of the Caucasus.   E. H. C. L.

NORSK TIDSSKRIFT FOR SOVAESEN.

8TH ANNUAL SERIES, VOLUMES 1 AND 2. The review at Spithead. Merchant fleets of different countries (from *Nautical Magazine*). The English fleet manœuvres in August, 1889. Stranding of the German men-of-war at Apia. On the fouling of iron and steel vessels (from *Nautical Magazine*). French navy. Minor notices: The course of the abandoned American schooner W. S. White; The International Maritime Congress; The latest English armor-clad vessel; Largest sailing vessels; New Italian torpedo-boats; Additional remarks on the English fleet manœuvres.

VOLUME 3. Maxim-Nordenfeldt automatic gun. Luminous night-clouds. Additions to the merchant fleets of different countries. Short visits to the dockyards at Copenhagen, Kiel, and Karlskrona. The United States navy. Derelict vessels. Minor notices: The

Italian cruiser Piemonte; Speed trials in the English fleet; The Medea type of vessel; The cruisers Blake and Blenheim

E. H. C. L.

PROCEEDINGS OF THE INSTITUTION OF CIVIL ENGINEERS.

VOLUME XCVIII. Armor for ships. The treatment of steel by hydraulic pressure, and the plant employed for the purpose. Experiments on a steam engine, the cylinder of which was heated externally by gas flames. Investigation of the heat expenditure in steam engines. The improvement of the river Avon below Bristol. Experiments on the strain in the outer layers of cast-iron and steel beams. The cyclical velocity. Variations of steam and other engines. Balanced slide valves. On the new programme for shipbuilding.

THE RAILROAD AND ENGINEERING JOURNAL.

NOVEMBER, 1889. Engineering in naval warfare. Irrigation. The world's shipping. Cable towing on canals. A new Chilian battle-ship. The development of armor. United States naval progress.

A review of the report of the trial of the Charleston, and the tests of the guns of the dynamite cruiser Vesuvius.

French criticism of our ships.

DECEMBER. On stresses produced by suddenly applied forces and shocks. A propeller for light-draft boats. An English view of the American navy.

An extract from the *Army and Navy Gazette*, commenting on the remarkable strides which have been taken by the United States in the rehabilitation of the navy. Commenting on the arrival of the squadron of evolution in European waters, it states that it is not so much that the United States have made a very good beginning towards building of a modern navy, but there have been developed in the country facilities of every kind for the creation of that navy without outside aid. This can hardly be said of any other power except France and Great Britain. Not only has much been done in the way of constructing vessels which are as good as anything, but arrangements have been made by which they will be able shortly to create entirely from their own resources every modern implement of war, including steel-clad battle-ships of the heaviest tonnage, with their guns and armor.

United States naval progress.

The launch of the San Francisco, the report of the board to report a plan for the development of the League Island Navy Yard, the work now in progress at the Washington gun factory, and the report of the board upon the question of consolidating the naval establishment at Newport.

The development of armor. A French naphtha launch engine.

JANUARY, 1890. The development of armor, by Lieutenant Califf, U. S. army. Quadruple expansion engines for S. S. Singapore. A new Russian battle-ship.

The latest addition to the Russian navy is the armored battle-ship Nicholas I., which was begun in 1886, and was built by the Franco-Russian Company under contract. The vessel is of steel, of Russian manufacture, while the

armor-plates were made at the Admiralty works at Ijorski.   The chief dimensions of the Nicholas I. are as follows : Height between perpendiculars, 333⅓
feet ; length over all, 347½ feet; breadth, 67 feet ; average draft, 23 feet ;
displacement, 8440 tons.

Propelling on light draft.   The Burgin high-speed engine. United
States naval progress; a synopsis of the annual report of the Secretary of the Navy.   Notes on the New York Naval Reserve.

J. K. B.

REVISTA MILITAR DE CHILE.

No. 36, SEPTEMBER, 1889.   Extension of the Central Railroad of
Chili.   Arms in use in different countries.   On the defense of the
Chilian coast.   Notes on the " Great Captain," Fernandez de Cordova.
On the employment of cavalry in raids. The uses of modern artillery.

No. 37, OCTOBER.   Examinations in military schools.   Studies on
the de Bange system.   Extension of the Central Railroad of Chile
(continued).   Treatise on military hygiene.   On the employment of
cavalry in raids (continued). The uses of modern artillery (continued).

No. 38, NOVEMBER.   The National Guard.   On open order in
military operations.   On the establishment of Le Creusot.

No. 39, DECEMBER.   On mobilization of troops.   The Pfund-
Schmidt torpedo.   The electric light in searching for wounded.

J. B. B.

REVISTA MARITIMA.

No. 2, AUGUST, 1889.   Naval reforms.   Use of oil in storms at sea.
Sea-going torpedo-boats.   Notes on Brazilian naval history.   The
dynamite cruiser Vesuvius.   Naval warfare.   On explosives.

SEPTEMBER, 1889.   The ports of Liverpool and Birkenhead, with
plates and descriptions of docks and basins.   Regulators for arc
lamps.   Historical study of steam navigation.   The French naval
manœuvres.   On coaling men-of-war (translation).   The Society
Islands and the natives of Polynesia.

OCTOBER, 1889.   Arctic exploration.   On penetration of armor-
plates.   Notes from observations taken at Krupp's works.   Liverpool
and Birkenhead (continued).   Study on the composition of fleets.
The Hicks machine for throwing shell charged with high explosives
(translation).   The Society Islands and the natives of Polynesia
(continued).

NOVEMBER, 1889.   Combined military and naval operations.
Arctic exploration (continued).   Liverpool and Birkenhead (continued).   On naval mobilization (reproduction of portions of article
by Lieut. S. A. Staunton, U. S. N.)   Geography of the sea.   The
evolution of the torpedo-boat.   Boilers in the new English men-of-
war.   Trials of the Spanish submarine torpedo-boat Peral (with
illustration).

DECEMBER, 1889.   Combined military and naval operations (continued).   Liverpool and Birkenhead (concluded).   On the penetration

of armor-plates (concluded). The question of heavy guns in England. The Society Islands and the natives of Polynesia (concluded). Photographing projectiles in motion (translation).   J. B. B.

REVUE DU CERCLE MILITAIRE.

OCTOBER 6. Execution of counter-attacks by troops of all arms. The "Sud-Oranais" expedition of 1881, Algeria (continued). The military exhibit of 1889 (continued).

OCTOBER 13. The Dutch army.

OCTOBER 20. Regulations for infantry drill. The military exhibit.

OCTOBER 27. Regulations for infantry drill (end). Two days at the school of firing at Le Mons.

NOVEMBER 7. Regulating the fire of a field battery by the mere observation of fuzing shots. The "Sud-Oranais" expedition (ended). Remount of officers and requisition for horses in Italy.

NOVEMBER 10. Strategic transportations.

NOVEMBER 17. Infantry tactics. Battle formation of the company, the battalion, and regiment.

DECEMBER 1. The moral education of the Russian soldier. The defense of Plevna (with cuts in the text)

DECEMBER 8. The defense of Plevna (continued). The smokeless powder: its influence upon tactics.

DECEMBER 15. The defense of Plevna (ended).

JANUARY 2, 1890. Renewing the supply of ammunition to the artillery. Moral education of the soldier, and particularly of the trooper.   J. L.

REVUE MARITIME ET COLONIALE.

OCTOBER, 1889. The cohorts of the Legion of Honor (end). Note on the use of the deflector for regulating the compass at sea. Verifying the compass route of the ship by means of luminous guiding marks (repères). An ingenious device which allows the helmsman, the quartermaster, and the watch officer to verify at night the ship's course, at one glance and without reading the points of the compass.

NOVEMBER, 1889. Action of heated iron hulls upon the deviations of the compass. Depopulation of the sea, and the advisory board on sea fisheries. Historical studies on the military marine of France (continued). The end of large fleets. Oceanography (statics).

DECEMBER, 1889. The tides in the Lower Seine, by Adm. Cloué. The Cape Horn mission (continued). The hurricane of March, 1889, in Samoa. Oceanography (statics) (continued).   J. L.

THE SCHOOL OF MINES QUARTERLY.

NOVEMBER, 1889. Incandescent lamps. The performance of certain multiple expansion marine and pumping engines.

A series of tables compiled for the use of investigators on the boilers, engines, pumps and condenser of various marine and pumping engines, with their dimensions and performances.

Summary of useful tests with the blow-pipe. The total solar eclipse of December 22, 1889. J. K. B.

THE STEAMSHIP.

OCTOBER, 1889. Electric lighting. The corrosion and fouling of ships. Anti-fouling compositions. Feed-water heating. Morton's reversing valve-gear. Note on Bremme's valve-gear. The new cruiser Barham.

The engines of this vessel have the distinction of being the lightest in the service for the power to be developed, their weight being less than those of the Barrosa, which has only to indicate 3000 horses, while the Barham has engines of 6000 horse-power. The lightness of the machinery, which is a repetition of that of the Bellona, has caused much comment among naval engineers. The standards which usually support the vertical cylinders have been dispensed with and exceedingly light framing substituted. This framing is, however, tied together by longitudinal and diagonal girders. The Barham is without the protection of a double bottom, and in order to economize weight, the plummer blocks are placed directly upon the bottom plating of the vessel without the intervention of the usual head plates. To give stability to the engines, the cylinder of each set are connected together by stops passing through the dividing longitudinal bulkhead. The engines on the port and starboard sides respectively are also tied by girders to the protective deck above. As the Barham is intended for the speed of 19½ knots per hour, the engines are especially designed for quick running, the power developed being secured by the exceptional number of revolutions—230 per minute.

NOVEMBER, 1889. Notes on Bremme's valve-gear. Forced draft.

A paper read before the Institute of Marine Engineers, giving the various systems of artificial draft that have been tried, together with a discussion of their relative values.

Morton's reversing valve-gear. Bilge pump.

DECEMBER, 1889. Broken shafts.

A discussion of the causes of breakage in the shafts of sea-going steamers. To avoid breakdown the author recommends (1) an efficient governor, (2) a flexible coupling, (3) built crank-shafts; and when great power has to be transmitted, hollow steel screw shafts instead of solid wrought-iron ones.

The coaling of steamships. The Pamphlett-Ferguson distilling apparatus. A Canadian ship railway. Launch of the Australian cruiser Phœnix. The obliquity of connecting rods. Smith & Cowan's patent marine boiler.

JANUARY, 1890. The Elsmore copper depositing process. The modern marine boiler.

A paper read before the Junior Engineering Society by Mr. Lilly, on the design and construction of modern marine boilers.

The twisting moment diagram of a triple expansion engine. Means of increasing the speed of steam vessels. Radial valve-gears.

J. K. B.

THE STEVENS INDICATOR.

OCTOBER, 1889. Experimental mechanics as developed in foreign technical schools. Testing machines. Efficiency. On the use of ropes for prong brakes.

JULY, 1889. Energy-relation of gas and electric lighting. The charging of blast furnaces. The mechanics of the injector. Examination of olive oil for adulteration. A chimney paradox.

J. K. B.

AMERICAN SOCIETY OF MECHANICAL ENGINEERS (TRANSACTIONS).

VOLUME X, 1889. On the use of the compound engine for manufacturing purposes, the relative areas of the cylinders in the compound engine and the regulation of the pressure in the receiver. Foundry cupola experience. Electric welding. The distribution of internal friction of engines. Some tests of the strength of cast iron. The cost of power in non-condensing engines. The flow of steam in a tube. A simple calorimeter. The mechanics of the injector. The identification of dry steam. The friction of piston packing rings in steam cylinders. The comparative cost of steam and water power. Cylinder ratios of triple expansion engines. Improved motion device for engine indicators. The piping of steel ingots. Formulas for saturated and superheated vapors. Longitudinal riveted joints of boiler shells. Steam consumption at various speeds. Use of crude petroleum in steam boilers. Some properties of vapor and vapor engines.

J. K. B.

AMERICAN SOCIETY OF MINING ENGINEERS (TRANSACTIONS

Aluminium in cast iron.

A paper prepared to substantiate the statements made in an article upon the same subject read before the American Association for the Advancement of Science, in August, 1888, together with the chemical analyses of the test bars belonging to the casts reported in that paper. J. K. B.

TRANSACTIONS OF THE CANADIAN SOCIETY OF CIVIL ENGINEERS.

VOLUME III, PART I. The Panama Canal. Cantilever bridges. The development of the locomotive. The Esquimault Graving Dock Works, British Columbia.

J. K. B.

TRANSACTIONS OF AMERICAN SOCIETY OF CIVIL ENGINEERS.

SEPTEMBER AND OCTOBER, 1889. A discussion on the construction of the Sibley bridge over the Missouri River. Cylindrical wheels and flat-topped rails for railways.

J. K. B.

THE UNITED SERVICE.

VOLUME III, No. 1. Our coast defenses. Modern naval educa-, tion. The United States Revenue Cutter service. Comity in the mess. The army as a home. Some cavalry leaders I have known. Among our contemporaries.

No. 2. Glances at the wars of to-morrow. Curious phases of naval warfare. The United States Revenue Cutter service. Our coast defenses. Operations before and fall of Atlanta. A grand palaver.

No. 3. The Indian as a soldier. The modern law of storms. The United States Revenue Cutter service. Great commanders of modern times. Among our contemporaries.

UNITED SERVICES GAZETTE.

NOVEMBER 2, 1889. Torpedo, gunnery, hydraulic and pilotage courses. Royal Naval School. Dockyard economies. The Vesuvius.

NOVEMBER 9, 1889. Imperial federation. The Victoria's 110-ton guns. Steel guns for the American navy. India since the mutiny. The American " New Navy."

NOVEMBER 16, 1889. Lord George Hamilton and Mr. Stanhope on the services. Our monster ordnance. Sights and laying. Engine-room artificers, R. N.

NOVEMBER 23, 1889. Mr. Forwood on the navy. The twelve-pounder. The stoker question.

NOVEMBER 30, 1889. The value of case shot. Sir Edward Watkin on the Channel tunnel. Lord George Hamilton on the navy. Launch of war vessels. The Australian defenses. Gun trials of the Camperdown. Naval uniforms.

DECEMBER 7, 1889. Naval uniforms. Modern drill and tactics. An American view of the late naval manœuvres. Imperial federation. The Portsmouth Royal Naval College.

DECEMBER 14, 1889. Vice-Admiral Sir Anthony Hiley Hoskins, K. C. B. H. M. S. Victoria. The Royal Navy. Engine-room artificers and their grievances.

DECEMBER 21, 1889. The repair of H. M. S. Amphion. The Royal Navy. The naval uniform question. The right of way at sea.

DECEMBER 28, 1889. Canada and the United States. Arrival of the Sultan. The Sharpshooter class. The Maxim gun.

JANUARY 11, 1890. Machine guns. The defenses of St. Helena and Cape Colony. Naval retrospect, 1889.

JANUARY 25, 1890. The tactics of coast defense. Court-martial. Out of sight, out of mind. Protective and anti-fouling composition.

FEBRUARY 1, 1890. Modern gunpowder as a propellant. Naval uniforms. Steam reserve economies.

LE YACHT.

OCTOBER 12, 1889. Editorial on the slow promotion of officers in the French navy; case of the ensigns. The French five-masted sailing ship France; sail plan and longitudinal plan amidships. The merchant marine exhibit at the Paris Exhibition.

OCTOBER 19. The coming International Marine Congress. The new cruisers of the United States; sketch of the Baltimore. Judges' decision in the late British manœuvres. Notes from foreign ship-yards. The merchant marine exhibit at the Paris Exhibition.

OCTOBER 26. Naval policy of France and foreign nations for the year. Argument against torpedo-boats. Review of the merchant marine exhibit at the Paris Exhibition.

NOVEMBER 2. Compound boilers of the steamer Calorie plan. Description of the Collingwood; profile and deck plan. Launch of the Greek cruiser Spetsia.

NOVEMBER 9. The different shapes given screw propellors. Naval notes. Comparison of promotion of officers in England, Russia, Germany, and France. Description and plan of Mr. Bennett's yacht Sereda.

NOVEMBER 16. Editorial on the resignation of Amiral Krantz, Minister of Marine. Engines and boilers at the Paris Exhibition. The British battle-ship Royal Sovereign. Review of the merchant marine.

NOVEMBER 23. Engines and boilers at the Paris Exhibition. Notes from foreign ship-yards. Effect of subsidies on the merchant marine.

NOVEMBER 30. Launch of the French cruiser Alger. Sketch of the British battle-ship Benbow.

DECEMBER 7. Editorial on the appropriation of fifty-eight millions of francs. British protected cruisers of the first class. Review of the merchant marine.

DECEMBER 14. Editorial on the French navy. French submarine boats. A new four-masted sailing ship Le Nord.

DECEMBER 21. New ships to be constructed (editorial). The Chinese protected cruiser Chih Yuen; sketch.

DECEMBER 28. Description and sketch of the cruiser Vesuvio of the Italian navy. Notes from foreign ship-yards. Review of the merchant marine.

JANUARY 4, 1890. A review of the increase in European navies, 1889–90. Report of the Secretary of the U. S. Navy. A. C. B.

ANNUAL REPORT OF THE SEC. AND TREAS. OF THE U. S. NAVAL INSTITUTE.

To the Officers and Members of the Institute.

Gentlemen :—I have the honor to submit the following report of the affairs of the Institute for the year ending December 31, 1889.

ITEMIZED CASH STATEMENT.

RECEIPTS.

Items.	First Quarter.	Second Quarter.	Third Quarter.	Fourth Quarter.	Totals.
Advertisements............	$50.00	$30.00	$110.00	...	$190.00
Dues	1153.00	574.78	168.00	$90.25	1986.03
Subscriptions	88.25	114.25	215.60	136.75	554.85
Sales....................	235.77	157.70	3.80	102.80	500.07
Life-membership fees.....	...	30.00	...	60.00	90.00
Interest on bonds.........	44.58	18.00	18.25	18.00	98.83
Binding, extra............	19.60	3.00	3.00	1.00	26.60
Sundries	...	1.00	...	...	1.00
Totals................	$1591.20	$928.73	$518.65	$408.80	$3447.38

EXPENDITURES.

Items.	First Quarter.	Second Quarter.	Third Quarter.	Fourth Quarter.	Totals.
Postage, freight, etc.......	$42.65	$25.83	$25.41	$39.30	$133.19
Stationery	80.36	29.24	1.50	12.13	123.23
Messenger...............	97.00	105.00	105.00	112.00	419.00
Branch expenses	1.77	...	...	...	1.77
Extra binding	...	10.85	...	...	10.85
Purchase of back numbers.	...	...	...	...	...
Publication	735.47	220.58	600.00	687.24	2243.29
Office expenses...........	9.97	3.62	12.00	6.72	32.31
Subscript'n Army and Navy Register..............	5.00	...	...	...	5.00
Purchase of bonds	123.75	...	...	...	123.75
Rebate on dues...........	3.00	...	9.00	...	12.00
Totals................	$1098.97	$395.12	$752.91	$857.39	$3104.39

SUMMARY.

Balance of cash unexpended for year 1888........................... $715.28
Total receipts for 1889... 3447.38

Total available cash for 1889.....................................$4162.66
Total expenditures for 1889....................................... 3104.39

Cash unexpended, January 1, 1890............................:..............$1058.27
Cash held to credit of Reserve Fund 198.87

    True balance of cash on hand, January 1, 1890................ $859.40
Bills receivable for sales of No. 51.............................. 75.00
  "   "   " dues, 1889.................................... 633.30
  "   "   " back dues 204.00

    Total assets, January 1, 1890..............................$1771.70
Bill outstanding for No. 51 ....

RESERVE FUND.

List of bonds deposited for safekeeping in the Farmers' National Bank of Annapolis, Md.

United States 4 per cent registered bonds.......................... $900.00
District of Columbia 3.65 per cent registered bonds................ 1000.00
District of Columbia 3.65 per cent coupon bonds.................... 950.00

                                                 $2850.00
Cash in bank uninvested.. 198.87

    Total Reserve Fund.......................................$3048.87
Annual interest on bonds 107.17

Number of new life members......... 3

During the year one District of Columbia bond, 3.65 per cent, was purchased for $123.75. Another has been ordered, and as soon as received the coupon bonds will be sent to the Treasurer of the United States, ex-officio Commissioner of the District of Columbia sinking fund, and registered in the name of the United States Naval Institute.

MEMBERSHIP.

The membership of the Institute to date, January 1, 1890, is as follows : Honorary members, 6 ; life members, 96 ; regular members, 573 ; associate members, 159 ; total number of members, 834.

The total number of members shows a slight decrease. This is owing to the fact that a number of names which have been carried on the books for some years without profit to the Institute have been dropped. All the members now on the books are in good standing, and in the cases where dues are still owing it is because the members are on distant service.   The actual circulation has increased, as may be seen from the following :

Members......... 834
Subscriptions...... 263
Exchanges................. 63
Average sales per quarter 151

Total circulation per quarter......... 1311

PUBLICATIONS ON HAND.

'The Institute had on hand at the end of the year the following copies of back numbers of its Proceedings:

Whole Nos.	Plain Copies.	Bound Copies.	Whole Nos.	Plain Copies.	Bound Copies.
No. 1	202	...	No. 27.	294	27
2	205	...	28	5	...
3	63	...	29	226	27
4	153	...	30.	257	2
5	126	...	31.	30	54
6	9	...	32	6	173
7	15	...	33	16	164
8	41	...	34	55	7
9	47	...	35	106	66
10	5	...	36	253	25
11	220	...	37	171	21
12	61	...	38	249	2
13	4	...	39	146	2
14	7	...	40	25	111
15	2	...	41	247	16
16	224	...	42	132	13
17	4	...	43	286	3
18.	95	...	44.	272	10
19	115	...	45	210	18
20	131	...	46	217	18
21	235	...	47	198	18
22	278	...	48	187	18
23	180	...	49	216	18
24	202	...	50	204	17
25	1145	42	51	229	18
26	207	77			

2. Vol. X., Part 1, bound in half morocco.
1.   "   "   2,   "   "   "
1. Vol. XIII, "   2,   "   "   "
1. No. 34,   "   "   "
5.   "   "   "   calf.
6.   "   "   full sheep.

The archive set complete, Vol. I. to Vol. XIII. inclusive, bound in full turkey.

All business communications should be addressed to the Secretary and Treasurer U. S. Naval Institute, Naval Academy, Annapolis, Md., and all postoffice orders, checks or drafts should be drawn in favor of the Secretary and Treasurer.

Very respectfully,

RICHARD WAINWRIGHT, *Lieut., U. S. N.,*

Secretary and Treasurer.

ANNAPOLIS, MD., *January* 1, 1890.

THE PROCEEDINGS

OF THE

UNITED STATES NAVAL INSTITUTE.

| Vol. XVI., No. 2. | 1890. | Whole No. 53. |

U. S. NAVAL INSTITUTE, ANNAPOLIS, MD.

REPORT OF POLICY BOARD.

NAVY DEPARTMENT,
WASHINGTON, *January* 20, 1890.

Sir: In submitting a report under your instructions of July 16, it is deemed necessary in the beginning to state clearly, as they appear to the Board, the considerations demanding the maintenance of a strong naval force by the United States. Thence the limits of a sufficient force may be evaluated, and its different elements determined with reference to the services shown to be required of it.

It need not be pointed out that there are several ways of approaching this subject, and especially of estimating in gross the magnitude of the force required. The following argument, besides covering the broad questions, defines the duties of the more important types and supplies certain of their controlling features.

It may be stated that a navy is essential (1) for waging war and (2) to prevent war; and the second purpose may be far more important than the first. The Board in making this statement does not overlook the many important duties which a navy has to perform in time of peace, but a navy which is adjusted to meet the requirements of the country in time of war will, undoubtedly, be ample to perform the duty required of it in time of peace; and it may be rigidly accepted that there is no real ultimate economy in the maintenance of vessels apparently adapted to the more economical performance of the latter duty, if ill adapted to the former. The magnitude of the naval force to be maintained by a government should be adjusted on the one hand to the chances of war, and the magnitude of the naval force which a war would bring against it; and, on the other hand, it should be commensurate with the wealth of the country and the interests at stake in case of war. If the chances of war are slight, and the interests to be guarded are unimportant, certainly the naval force to be maintained need not be great; if the chances of war are great, then the naval force which should be maintained ought to be limited by nothing but the limits of the nation's wealth; for, in case of war, its life is at stake. Whatever may be the chances of war, if the interests to be guarded are great, then the naval force to be maintained should also be great.

In these cases, it is assumed that the protection required is best given by a naval force. For the United States, it may confidently be asserted that the chances of war are much less than for most European nations, or at least that the chances of war with any nation comparable with this in wealth and power are much less than the chances of war among the nations of Europe. The isolated position of this country removes many incentives to war.

We fear no encroachments upon our territory, nor are we tempted at present to encroach upon that of others. Our territory does not obstruct the free passage to the sea of the commerce of any nation, nor is our own obstructed in any similar way. We have no colonies, nor any apparent desire to acquire them, nor will this desire probably arise until the population of this country has overflowed its vast limits, or its great resources become strained to maintain it.

At present our foreign commerce is carried in foreign vessels, and our manufacturers compete with those of other nations in but few markets, when we consider the manufacturing resources of this country.

All these reasons combine to make the United States self-contained to a greater degree than any other important nation, and, as a consequence, we are brought into conflict with the interests of other nations to the least possible extent. For this reason, the chances of war would seem to be at a minimum. But there are not wanting indications that this comparative isolation will soon cease to exist, and that it will gradually be replaced by a condition of affairs which will bring this nation into sharp commercial competition with others in every part of the world. Even now our commercial relations with our nearest neighbors are clamoring for modification both by sea and land, and in the adjustment of our trade with a neighbor we are certain to reach out and obstruct the interests of foreign nations.

The time cannot be distant when we shall compete in earnest with others for the vast and increasing ocean-carrying trade; the time is near at hand when our own people will find it profitable to carry at least their own goods to foreign markets.

The construction of a canal to form an ocean route between the Atlantic and Pacific will place this nation under great responsibility which may be a fruitful source of danger.

While these changes in our national affairs and relations which loom up in the near future all tend to contribute to the wealth and greatness of the nation, they will also bring with them responsibilities and chances of war which we should prepare to meet. When we consider for a moment the naval force which might be brought against us, in case of war, even by a second-rate power, we are astonished at our own weakness and total lack of preparation. Never in the history of the country have we been so unprepared as now to maintain our rights upon the seas, or to defend our own shores. In years gone by, our ships were as good as any in the world, but while for a quarter of a century we have disregarded all advances in naval warfare, others have steadily pursued a policy of building ships embodying the improvements which have grown in such numbers and character that the modern man-of-war in no way resembles the fighting ships of fifty years ago. Not only, then, have we a navy which is insignificant in numbers, in comparison with those of other nations, but even those vessels which we possess are not capable of meeting their great armored ships. Fifty years ago a ship and her armament could be constructed in a few months,—even in our late war it was possible to improvise a fleet of considerable power from our merchant shipping,—but the fighting ships of to-day require years to construct, and demand for the work such great mechanical appliances that a nation without such ships ready built, in time of war, is certain to suffer defeat. This is, unquestionably, the condition of the United States at this moment.

When we consider the wealth of our country, or our ability to maintain a navy, and the interests at stake in case of war, we are again forced to admit that our navy is insignificant and totally disproportionate to the greatness of the country and to the task which would certainly fall to it in case of war.

If we omit entirely from consideration the property along our coast which is now exposed to destruction, and consider only the exports and imports which are exposed upon the high seas, we find that the value of such property is annually $1,500,000,000. This does not include our coastwise trade, which would double these figures; and this trade would also be exposed to destruction in case of a naval war. While more than 75 per cent of this vast foreign traffic is carried in foreign ships, the exports represent a part of the nation's wealth and would be subject to capture. Nor does the loss of this property adequately represent the loss to the country which would result from its capture in whatever ships it might be carried. The enormous home industries and inland transportation directly dependent upon the continuance of these exports would be paralyzed.

It matters not whether this commerce is carried in foreign ships; if it were to cease, it would produce wide-spread misery. It is not, then, alone sufficient that our sea-ports should be protected against bombardment—they must not be blockaded—our commerce must be free to enter and leave at its accustomed ports. The Board feels that it cannot emphasize this too strongly. If the port of New York alone were blockaded for even three months, it would produce greater confusion and loss than would be directly inflicted by a bombardment of the city, even in its present unprotected condition.

When compared with the foreign exports and imports of other nations, we find those of the United States nearly equal to one-half of those of Great Britain, three-fourths of those of France, five times those of Spain, more than three times those of Italy, and more than three times those of Russia.

The total exports and imports of Great Britain and of the United States have steadily increased during the past ten years, while those of other nations have decreased. It may, then, be accepted as a reasonable belief that the commerce of this country must continue to increase, and that the time is not far distant when it will surpass those of the only two nations which now exceed it. If, then, it be deemed prudent to maintain a naval force capable of protecting this vast and increasing traffic, the United States should possess no insignificant navy.

While the Atlantic and Pacific furnish a practically insurmountable barrier to an invading army, our extended coast-line is a constant temptation to an enemy to assail us from the sea. When we consider the simplicity of the means required to resist an invasion, and remember the greatness of our population from which armies are to be recruited, and the railway and other facilities for concentrating them upon any threatened points of our coast, we must be convinced that only a powerful army would venture to invade our territory. The vast means of ocean transportation demanded for any considerable army, and the difficulties to be overcome in keeping up its supplies, would, manifestly, render it impossible to invade our country. We have, then, only to fear a naval attack, which may vary from the marauding attack of a few cruisers upon our shores or commerce, up to the organized and destructive attack of a well-appointed fleet of armor-clads. How shall we best protect ourselves against these? Shall we place our main dependence upon fortifications for our principal ports and permit ourselves to be shut up within them, and suffer our vast commerce to be driven from the seas? or shall we be prepared to maintain our rights at sea, to keep our ports open to our trade, and, incidentally, protect the ports?

While feeling reasonably secure against any invasion of our territory by an enemy, we should redouble our efforts to protect ourselves against the one certain form of attack in case of war, and the only complete protection is to be found in an efficient and sufficient navy.

Considering, then—

(1) That we are a nation possessed of great and increasing wealth;

(2) That much of this wealth must pass to and fro across the ocean or along our coast, where it is exposed to capture or destruction in case of war;

(3) That we are now totally unprepared—even against second-rate powers—to protect our commerce, to prevent the blockade of our ports, or to maintain our rights and honor away from home;

(4) That these objects can only be secured by a navy;

(5) That while we are now at peace with all nations, this fortunate condition of affairs may not always continue:—it is evident that we should proceed with all possible dispatch to provide a navy of such a character and magnitude as will efficiently serve these purposes.

To do this is a naval problem which depends for its complete solution upon several conditions, which, in turn, are fixed or known with a moderate degree of exactness.

The Board considers these conditions under three heads:

The *first* may be stated as the naval strength of our possible enemies, which is, in all cases, so far as concerns the naval strength of individual nations, known with precision; but the possibility of combinations, either for or against us, introduces an element of uncertainty which, under given circumstances, may be estimated but cannot be anticipated.

The *second* condition is the proximity of the respective bases of coal and ammunition supplies to the area which we might be called upon to attack or defend. The imperative necessity of an ample supply of coal and ammunition to insure the maximum efficiency of a modern fighting ship renders it certain that the nation which is best supplied in these respects will have an enormous advantage.

Third, the extent of the area to be attacked or defended. The greater the area or coast-line to be defended against a given force, the greater will be the force required; whereas, in many cases the force required to attack may be much smaller than the defending force. A few cruisers may attack and greatly injure the commerce of a nation having many cruisers in search of the marauders. A few armor-clads may attack and destroy public stores or dock-yards, notwithstanding a much greater defending force. Consequently, if we know the naval strength of our possible enemy, his facilities for replenishing his supply of coal and ammunition as compared with our own facilities, both acting in any given part of the world, and give due weight to our respective positions in regard to the areas to be attacked or defended—that is, whether we are defenders with a convenient base, or the reverse—we shall have a reliable basis upon which to estimate the naval force which should be maintained by the United States.

First, then, if other conditions were equal, we should maintain a navy at least equal in strength to the most powerful navy in the world, on the theory that we might have to fight such a nation. But the second condition, viz., proximity of bases of supply, greatly modifies this estimate as far as the magnitude of the navy required for the protection of our interests in the western hemisphere is concerned, and these are our greatest interests; for in this part of the world we should be acting near our own base, while most European nations would be separated several thousands of miles from any base of supply—a fact which would serve to exclude some of the most powerful ships in the world from acting against us, provided our naval strength was sufficient to prevent our enemy from seizing some of the islands along our coast and establishing the necessary bases, or taking possession of Long Island Sound, Chesapeake Bay, or other suitable body of water where colliers could anchor in security.

In those cases where nations have bases of supply in proximity to our coasts, vessels of very moderate coal endurance could act against us; they would, however, be forced to defend their bases and protect all supplies sent to them. The proximity to our shores of naval bases of supply, or their existence even at more distant points which would in any degree assist an enemy, is a consideration of the gravest import to us, both as regards our national safety and in deciding the naval force we should maintain. It would, therefore, in case of war, lead to an immediate struggle between the possessors of such naval bases near our coast for their protection, and ourselves for their destruction. We should be prepared with a naval force adequate to such work, and only the most powerful armor-clads would suffice. Whatever force our enemy could fairly be expected to assign to such duty, we should be able to destroy, beyond a doubt. To fail in this would multiply the destructive force of our enemy, and leave us helpless to resist direct attacks upon our own seaboard.

We find, then, in this consideration, a limit fixed for our naval strength, which, it is probable, we need not exceed, and with less than which we cannot rest secure.

To be able in case of war to destroy completely, at its outbreak, every base of supplies belonging to our enemy which should be in proximity to our country, and at the same time to protect the converging highways of our commerce, both foreign and coastwise, would represent the principal demand to be made upon our navy for purposes of defense, and, hence, for any purpose.

By destroying all proximate bases of supply we force the enemy back, and exclude from the contest all vessels of limited endurance. The following table exhibits the endurance of the principal armor-clads and cruisers of Europe, now built or building, classified according to endurance. The table includes only such vessels as could act against us were we provided with the naval force above referred to. With our present naval strength, every vessel of an enemy, without respect to endurance, armor, or armament, would be effective against us; for where such enemy does not already possess near bases of supply, they could easily be established along our coasts at points inaccessible to a land force, and our whole coast completely invested.

[Numbers of table give the *total* number of ships having at least the endurance given at head of column.]

ARMORED SHIPS.

Nation.	Effective battle-ships.							
	Endurance at 10 knots exceeding—				Maximum endurance exceeding—			
	7,000	6,000	5,000	4,000	12,000	10,000	8,000	6,000
Great Britain :								
Complete	8	12	15	20	6	7	15	20
Building........	10	10	10	10	..	10	10	10
France :								
Complete	..	..	4	19	..	..	..	19
Building........	..	..	2	4	..	..	..	4
Spain :								
Complete	..	..	..	1	..	..	..	1
Building........	..	..	..	..	..	..	..	..
Italy :								
Complete	2	2	2	4	2	2	2	4
Building........	3	3	3	6	3	3	3	6
Germany :								
Complete	..	..	..	4	..	..	..	4
Building*........	..	..	..	..	..	..	..	..
Russia :								
Complete	..	..	3	3	..	..	3	3
Building........	..	..	7	7	..	..	7	7

Nation.	Effective armored cruisers.							
	10 knots endurance exceeding—				Maximum endurance exceeding—			
	12,000	10,000	8,000	6,000	15,000	12,000	9,000	6,000
Great Britain :								
Complete	..	..	7	9	..	9	9	12
Building........	..	..	..	..	..	..	..	..
France :								
Complete	..	..	..	..	..	..	..	..
Building........	..	..	..	5	..	..	5	5
Spain :								
Complete	..	..	..	..	..	..	..	..
Building........	6	6	6	6	6	6	6	6
Italy :								
Complete	..	..	..	..	..	..	..	..
Building........	..	..	..	..	..	..	..	..
Germany :								
Complete	..	..	..	..	..	..	..	..
Building*.......	..	..	..	..	..	..	..	..
Russia :								
Complete	..	..	..	..	..	..	3	3
Building........	..	..	2	2	2	2	2	2

* Four first-class armor-clads are being laid down, of 9000–10,000 tons; no information as to endurance.

[Numbers of table give the *total* number of ships having at least the endurance given at head of column.]

EFFECTIVE PROTECTED CRUISERS.

Nation.	10 knots endurance exceeding—				Maximum endurance exceeding—			
	12,000	10,000	8,000	6,000	15,000	12,000	9,000	6,000
Great Britain :								
Complete.......	..	..	9	23	4	9	23	47
Building.......	2	11	40	40	11	40	40	40
France :								
Complete.......	..	..	..	4	..	..	4	4
Building........	..	..	..	5	..	..	5	9
Spain :								
Complete.......	1	1	1	1	1	1	1	1
Building........	..	3	3	3	..	3	3	3
Italy :								
Complete.......	1	1	1	9	1	1	9	9
Building........	..	..	..	6	..	..	6	6
Germany :								
Complete.......	..	..	..	..	..	..	..	..
Building*.......	..	..	10	10	..	10	10	10
Russia :								
Complete.......	..	..	..	3	..	..	3	3
Building........	..	..	..	..	..	..	..	..

The above table does not take account of coast-defense vessels nor of armored ships constructed prior to 1870. Some of their older armored ships are having new engines put in by the *English*, and thus becoming armored cruisers. *No data on this.*

No vessels of less than 1000 tons have been considered with protected cruisers.

Armored ships constructed between 1868 and 1875 having less than 5000 miles total endurance have been considered coast-defense vessels. Half a dozen English ships under this head might still be considered effective battle-ships.

Similar ships of other navies (not considered above) are practically obsolete.

Assuming that a vessel, or a small number of vessels, cannot operate at a greater distance from a base than one-third of her endurance,—and even at this distance some dependence would have to be placed upon coaling at sea from captured vessels or from colliers, —it may be seen that a large proportion of these ships cannot attack us with a base of supply so distant as Europe. Probably a fleet of overwhelming strength could bring colliers with it, and detach a sufficient force to protect them in some sheltered place on the coast where coal could be taken on board, and thus to a limited extent make up for the lack of a protected base of supplies. This contingency has been considered by the Board.

Great Britain will soon possess seventeen battle-ships having the necessary endurance to act from a home base, nine armored cruisers, and sixty-three protected cruisers.

Of similar vessels, France will soon possess five armored cruisers and nine protected cruisers.

Spain will soon have six armored cruisers and four protected cruisers.

Italy will have five battle-ships and six protected cruisers.

Germany will have ten protected cruisers.

Russia will have five armored cruisers and three protected cruisers.

With coaling stations near our coasts, the conditions would be entirely changed, and Great Britain under these new conditions will soon be in possession of the following vessels which could operate against the United States : thirty battle-ships, twelve armored cruisers, and eighty-seven protected cruisers.

France, in like manner, will soon be in possession of the following vessels which could operate against the United States : twenty-three battle-ships, five armored cruisers, and thirteen protected cruisers.

Spain will have six armored cruisers and four protected cruisers.

Without bases of supply on this side of the Atlantic, the three first-named nations combined would have one hundred and thirteen modern ships of all classes which could operate against the United States. Of this number, but seventeen are battle-ships.

With their present bases of supply intact, these same three nations possess one hundred and eighty modern ships, of which fifty-three are battle-ships which could operate against this country.

As above stated, the ships here enumerated as effective ships comprehend only ships of recent construction and great power. If all England's battle-ships were included, the number given for that nation would have to be placed nearer fifty than thirty; hence the manifest need of destroying such bases at the outset. So great would be the advantage of an enemy in protecting such stations, and so much also would it diminish his power if they were destroyed, that we might be assured of a desperate struggle over their possession.

No more forcible illustration can be found of the importance of these bases of coal and ammunition supply, not to mention the facilities which they also furnish for repairing vessels, than in the fact that without them, were we possessed of an adequate force, no nation could send an ironclad against us, except England and Italy. This statement is subject to modification in case of an alliance which might render available bases of supply to a nation otherwise without them.

To protect our ocean highways of commerce involves the protection of the approaches to all our commercial ports, and consequently their protection against blockades or bombardment. Less than this we should not be satisfied with, and to accomplish these objects fixes the limits of the naval strength required by this country. In basing the necessary strength of our navy upon these requirements, the Board has not lost sight of the possible alliances which might be formed by either side, and, considering the political relations existing between European nations, it may fairly be assumed that the alliances we would be able to form would more than counterbalance those formed against us.

In this assumption there is, however, an element of uncertainty which prudence would suggest should discount any such assumed advantages. The most powerful navy of these three nations is that of Great Britain, and her effective force is now thirty battle-ships and one hundred cruisers; but the conditions will not arise when that or any other nation possessing an important commerce can detach all her effective navy from her own coast for distant operations.

On the other hand, a considerable preponderance of defensive force must exist to accomplish the purposes above set forth, and especially in heavy ships suitable to the attack of fortified places within a thousand miles of our nearest naval depot.

The Board is therefore of the opinion that the following vessels, to be hereafter described, should be added to the navy of the United States:

Battle-ships of great coal endurance	10
Battle-ships of limited coal endurance	25
Cruisers of 4000 tons and over	24
Torpedo cruisers of about 900 tons	15
Special cruisers for China service of about 1200 tons	5
Rams	10
Torpedo-depot and artificer's ships	3
Ships of all classes	92

Besides one hundred first-class torpedo-boats and numerous second-class torpedo-boats carried by battle-ships of great endurance and the larger cruisers and torpedo-depot ships.

In general terms, the battle-ships of great coal endurance would constitute the basis of a fleet which might be detached, in whole or in part, for distant service and for the purpose of cruising against the enemy; having the power to remain at sea during a long period, and to attack points on the other side of the Atlantic.

The Board deems it absolutely essential that we should possess this number of battle-ships of such endurance, for it is evident that a policy of protection without the power to act offensively, even to carrying war to the very doors of an enemy, would, at the present time, double the force with which we would have to contend. For, an enemy, knowing it was possible for such a fleet to appear upon its own coast, would be obliged to assign a superior force for its protection, thus greatly diminishing that to be sent against us.

The battle-ships of limited endurance would constitute a fleet of heavily armed and armored ships, having the necessary endurance to act at any point along the Atlantic or Pacific coasts or in the Gulf of Mexico, and within a thousand miles from a depot of supplies. They would serve the purpose of keeping our ports open and destroying an enemy's bases of supplies within a thousand miles of our coast.

While the great coal and ammunition endurance required in the first-class battle-ships necessitates a vessel of 10,000 tons displacement, with a draught of 25½ feet of water, battle-ships of endurance limited as above defined, carrying even a more powerful main battery, can be constructed upon a displacement of 8000 tons, and a draught of 23½ feet.

The limited depth of water to be found at the entrance to the harbors of the Atlantic and Gulf ports makes it necessary that even our first-class battle-ships should be constructed on the least possible draught.

Three classes of these battle-ships of limited endurance are recommended, all having the same general characteristics of speed and manœuvring power, in order that they might act together as a unit or in squadrons to the greatest possible advantage. This is a consideration which the Board deems of the utmost importance, as it would give such a fleet an advantage over any fleet now in existence. At the same time the diminished draught of water of the smaller vessels permits them to enter ports along our Southern coast, which the larger vessels could not enter. The main object of all these battle-ships is the protection of our own coast.

The Board deems it unnecessary to indicate further in this report the disposition to be made of these vessels to accomplish this object.

In considering the general type of these vessels, the Board has naturally investigated carefully the merits and demerits of the essentially American monitor type. As a result it finds that this type of vessel is adapted only to smooth-water service, and that the conditions of its efficient working are therefore at variance with a large part of the duty demanded. As it is believed that a very wide misconception exists on this subject both within and without the navy, the leading features of the investigation are given in the Appendix B.

The 22-knot protected cruisers possess the greatest endurance, and are intended to capture or destroy the fastest merchant vessels in the world, while they will also be capable of fighting vessels of their own class.

The cruisers of 20 knots speed are to serve the same general purpose, being, however, better capable of fighting, having the same battery, and great, though less, endurance, with greater protections, on a less displacement.

The armored cruisers have the fighting power and protection still further developed, and they would probably overpower any vessel of the cruiser type now built or projected.

The Board has carefully considered the question of protection for the guns and crews in all these cruisers. The development of rapid-fire guns, especially of moderate caliber, and the large number now carried, renders it absolutely necessary that the gun's crews should be protected, or they could not remain at their guns. It is believed that within certain limits the weight assigned to such protection is far more useful than the same weight in additional guns.

It is, perhaps, unnecessary to state that a certain proportion of these cruisers of all types would find their proper field of usefulness in destroying the commerce of our enemy, while the remainder would form necessary constituents of the squadrons, especially in offensive operations. Commerce upon the high seas is the most sensitive of all interests in time of naval war. It is difficult to protect and easy to destroy.

The Board does not consider it well to include any vessels of the Vesuvius type in the programme, as being of uncertain value in naval warfare. At the same time it has no wish to discountenance experiment in naval weapons such as the dynamite guns or the submarine torpedo-boat. On the contrary, it would appear wise for a nation of

such undoubted inventive ability, possessing no large naval establishment to be depreciated by the successful development of such
weapons, to devote a certain amount of attention to them. Their
consideration, however, is rather beyond the province of the Board.

The tonnage, cost, and proportion of ships of different types in the
proposed programme are as in the following table:

Naval Programme.

No.	Type.	Displacement of each.	Aggregate displacement of type.	Approximate total cost.
		Tons.	Tons.	
10	First-class battle-ships of great endurance.....................	10,000	100,000	$56,400,000
8	First-class battle-ships of limited endurance.....................	8,000	64,000	39,890,000
12	Second-class battle-ships of limited endurance................	7,100	85,200	52,200,000
5	Third-class battle-ships of limited endurance.....................	6,000	30,000	18,000,000
10	Rams.........................	3,500	35,000	18,700,000
9	Thin-armored cruisers...........	6,250	56,250	28,800,000
4	Protected cruisers..............	7,500	30,000	15,760,000
9	do.	5,400	48,600	25,200,000
2	do.	4,000	8,000	4,200,000
5	Special cruisers, about	1,200	6,000	2,400,000
15	Torpedo cruisers, about.........	900	13,500	7,000,000
3	Torpedo-depot and artificer ships.	5,000	15,000	6,500,000
92	Ships of all classes..............	..	491,550	$275,050,000
100	120-foot torpedo-boats..........	65	6,500	6,500,000
				$281,550,000

Including new vessels completed, building, or already authorized—
chiefly in the cruiser classes—the proposed naval establishment of
the United States would consist of the following number of vessels
of each class, with approximate tonnage and cost.

The Maine and 7500-ton "armored cruiser" authorized are
classed with the Texas, as third-class battle-ships, as being sufficiently exact for our present purposes. The coal endurance of the
Maine, indeed, is less than that of many recent battle-ships, and her
sail power is a supplement of doubtful value.

*Proposed naval establishment of the United States, including vessels
already built or appropriated for.*

No.	Type.	Displacement of each.		Aggregate displacement of type.	Approximate total cost.
		Tons.		Tons.	
10	First-class battle-ships of great endurance......................	10,000		100,000	$56,400,000
3	Third-class battle-ships of great endurance....................	6,300 6,650 7,500	}	20,450	11,000,000
8	First-class battle-ships of limited endurance......................	8,000		64,000	39,890,000
12	Second-class battle-ships of limited endurance......................	7,100		85,200	52,200,000
5	Third-class battle-ships of limited endurance......................	6,000		30,000	18,000,000
6	Harbor-defense monitors.......	(4) 3,815 (1) 4,000 (1) 6,060	}	25,320	25,000,000
1	Cruising monitor................	3,800		3,800	1,900,000
11	Rams.........................	(10) 3,500 (1) 2,000	}	37,000	19,500,000
9	Thin-armored cruisers...........	6,250		56,250	28,800,000
4	First-class protected cruisers......	7,500		30,000	15,760,000
10	" " "	5,400		54,000	28,000,000
12	Second-class protected cruisers....	3,000 to 4,500		45,600	22,500,000
6	Third-class protected cruisers.....	1,700 to 3,190		11,100	5,500,000
10	Gun-vessels and dispatch-boats...	850 to 1,500		11,450	4,500,000
16	Torpedo cruisers (including Vesuvius).......................	About 900		14,500	7,500,000
3	Torpedo-depot and artificer ships .	5,000		15,000	6,500,000
126	Ships of all classes:..............	......		603,470	$342,950,000
101	120-foot torpedo-boats...........	65		6,565	6,565,000
					$349,515,000

It will be observed, comparing the above tables, that the grand
total of cost includes the sum of $67,965,000 already expended or
appropriated for vessels, chiefly of the cruiser class, now completed
or in course of construction. .

The following summary gives the distribution of the elements of
the fleet, including all vessels built, building, appropriated for, and
recommended by the Board.

Battle-ships of great endurance.—Thirteen vessels of an aggregate
tonnage of 120,450 and total cost of $67,400,000.

Battle-ships of limited endurance.—Twenty-five vessels of an
aggregate tonnage of 179,200 and total cost $110,090,000.

Harbor defense and rams.—Seventeen vessels of an aggregate tonnage of 62,320 and total cost $44,500,000.

Cruisers of all classes, including gun-vessels and dispatch-boats, sixty-eight vessels, of an aggregate tonnage of 225,500 and total cost $114,460,000.

Torpedo-depot and artificer ships.—Three vessels of an aggregate tonnage of 15,000 and cost $6,500,000.

Sea-going torpedo-boats.—One hundred and one of an aggregate tonnage of 6565 and cost $6,565,000.

Reliance is placed on the development of an auxiliary navy of fast, well-built merchant steamers to supplement the rather small proportion of cruisers contemplated in the programme.

In attempting to outline any scheme for the execution of the proposed naval programme, we are met at the start by very unsatisfactory conditions as to armor supply; and the most pressing need of the navy is armored ships. At present there exists in the United States but one establishment which will, in a short time, be able to furnish armor of the character needed.

On the 1st of June, 1887, a contract was signed with the Bethlehem Iron Company, of Bethlehem, Pa., for gun steel and armor, including about 5400 tons of vertical armor and bolts. The necessary plant was to be ready within two and a half years of the date of contract, and all the armor to be delivered within two years thereafter, or early in 1892. All this armor is needed for ships under construction, and meanwhile there is required about 3000 tons, not yet contracted for, for ships building or authorized. The Board is of opinion that if this plant continues to be the only source of supply, no large quantity of armor for new vessels can be delivered before 1895. The existing lack of ability to furnish armor in large quantities is entirely due to the absence hitherto of a sufficient demand for it. Were our steel-makers assured sufficient orders, the plants necessary for the manufacture of heavy armor would certainly be created.

Upon this ground, therefore, the Board believes that it would be wise for appropriations to be made at once for a considerable number of heavily armored ships, which should, however, be laid down only as fast as the Department is assured that the developing resources of the country will admit of their rapid completion. On the other hand, should a policy be adopted of appropriating only small sums for armor and armament, it is feared that the necessary stimulus to enlarging present and creating new plants will not be given, and in such case, since the armor of internal bulkheads and redoubts of armored ships built with fair dispatch will be wanted certainly within two years of the laying of their keels, the Board is of the opinion that the requisite armor will not be furnished in time for ships whose construction is begun in 1890.

In the type of battle-ships of great endurance and of thin-armored cruisers submitted, a large proportion of the armor protection is afforded by plates of five inches or less in thickness, but this, it is believed, could be readily manufactured in this country within the necessary time.

Besides the condition of armor supply, a certain restriction is imposed by the working facilities of the country for ship and engine building of the kind required. The slow progress which has been made since the first vessels of the new navy were appropriated for is a warning that expectations should be moderate for some years to come. The average annual output for the navy for the last six years is only 4000 tons, and the present rate may be placed as not more than 7500 tons. Any adequate progress will therefore be accompanied by great expansion of the ship-building industries of the country.

The Board has carefully considered the facilities possessed by the ship-building establishments, including navy-yards now in existence or soon to commence work, together with the supply of skilled labor available. The practical limitation is found in the latter element. Were the urgency declared and very high prices offered for ship-building labor, the demand could probably be rapidly filled by drawing extensively on labor employed in other occupations. Time is necessary for even the stimulated development of the industry in any healthy manner.

Again, it is now very generally recognized that the best naval ship-building policy lies in concentrating all available energy upon such a number of ships as can be completed in a reasonable time, and the efficiency of their type made available as soon as possible, rather than in laying down at one time a greater number of vessels requiring a much longer time to complete, during which none are available, and the efficiency of the type is depreciating.

From all these considerations the Board is of opinion that about 70,000 tons may profitably be laid down in the next two years, the distribution among the different types depending upon the capacity of the various ship-building establishments of the country, some of which are adapted only to the building of unarmored ships. As much as possible of this aggregate tonnage, however, should be in battle-ships.

The Board respectfully recommends that designs be prepared in anticipation of appropriations, as much time has been lost heretofore from this cause, and it has frequently been necessary to adapt the designs to the appropriations, which is manifestly wrong. Should such a policy be followed in the design of the battle-ship recommended, it would be certain to result unsatisfactorily.

Commencing with 14 per cent of the programme to be laid down during the next two years, under the conditions stated, it is believed that the whole work may be carried on at the following rate:

Fourteen per cent to be laid down the first and second years; 10 per cent to be laid down the third and fourth years, and 12 per cent to be laid down each succeeding year, and the programme can be completed in fourteen years.

It is evident that at the end of the third year 24 per cent of the total would be under construction or completing.

It is herein contemplated and urged that the four principal navy-yards, viz., New York, Boston, Norfolk, and Mare Island, will be in the least practicable time completely equipped for constructing the hulls and machinery of all the classes of vessels and their equipment, and with largely increased facilities for docking.

The Board also recommends that the League Island navy-yard be equipped as soon as possible for the construction of the hulls of, vessels, believing that such work may be advantageously done there without the expense and delay consequent on equipping this yard for all classes of work. This may be the better done, as it is believed that the Department will find it advantageous to place contracts for much machinery separately from the hulls, as is very generally done in foreign countries.

In consideration of the undetermined value of many accepted and other proposed means of naval attack and defense, as included in the types of ships in use and proposed, and of the very large expenditure proposed for the building of the fleet on this basis, the Board has to recommend that a wide series of artillery and torpedo experiments be made: to include the effect of common and high-explosive shells on coal, woodite, cellulose, and other water-excluding materials and devices contained in compartments; the attack of inclined armor and decks; the attack of thin-armored sides, and the best form and material of shells to carry various explosives through them; the composition of such explosives; the efficiency of compressed paper as backing for thick and thin armor and for deck armor; torpedo nets and other forms of torpedo defense. To begin these experiments, $100,000 should be appropriated for immediate use.

The Board has also to call attention to the necessity of increased coaling facilities both at our navy-yards and coaling stations, and for coaling ships at sea. For this latter purpose it is not considered necessary to build special colliers, provided suitable appliances can , be determined upon and readily applied to ordinary colliers, as is believed may be done.

GENERAL CONSIDERATIONS RELATING TO THE DESIGNS.

What follows is intended to present the general features of the principal types of the vessels recommended in accordance with the present state of war-ships designed in the light of recent developments in the elements of naval attack and defense, and especially of the steel armor-piercing shell and of shells charged with high explosives.

As these latter elements have controlled certain of the prominent features of many of these and of other recent designs, while experience with them is almost entirely foreign to our navy, it is well to reiterate here the necessity of early and extensive experiment, both as to the nature of such attacks and the efficiency of the means taken to receive them.

Meanwhile, as the results of such experiments cannot be expected to condemn positively existing means of defense on the one hand, or on the other to develop a perfectly satisfactory defense on the same weight, but only to evaluate more closely the present methods or produce others of somewhat greater efficiency, construction may be proceeded with in the confidence of possessing such features at least as efficiently as in the most recent constructions of other nations.

As a general naval progress develops improvements in the means of warfare, they should be embodied in these vessels as they are successively laid down, to the end that each ship shall represent the most approved ideas in construction at the time she is completed. Yet a fairly complete scheme on paper, for a fleet of present construction, adapted to the conditions before laid down, cannot fail to furnish a most useful basis for some years, besides affording the means of determining the approximate aggregate cost and applying directly to such vessels as are to be immediately commenced.

In order that the scheme as prepared may be so used with confidence, it is considered well to lay down briefly the general basis
adopted in the designs for each of the main elements, such as hull,
armor, guns, machinery, etc.

Hulls.—The best construction of mild steel as at present used
is contemplated with the following general differences.   Machine
riveting should be used in all framing and elsewhere, whenever possible, and the rivet material* should throughout be stronger than
now used in the navy.   Also, most of the material within the shell of
the ship, such as frames, bulkheads, plating—other than stringers—
of decks, all supports to redoubts and guns, and such parts normally
in compression, should be of material of from 10 to 20 per cent
greater tensile strength and elastic limit than now used.

As little wood-work should be used as may be necessary for the
comfort of the people and the dry stowage of supplies.   Cementing
is, generally, by admiralty composition.   The weights assigned to
hull and fittings admit of scantlings equal to or exceeding those
of recent analogous European constructions, and are, throughout,
ample with careful distribution of material and fastenings and close
attention to wood-work and fittings.

In all cases there are excluded from these weights all material
specially relating to the protection of the hull and gun positions by
heavy armor, such as plating, framing, etc., behind armor, very commonly included in the weight of hull, but here considered as rather
belonging to the protection.   An exception is made of the framing
to thin armor of casemates or belts.   All heavy plating for gun
protection is likewise excluded.

Armor and protective material.—All steel armor has been contemplated.   The construction of the heavy belts and bulkheads is
with the armor on thin backing against heavy plating in two thicknesses, supported by a close cellular structure and inner mantlet, the
numerous bolts to stand on sleeves within the cellular structure.

The material of the backing is teak, subject to experiment as to
the advantages of compressed paper in large plates of the required
thickness, which may be less than that of the teak as designed.   The
thin armor of 4 to 5 inches is to be worked without backing on heavy
transverse frames, doubled at the butts, with a deep longitudinal
along the middle of the plates.

---

* The advantage of chrome and nickel steels for this purpose, as possessing
greater toughness for the same strength than carbon steel, should be investigated, though the changes contemplated here relate to the ordinary material.

All armor and protective decks are to be worked in single thickness on deep beams of channel or Z section, with a water-tight jacket of 10-pound plating on the lower flanges of the beams between the wing bulkheads. Where sufficient longitudinal support is not afforded by bulkheads, it is to be given by carlings in short lengths between the beams.

A prominent feature in most of the designs is the use of special water-excluding material in the wings. The material known as *woodite* has been adopted for this purpose, subject to the experimental determination of a more suitable substance. Woodite is a light material, composed essentially of cork and rubber. As now made, it lacks elasticity, and has no leak-stopping quality, but fills about 95 per cent of the volume containing it, leaving practically no room for entering water except where the projectile has passed. It may be mined away by shell exploding within it, but, having consistency, will not run out of a shot-hole like cellulose. It is largely relied upon in these designs in connection with ordinary coal and patent fuel over heavy decks for the maintenance of ample stiffness in vessels not heavily armored, and along the unarmored water-line of belted ships, against all forms of gun-attack. Its weight has been figured at 12 pounds to the cubic foot of gross volume of compartments as filled. If packed in tin boxes, as may be found desirable, a somewhat greater weight will be required.

In the battle-ship and protected cruisers the longitudinal bulkhead of the woodite belt is water-tight all fore and aft, and subdivision within is by transverse bulkheads alone throughout the coal bunkers, so as to avoid the serious list consequent on longitudinal subdivision in these parts of ordinary protected ships under conditions of water-line damage when deeply laden. *Cellulose* is to be used in the cofferdams to hatches in heavy decks.

Protection is afforded the 12- and 13-inch guns by turrets differing somewhat from the common form in being lower in front than elsewhere, the port openings extending into the turret crowns and being covered by 3-inch aprons carried on the guns. Barbettes are used only with the 10-inch guns on the smaller battle-ship of limited endurance, in order to keep the size of the ship within certain limits.

In the vessels designed to take any large amount of gun punishment, the guns of 5-inch and over, not protected by heavy armor, are given completely localized protection, being mounted in low barbettes of 2½ to 3½ inches thickness, with equivalent complete shields, entered by door in rear, carried over them, turning with the piece. The connection of movable shield to fixed barbette is to be such as to avoid jamming on moderate injury near the joint, while being tight enough to prevent neighboring explosions of heavy shell —including those containing moderate amounts of high explosive— and the blast of adjacent guns, from being felt within.

In such vessels without a casemate, the barbettes are continued to the armor deck by thick cone and tube bases, with external radial frames, and lightly connected by plate rings and brackets at the junctions with other decks. The tubes open into passages below the armor deck, and the ammunition is to be passed up through the tubes and cones, the guides, etc., being supported as far as possible without connection to them.

In the ordinary protected cruisers, complete shields are fitted to the 8-inch guns above the deck, turning with them. Allowance is made for power—preferably electric—for turning the 8-inch guns and shields. The 5-inch guns are protected by thick plating over the sponsons, extending for a short distance on the side, with open ports covered by inner wing-shields on the gun.

Guns.—After long discussion it was determined that the largest guns to be carried should be of 13-inch caliber, 35 calibers long, and weighing about 60 tons, to be mounted on the largest ships of limited endurance. The greater number of the heavy guns, including those for the battle-ship of great endurance, to be 12-inch, 35 calibers, weighing about 50 tons.

These guns are all mounted in the middle line in turrets above redoubts with single loading positions fore and aft.

In all the large cruisers a certain amount of heavy shell fire and armor-piercing capacity is afforded by the two 8-inch bow and stern guns, while the greater part of the volume of fire is assigned to the 5-inch guns.

One type of large rapid-fire gun, 5 inches caliber and 3 tons weight, is used throughout, as being considered the largest caliber admitting of reasonably convenient fixed ammunition, and well adapted to the attack of unarmored or lightly protected sides. Two types of smaller rapid-fire guns are used—the 6-pounder and 1-pounder—of which, in all cases, four of the latter are mounted for boat attack on the first deck above water, on screw traversing mounts, to stow entirely within the ship's side without projections.

In all the battery plans, the governing considerations have been to obtain a large proportion of the total fire on all bearings, and, in general, equally forward and abaft the beam, while retaining the greatest through a considerable angle of the broadside, and that the lightly protected guns should be distributed as widely as possible, and not more than two in the same line to be disabled by a single projectile. It is believed that these objects have been well attained in the plans submitted.

Machinery.—The machinery throughout is designed for three-cylinder, triple-expansion, with a boiler pressure of 160 pounds; the expansion gears to admit of realizing a combined mean effective pressure by card at the highest power of generally 41 pounds per square inch of low-pressure pistons. In all the vessels of great endurance, and in the ram, there are four sets of vertical engines, two driving each screw. Such an arrrangement in these ships admits of much better proportions with vertical engines accompanied by smaller rubbing velocities and better lubrication, greater reliability, less first cost, and above all, greater economy at low cruising speeds with the forward engines detached, the after engines alone driving the screws and using a good head of steam with most economical expansion. While such engines weigh more by 6 to 7 per cent than single engines of equally good proportions, there is not so much excess weight over engines of the proportions generally necessary, and the slightly greater weight is much more than compensated for by the advantages gained.

The boilers are of cylindrical types in general use. Forced draught by closed stoke-holes is contemplated. Except in the vessels of limited endurance, there are about 1.8 square feet of heating surface per indicated horse-power at maximum power. In the vessels of limited endurance, about 1.95 square feet of heating surface is allowed under the same conditions. As in all the turret-ships a liberal allowance of steam is made for turret and other auxiliary machinery, a greater proportion of heating surface exists when such auxiliaries are not in use. With such large boiler powers very good speeds may be continuously and economically maintained at sea under conditions producing no injury to the boilers, as shown by the appended Table A.

A comparative statement will show the greater weight necessary for the comparative comfort and economy effected in the machinery of these designs, as compared with recent foreign designs. The new British battle-ships of about 14,000 tons, 13,000 indicated horse-power, and 17¼ knots speed, are allowed a weight of 1100 tons for propelling and auxiliary machinery. In the proposed design of United States battle-ship of 10,000 tons, 11,000 indicated horse-power, and 17 knots speed, the weight assigned to propelling machinery, including donkey boiler and stores, is 1215 tons, although it should be stated that the performance predicated of the latter design is not so good as that of the former. The Board is better satisfied with the balance of qualities afforded by the basis of working of the machinery as adopted, at least under existing conditions.

Endurance.—The coal endurance at 10 knots is based on a consumption for all purposes at the rate of 1.8 pounds per indicated horse-power in vessels fitted with double engines on each screw, and 1.9 pounds in the others. The steaming distances under different conditions are given in Table A.

The ammunition endurance at the load displacement has been determined in view of the following considerations:

(1) The intended service of the vessel: being greater, other things being equal, the greater the endurance of coal and supplies.

(2) The size and nature of the piece, and the rapidity of its service.

(3) The degree of protection afforded.

(4) The position and train or arc of bearing.

(5) The number of pieces of the same kind in battery.

The basis has been an allowance of eighty rounds per gun for the 12-inch guns of the design of battle-ship of great coal endurance. The endurance of provisions and small stores on the load displacement is generally for 50 per cent longer than the total coal endurance at 10 knots, being for the cruisers three months, for the battle-ship two months, and for the ships of limited endurance one month.

The water supply is based upon the considerations of allowing at once for considerable derangement of the distilling apparatus and for the water to become thoroughly potable when used. It is thus practically alike for all the vessels, and the tank capacity is fixed at one gallon per man for 15 days. As the tanks are not, however, kept normally full, the corresponding weight is taken to include water carried for sanitary purposes.

Cost.—The costs given will be distinguished from the ordinary contract price of bull and machinery, being for final costs for the best work of the character contemplated on what is considered to be a fair basis for such constructions in the immediate future. As much of the work is larger than has yet been undertaken, and the circum-

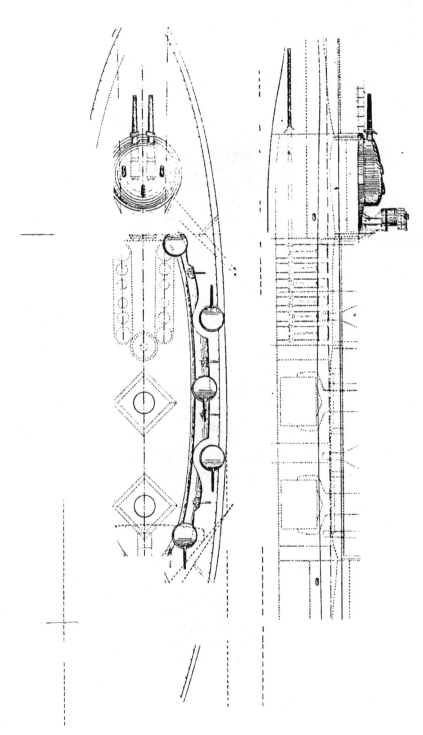

Maximum S.

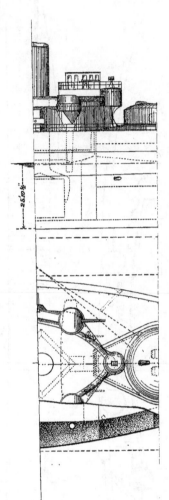

ARMAMEN

4—12 *inch B.*

10—5 *inch R. F*

12—6 *pounder*

6—1 *pounder*

2—37 *mm. R.*

6—*Torpedo T*

stances under which such work may have to be done are very uncertain, some latitude must be allowed in this respect.

Plans.—The plans include only the features peculiar to the type, and are not filled in as to details of subdivision, equipment, etc.

TABLE A.

Type of ship.	Load displacement in tons.	Performance.						Estimated endurance.			
		Forced draught.		Full sea power with air pressure of one-half inch of water in fire-rooms.		Natural draught.		At 10 knots on coal carried at load draught.	Total at 10 knots	Total at most economical speed.	Total at full sea power.
		I. H. P.	Speed	I. H. P.	Speed	I. H. P.	Speed				
First-class battle-ship of great endurance	10,000	11,000	17.0	8,250	15.7	7,070	15.0	5400	10,800	13,000	2300
First-class battle-ship of limited endurance......	8,000	7,500	15.8	6,100	15.1	5,225	14.1	2770	4,600	5,200	1100
Second-class battle-ship of limited endurance..	7,100	6,750	15.8	5,485	14.9	4,700	14.3	2650	4,600	5,200	1100
Third-class battle-ship of limited endurance	6,000	6,500	15.8	5,280	15.2	4,525	14.5	2550	4,600	5,200	1100
Thin-armored cruiser	6,250	9,600	19.0	7,200	17.9	6,170	16.9	7000	10,700	13,000	2100
7500-ton protected cruiser..........	7,500	20,250	22.0	15,150	20.8	13,000	20.0	8600	15,000	18,000	2100
5400-ton protected cruiser	5,400	12,000	20.0	9,000	19.0	7,700	18.4	6700	11,200	13,000	1850
Torpedo depot and artificer's ship..	5,000	10,750	20.0	8,060	18.9	6,910	18.2	7500	11,200	13,000	2000
Ram.............	3,500	9,700	20.25	7,270	19.1	6,230	18.5	2000	4,800	5,500	1100

DESIGN OF BATTLE-SHIP OF GREAT COAL ENDURANCE.

Certain recent foreign designs and constructions suffice to show the type of ship having the qualities of attack, defense, endurance, and seaworthiness demanded for the battle-ship, as the term is generally understood. The introduction of shells containing a considerable amount of explosives, such as gun-cotton and melinite, has rendered many of the essential features of most of these vessels highly objectionable, and particularly so in the case of the protection afforded the heavy guns by redoubts entirely above the hull armor, leaving large spaces between the redoubt floors and the armor deck for the most advantageous action of high-explosive shell. To carry the increased weight demanded to continue the great-gun protection down to the hull armor, and otherwise provide for the protection

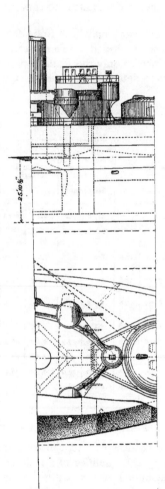

25.10½″

ARMAMEN

4—12 *inch B.*
10—5 *inch R. F*
12—6 *pounder*
6—1 *pounder*
2—37 *mm. R.*
6—*Torpedo T*

INSERT

FOLD-OUT

OR MAP

HERE!

against high explosives of the numerous guns of the powerful secondary batteries now so justly esteemed, while retaining other qualities the same, requires a very considerable increase of displacement.

Simultaneously with the introduction of high-explosive shell-fire, the value of protection by armor of moderate thickness, as relied on to keep out projectiles entirely, has been depreciated by the development of the forged steel, armor-piercing shell, while a new value has been given to thin armor of 4 to 6 inches in causing the explosion before penetration of shell containing high explosives, and incidentally affording a measure of protection against the largely increased volume of fire of small caliber. There is this difference, however, that whereas complete protection was expected of thick armor, accordingly worked below the armor deck, the thin armor is worked entirely above such decks, the chief protection against armor-piercing shell being afforded by thick submerged decks, with water-excluding material above them to insure buoyancy and stability when the thin armor is pierced. As the armor-piercing shell can carry but a small amount of explosive of any kind, it will not mine out any large quantity of such material.

A limit to the vertical extent of thin armor is imposed by the consideration that it should be only so high on the side that an armor-piercing shell which has entered on one side and been deflected by the armor deck should leave the off side above the thin armor, otherwise the tumbling shell with reduced velocity might carry out a large area of the resisting side. It may be remarked that the thin armor of casemates above thick armor belts is specially subject to this risk, except that the armor deck is much less liable to be struck at all, the majority of armor-piercing shell being relied on to go through both sides cleanly.

The armor-piercing shell then has depreciated armor of moderate thickness as commonly worked, and demands considerably increased thickness for the fairly complete protection of the water-line region against penetration by any single projectile, as sought for in the battle-ship of great size.

But this is offset by the influence of high-explosive shell in demanding protection over a greater length of water-line. This has resulted, in the most recent British designs, in retaining the thickness of belt armor at 18 inches over a very much longer belt than formerly.

The general effect of the development of the armor-piercing and high-explosive shells and extension of secondary battery has been to require a very great increase of displacement, with all that is entailed, while increasing the speed but little over the previous designs—and that chiefly by great and perhaps extravagant forcing of the boiler-power—and affording rather inadequate thickness of water-line armor for the increased value of the vessel. Neither is the great-gun protection satisfactory in the vessels of good free-board, being by bar-bettes, leaving the guns and straps entirely exposed; whereas, when they are protected by turrets, the free-board has to be too low for satisfactory work in a seaway.

To such vessels, then, must be urged the objections of inordinate size (14,000 tons) and cost, excessive draught of water (mean load-draught, 27 feet 6 inches, and extreme, at deep load 30 feet), and unsatisfactory protection of great guns and water-line with sufficient free-board, or of water-line with too low free-board with satisfactory protection of the great guns.

With a view of accomplishing the duty required of this type of war vessel on a much reduced displacement and draught of water (and the draught of the British type is practically prohibited by our harbors), the Board submits a design of somewhat novel character, but free from many of the objections of the British type. Protection to buoyancy and stability is afforded by thin-armored sides with thick submerged deck well covered with water-excluding material, on the principles before laid down.

The great-gun positions are well separated and protected by heavy armored redoubts and turrets, and the numerous 5-inch guns are given complete local protection of considerable thickness. The endurance of coal and all supplies except ammunition exceeds that of any equally powerful ships of this class, and the machinery is of type and size to be economical and reliable. It is believed that to obtain equally satisfactory results by thick-armor protection to the water-line region will involve about 1200 tons greater displacement with correspondingly increased draught of water and more than correspondingly increased cost.

Following is a general description of the design:

PRINCIPAL DIMENSIONS.

Length over all .... 349′ 2″
Length on load water-line .... 340′ 2″
Length between perpendiculars.................................. 326′ 6″

Beam extreme*...	71'	6''
Beam on load water-line.....................................	69'	9¾''
Draught mean..	25'	4½''
Displacement in tons..	10,000	
Free-board to top of berth-deck plank at side................	5'	
Free-board to top of main-deck plank at side.................	13'	
Maximum indicated horse-power (forced draught)............	11,000	
Maximum speed in knots.....................................	17	
Metacentric height..	5'	
Metacentric height after complete riddling of water-line.......	2'	6''

HULL AND ITS PROTECTION.

To have a double bottom well fore and aft and to be closely sub-divided up to the berth-deck. The engines to be in four compart-ments, and the boilers in four compartments, the latter separated across the middle line by longitudinal bulkheads containing the magazines for fixed ammunition, with complete passage above and below throughout boiler and engine spaces, narrowed in wake of the engines. Longitudinal wing-bunkers along the boilers and engines to contain fore-and-aft passages communicating with the middle-line passage, and the hand-up of 5-inch ammunition to be through the thick tubes opening into these wing passages. There are thus eight longitudinal skins or bulkheads throughout the machinery spaces, and close transverse subdivision. Similarly close subdivision to be carried to the ends of the ship so far as the stowage, etc., will permit. Protection against water-line damage to be afforded by a wide belt of 5-inch armor extending the whole length of the vessel from a maximum depth of five feet below the load water-line to the under side of the berth-deck, 4 feet 9 inches above it, raised forward to include the bow torpedo-tubes. From the lower edge of this armor a 3-inch armor-deck shall extend from side to side, arched as shown on the plan and domed up in wake of the redoubts. Glacis to be worked around the junction with redoubt armor, and also to boiler-hatches, as shown. These and the engine and ventilating hatches to be continued to the berth-deck by plating of about 1½-inch thick-ness, and armor gratings to be fitted in the openings in the armor-deck. A belt of woodite to extend along the side, between the armor and berth decks the whole length of the ship, to be 12 feet thick at the water-line amidships and 8 feet at the ends, as shown on

* To accomplish a reduction in draught a special form has been given this vessel, involving less beam on the water-line than below it.

the plans. The woodite on each side to be in compartments outboard
of a water-tight sloping fore-and-aft bulkhead. Fuel to be stowed
over the deck between these bulkheads, the subdivision in these
parts being transverse only, and as shown in plan.

CONNING TOWERS.

Two conning towers to be carried. The forward one to be circular,
6 feet 11 inches internal diameter, walls 5 feet 8 inches high and 14
inches thick, crown 2 inches, and base 1½ inches. To stand on a
circular armored shaft from rear of redoubt, 8 feet outside diameter
and 7 inches thick. The after conning tower to be circular, but only
5 inches thick, with three-fourths of an inch crown, and supported
by a 3-inch circular shaft, 6 feet 6 inches outside diameter.

Both towers to be entered by shafts from below, and the shaft to
afford an armored hatch to superstructure deck for hand up of fixed
ammunition.

MAIN ARMAMENT AND PROTECTION.

The great-gun positions to be two in number, one forward and one
aft in the middle line. Each position to be for two 12-inch, 50-ton,
breech-loading rifles in turrets, with complete train around bow and
stern, respectively, to well forward and abaft the beam as shown.
Height of horizontal axis of gun above water-line, 18 feet.

Turrets to be 22 feet inside diameter with armor of 16 inches on
backing of 6 inches, with plating behind armor of 1½ inches, stiffened
by vertical frames and inner mantlet of $\frac{7}{16}$ inch. Turret crown 2½
inches, with three sighting hoods, and hatch with armor grating.
Aprons of 3 inches carried on guns over ports.

Turrets to stand on and in armored redoubts as shown. Thickness
of redoubt armor, 16 inches, tapering to a minimum of 11 inches behind
the 5-inch armor belt. The armor on backing of 6 inches, with
plating behind armor of 1½ inches, with 12-inch stiffeners and 17½
pounds mantlet. Stiffeners and mantlet omitted to armor of rear.
Thickness of crown of rear of redoubt 2½ inches. The armor to be
worked in two widths, except to rear, joining at platform within
redoubt. Turret and redoubt supports to include magazines and
shell-rooms for main battery.

Guns to load in a single fore-and-aft position, elevated and run in.
Eighty rounds to be carried for each gun at load displacement.

SECONDARY BATTERY AND ITS PROTECTION.

Between the great-gun positions, a central superstructure to extend as shown on the plans. Ten 5-inch rapid-firing guns to be carried on main and superstructure decks as shown in low 3-inch barbettes, with complete shields of 3 inches or equivalent over the gun and turning with it, the barbettes to extend to the wing-passages below armor-deck by cone and tube bases; cones and tubes above berth-deck 2 inches thick, and tubes below berth-deck 1 inch thick. All as before described.

Twelve 6-pounder, rapid-firing guns to be carried, eight within the superstructure and four on bridges above the conning towers.

Allowance is made for thick plating and shields for these guns. Six 1-pounder, rapid-firing guns to be carried, four as against boat attack on the berth-deck and two in the tops.

Two 37-millimeter, revolving cannon to be carried in the tops.

The ammunition allowance is as follows: One hundred and fifty rounds for each 5-inch, rapid-firing gun; 400 rounds for each 6-pounder, rapid-firing gun; 700 rounds for each 1-pounder, rapid-firing gun; 1500 rounds for each 37-millimeter, revolving cannon.

TORPEDO OUTFIT.

To consist of six tubes—two fixed above water behind the 5-inch armor in the bows, and four under water in the sides near great-gun positions.

MOTIVE POWER.

Boilers.—Under forced draught to supply steam of 160 pounds for main engines and auxiliaries.

The main boilers are four in number, 15 feet 2 inches diameter and 20 feet long, of double-end type, with eight furnaces and separate combustion chambers. Each pair of boilers to be served by one stack. Aggregate heating surface, 21,500 square feet. Corresponding indicated horse-power at 1.8 square feet, 11,945 indicated horse-power.

A donkey boiler to be carried on the armor deck; to have 30 square feet of grate and about 950 square feet of heating surface.

Engines.—To be four in number, two on each shaft. Two to be of the vertical three-cylinder, triple-expansion type. Cylinders 30½, 45, and 68 inches diameter; stroke, 39 inches. Revolutions per minute, 93¾ at maximum speed of 17 knots. Corresponding piston speed, 609 feet per minute.

COAL ENDURANCE.

The coal carried at load draught is 675 tons, and total coal capacity is estimated at 1350 tons. On the coal at load draught the steaming distance at 10 knots is 5400 knots. With full bunkers at 10 knots, 10,800 knots ; and at most economical speed of about 8 knots, about 13,000 knots. The steaming distance at full sea power and speed of 15.7 knots is 2300 knots.

EQUIPMENT AND OUTFIT.

A single military mast is carried, with two tops. Eleven boats are carried and two 60-foot torpedo-boats. The boats are chiefly handled by derrick on mast. Provision is made for defense by torpedo nets and booms. The complement of officers and men is fixed at five hundred.

TABLE OF WEIGHTS AND DISTRIBUTION OF DISPLACEMENT.

The statement of weights and their distribution among the various elements of efficiency is as follows:

	Tons.
Hull proper and fittings	3,050
Hull protection (5-inch side armor, 620 ; 3-inch armor deck with glacis and hatches, 910 ; water-excluding material, 270)	1,800
Conning towers with armored shafts and connections	90
Two great-gun positions (turret and guns, 430 ; redoubt, 525 ; ammunition and equipment, 105 = 1060)	2,120
Secondary battery, gatlings, and small arms, with ammunition, torpedo outfit, and search lights (242), and protection (212)	454
Propelling machinery, donkey, and spare parts and stores	1,215
Hydraulic and auxiliary machinery, including electric lighting and dynamos	100
Coal at load displacement	675
Equipment and outfit	500
Total weight and displacement	10,004

ESTIMATED COST.

The elements of cost are itemized as follows :

Hull and armor	$3,515,000
Propelling and auxiliary machinery, including donkey boiler and spare parts	1,050,000
Ordnance and torpedo outfit	850,000
Equipment outfit and stores	225,000
Estimated total cost	$5,640,000

BATTLE-SHIPS OF LIMITED COAL ENDURANCE.

It has been hereinbefore demonstrated that the most important and immediate area of naval action, in the event of war with the European nations most powerful on the sea, is included within a line drawn a thousand miles from the Atlantic coast, and extending from the Gulf of St. Lawrence on the north, to the Windward Islands and the Isthmus on the south.

To destroy at the outset all bases of supply and maintain our complete supremacy within this area would insure a comparatively perfect defense of the Atlantic coast and its trade. In addition, to provide for the raids of swift cruisers and the concentrated attack of an enemy's squadron at specially important points, after evading our cruiser scouts, demands a force localized and adapted to local requirements, and yet capable of rapid concentration. Thus a few suitable guard-ships would suffice for the complete protection of an important point against any small force of cruisers, and, in conjunction with fortifications, mines, and torpedo-boats, would prevent bombardment until, by the rapid concentration of a powerful force of our own from adjacent points, the investing squadron could be attacked in force and to great advantage.

To perform this duty, a type of vessel is required of great fighting power and adequate protection, able to take the sea without regard to weather, and fight her guns at sea with the ordinary battle-ships, possessing considerable sea speed, and of draught suitable for comfortable work in our harbors and their approaches. The endurance of all supplies may be limited, and the coal supply should be sufficient for independent action within about 1000 miles, and therefore for a total endurance of not less than about 3000 miles.

Three types of vessels have been designed to meet these requirements, alike in all essentials except great-gun power and draught of water, and differing slightly in protection. All have a maximum speed of 15¾ knots and a sea speed of about 15 knots, with a coal endurance of 4600 knots at 10 knots, and 5200 knots at most economical speed, while at full sea speed they can steam 1100 knots.

Their arrangements are similar, except that the smaller type has a preponderance of bow over stern fire, and her after great guns are mounted in barbette, while the others are in turrets. All are mounted in the middle line, and between the two great-gun positions is a central superstructure and a secondary battery, practically alike for all.

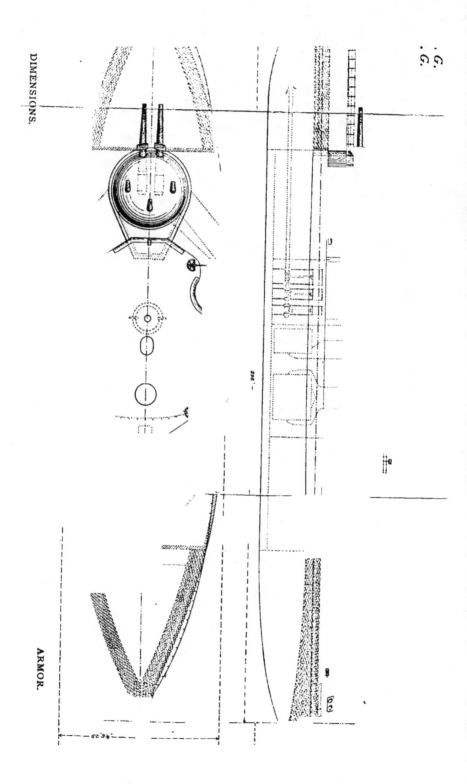

DIMENSIONS.

ARMOR.

Hull protection is afforded by belts of heavy armor of length exceeding one-half the length of the ship, terminating in athwartships bulkheads, the whole including about 72½ per cent of the water-line area. Over the belt and forward and abaft the bulkheads, heavy armor decks are worked, associated with wing coal bunkers over the belt, and wing belts of water-excluding material forward and aft. An additional feature of the protection is a casemate of 4-inch armor extending above the belt between the gun positions for about one-third the length of the ship on the side.

This arrangement of protective material is not considered as necessarily the best or only suitable one. For instance, a preferable arrangement in some respects would be to shorten the belts of thick armor, and terminate it in oblique triangular bulkheads, dispense with the casemate, giving local protection to the supports of 5-inch guns, etc., and continue in the water-line protection to the ends of the ship with·thin armor as in the design of battle-ship of great endurance. The present arrangement was adopted as being in greater general accordance with former and recent foreign designs, and perhaps less questionable than the alternative suggested.

The first type of these vessels is of 8000 tons displacement, and 23 feet 3 inches mean load draught, mounting four 13-inch (60-ton) breech-loading rifles in turrets, with thick armor protection of 17 inches.

The second type is of 7100 tons displacement, and 22 feet 6 inches mean load draught, mounting four 12-inch (50-ton) breech-loading rifles in turrets, with thick armor protection of 16 inches.

The third type is of 6000 tons displacement, and 21 feet mean load draught, mounting two 12-inch (50-ton) breech-loading rifles forward in turret, and two 10-inch (26-ton) breech-loading rifles aft in barbette. Thick armor protection of hull and 12-inch guns, 15 inches; of 10-inch guns, 12½ inches.

Having the same speed, endurance, and general qualities, they are specially well adapted to act together in squadron.

DESIGN OF FIRST-CLASS BATTLE-SHIP OF LIMITED COAL ENDURANCE.

Principal Dimensions.

Length over all .... 314′ 0″
Length on load water-line... 306′ 0″
Length between perpendiculars .... 296′ 0″

Beam extreme .... 67' 9½"
Mean draught.. 23' 3"
Displacement in tons.. 8000
Free-board to top of berth-deck plank at side.............. 3' 1"
Free-board to top of main-deck plank at side.............. 10' 6"
Maximum indicated horse-power (forced draught).............. 7500
Maximum speed in knots...................................... 15.8
Estimated metacentric height .... 5' 0"

HULL AND ITS PROTECTION.

To have a double bottom, extending from the armor shelf, from side to side, well fore and aft. The engines to be in two compartments and the boilers in two, separated across the middle line by double longitudinal bulkheads, affording upper and lower passages, and containing the main drainage and pumping systems. Longitudinal coal bunkers extend outboard along the length occupied by boilers and engines. There are thus ten longitudinal skins or bulkheads throughout the machinery spaces, with close transverse subdivision. Similarly close subdivision to be carried to the ends of the ship, so far as the storage, etc., will permit.

Protection against water-line damage to be afforded by a belt of 17-inch armor for a length of 164 feet, extending from a depth of 4 feet 6 inches below the load water-line amidships, and 3 feet 6 inches at the ends, to a uniform height of 2 feet 10 inches above; to terminate in athwartship bulkheads of 14-inch armor. An armor deck of 2¾ inches to be worked over the belt, and a submerged deck of 3 inches from the armor bulkheads to the ends of the vessel, with stability wings of water-excluding material to the berth-deck, 6 feet thick at the water-line.

Over the belt amidships, a length of 95 feet on each side between the berth and main decks to be plated with 4 inches, terminating in oblique bulkheads of 5 inches to the redoubts. Wing coal bunkers to extend along the sides of this casemate, and torpedo-tubes to be fitted in the corners.

Glacis to be worked around the junction of armor deck with redoubt armor. All hatches in this deck to have coffer-dams filled with cellulose to a height of 3 feet.

The decks to have no sheer. Crown of armor deck over belt, 3 inches; of main deck, 9 inches. Clear height of between decks at side, 6 feet 3½ inches; amidships, 6 feet 9½ inches.

CONNING TOWER.

A conning tower to be carried on the superstructure forward; to be circular in plan, 6 feet 11 inches internal diameter, armor 5 feet 8 inches high and 14 inches thick, crown 2 inches, and base 1½ inches.

To stand on a circular armored shaft from rear of forward redoubt, 8 feet outside diameter, and 7 inches thick.

The tower to be entered by the shaft from below, and the shaft also to afford an armored hatch to superstructure deck for hand-up of fixed ammunition.

MAIN ARMAMENT AND ITS PROTECTION.

The great-gun positions to be two in number, one forward and one aft, in the middle line. Each position to be for two 13-inch 60-ton breech-loading rifles in turrets, with complete train around bow and stern to well abaft and forward the beam respectively as shown. Height of horizontal axis of guns above water-line, 16 feet.

Turrets to be 24 feet internal diameter, with armor of 17 inches, worked on backing of 6 inches, with plating behind armor of 1½ inches, stiffened by vertical frames and with inner mantlets of $\frac{7}{16}$ inch. Turret crown, 2½ inches, with three sighting-hoods and hatch with armor gratings. Deflective aprons of 3 inches carried on guns in front of the ports.

The turrets to stand over and in armored redoubts which extend to the armor deck. The thickness of armor on the redoubt to be 17 inches, except where protected by the casemate of thin armor, where it is to be 12½ inches. The armor to be worked on backing 6 inches thick; the plating behind to be 1½ inches, stiffened by vertical frames, and with an inner mantlet of $\frac{7}{16}$ inch.

The stiffeners and mantlet to be omitted at the rear of the redoubt in wake of the loading apparatus; the sloping crown over the rear of the redoubt to be 2 inches thick.

The turret and redoubt supports to include magazines and shell-rooms for main battery.

The guns to load in a single fore-and-aft position, elevated and run in.

The ammunition supply for each 13-inch gun to be forty-five rounds at load displacement.

SECONDARY BATTERY AND ITS PROTECTION.

Between the great-gun positions a central superstructure to extend as shown on the plans. Four 5-inch guns to be carried on the upper deck in low barbettes of 3-inch thickness, and with shields over guns of 3 inches, or the equivalent, which turn with the guns.

The smaller guns of the secondary battery to be six 6-pounder rapid-firing guns, four 1-pounder rapid-firing guns, two 37-millimeter revolving cannon.

The 6-pounder guns are carried on the superstructure deck, and protected by shields on the guns.

The 1-pounder guns are carried on the berth deck, being intended primarily for defense against torpedo-boat attacks by night, and are protected by extra thick plating in wake of their ports and small shields on the guns.

The revolving cannon are carried in the fighting top, and are to have a train of 360° each.

The ammunition allowances are as below:

	Rounds.
For each 5-inch rapid-firing gun	120
For each 6-pounder rapid-firing gun	300
For each 1-pounder rapid-firing gun	600
For each 37-millimeter revolving cannon	1200

TORPEDO OUTFIT.

To consist of six tubes above water, two fixed at the bow, diverging at 5 degrees and to be worked simultaneously, and four training at corners of casemate. Twelve torpedoes to be carried.

EQUIPMENT AND OUTFIT.

A single military mast is fitted with one fighting-top. A derrick to be fitted for handling boats. The supply of boats and rafts to be only sufficient for the ordinary service of the ship and for life-saving purposes.

The complement, officers and men, is fixed at three hundred and fifteen.

MOTIVE POWER.

Boilers.—Under forced draught to supply steam of 160 pounds pressure for main engines and auxiliaries.

The main boilers to be four in number. Two of them to be 16 feet in diameter and 18 feet 6 inches long, double-ended, with eight furnaces and a common combustion chamber. Two of them to be 16 feet in diameter and 10 feet long, single-ended, with four furnaces and a common combustion chamber.

The boilers to be arranged in two water-tight compartments, one on either side of the middle line, the double-ended boiler forward and the single-ended boiler aft in each boiler-room.

The aggregate heating surface of main boilers to be about 16,600 square feet, and the corresponding indicated horse-power required of main and auxiliary engines, 8500.

Engines.—The main propelling engines to be two in number, one on each shaft, and of the vertical, three-cylinder, triple-expansion type. Cylinders to be 33½, 50½, and 75¾ inches in diameter, respectively, and the stroke, 43 inches.

The revolutions corresponding to the proposed maximum speed of 15.8 knots to be ninety-four per minute, when it is expected that the main engines will develop about 7500 indicated horse-power. The piston speed at the proposed maximum revolutions will be 674 feet per minute.

COAL ENDURANCE.

The coal supply at load draught is 300 tons, and the total coal capacity is estimated at 500 tons. On the coal at load draught the steaming distance at 10 knots is 2770 miles. On the total coal capacity the steaming distance at 10 knots is 4600 miles. At full sea speed of 15.1 knots the steaming distance on the total coal capacity is 1100 miles.

ASSIGNMENT OF WEIGHTS.

		Tons.
Hull proper and fittings..		2,200
Hull protection :		
Belt armor..	1,122	
Deck armor and glacis.....................................	673	
Casemate armor..	168	
Water-excluding material.................................	53	
		2,016
Conning tower and armored shaft.....		70
Two great-gun positions—Turret and guns, 534; redoubt, 393; ammunition and equipment, 78—each 1,005, equal............		2,010
Secondary battery, gatlings, and small arms, with ammunition, torpedo outfit, and search lights (109), and protection (51)................		160
Propelling machinery, with stores and spare parts....................		865
Auxiliary machinery, including electric lights outfit		110
Coal at load displacement...		300
Equipment and outfit...		298
Total...		8,000

ESTIMATED COST.

Hull and armor...	$3,318,650
Propelling and auxiliary machinery.............................	717,350
Ordnance and torpedo outfit......................................	862,500
Equipment outfit and stores.......................................	88,000
Total cost..........	$4,986,500

·SECOND-CLASS BATTLE-SHIP OF LIMITED COAL ENDUR- ANCE, ETC.

[No plans are furnished for this ship, which is to be precisely similar in type to the first-class battle-ship of 8000 tons.]

Length over all... 3c8′ 0″
Length on load water-line...................................... 300′ 0″
Length between perpendiculars................................. 290′ 0″
Beam extreme.. 64′ 6″
Mean draught.. 22′ 6″
Displacement in tons 7100
Free-board to top of berth-deck plank at side .... 3′ 0″
Free-board to top of main-deck plank at side.................... 10′ 4″
Maximum indicated horse-power (forced draught)................ 6750
Maximum speed in knots....................................... 15.8
Estimated metacentric height.................................. 4′ 6″

HULL AND ITS PROTECTION.

To have a double bottom running well fore and aft, and extending to the armor shelf on each side. The engines to be in two compartments, and the boilers in two, separated across the middle line by double longitudinal bulkheads, affording upper and lower passages and containing the main drainage and pumping systems.

Longitudinal coal bunkers extend outboard along the length occupied by boilers and engines.

There are thus ten longitudinal skins or bulkheads throughout the machinery spaces, with close transverse subdivision. Similarly close subdivision to be carried to the ends of the ship, so far as the requirements of storage, etc., will permit.

Protection against water-line damage to be afforded by a belt of 16-inch armor for a length of 160 feet amidships, extending from a depth of 4 feet 6 inches below the water-line amidships, and 3 feet 6 inches below the water-line at the ends, to a uniform height of 2 feet 8 inches above the water-line. To terminate in athwartship bulkheads of 14-inch armor. An armor deck 2⅝ inches thick to be worked over the belt, and submerged armor decks of 2¾ inches to be worked from the armor bulkheads to the ends of the vessel, with stability wings of water-excluding material extending from the submerged deck to the berth deck. These to be 6 feet thick on each side. Over the belt amidships, for a mean length of 90 feet on each side, between the berth and main decks, to be plated with 4-inch armor, terminating in oblique bulkheads to the redoubts 4 inches thick. Wing coal bunkers to extend along the sides of this armored casement, and a torpedo-tube to be fitted in each corner where the diagonal bulkheads join the sides.

Glacis to be worked around the junction of the armor deck with the redoubt armor. All hatches in this deck to have coffer-dams filled with cellulose to a height of 3 feet.

The decks to have no sheer. Crown of armor deck over belt, 3 inches, of main or upper deck, 9 inches. Clear height of between decks at side, 6 feet 3½ inches, and amidships, 6 feet 9½ inches.

CONNING TOWER.

A conning tower to be carried on the superstructure forward. To be circular in plan, 6 feet 11 inches in diameter inside. The walls to be 5 feet 8 inches high and 13 inches thick. The crown to be 1¾ inches, and the base-plate, 1½ inches. To stand on a circular armored shaft from rear of forward redoubt, 7 feet 9 inches in external diameter and 6½ inches thick.

The tower to be entered by the shaft from below, and the shaft also to afford an armored hatch to the superstructure deck for the passage of fixed ammunition.

MAIN ARMAMENT AND ITS PROTECTION.

The great-gun positions to be two in number, one forward and one aft, in the middle line.

Each position to be for two 12-inch 50-ton breech-loading rifles, with complete train around bow and stern, respectively, to well abaft and forward the beam, as in the 8000-ton battle-ship.

The height of the horizontal axis of the guns above the water-line to be 15 feet 3 inches.

The turrets to be 22 feet 6 inches internal diameter, with armor of 16 inches worked on wood backing 6 inches thick, and plating behind, 1½ inches thick. The latter is stiffened by vertical frames, and inside of all is a mantlet $\frac{7}{16}$ inch thick. The crown of the turrets is 2½ inches thick, with three sighting-hoods and hatch with armor gratings. Deflective aprons of 3 inches to be carried on the guns in front of the ports.

The turrets to stand over and in armored redoubts which extend to the armored deck. The thickness of armor on redoubts to be 16 inches, except where protected by the casemate of thin armor, where it is to be 12 inches. In the extreme rear the redoubt is to have but 10 inches of armor.

The redoubt armor is to be worked on 6-inch backing; the plating behind to be $1\frac{1}{2}$ inches, stiffened by vertical frames, and with an inner mantlet of $\frac{7}{16}$ inch.

The stiffeners and mantlet to be omitted at the rear of the redoubt in wake of the loading apparatus. The sloping crown over the rear of the redoubt to be 2 inches thick.

The turret and redoubt supports to include magazines and shell rooms for the main battery.

The guns to load in a single fore-and-aft position, elevated and run in.

The ammunition supply for each 12-inch gun to be forty-five rounds at load displacement.

THE SECONDARY BATTERY AND ITS PROTECTION.

Between the great-gun positions a central superstructure to extend as shown on the plans. Four 5-inch guns to be carried on the upper deck in low barbettes of 3 inches thickness, and with shields over the guns of 3 inches thickness or the equivalent, which are to turn with the guns.

The smaller guns of the secondary battery to be four 6-pounder rapid-firing guns, ten 1-pounder rapid-firing guns, and two 37-millimeter revolving cannon.

The 6-pounder guns to be carried on the superstructure deck, and protected by shields on the guns.

Four of the 1-pounder guns to be carried on the berth deck, being intended primarily for defense against torpedo-boat attack by night, and to be protected by extra thick plating in wake of their ports, and small shields on the guns.

The remainder of the 1-pounder guns to be carried on the superstructure deck, and protected by small shields on the guns.

The revolving cannon to be carried in the fighting-top, and to have a train of 360° each.

The ammunition allowances are as below:

	Rounds.
For each 5-inch rapid-firing gun	120
For each 6-pounder rapid-firing gun	300
For each 1-pounder rapid-firing gun	600
For each 37-millimeter revolving cannon	1200

TORPEDO OUTFIT.

To consist of six tubes above water, two fixed at the bow, diverging at 5 degrees, and to be worked simultaneously, and four training at the corners of the armored casemate. There are to be twelve torpedoes carried.

EQUIPMENT AND OUTFIT.

A single military mast, to be fitted with one·top. A derrick to be fitted for handling boats. The supply of boats and life-rafts to be only sufficient for the ordinary service of the ship and for life-saving purposes.

The complement, officers and men, is fixed at three hundred and ten.

MOTIVE POWER. '

' *Boilers.*—Under forced draught to supply steam of 160 pounds pressure for main engines and auxiliaries. The main boilers to be four in number. Two of them to be 15 feet 4 inches in diameter by 18 feet 6 inches long, double-ended, with eight furnaces and a common combustion chamber. Two of them to be 15 feet 4 inches in diameter by 10 feet long, single-ended, with four furnaces and a common combustion chamber. The boilers to be arranged in two water-tight compartments, one on either side of the middle line, the double-ended boiler forward and the single-ended boiler aft in each boiler-room.

The aggregate heating surface of main boilers to be about 15,100 square feet, and the corresponding indicated horse-power required of main and auxiliary engines, 7750.

·*Engines.*—The main engines to be two in number, one on each shaft, and of the vertical, three-cylinder, triple-expansion type. The cylinders to be 32, 48, and 71½ inches in diameter, respectively, and the stroke, 41 inches. The revolutions, corresponding to the proposed maximum speed of 15.8 knots, to be ninety-nine and one-quarter per minute, when it is expected that the main engines will develop about 6750 indicated horse-power. The piston speed at the proposed maximum revolutions will be 678 feet per minute.

COAL ENDURANCE.

The coal supply at load draught to be 265 tons, and the total coal capacity is estimated at 460 tons. On the coal at load draught the steaming distance at 10 knots is 2650 miles. On the total coal capacity the steaming distance at 10 knots is 4600 miles. At full sea speed of 14.9 knots the steaming distance on the total coal capacity is 1100 miles.

ASSIGNMENT OF WEIGHTS.

	Tons.
Hull proper and fittings............................	1,976

Hull protection :

Belt...	1,024	
Deck armor and glacis....................................	620	
Casemate armor..	148	
Water-excluding material................................	52	
		1,844

Conning tower and armored shaft	62
Two great-gun positions: Turret and guns, 433 ; redoubt, 350 ; ammunition and equipment, 62 = 845	1,690
Secondary battery, gatlings, and small arms, with ammunition, torpedo outfit, and search-lights (108), and protection (50)...................	158
Propelling machinery, with stores and spare parts	775
Auxiliary machinery, including electric lighting......................	85
Coal at load displacement	265
Equipment and outfit...	252
Total	7,107

ESTIMATED COST.

Hull and armor..	$2,950,000
Propelling and auxiliary machinery	640,000
Ordnance and torpedo outfit...................................	700,000
Equipment outfit and stores....................................	85,000
Total...	$4,375,000

THIRD-CLASS BATTLE-SHIP OF LIMITED COAL ENDURANCE.

Principal Dimensions, etc.

Length over all..	298′	0″
Length on load water-line	290′	0″
Length between perpendiculars.................................	280′	0″
Beam extreme..	61′	6″
Mean draught...	21′	0″
Displacement in tons..	6000	
Free-board to top of berth-deck plank at side	2′	9″
Free-board to top of main-deck plank at side....................	10′	2″
Maximum indicated horse-power (forced draught)	6500	
Maximum speed in knots.......................................	15.8	
Estimated metacentric height..................................	4′.8	

HULL AND ITS PROTECTION.

To have a double bottom running well fore and aft, and extending to the armor shelf on each side. The engines to be in two compartments,

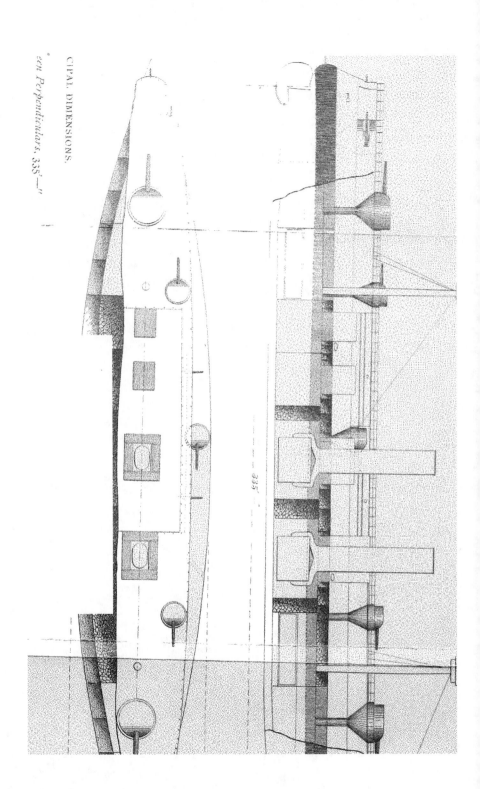

CIPAL DIMENSIONS.

een Perpendiculars, 335'——"

and the boilers in two, separated across the middle line by double longitudinal bulkheads, affording upper and lower passages, and containing the main drainage and pumping systems.

Longitudinal coal bunkers extend outboard along the length occupied by boilers and engines.

There are thus ten longitudinal skins or bulkheads throughout the machinery spaces, with close transverse subdivision.

Similarly close subdivisions to be carried to the ends of the ship, so far as the requirements of stowage, etc., will permit.

Protection against water-line damage to be afforded by a belt of 15-inch armor for a length of 155 feet amidships, extending from a depth of 4 feet 6 inches below the water-line amidships, and 3 feet 6 inches below the water-line at the ends to a uniform height of 2 feet 6 inches above water-line. To terminate in athwartship bulkheads of 13-inch armor.

An armor deck 2½ inches thick to be worked over the belt, and submerged armor decks of 2¾ inches to be worked from the armor bulkheads to the ends of the vessel, with stability wings of water-excluding material extending from the submerged deck to the berth deck. These to be 6 feet thick on each side.

Over the belt amidships, for a mean length of 100 feet, on each side, between the berth and main decks, to be plated with 4-inch armor, terminating in oblique bulkheads to the redoubts, 4 inches thick. Wing coal bunkers to extend along the sides of this armored casemate, and a torpedo-tube to be fitted in each corner, where the diagonal bulkheads join the sides.

Glacis to be worked around the junction of the armor deck with the redoubt armor. All hatches in this deck to have coffer-dams filled with cellulose to a height of 3 feet. The decks to have no sheer. Crown of armor deck over belt, 3 inches; of main or upper deck, 9 inches. Clear height of between decks at side, 6 feet 3½ inches, and amidships, 6 feet 9½ inches.

CONNING TOWER.

A conning tower to be carried on the superstructure forward. To be circular in plan, 6 feet 11 inches in diameter inside. The walls to be 5 feet 8 inches high and 12 inches thick. The crown to be 1½ inches, and base-plate, 1¼ inches. To stand over a circular armored shaft, from the rear of forward redoubt, 7 feet 9 inches in diameter and 6¼ inches thick. The tower to be entered by the shaft from below, and the shaft also to afford an armored hatch to the superstructure deck for the passage of fixed ammunition.

MAIN ARMAMENT AND ITS PROTECTION.

The great-gun positions to be two in number, one forward and one aft, in the middle line.

The forward position to be for two 12-inch 50-ton breech-loading rifles in a turret, and the after position for two 10-inch 26-ton breech-loading rifles in a barbette.

These guns to have complete train around bow and stern respectively to well abaft and forward the beam, as shown.

The height of the horizontal axis of the forward guns above the water-line is 14 feet 6 inches, and of the after guns is 13 feet 6 inches. The turret to be 22 feet 6 inches internal diameter, with armor of 15 inches worked on wood backing 6 inches thick, and plating behind armor $1\frac{1}{2}$ inches thick. The latter is stiffened by vertical frames, and inside of all is a mantlet $\frac{7}{16}$ of an inch thick. The crown of the turret is $2\frac{1}{2}$ inches thick, with three sighting-hoods and hatch with armor gratings. Deflective aprons of 3 inches are carried on the guns in front of the ports.

The turret to stand over and in an armored redoubt which extends to the armor deck. The redoubt armor to be 15 inches thick in front, and 10 inches on sides and rear where protected by the armored casemate.

The redoubt armor is to be worked on 6-inch backing. The plating behind is to be $1\frac{1}{2}$ inches, stiffened by vertical frames, and with an inner mantlet of $\frac{7}{16}$ of an inch. The turret and redoubt supports to include magazine and shell-room for the 12-inch guns.

The 12-inch guns to load in a single fore-and-aft position, elevated and run in. The ammunition supply for each 12-inch gun to be forty-five rounds at load displacement.

The barbette aft to be 17 feet inside diameter, with armor of $12\frac{1}{2}$ inches, suitably diminished where protected by the casemate, worked on 6-inch backing, and $1\frac{1}{2}$-inch plating behind. The latter to be stiffened by vertical framing, and a seven-sixteenth mantlet worked inside of all.

The crown, gun-shields, etc., to be 3 inches thick, or the equivalent, and a suitable working tower is to be fitted.

The guns are to load at any train.

The ammunition supply for each 10-inch gun to be fifty rounds at load displacement.

SECONDARY BATTERY AND ITS PROTECTION.

Between the great-gun positions a central superstructure to extend as shown on the plans. Four 5-inch guns to be carried on the upper deck in low barbettes of 3 inches thickness, and with shields over the guns of 3 inches thickness, or the equivalent, which are to turn with the guns.

The smaller guns of the secondary battery to be four 6-pounder rapid-firing guns, ten 1-pounder rapid-firing guns, and two 37-millimeter revolving cannon.

The 6-pounder guns are carried on the superstructure deck, and protected by shields on the guns.

Four of the 1-pounder guns are carried on the berth deck, being intended primarily for defense against torpedo-boat attacks by night, and are protected by extra thick plating in wake of their ports and by small shields on the guns.

The remainder of the 1-pounder guns are carried on the superstructure deck and protected by small shields.

The revolving cannon are carried in the fighting-top, and are to have a train of 360 degrees each.

The ammunition allowances are as below :

	Rounds.
For each 5-inch rapid-firing gun	120
For each 6-pounder rapid-firing gun	300
For each 1-pounder rapid-firing gun	600
For each 37-millimeter revolving cannon	1200

TORPEDO OUTFIT.

To consist of six tubes above water, two fixed at the bow, and diverging at 5 degrees, and to be worked simultaneously, and four training at the corners of the armored casemate.

There are to be 12 torpedoes carried.

EQUIPMENT AND OUTFIT.

A single military mast is fitted with one top. A derrick to be fitted for handling boats. The supply of boats and life-rafts to be only sufficient for the ordinary service of the ship and for life-saving purposes. The complement, officers and men, is fixed at three hundred.

MOTIVE POWER.

Boilers.—Under forced draught to supply steam of 160 pounds pressure for main engines and auxiliaries. The main boilers to be four in number. Two of them to be 15 feet in diameter by 18 feet 6 inches long, double-ended, with eight furnaces and a common combustion chamber. Two of them to be 15 feet in diameter by 10 feet long, single-ended, with four furnaces and a common combustion chamber.

The boilers to be arranged in two water-tight compartments, one on either side of the middle line, the double-ended boiler forward and the single-ended boiler aft in each boiler-room.

The aggregate heating surface of main boilers to be about 14,100 square feet, and the corresponding indicated horse-power required of main and auxiliary engines, 7250.

Engines.—The main engines to be two in number, one on each shaft, and of the vertical, triple-expansion, three-cylinder type. The cylinders to be 31, 46¾, and 69¾ inches in diameter, respectively, and the stroke, 39½ inches. The revolutions corresponding to the proposed maximum speed of 15.8 knots to be one hundred and one per minute; when it is expected that the main engines will develop about 6500 indicated horse-power. The piston speed at the proposed maximum revolutions will be 665 feet per minute.

COAL ENDURANCE.

The coal supply at load draught is 230 tons, and the total coal capacity is estimated at 425 tons. On the coal at load draught the steaming distance at 10 knots is 2550 miles. On the total coal capacity the steaming distance at 10 knots is 4600 miles. At full sea speed of 15.2 knots the steaming distance on the total coal capacity is 1100 miles.

ASSIGNMENT OF WEIGHTS.

		Tons.
Hull proper and fittings		1,725
Hull protection :		
Belt	925	
Deck armor and glacis	545	
Casemate armor	158	
Water-excluding material	50	
		1,678
Conning tower and armored shaft		58
Forward great-gun position (turret and guns, redoubt, ammunition, and equipment)		795
After great-gun position (guns, barbette, ammunition, and equipment)		320
Secondary battery, gatlings, and small arms, with ammunition, torpedo outfit, and search lights (108), and protection (50)		158

	Tons.
Propelling machinery, with stores and spare parts	725
Auxiliary machinery, including electric lighting	75
Coal at load displacement	230
Equipment and outfit	235
Total	5,999

ESTIMATED COST.

Hull and armor	$2,488,5co
Propelling and auxiliary machinery	549,600
Ordnance and torpedo outfit	488,500
Equipment outfit and stores	81,000
	$3,607,600

CRUISERS.

The numerous duties expected of this general type in war may be stated as:

(a). When acting singly or in small cruising squadrons—

(1) To destroy commerce.

(2) To convoy commerce.

(3) To generally protect commerce by clearing the trade routes of the enemy's cruisers.

(4) To ravage the enemy's coasts at unprotected or weakly defended points.

(5) To carry on the smaller naval wars in distant countries, effect reprisals, etc.

(b). When acting directly or indirectly with a squadron of battle-ships—

(6) To act as scouts, keep touch with the enemy, mark and give prompt information of his movements, cut off disabled vessels, harass his weaker ships as opportunity offers, and especially his supply ships.

(7) To keep touch between the squadron and its base of supplies, and protect the route of supplies.

(8) In squadron under way, to form the front, rear, and wings of the fleet, making first contact with an enemy, and engaging them until their force becomes well defined and order of battle may be laid down. To engage the lighter ships, during action, and, as far as their fighting power will permit, act as reserves to strengthen weak places in the line, and generally fulfill certain tactical requirements connected with the line of battle.

(9) In squadron blockading, by their great numbers and activity, to insure the efficiency of the blockade, to serve as bases of supplies for the numerous lighter vessels, such as sea-going torpedo-boats required for this service, to pursue and destroy or capture similar ships of the enemy which may pass the lines.

(10) In squadron blockaded, by their number and activity, to deceive and disconcert the blockading force and mark the opportunities for sally. To escape singly or in small numbers, and, by ravaging the enemy's coast and destroying commerce and attacking the supply routes of his squadron, to effect the diversion of a superior force from the blockade or other sources.

While all vessels of the cruiser type should be more or less well adapted to all these services, the necessary development of special fitness for one or more particular services results in the creation of numerous types, differing widely in size and power.

Certain broad features are believed to obtain universally with this class, viz.:

(1) The use of steam of high pressure in triple expansion, with machinery of types specially adapted to the wide range of speed and fuel economy demanded of all vessels of this class, has rendered it undesirable to assign any of the displacement to sail power. Under present conditions a better return will be given by applying such weight to other qualities, and particularly to the coal endurance. But it becomes all the more desirable for the country to obtain and strongly defend coaling stations at certain points, and especially one in mid-Pacific. The attention of the Department is also directed to the necessity of immediate careful study of fuel consumption in all distinct types of vessels added to the navy, and especially in cruisers, and in addition to the supply and general use of the fuel best adapted to present conditions.

(2) It is believed that a certain line of size is drawn at about 3000 tons displacement for vessels intended to act singly, by the impossibility of combining the necessary speed and endurance in smaller vessels, and of fitting the very necessary feature of a satisfactory double bottom, and by the very serious loss of speed in a moderate sea-way. Smaller vessels of the class should not exceed 1200 tons displacement, with very high smooth-water speed, and armed only

with the lightest weapons. Being incapable of efficient single action, their endurance may be less than the 9000 miles at 10 knots, which it is believed all the larger cruisers should possess. Their place is with the fleet or as part of the torpedo-boat defense.

(3) The minimum smooth-water speed of the larger vessels should not be less than 19 knots, otherwise sufficient mobility will not be possessed for many of the services required. Similarly, an endurance of 9000 miles at 10 knots may be accepted as a minimum for independent action, and should generally be considerably, and in some vessels largely, exceeded.

, (4) Inasmuch as it is now possible to prepare a steel bottom against fouling to give perfectly satisfactory results for continuous cruising for a time from three to five times the coal endurance, for a time, indeed, so long that it becomes questionable whether, as dependent on prizes alone for continuance of fuel supply, many vessels, even of the cruiser class, will require to remain longer out of the home ports, it is clear that sheathing such ships with wood and copper must be classed with sail-power as not affording an adequate return of efficiency. Sheathing must then be regarded as a concession to the peace duties of war-ships, though it should be noted that it is by no means a complete remedy for the evil of fouling, sheathed ships lying for several months in harbor, as very commonly happens in time of peace, being found fouled as badly as can be and suffering the same loss of speed and fuel economy as an unprotected iron hull. It so happens that below the limiting minimum displacement of 3000 tons, numerous types of vessels may be built of moderate speed and endurance, more or less complete sail-power, hulls sheathed with wood and coppered against fouling—and thus also rendered better able to take the ground under ordinary circumstances than a steel hull with a double bottom—vessels very well adapted to the economical performance of the peace duties of a navy; but it is now very generally conceded that such constructions are essentially extravagant as having no place or function in time of war with powerful nations.

The Board has hereinbefore declared its opinion that true ultimate economy exists in building for war purposes only, the trifling apparent economy resulting from building vessels essentially adapted to the peace duties of the navy being out of all proportion to the depreciation in value of all such material in time of war, for which the navy essentially exists, while the training of the personnel in the one class of ships is not conducive to the highest efficiency in the other.

In connection with its attitude on the sheathing question, the Board desires to impress on the Department, as a measure of efficiency and economy, the necessity of the highest attainable standard in the matter of protective and anti-fouling compositions and their application to the bottoms of ships. ˙ It is believed to be a matter of the greatest importance, demanding the most intelligent special study and accumulated experience, and not to be advanced by detached and disconnected experiments. The progress made of late years in these matters leads to the belief that much more will soon be accomplished, and the Department is urged to take such measures as will assist the development of the subject, and especially insure a high efficiency in all such work done for the navy.

In view of the above-defined propositions, and of the types of cruisers existing in, or about to be added to, the navy, the Board has recommended the construction of no additional cruisers of less than 4000 tons, except certain torpedo-cruisers having the qualities above defined for cruisers of small displacement, and certain gun-vessels for river service. All the cruisers recommended have steel hulls unsheathed, no sail power, a minimum smooth-water speed of 19 knots, and, with the exception of the torpedo-cruisers, a minimum endurance of 10,000 knots at 10 knots, and machinery of a type admitting at once of great power and great economy at low speeds.

The Board desires to draw special attention to the design of cruiser with thin-armored belt and complete local protection to the guns, as representing the outcome of applying the arguments advanced on the design of battle-ship of great coal endurance to the case of a cruiser. It is believed that in this type lies the germ of much development. Thus the universal use of high-explosive shell may render it necessary to build all large war-vessels, of whatever class, with protection of this kind, only a somewhat greater thickness of belt armor being demanded for battle-ships, as the best shell to carry high explosives from guns of large calibers through such armor becomes developed. And again, such vessels, with perhaps others of somewhat greater size and heavy gun-power, having so much capacity to give and receive punishment, could in squadron (as it were a fleet of light, fast battle-ships) overcome the defenses of many important ports. In other words, for certain purposes their individual value would be specially increased by concentration into squadrons.

They have been designed to fulfill specially requirements *a* 3, *a* 4, and *b* 8.

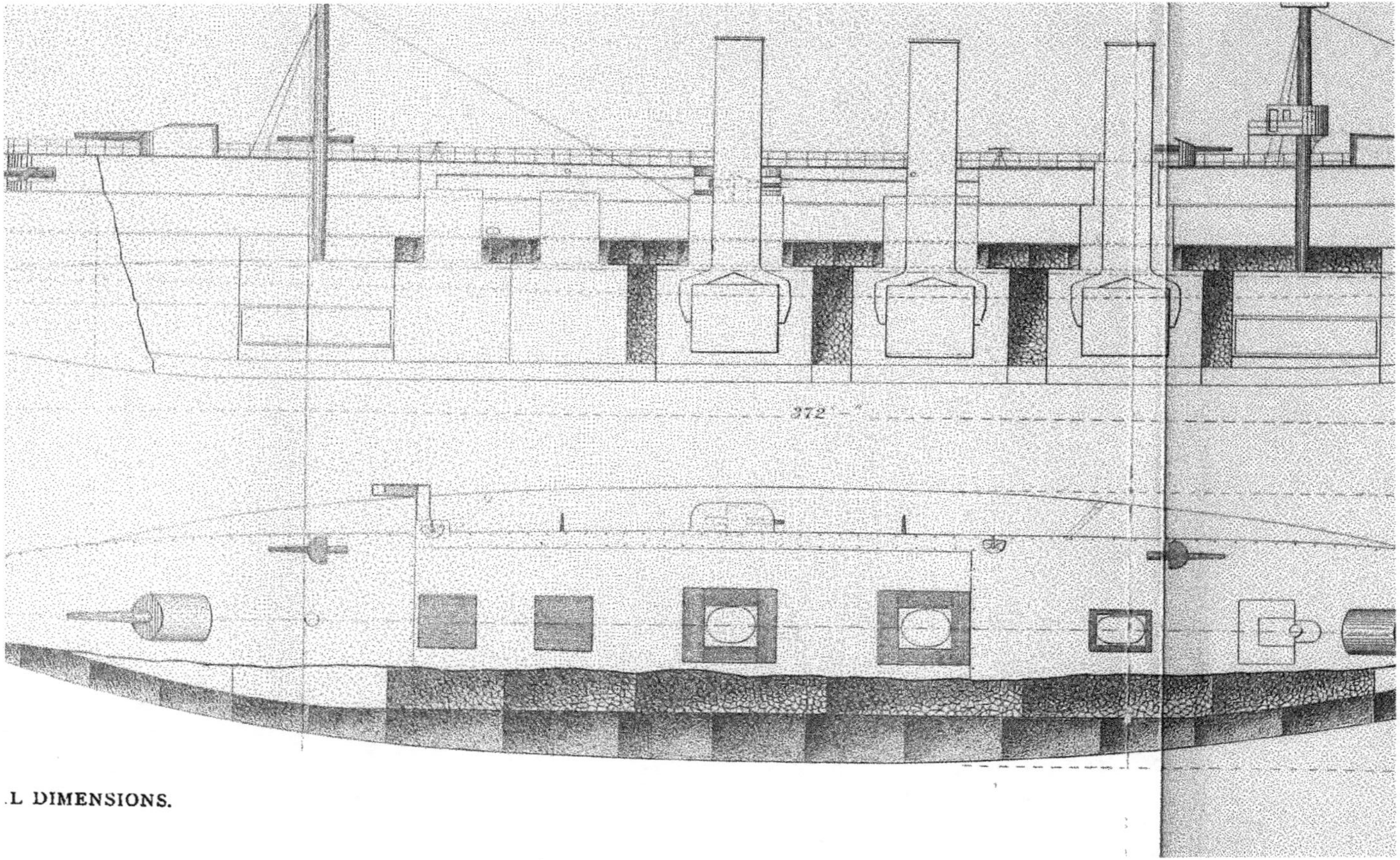

.L DIMENSIONS.

The first-class protected cruiser is specially adapted to the requirements of divisions *a* 1, *a* 3, *a* 4, *b* 6, and certain of the duties of *b* 9 and *b* 10. She has the unexcelled speed of 22 knots and coal endurance of 15,000 miles. Her gun-power is good, equal to that of the largest protected cruisers built or building by other nations, and her protection by armor deck and stability belt is such as to enable her to stand considerable punishment.

To obtain these qualities it has been necessary to adopt the somewhat large displacement of 7500 tons, which incidentally assures the maintenance of very high speed at sea in all weathers, and makes her capable of overtaking any merchant steamer now completed or likely to be built in the near future, while able to match or overpower any man-of-war approaching her in speed.

Two types of second-class protected cruisers are recommended. The first, of 5400 tons, may be considered the standard type of protected cruiser, being equally well adapted to each of the duties required of this class, except that from her size and cost such vessels may not be sufficiently numerous. Her gun-power and defensive qualities both exceed those of the first-class cruiser.

The smaller type, of about 4000 tons, fairly represented by such vessels as the Baltimore, with the addition of somewhat greater capacity to receive water-line injury, and a reduction of great-gun power, is less well adapted to independent action, but almost equally useful with a fleet, while their reduced cost assists their numbers.

For services in the cruiser class, involving only small powers of offense and defense, with good sea speed and endurance, unless the cost of the cruiser fleet is to be much increased, reliance must be had on the development of an auxiliary navy of merchant steamers. To this end it will be necessary to offer suitable inducements for owners of vessels in certain trades to build them up to a certain limit of speed and under certain requirements as to subdivision and local strength for light-gun positions. The advantage, both to the Government and to owners in trades requiring vessels of fair speed, is so obvious that the Board considers no elaborate arguments necessary to demonstrate the wisdom of such an arrangement, the conclusion of which is contemplated in the fleet proposed.

THIN-ARMORED CRUISERS OF 6250 TONS.

Principal Dimensions, etc.

Length over all.. 345′ 0″
Length on load water-line..................................... 344′ 0″
Length between perpendiculars................................ 335′ 0″
Beam extreme.. 57′ 6½″
Mean draught... 22′ 3″
Displacement in tons... 6250
Free-board to top of berth-deck plank at side................. 5′ 0″
Free-board to top of main-deck plank at side................. 13′ 0″
Maximum indicated horse-power (forced draught).............. 9600
Maximum speed in knots....................................... 19
Estimated metacentric height................................. 4′ 0″

HULL AND ITS PROTECTION.

To have a double bottom running well fore and aft and to the turn of the bilge on either side. The engines to be in four compartments, and the boilers in two, with thwartship coal bunkers. Wing coal bunkers to be fitted in addition, and the subdivision forward and aft to be as close as the requirements of stowage, etc., will permit. To have a long poop and forecastle.

The buoyancy and stability are to be protected—

(1) By a complete belt of armor extending from the under side of berth-deck plank to 4 feet 6 inches below the water amidships, rising to 3 feet 6 inches below the water forward and aft and thence down to the ends of the ship. This belt to be 5 inches thick for one-third the length amidships, and 4 inches thick forward and aft. This armor to be worked on frames without backing.

(2) By an arched armored deck reaching the water-line at its crown amidships and the bottom of the armor belt at sides. To be 2¼ inches thick at the center and 2½ inches thick at the sides.

(3) By a complete belt of water-excluding material between the armor and berth decks. This is worked outside of sloping water-tight wing bulkheads between the armor deck and the berth deck as shown. The belt to be about 9 feet thick at the water-line amidships, and to taper gradually with the curve of the water-line until its thickness is 6 feet, this thickness being retained for the rest of the length. The belt to be divided at suitable intervals by partial transverse bulkheads, as shown.

(4) Within the wing bulkheads of the belt, coal to be stowed over the armor deck as shown, and light water-excluding stores forward and aft. The coal bunkers above the armor deck to have transverse bulkheads only.

Glacis to be worked around openings in the armor deck for smoke-stack, etc., and armor gratings and coffer-dams as required.

CONNING TOWER.

A conning tower, with sides of not less than 6 inches, to be fitted on the bridge or in other suitable commanding position.

ARMAMENT AND ITS PROTECTION.

The battery to consist of two 8-inch breech-loading rifles, ten 5-inch rapid-firing guns, six 6-pounder rapid-firing guns, eight 1-pounder rapid-firing guns, and four 37-millimeter revolving cannon.

The 8-inch guns are mounted in the middle line, one on the poop and one on the forecastle, in partial barbettes $3\frac{1}{2}$ inches thick. A shield covers the rear of the gun and the crew. This shield is $3\frac{1}{2}$ inches thick in front and its equivalent elsewhere. Beneath these barbettes are worked protective cones with tubular bases each 2 inches thick. The tubes lead below the armor deck, and ammunition is to be supplied through them. Four of the 5-inch guns are mounted on poop and forecastle, and two in the waist, similarly to the 8-inch guns, but with barbette and shield 3 inches thick or the equivalent. The cone and tube are 2 inches thick or the equivalent.

The remaining four 5-inch guns are mounted on the main deck forward and aft, in recessed ports with thick plating and closed shields, in rear equivalent to 3 inches, and with cone and tube beneath the position as at the other mounts. The smaller guns are mounted as shown on the plans, being protected by small shields on the guns or by extra thick plating in wake of ports.

The ammunition allowances are as below:

	Rounds.
For each 8-inch breech-loading rifle	80
For each 5-inch rapid-firing gun	160
For each 6-pounder rapid-firing gun	500
For each 1-pounder rapid-firing gun	750
For each 37-millimeter revolving cannon	2000

EQUIPMENT AND OUTFIT.

Two military masts, to be fitted with fighting-tops. Derricks to be fitted for handling boats, including two 60-foot torpedo vidette boats. The complement of officers and men is fixed at four hundred.

MOTIVE POWER.

Boilers.—Under forced draught to supply steam of 160 pounds pressure for main engines and auxiliaries. The main boilers to be four in number, 14 feet 7 inches in diameter by 18 feet 6 inches long, double-ended, with six furnaces and a common combustion chamber.

The boilers to be arranged in two water-tight compartments separated by a transverse coal bunker. Aggregate heating surface of main boilers to be about 17,300 square feet, and corresponding indicated horse-power, 9600.

Engines.—The main engines to be four in number, two on each shaft, and of the vertical, triple-expansion, three-cylinder type.

The cylinders to be 26, 38, and 58¼ inches in diameter respectively, and the stroke, 33 inches. The revolutions corresponding to the proposed maximum speed of 19 knots to be one hundred and thirty-three per minute, when it is expected that the main engines will develop about 9600 indicated horse-power. The piston speed at the proposed maximum revolutions will be 732 feet per minute.

Donkey.—A donkey boiler to be carried on the armor deck. To have about 20 square feet of grate surface, and about 620 square feet of heating surface.

Coal endurance.—The coal supply at load draught is 625 tons, and the total coal capacity is estimated at 950 tons. On the coal at load draught the steaming distance at 10 knots is 7000 miles. On the total coal capacity the steaming distance at 10 knots is 10,700 miles. At full sea speed of 17.9 knots the steaming distance on the total coal capacity is 2100 miles.

ASSIGNMENT OF WEIGHTS.

		Tons.
Hull proper and fittings, including 6-inch conning tower (15 tons) and extra plating on berth deck....................................		2,400
Hull protection :		
Armor...	460	
Armor deck..	600	
Water-excluding material.................................	200	
		1,260
Armament (326) and its protection (312)............................		638
Machinery, donkey, stores, and spare parts		913
Coal at load displacement		625
Equipment and outfit...		415
Total...		6,251

ESTIMATED COST.

Hull and protection...	$2,085,300
Machinery..	623,000
Ordnance...	371,800
Equipment..	120,000
	$3,200,100

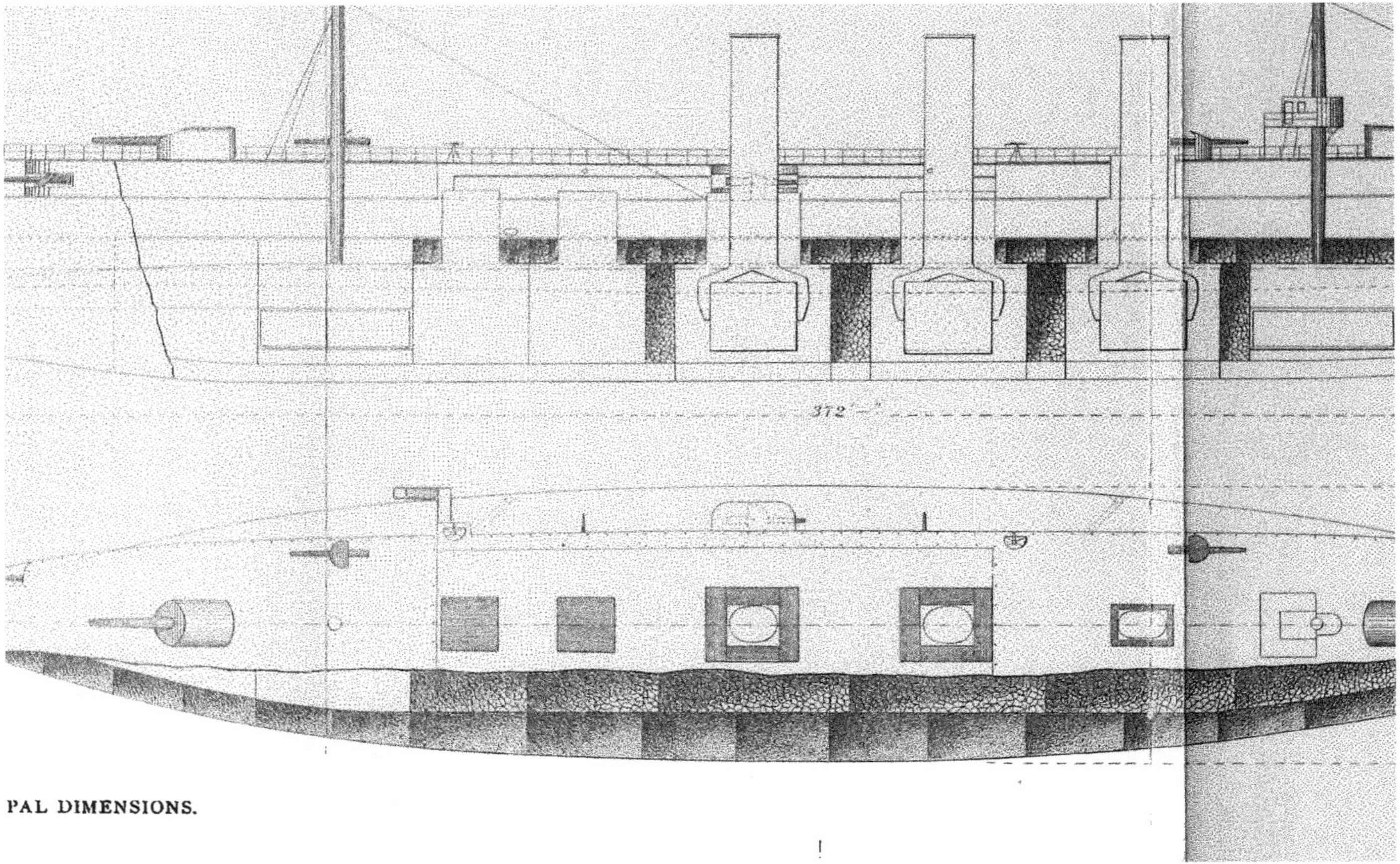

PAL DIMENSIONS.

FIRST-CLASS PROTECTED CRUISER OF ABOUT 7500 TONS.

Principal Dimensions, etc.

Length over all...	382′	6″
Length on load water-line.................................	381′	0″
Length between perpendiculars............................	372′	0″
Beam extreme........	58′	7″
Mean draught...	23′	1½″
Displacement in tons	7500	
Free-board to top of berth-deck plank at side............	5′	9″
Free-board to top of main-deck plank at side.............	13′	9″
Maximum indicated horse-power (forced draught)............	20,250	
Maximum speed in knots	22	
Estimated metacentric height.............................	4′	5″

HULL AND ITS PROTECTION.

To have a double bottom running well fore and aft and to the turn of the bilge on either side. The engines to be in four compartments and the boilers in three, with thwartship coal bunkers. Wing coal bunkers to be fitted in addition, and the subdivision forward and aft to be as close as the requirements of stowage, etc., will permit. To have a long forecastle and ordinary poop.

The buoyancy and stability to be protected—

(1) By an armor deck with flat crown 1 foot above the water and 2 inches thick. This is to slope to the side with two slopes, the inner of 20 degrees and the outer of 32 degrees. Plating on both slopes to be 4 inches in thickness. The edge of the armor deck to be 5 feet below the water amidships.

(2) By a belt of water-excluding material worked outside of sloping water-tight wing bulkheads between the armor deck and the berth deck as shown. The belt to be 10 feet thick at the water-line amidships, tapering with the water-line, as shown, to a thickness of 6 feet, which is continued forward and aft, and divided at suitable intervals by partial transverse bulkheads.

(3) Within the wing bulkheads of belt, coal to be stowed, as shown, above the armor deck in wake of the machinery space and light and water, excluding stores forward and aft of the coal as far as practicable. The coal bunkers above the armor deck to be subdivided by transverse bulkheads only.

Glacis to be worked around openings in the armor deck for smokestacks, etc., and armor-gratings and coffer-dams as required.

CONNING TOWER.

A conning tower, with sides of not less than 6 inches thickness, to be fitted on the bridge or in other suitable commanding position.

ARMAMENT AND ITS PROTECTION.

The battery to consist of two 8-inch breech-loading rifles, ten 5-inch rapid-firing guns, six 6-pounder rapid-firing guns, eight 1-pounder rapid-firing guns, and four 37-millimeter revolving cannon.

The 8-inch guns to be mounted in the middle line, one on the poop and one on the forecastle, as shown. They are to be completely shielded by closed shields, as shown, which are to work with the guns. These shields are to have a front of 2 inches, sides of 1½ inches, and rear of 1 inch. The sloping part of the crown in front to be 1½ inches thick, and the remainder of the crown, ¾ inch. These shields are to be large enough to contain the whole gun's crew. Allowance of weight is made for working these guns by power.

The 5-inch guns are to be mounted as shown, two in the waist, two on the forecastle, two on the poop, two on the main deck under forecastle, and two on the main deck under poop.

The guns on poop and forecastle to be protected by shields on the guns of 2 inches in front and the equivalent elsewhere. The remainder of the 5-inch guns are protected by plating around their ports, 2 inches in thickness, with equivalent small shields on the guns to cover the ports. The smaller guns to be mounted as shown on the plans, the 6-pounders being protected by shields on the guns.

The ammunition allowances are as below :

	Rounds.
For each 8-inch breech-loading rifle	70
For each 5-inch rapid-firing gun	130
For each 6-pounder rapid-firing gun	500
For each 1-pounder rapid-firing gun	750
For each 37-millimeter revolving cannon	2000

TORPEDO OUTFIT.

To have six torpedo-tubes. Two in the bow are fixed, diverging with the angle between them 5 degrees, and arranged to work simultaneously. The remaining four are training tubes, two on each broadside, one forward and one aft. Fifteen torpedoes to be carried.

EQUIPMENT AND OUTFIT.

Two military masts, to be fitted with fighting-tops. Derricks to be fitted for handling boats, including two 60-foot vidette boats. The complement of officers and men is fixed at four hundred and thirty.

MOTIVE POWER.

Boilers.—Under forced draught to supply steam of 160 pounds pressure for main engines and auxiliaries. The main boilers to be eight in number, 15 feet in diameter by 18 feet 6 inches long, double-ended, with six furnaces and a common combustion chamber. The boilers to be arranged in three water-tight compartments, separated by transverse coal bunkers. Aggregate heating surface of main boilers to be about 36,500 square feet, and corresponding indicated horse-power, 20,250.

Engines.—The main engines to be four in number, two on each shaft, and of the vertical, triple-expansion, three-cylinder type. The cylinders to be $33\frac{1}{2}$, $50\frac{1}{2}$, and $75\frac{3}{4}$ inches in diameter respectively, and the stroke, 40 inches. The revolutions corresponding to the proposed maximum speed of 22 knots to be one hundred and thirty-six per minute, when it is expected that the main engines will develop about 20,250 indicated horse-power. The piston speed at the proposed maximum revolutions will be 907 feet per minute.

Donkey.—A donkey boiler to be carried on the armor deck. To have about 22 square feet of grate surface, and about 690 square feet of heating surface.

COAL ENDURANCE.

The coal supply at load draught is 900 tons, and the total coal capacity is estimated at 1600 tons. On the coal at load draught the steaming distance at 10 knots is 8600 miles. On the total coal capacity the steaming distance at 10 knots is 15,000 miles. At full sea speed of 20.8 knots the steaming distance on the total coal capacity is 2100 miles.

ASSIGNMENT OF WEIGHTS.

	Tons.
Hull and fittings (including conning tower)	2,800
Protection :	
Armor deck......880	
Water-excluding material......190	
	1,070
Propelling machinery, donkey, stores, and spare parts	1,815
Coal at load displacement	900
Armament, its protection and ammunition	362
Equipment, outfit and stores	550
Total	7,497

ESTIMATED COST.

Hull and protection..$2,084,000
Machinery 1,260,000
Armament and ammunition...................................... 400,000
Equipment and outfit... 197,500

Total...$3,941,500

SECOND-CLASS PROTECTED CRUISERS OF 5400 TONS.

Principal Dimensions, etc.

Length over all... 337′ 6″
Length on load water-line....................................... 335′ 6″
Length between perpendiculars.................................. 324′ 0″
Beam extreme.. 53′ 6″
Mean draught.. 20′ 7″
Displacement in tons 5400
Free-board to top of berth-deck plank at side.................... 5′ 6″
Free-board to top of main-deck plank at side .... 13′ 6″
Maximum indicated horse-power (forced draught) 12,000
Maximum speed in knots... 20
Estimated metacentric height................................... 4′ 25″

HULL AND PROTECTION.

To have a double bottom running well fore and aft and to the turn of the bilge on either side. The engines to be in four compartments and the boilers in two, with athwartship coal bunkers. Wing coal bunkers to be fitted in addition, and the subdivision forward and aft to be as close as the requirements of stowage, etc., will permit. To have a long forecastle and ordinary poop (a preferable arrangement for this ship may be with central superstructure and free-board forward and aft to main deck only).

· The buoyancy and stability to be protected:

(1) By an armor deck with flat crown 1 foot above the water-line and 2½ inches thick. This to slope to the side with two slopes, the inner of 22 degrees and the outer of 36 degrees. Plating on both slopes is to be 4½ inches thick. The edge of the armor deck to be 4 feet 6 inches below the water amidships.

(2) By a belt of water-excluding material worked outside sloping water-tight wing bulkheads between the armor deck and the berth deck, as shown. The belt to be 10 feet thick at the water-line amidships, tapering with the water-line, as shown, to a thickness of 5½ feet, which is continued forward and aft, and divided at suitable intervals by partial transverse bulkheads.

PRINCIPAL DIMENSIONS.

Length between Perpendiculars, 324'—"

Beam Extreme, ... 53' 6"

Mean Draught, ... 20' 7"

Load Displacement in Tons, . . . 5400

Maximum I. H. P., ... 12,000

Maximum Speed in Knots, . . . 20

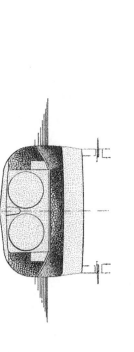

(3) Within these wing bulkheads coal to be stowed above the armor deck, as shown, in wake of the machinery space, and light water-excluding stores forward and aft. The coal bunkers above the armor deck to be subdivided by transverse bulkheads only.

Glacis to be worked around openings in the armor deck for smokestacks, etc., and armor-gratings and coffer-dams as required.

CONNING TOWER.

A conning tower with sides of not less than 6 inches in thickness to be fitted on the bridge or in other suitable commanding position.

ARMAMENT AND ITS PROTECTION.

The battery to consist of two 8-inch breech-loading rifles, twelve 5-inch rapid-firing guns, six 6-pounder rapid-firing guns, four 1-pounder rapid-firing guns, four 37-millimeter revolving cannon.

The 8-inch guns are to be mounted in the middle line, one on the poop and one on the forecastle, as shown. They are to be completely shielded by close shields, as shown, which are to work with the guns. These shields are to have a front of 2 inches, sides of 1½ inches, and rear of 1 inch, the sloping part of the crown in front to be 1½ inches thick, and the remainder of the crown, three-quarter inch. These shields are to be large enough to contain the whole gun's crew. Allowance of weight is made for working these guns by power.

The 5-inch guns are to be mounted as shown, two on the poop, two on the forecastle, and the remaining eight in recessed ports on the main deck, four forward and four aft, the guns on poop and forecastle to be protected by shields on the guns of 2 inches in front and the equivalent elsewhere, the 5-inch guns on the main deck to be protected by 2-inch plating around their ports, with equivalent small shields on the guns to cover the ports.

The smaller guns are to be mounted as shown on the plans, the 6-pounders being protected by shields on the guns. The ammunition allowances are as below :

	Rounds.
For each 8-inch breech-loading rifle	70
For each 5-inch rapid-firing gun	125
For each 6-pounder rapid-firing gun	500
For each 1-pounder rapid-firing gun	750
For each 37-millimeter revolving cannon	2000

TORPEDO OUTFIT.

To have six torpedo-tubes. Two in the bow are fixed, diverging with the angle between them 5 degrees, and arranged to work simultaneously. The remaining four are training tubes, two on each broadside, one forward and one aft. Fifteen torpedoes to be carried.

EQUIPMENT AND OUTFIT.

Two military masts, to be fitted with fighting-tops. Derricks to be fitted for handling boats, including two 60-foot vidette torpedo-boats. The complement, officers and men, to be fixed at four hundred.

MOTIVE POWER.

Boilers.—Under forced draught to supply steam of 160 pounds pressure for main engines and auxiliaries. The main boilers to be four in number, 15 feet 2 inches in diameter by 20 feet long, double-ended, with eight furnaces and two combustion chambers. The boilers to be arranged in two water-tight compartments, separated by transverse coal bunkers. Aggregate heating surface of main boilers to be about 21,500 square feet, and corresponding indicated horse-power, 12,000.

Engines.—The main engines to be four in number, two on each shaft, and of the vertical, triple-expansion, three-cylinder type. The cylinders to be $27\frac{1}{4}$, 41, and $61\frac{1}{4}$ inches in diameter respectively, and the stroke, $35\frac{1}{2}$ inches. The revolutions corresponding to the proposed maximum speed of 20 knots to be one hundred and forty per minute, when it is expected that the main engines will develop about 12,000 indicated horse-power. The piston speed at the proposed maximum revolutions will be eight hundred and twenty-eight feet per minute.

Donkey boiler.—A donkey boiler to be carried on the protective deck. To have about 20 square feet of grate surface, and about 620 square feet of heating surface.

COAL ENDURANCE.

The coal supply at load draught is 540 tons, and the total coal capacity is estimated at 900 tons. On the coal at load draught the steaming distance at 10 knots is 6700 miles. On the total coal capacity the steaming distance at 10 knots is 11,200 miles. At full sea speed of 19 knots the steaming distance on total coal capacity is 1850 miles.

ASSIGNMENT OF WEIGHTS.

		Tons.
Hull and fittings (including conning tower)		2,115
Protection :		
Armor deck	800	
Water-excluding material	140	
		940
Propelling machinery, donkey, stores, and spare parts		1,049
Coal at load displacement		540
Armament, its protection and ammunition		384
Equipment, outfit and stores		380
Total		5,408

ESTIMATED COST.

Hull and protection	$1,542,500
Machinery	715,620
Armament and ammunition	418,000
Equipment and outfit	124,740
Total	$2,800,860

SECOND-CLASS PROTECTED CRUISERS OF ABOUT 4000 TONS

No plans are submitted of this type, of which several examples have been and are about to be added to the navy.

Future vessels of this class, however, should have the special feature of the stability belt, as in the larger types. Their speed should not be less than 19 knots in smooth water, nor their coal endurance less than 9000 miles at 10 knots, as before laid down for the whole class of large cruisers. Their battery should consist of two 8-inch guns at the ends of a long central superstructure, the freeboard forward and aft being to the main deck only. In and on the superstructure at least eight 5-inch and other smaller rapid-fire guns should be mounted. All the guns should be protected as in the other types of protected cruisers.

TORPEDO CRUISERS OF FROM 700 TO 1000 TONS.

This type is the chief representative of the class of small cruisers, and should be developed along with the sea-going torpedo-boat. Numerous representatives of this class have been recently added to foreign navies, such as the Sharpshooter class in the British and the Tripoli class in the Italian navies, while in all European navies there are representatives of smaller types in which the same objects are sought for, but we believe by no means so efficiently obtained.

No type plans are presented of this class, but the following general requirements may be laid down :

The displacements should not be less than 700 nor more than 1000 tons, approaching the higher limit, with lengths of from 230 to 260 feet. The least smooth-water speed which can be considered satisfactory is 22 knots, with a total endurance of about 3000 miles at 10 knots, the coal carried at normal load displacement to be not less than 60 per cent of the total capacity. The gun armament should consist of two 4-inch and at least six 3-pounder rapid-fire guns, the latter being considered large enough for sure execution against seagoing torpedo-boats. The torpedo armament should consist of two bow tubes fixed at an angle of 5 degrees, working simultaneously, and moderately protected by extra plating; and two training tubes elsewhere. It is suggested and believed to be feasible that the torpedo-tubes should be mounted in pairs on the disappearing principle, to be exposed only immediately before and after firing. Every means of incidental protection to the hull, by coal and otherwise, should be utilized, and direct protection afforded in wake of the machinery, so far as the displacement will admit. The subdivision should be very close, and great "emergency" pumping power should be supplied by ejectors in each compartment.

SPECIAL LIGHT-DRAUGHT GUN-VESSELS OF ABOUT 1200 TONS.

These vessels are intended for river service, and are, of necessity, of special type, though it has not been considered necessary to present plans. Their normal load draught should not exceed 9 feet, with a length of about 230 feet. Maximum speed not less than 18 knots, and coal endurance of about 3000 knots at 10 knots. Gun battery should consist of two 4-inch and four 1-pounder rapid-fire guns, four 37-millimeter revolvers, and four 45-caliber machine guns. They should have a high free-board with complete berth deck, and raised quarterdeck and sunk forecastle, affording accommodation on occasions for a very large complement. They should be driven by twin screws with horizontal engines and locomotive boilers, the machinery being under an arched water-tight deck, extending to about 12 inches below water at sides. Subdivision below water should be close, and great emergency pumping power should be supplied by ejectors in each compartment.

TORPEDO-DEPOT AND ARTIFICER'S SHIPS.

The necessity of economizing space and weight on modern warships prevents the installation on board each ship of many of the tools and appliances frequently needed for the repairs of machinery, boilers, etc. When, however, a large number of such ships, constituting a squadron, are employed together, the necessity of supplying adequate means of repair becomes so great that it is advisable to have each such squadron accompanied by a vessel of special design, fitted out with all the appliances requisite for the ordinary repairs of large machinery, and having as a part of her complement a large force of skilled mechanics and artificers. Such vessels are of special importance to the United States, which possesses no foreign dockyards or fortified stations where repairs can be executed, and whose cruising squadrons, in time of war, must be self-supporting in every way.

To supply the deficiency of torpedo-boat service with the offensive is another and hardly less important function of this type of vessel, which carries on deck a number of boats, with facilities for rapidly and easily handling them, and whose workshop appliances are equal to the care of the large amount of mechanism of the boats and their torpedoes and appliances.

Experience has shown that besides their peculiar field of offensive action, torpedo-boats, by their great mobility, are invaluable in blockading or investing a port or harbor. In spite of great increase of size, the "sea-going torpedo-boat" cannot accompany a squadron on any distant operations, while further increase of size would take away from the torpedo-boat many of its essential features. Required to make its own course, therefore, the torpedo-boat is confined to coast and harbor defense or to offensive operations in waters adjacent to its ports, and great advantage in this respect commonly lies with the defense. Battle-ships generally carry two vidette torpedo-boats, but may not conveniently carry more, and these are very necessary for the immediate service of the costly ironclad. In addition, the vessel carries stores of mines, gun-cotton, etc., suited to mining and countermining operations.

With these qualities it is found possible to combine high speed and great endurance, moderate protection, and a light battery of rapid-fire guns, without demanding a very large displacement. The only fair representative of this class in existence is the Vulcan, building for the British Navy. In the vessels recommended it has not been considered advisable to supply so much deck protection as in the British vessel, while retaining all the essential qualities and equal speed and endurance on a somewhat less displacement.

There can be no question of the very great value of such a vessel for the duties described, but for lack of the fleet for which she essentially exists, such constructions, of course, must be postponed for some years.

TORPEDO-DEPOT SHIP OF 5000 TONS.

Principal Dimensions, etc.

	Feet.
Length over all	358
Length on load water-line	357
Length between perpendiculars	346
Beam extreme	52
Mean draught	18⅔
Displacement in tons	5000
Free-board to top of berth-deck plank at side	4
Free-board to top of main-deck plank at side	13
Free-board to top of spar-deck plank at side	22
Maximum indicated horse-power (forced draught)	10,750
Maximum speed in knots	20
Estimated metacentric height	$3\frac{5}{12}$

HULL AND ITS PROTECTION.

To have a double bottom of moderate extent. The engines to be in four compartments and the boilers in three. The ship to be as thoroughly subdivided as the peculiar nature of its service will permit. The machinery, tools, etc., to be carried on both berth and main decks amidships, which should each be 8 feet in the clear for that purpose. It will probably be necessary to place some fittings in wells of the depth of both decks. To have a complete spar-deck. Buoyancy to be protected by—

(1) A curved water-tight deck, 1 inch thick, which is to be (amidships) 5 feet below the water at sides, and 6 inches below the water in the middle line.

(2) A stability belt of water-excluding material worked as in the cruisers.

(3) Coal stowage above the water-tight deck, as in the cruisers.

ARMAMENT.

The armament to be light, consisting of light guns with shield protection only. Number and size of guns to be determined, but total weight of armament, its protection, and ammunition, not to exceed 80 tons.

EQUIPMENT, ETC.

To carry three military masts, with two derricks on each for handling boats. The complement to be as may be determined, but should be unusually large.

MOTIVE POWER.

Boilers.—Under forced draught to supply steam of 160 pounds pressure for main engines and auxiliaries. The main boilers to be nine in number, all of the long low type. Six of them to be 11 feet in diameter by 19 feet long, and three of them to be 10 feet in diameter by 18½ feet long. Aggregate heating surface of main boilers to be 19,000 square feet, and corresponding indicated horse-power, 10,750.

Engines.—To have four sets of vertical, three-cylinder, triple-expansion engines, working two screws. Cylinders to be 27, 40½, and 61 inches in diameter respectively, and the stroke, 31 inches. The revolutions corresponding to the proposed maximum speed of 20 knots to be one hundred and forty-five per minute, when it is expected that the main engines will develop about 10,750 indicated horse-power. The piston speed at the proposed maximum of revolutions will be 750 feet per minute.

COAL ENDURANCE.

The coal supply at load draught to be about 600 tons, and the total coal capacity, at least 900 tons, and as much greater as practicable. On the coal at load draught the steaming distance at 10 knots is 7500 miles. On the total coal capacity the steaming distance at 10 knots is 11,200 miles. At full sea-speed of 18.9 knots the steaming distance on the total coal capacity is 2000 miles.

TORPEDO-BOATS.

Fourteen torpedo-boats to be carried, each of a length of not less than 60 feet. Particular attention to be paid to their stowage and efficient handling.

PRELIMINARY ASSIGNMENT OF WEIGHTS.

Owing to the novel type of this ship it is impossible to make more than an approximate assignment of weights.

	Tons.
Hull and fittings...	2,035
Protection (water-tight deck, 230; water-excluding material, 150)......	380
Propelling machinery, etc...	1,010
Ordnance..	80
Equipment.....,...	450
Normal coal-supply...	600
Fourteen 12½-ton torpedo-boats.....................................	175
Machinery, tools, and stores for work on board.......................	150
Available for torpedo gear, mines, etc...............................	120
Total..	5,000

RAM.

No blow which can be inflicted by any weapon other than a mine or torpedo carrying very large quantities of high explosive can compare in effect to that of a vessel herself moving at a high rate of speed. This has long been recognized, and practically all vessels of war have their bows more or less specially strengthened and shaped for the purpose of ramming. The little experience developed as to the use of the ram leads to the belief that while attempting the operation the other offensive functions of the vessel are best temporarily suspended, and that in ordinary battle-ships and cruisers the presence of other weapons involves qualities not favorable to certainty and efficiency in ramming.

Many propositions have been made looking to the differentiation of the ram from the other ordinary weapons, especially the gun, in the construction of vessels in which all other weapons are, at least, subordinated to the spur. But little has been actually accomplished in this direction, partly from the inherent difficulties and partly because it has been considered doubtful economy to add such vessels to any but a great fleet.

The inherent difficulty lies chiefly in the fact that a number of very antagonistic qualities must be combined in a high degree of efficiency, and especially those of speed, handiness, and capacity to take punishment from the gun. It will be evident that the limits of displacement are very narrow which will admit of a satisfactory combination of those qualities, and it is questionable if any such vessel can be at all satisfactory against more than a single antagonist in daylight. In other words, if out of her proper sphere, or if overmatched at all by more than a single ship, the pure ram loses efficiency very rapidly. For this reason alone a certain amount of gun power is imperative; and, in the shape of rapid-fire guns compara-

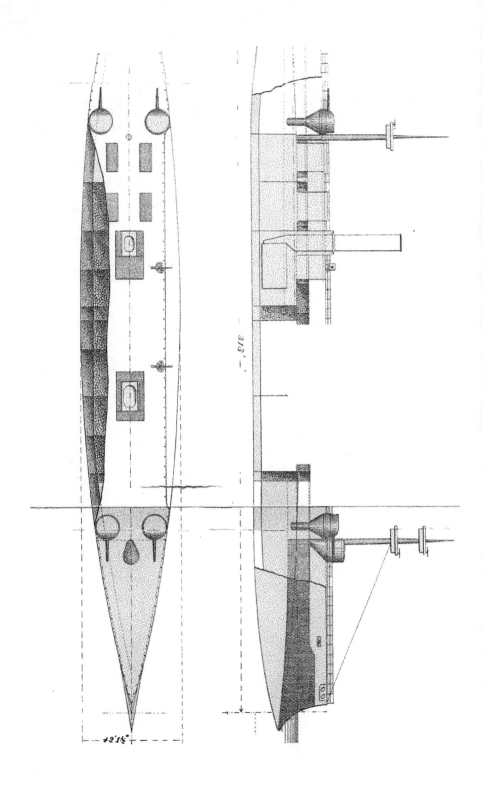

tively well protected and handled independently of the conning tower, is believed to afford adequate return under all circumstances, if only in reducing the enemy's small-gun fire and attacking the unprotected water-line.

Again, in any such vessel it will be evident that the conditions are peculiarly favorable to high efficiency of bow torpedo-tubes, while there seems no valid reason why their presence and use should disconcert the main attack, while it is believed that it may often render it unnecessary and yet more frequently afford favorable conditions. In any case, if they are found to be confusing to the conning tower, they can be served independently of it. Provision must be made for all crews of independent weapons to keep a good lookout and signal to those engaged in their service below to prepare themselves for the shock in time.

Above all it should be noted that, however fine a weapon of this character, the result is peculiarly dependent on the human element in the conning tower.

A design has been prepared which it is believed fulfills the conditions laid down. On a displacement of 3500 tons, a speed of 20½ knots is combined with capacity to turn considerably quicker and in less space than any suitable antagonist. The gun-power is confined to rapid-fire guns entirely abaft the conning tower, but with good command ahead, the larger caliber being in well protected positions. Bow torpedo-tubes are fitted under the protection of the bow armor. The arrangement of protective material is novel and satisfactory against a single enemy, as alone it is possible to make it in such a vessel, but offers fair resistance from all quarters. The light draught renders the vessel peculiarly serviceable in harbor and coast defense, and in the attack by surprise of vessels in port. By suitably arranging the coal stowage the same vessels may form part of a squadron of limited endurance as well as acting in local defense, the increased displacement due to the greater average coal supply still admitting of excellent qualities for the purpose intended; only, when acting within a very limited area, the specially valuable qualities of speed and handiness are improved by loading coal to 300 tons only, stowed entirely above the armor-deck.

RAM OF 3500 TONS.

Principal Dimensions, etc.

Length over all..	330′ 0″
Length on load water-line................................	321′ 6″
Length between perpendiculars................................	313′ 0″
Beam extreme..	42′ 1½″
Mean draught..	18′ 2½″
Displacement in tons...	3500
Free-board to top of main-deck plank at side....................	5′ 0″
Free-board to top of poop and forecastle plank at side...........	12′ 0″
Maximum indicated horse-power (forced draught)	9700
Maximum speed in knots.......................................	20¼
Estimated metacentric height.................................	5′ 0″

HULL AND ITS PROTECTION.

To have a double bottom and as great subdivision as possible, especially forward. The engines to be in four compartments and boilers in three.

The poop and forecastle to extend about 80 feet and be connected by a central parallel-sided superstructure about 30 feet wide.

For a length of 70 feet forward, protection will be afforded by side armor of 4 inches, worked on heavy Z frames with an inner mantlet of 15 pounds. The armor to extend from 4 feet 9 inches above the load water-line to the line of armor-deck, which slopes down to the spur, and throughout the length of this armor will be 2 inches thick on the slopes and 1½ inches on the crown. The lower plates on side next the spur to be 1½ inches thick. The main deck over the armor will be plated with a wide stringer of 40 pounds and the remainder of 30 pounds. The side armor will terminate in a 3-inch bulkhead extending from armor to main deck. A belt of water-excluding material 4 feet wide to extend from this bulkhead to the bow, and between the longitudinal bulkheads of this belt shall be arranged the chain lockers, fresh-water tanks (built in), cable tiers, wet provisions, and in the bows, the torpedo room.

Abaft the armor bulkhead the protection to be continued by an armor-deck, with double-side slopes of 22 and 36 degrees, to be 2 inches thick on the crown and 3½ inches on the slopes. Over boilers amidships the crown of this deck to be about 6 inches above the load water-line, and over the engines, about 15 inches.

Depth below water at sides amidships to be about 4 feet 6 inches, ranging to about 3 feet 6 inches towards the ends until it begins to slope down to the spur. A splinter jacket of 10 pounds to be worked on the under side of the beams of this deck between the longitudinal wing bulkheads.

A belt of water-excluding material about 4 feet 6 inches thick amidships and about 4 feet towards the ends, to extend from armor to berth deck.

Glacis of 3 inches inclined at 30 degrees to be worked to hatches in the armor-deck. Shell-proof gratings to be worked in addition.

CONNING TOWER.

A conning tower to be carried above the forecastle-deck in the middle line. To be pear-shaped in plan. The face, a circular arc of rather more than 180 degrees to an inner diameter of 6 feet, plated 12 inches. The rear an arc of rather less than 180 degrees to an inner diameter of 3 feet, plated 3 inches. Straight sides of 6 inches joining the face and rear. Thicknesses to be joined by tapering the greater thickness into the less across the joint. Height of armor, 5 feet 8 inches. Thickness of crown, $1\frac{1}{2}$ inches; of floor, three-fourths of an inch. To be supported by a cone tube 2 inches thick to main deck, the vertex being in axis of rear circle of tower, which also forms a portion of the foremast. Below the main deck to be continued to the protective deck by a 1-inch tube, and further supported by two radial belt frames under the cone. To be lightly connected at the decks by a ring of plating and light brackets.

ARMAMENT AND ITS PROTECTION.

The battery to consist of four 5-inch rapid-firing guns, four 6-pounder rapid-firing guns, four 1-pounder rapid-firing guns, and three 37-millimeter revolving cannon.

The 5-inch guns to be mounted, two on the forecastle and two on the poop, as shown, in partial barbettes, with shields on the gun and cone protection underneath. Barbettes and shields to be 4 inches or its equivalent, and cone and tube protection, 2 and $1\frac{1}{2}$ inches. The smaller guns to be mounted as shown, with the protection of extra side-plating and heavy shields.

The ammunition allowances are as below:

	Rounds.
For each 5-inch rapid-firing gun	115
For each 6-pounder rapid-firing gun	300
For each 1-pounder rapid-firing gun	500
For each 37-millimeter revolving cannon	1000

TORPEDO OUTFIT.

To have two fixed bow tubes diverging, with the angle between them 5 degrees, and working simultaneously as one weapon.

MOTIVE POWER.

Boilers.—Under forced draught to supply steam of 160 pounds pressure for main engines and auxiliaries. The main boilers to be eight in number, and of the long, low, three-furnace type. Three of them to be 10 feet 9 inches in diameter and 19 feet long. Five of them, 10 feet 6 inches in diameter and 19 feet long. The boilers arranged fore and aft in three water-tight compartments, with two boilers in the forward boiler-room. Aggregate heating surface of main boilers to be 17,500 square feet, and the corresponding indicated horse-power, 9700.

Engines.—The main engines to be four in number, two on each shaft, and of the vertical, triple-expansion, three-cylinder type. The cylinders to be 24¾, 36½, and 55 inches in diameter respectively, and the stroke, 33 inches. The revolutions corresponding to the proposed maximum speed of 20.25 knots to be one hundred and forty-nine per minute, when it is expected that the main engines will develop about 9700 indicated horse-power. The piston speed at the proposed maximum revolutions will be 820 feet per minute.

COAL ENDURANCE.

The coal supply at load draught is 200 tons, and the total coal capacity is estimated at 500 tons. On the coal at load draught the steaming distance at 10 knots is 2000 miles. On the total coal capacity the steaming distance at 10 knots is 4800 miles. At the full sea speed of 19.1 knots the steaming distance on the total coal capacity is 1100. When acting in harbor or other local defense, 300 tons of coal stowed entirely above the deck is to be considered her maximum coal supply, the corresponding endurance being 3000 miles at 10 knots.

ASSIGNMENT OF WEIGHTS.

	Tons.
Hull and fittings, with conning tower (28)	1338
Hull protection :	
4 inches side, 3 inches bulkhead 105	
Armor deck and glacis 515	
Water-excluding material 120	
	740
Armament, ammunition, and protection	177
Machinery, stores, and spare parts	865
Coal at load displacement	200
Equipment	190
Total	3510

ESTIMATED COST.

Hull and protection..	$1,135,050
Machinery..	568,830
Ordnance...	113,300
Equipment..	54,000
Total....................................	$1,871,180

Very respectfully,

W. P. McCann,
Commodore U. S. Navy, President of Board.

R. L. Phythian,
Captain U. S. Navy, member.

W. T. Sampson,
Captain U. S. Navy, member.

Wm. M. Folger,
Commander U. S. Navy, member.

Willard H. Brownson,
Lieutenant-Commander U. S. Navy, member.

R. Gatewood,
Naval Constructor U. S. Navy, member.

Philip R. Alger,
Ensign U. S. Navy, Recorder of Board.

Hon. B. F. Tracy, *Secretary of the Navy.*

APPENDIX A.

ORDERS CONSTITUTING THE BOARD.

NAVY DEPARTMENT, *Washington, July* 16, 1889.

SIR: You are appointed president of a Board which will consider and report as to the policy which should be pursued by this Department in the construction of a fleet to meet the future wants of the United States:

(1) How many years should be allowed for the building of the fleet.

(2) What should be the number of vessels of which the fleet should be composed when completed, both for cruising and for coast-defense purposes.

(3) What classes of vessels should be built both for cruising and coast-defense purposes.

(4) What should be the size and general features of each class of vessels.

(5) What proportion of the entire fleet projected should be constructed annually.

(6) What additional percentage should be constructed annually.

(7) What number and classes of vessels should be asked for at the coming session of Congress.

(8) What will be the annual cost of construction and the aggregate cost of the fleet as recommended.

This Board will consist of yourself as president, and of Captains R. L. Phythian and W. T. Sampson, Commander W. M. Folger, and Lieutenant-Commander W. H. Brownson as members, and will convene at the Navy Department on the 23d instant.

You are authorized to call upon the bureaux of the Department for any information or assistance which you may require.

Very respectfully,   B. F. TRACY,
                                        Secretary of the Navy.

Commodore W. P. McCANN, U. S. Navy,
          Commandant Navy Yard, Boston, Mass.

NAVY DEPARTMENT, *Washington, August* 5, 1889.

SIR: In addition to your present duties you will report by letter to Commodore W. P. McCann for duty as a member of the Board of which he is president, organized for the purpose of considering and reporting as to the policy which should be pursued by this Department in the construction of a fleet to meet the future wants of the United States. When notified by that officer, you will proceed to Washington and report to him in person for the duty.

Very respectfully,   B. F. TRACY,
Secretary of the Navy.

Naval Constructor RICHARD GATEWOOD, U. S. Navy,
Newport, R. I.

APPENDIX B.

INVESTIGATION OF THE MONITOR TYPE.

In a vessel of low speed and very small free-board, such as the monitor as employed during the civil war, it is evident that a greater portion of the displacement can be assigned to armor and armament than in a vessel of comparatively high speed and free-board, though the advantage in this respect is somewhat reduced by the necessity of protecting by armor certain openings, such as smoke-pipe and ventilating hatches, in the absence of other provision against swash as afforded by the sides and subdivision of high-sided ships above the armor deck.

Some incidental advantage accompanies the small target afforded by the low free-board and gun positions, but this has been very much exaggerated in many minds, since damage in the higher unarmored parts of other well-designed armored ships is of little consequence to the vital qualities of the ships, except incidentally, as producing the risk of jamming turret or barbette guns by the explosion of shells within it. The greater target of higher armored parts is more than offset by the greater command of guns, etc., so afforded.

As the speed approaches that of the high-sided ships, the excess of weight available for increased armor and armament diminishes, from the greater fineness of form required. But with the very low free-board, a limit to speed, even in smooth water, is soon reached when the disturbance created by the vessel comes aboard, requiring a considerable fraction of the velocity of the ship to be imparted to considerable masses of water. Thus, such a vessel may not be driven even in smooth water to the same extent as a high-sided ship, and less power may be advantageously exerted in a form of given fineness. No experience of this character exists, but it may be anticipated, especially in the 4000-ton monitor of comparatively high power now under construction for the navy.

This disadvantage of the low free-board type becomes rapidly more pronounced as the water becomes rougher, and especially with the short seas of moderate weather and the choppy seas of large bays.

The low sides of the monitor render necessary enormous stiffness or metacentric height in order to insure even the small range of stability commonly necessary in a sea-way, and, deriving no advantage from the subdivision and presence of water-excluding materials, such as fuel, commonly found in the wings above the armor decks of high-sided ships, any water-line damage is of greater importance as affecting stability and reserve buoyancy. This consideration, together with the great effect of a moderate change of trim on the speed, renders necessary a longer belt of greater thickness, and discounts a portion of the apparent advantage in this respect in existing vessels of this type. Were, however, the monitor so large as to be able to carry armor completely proof against existing guns—as it may be pointed out was practically the case during the civil war—these considerations would not obtain at all, and it will be evident that the advantage of the type in this respect increases rapidly with increasing thickness of armor on a complete or very long belt, or, what is practically the same thing, with the size of the vessel.

The monitor, then, is found to possess valuable qualities where the requirements are smooth water, low speed, light draught, and great heavy-gun fire opposed to the attack of guns not overmatching the armor protection carried in the long or complete belt; and, within the limits of draught permissible for the work, these advantages increase rapidly with increasing size.

Opposed to the fire of guns overmatching her armor, the monitor can stand very little punishment. In a sea-way she has not speed enough to force a combat with any intact ship. But the great and cardinal defect of the monitor type in anything but smooth water remains to be stated, as consisting in the utter lack of steadiness accompanying the great stiffness required for stability and the consequent impossibility of doing any work with the guns. All ships have great stiffness in the direction of their length, and therefore closely follow the effective wave slope in pitching, this slope being, however, very much reduced by the considerable length of most ships as compared with that of all ordinary waves. On account of the great transverse stability of the monitor type, the same conditions obtain in their rolling as in all pitching, except that the beam and draught being small, the effective wave slope is very much steeper. These vessels in rolling, then, closely follow the wave slope, and in such a ship in the trough of waves having a period of eight seconds and a maximum slope of 10 degrees—very ordinary values—guns trained on the broadside would point at intervals of four seconds from 10 degrees above the horizon to 10 degrees below it. Ordinary vessels roll through larger angles, albeit in a greater time, but they undergo periods of comparative quiet when a gun can be laid with some degree of accuracy.

The monitor's artillery suffers the additional disadvantage of being so close to the water that it is only well up a fair-sized wave that another vessel can be seen, while the swash of water over the decks seriously interferes with the working of common turrets, although it will be observed that from so closely following the water surface much better weather is made than would at first be expected from the extreme lowness of free-board. For the same reason, the effective force on a person aboard being at each instant normal to the wave slope, the motion is much less apparent than in ordinary vessels, and a common short pendulum tending to set itself at right angles to the wave slope gives no indication of the motion of the ship with respect to the true vertical, so that the true roll of these vessels has been overlooked.

The surprise consequent on the comparatively good weather made by monitors in riding out heavy gales and afterwards in long voyages, and the very good work done by them in the attack of fortifications under conditions approaching those of their maximum efficiency, caused a reaction in naval opinion, originally unfavorable, until extravagant opinions were even expressed of their efficiency as cruisers for work on the high seas.

Under certain conditions of the attack and defense of harbors, the Board recognizes real advantages in the type. But it has no efficiency in protecting the coast and commerce from without the harbors, which, in the opinion of the Board, is the chief naval problem. Being further incapable of rapid mobilization and concentration under all conditions of weather, coast-defense by ships of this type would really amount to local defense of the principal harbors, when the enemy had been allowed, by the absence or defeat of our fleets on the high seas, to invest them.

That the same measure of defense can be obtained by very numerous and powerful sea-going fleets off the coast supplemented by monitor· and shore defenses of the principal harbors, as with the vessels proposed by the Board, is unquestionable, but we are of opinion that greater security, with far greater economy, can be obtained by the type of battle-ships of limited endurance, capable not only of affording local defense to the harbors, but of rapid mobilization and concentration under all circumstances, and of acting with perfect confidence on the high seas.

In the opinion of the Board, the six vessels of the monitor type already provided for afford a sufficient proportion of purely harbor-defense vessels.

THE PROCEEDINGS

OF THE

UNITED STATES NAVAL INSTITUTE.

| Vol. XVI., No. 3. | 1890. | Whole No. 54. |

[COPYRIGHTED.]

U. S. NAVAL INSTITUTE, ANNAPOLIS, MD.

NAVY BOATS.

BY ENSIGN ALBERT A. ACKERMAN, U. S. N.

PART I.

The writer has freely employed information with regard to coast life-boats, derived from the publications of the U. S. Life-Saving Service; the Royal Life-Boat Institution of Great Britain; "Life-Boats, Projectiles, and other means for Saving Life," by the late R. B. Forbes; 'The Life Boat," by Captain J. R. Ward, Chief Inspector of Royal Life-Boat Institution; "History of the Life-Boat and its Work," by Richard Lewis; and "Specifications for U. S. Life-Saving Stations and Surf Boats."

With regard to the use of life-boats aboard ship little has been written, and that by either landsmen or those unacquainted with the peculiar needs of a man-of-war. Still, the writer can claim to have developed but little that is new. It has been his effort to state the case plainly and justly, to form an unprejudiced opinion, and to bring out points of practical and direct application aboard ship. The temptation to touch on the general features of policy and organization has been irresistible; but once committed, their criticism has been actuated by a sincere desire for improvement. Since the need of change has been recognized by the appointment of a "Board of Reorganization," many of these points will have already received attention.

It was only the other day that one of the editorials of a great engineering journal, in speaking of technical education, said: " Men, especially young men of the student's age, are not machines; they want some living interest to give them a zest for their work." In the same way, some fragment of discretionary action must be left every man aboard ship. The mind of a " blue-jacket," to make him fit for his place, must be made to work, as well as his muscles. The " points of isolated responsibility " aboard ship are far less *responsible* in drill than they will be in action—far less isolated than when it may cost one's life to go from one to another.

Give the young officer, the seaman, the apprentice, a chance to try himself; let him learn how to wield responsibility, not under the direction of his superior with all the authority of the ship at his back, but alone, when he must gain his point through the force of his own individuality, correct his own mistakes, and, in general terms, make a man of himself.

It is believed that there is no surer way to cultivate self-reliance and resolution in a modern man-of-war's crew than by boat drills at sea and frequent landing through the surf.

There is no space on a modern man-of-war allotted to sentiment; qualities must be limited by *existing* essentials rather than precedents, no matter how honorable and efficient they may have been in another day and generation. Changes based on necessity must continue to be made so long as there is a navy; they are not a reflection on the customs of the past. Each generation has its own advance to make.

Apart from sails and spars, there is, perhaps, no adjunct of the old-time man-of-war around which tradition lingers more fondly or about which a sailor's proper pride is more exacting than the boats. It may be said that the service would be anything but pleased by the substitution of new and safer types lacking some of the features that have characterized men-of-war boats for the past hundred years. They are swift and handsome. No one has watched them threading their way among the mongrel craft of a foreign port but he has felt a patriotic glow of satisfaction in their trim appearance. Their speed and grace is an old story the world around; not a ship in the navy but has its pet boat, not a blue-jacket but has reminiscences of racing triumphs.

And yet how many ships dare lower a boat in rough weather? How many wrecked crews have been saved by navy boats? That there have been such instances make them all the more remarkable for skill and courage. Instead of the exultation at their victory in the more peaceful contests, would it not be a nobler thing, more consistent with the purpose of a navy, if they were less feeble in their struggles with relentless winds and sea?

Is the present state of affairs to continue, when hundreds of lives have been lost in a few years past, in wrecks on coasts *where the stress of sea and surf yet permitted a few stray men to swim ashore?* The officer is lucky indeed who has not in some way been closely associated with or present at a fatal boat accident. The list of navy dead from this cause is growing stealthily but surely. On every coast, in the smooth harbor waters and in the booming surf, good men's lives have been paid as the price of an assumed military appearance and smooth-water speed in ship's boats.

There are those who will deny this: some minds leap to a stern judgment of all connected with disaster, no matter how defective experience, strength, and facilities may be. The truth is that the sphere of usefulness of navy boats has been so long and inflexibly limited by their defects, that many do not understand that there is a wider field for them. And here it may be said that in no other department of the respective experiences that go to make up the efficiency of different naval officers, is there a greater variety than in that of the handling of boats. The duty of " boat officer " is generally assigned to certain grades, and to some extent it is fair to suppose that the practical experience gained is proportional to the time spent in these grades. Of course there are many brilliant exceptions; boat duty is a very different thing on different stations and ships. For an entire cruise, it may be sometimes said that the boat officers are no more than sentries guarding the boat's crew against the enticing dangers of the beach, rather than from those of the harbor. No disparagement of this duty is intended; its necessity is lamented, but most junior officers will declare it to be one of the most instructive, if most unpleasant, lessons in the control of their men. Not much, however, is learned of the management of boats.

In bad weather, boat service is limited to the necessities of the occasion, not only because it is not desired to inflict hardship on officers and crew, but because the boats are known to be unfit for such work. So, however valuable the experience might become, the actual danger involved forbids the attempt to gain it.

Of all the branches of this encyclopedic profession, seamanship lies closest to the heart. It requires self-reliance, decision, and courage to deal with the grand forces of nature. The exhibition of ignorance or error in its practice, no matter how meager the experience or scant the opportunities of learning, is guarded against by many a young officer with anxious dread.

There is too much of this, for, in consequence, many stand in danger of blame for lacking practical knowledge which it has never been their opportunity to gain.

Before deciding that navy boats are of necessity fit only for smooth weather and light loads, the qualities essential to all-around seaworthiness should be analyzed, and the attempt made to combine their essentials in a boat able to perform routine work and be carried at a cruiser's davits.

No higher degree of seaworthiness can be found than in the heavier life-boats employed along some parts of our coast, and especially that of Norfolk and Suffolk in England, where it is necessary to go off under sail to reach outlying wrecks. Many of these boats are improved variations of the celebrated Beeching life-boat, the first self-righting boat ever built, and which won the prize of 100 guineas offered by the Duke of Northumberland in 1849, for the best model of a life-boat. This fact did not make them entirely safe, however, for two were lost through the shifting of water ballast while under great press of sail. The water ballast had leaked out of its well into the air compartments.

The report made by this committee of award is undoubtedly the best original publication on the subject of life-boats suited to the coasts of Great Britain. It has been largely drawn upon in this article. It must be borne in mind, however, that the coast with which this committee dealt is thickly peopled, and many willing hands may be found for manning and launching heavy boats. There are frequent harbors or shelters from which the boat can be pulled out to sea without launching through the surf, and in some parts it is even possible for a powerful tug to tow the life-boat to the vicinity of the wreck. Another point with regard to this report is that while the coast presents a great variety of aspects, and the duty of the life-boat will vary in character almost as much, the committee could give but one prize—to the best *all-around* competitor. This, of course, makes it more interesting to naval people, though the use of the boat from the shore as a base was alone considered. There were 280 models and plans submitted; the Beeching boat received a mark of 84 out of a possible 100; the next two competitors, 78 and 77. Only 36 received over 60. It is apparent, therefore, how damaging this method of determining relative standing was to special or local merit, especially when it is known that many of the boats were built for localities where they are still used.

The values assigned the various qualities by this committee are as follows:

Qualities as a rowing boat in all weathers 20
Qualities as a sailing boat............. 18
Qualities as a sea-boat as to stability, safety, buoyancy forward
    for launching through a surf, etc...................................... 10
Small internal capacity for water up to thwarts....................... 9
Means of freeing boat of water readily......... 8
Extra buoyancy; its nature, amount, distribution, and mode of
    application 7
Power of self-righting... 6
Suitableness for beaching......... 4
Room for, and power of carrying, passengers......................... 3
Moderate weight for transport along shore 3
Protection from injury to the bottom 3
Ballast, as iron 1, water 2, cork 3... 3
Access to stem and stern......... 3
Timber heads for securing warps to..................................... 2
Fenders, life-lines, etc.. 1

                                                                        ——
                                                                        100

Had the committee also been obliged to mark according to merit as a *ship's boat*, it is easy to see that no boat could be built *apparently* as efficient as the Beeching boat, through the introduction of limiting qualities of size and weight, and the high values assigned to them.

The late R. B. Forbes, an authority on the various appliances for life-saving, published a small volume in 1872 on life-boats.

He pointed out that, however fine a boat that of the R. L. Institution's was for the English coast, it was entirely too heavy and costly for the long uninhabited reaches of our open coast. Again, he notes that local customs must be respected if anything is to be accomplished; "seamen brought up on the coast have their own ideas as to the kind of vehicle they require when they are to risk their lives

to save life and property." For the New Jersey coast he recommends " a flat bottom, broad stern, sloping sided, cedar, or light corrugated iron boat, with considerable sheer. To possess detachable
end air tanks, and a solid cork fender, fastened half way between the
gunwale and the load-line. The latter will help when water has been
shipped, will make capsizing difficult, and righting almost as much
so. . . .

    ' If built of light wood, thirty feet long, eight feet wide, and three
feet deep, it need not weigh over 1000 pounds. The chances are
that a boat of this weight can be got off and on through a surf, the
men being supplied with life-belts, when no boat called a life-boat,
and deserving the name, could be manned and got off *on the Jersey
coast by Jersey beachmen.* . . .

    ' The same may be said of the coast of Long Island. . . .

    ' At Nantucket, where the surf is heavy and the beaches frequently
steep, the surfmen prefer light wooden clincher-built boats with large
beam and considerable sheer, pulling long oars, single bank, and
steering like whale-boats. . . . With end air cases and stout cork
fenders, these boats will stand up with considerable water in them,
and in case of being stove in they can be got ashore with some risk
and difficulty. . . .

    " Near Chatham and Orleans on the Cape shore, the surfmen use
for their fishing and wrecking purposes full-built, flat, round-bottomed cedar-built boats, both ends alike.

    " They naturally feel confidence in them, and will risk their lives
in them when nothing would induce them to attempt to launch a
life-boat weighing a ton only. . . .

    " At Provincetown they have boats more like whale-boats, only
with more beam, fair sheer, and very much shoaler than the Orleans
boats—very light-built boats that pull five oars, and that can be picked
up bodily by five or six men and run off to meet the rollers. It
requires considerable urging to induce them to admit any extra
buoyancy, or even to put on their belts. . . . At Cape Ann the
seamen have nearly all been long accustomed to handling dories
pulling short cross-handled oars, the largest pulling four and the rest
two oars. A well managed dory will get off a rough beach dry when
no other boat can. They are useful for saving life at times by
carrying off a line, etc., but they are worthless for landing numbers of
clumsy passengers."

    Mr. Forbes further suggested a boat that would not antagonize the prejudices of the coast men, and for which he claimed durability, economy in the long run, and ability to command the confidence of the men who are to use them.

It is a "light corrugated iron boat, 25' x 6½' x 30", with 15" sheer; flat midship section, not very fine ends, centerboard, and to pull, at Nantucket, five oars single banked, elsewhere, six short oars double banked and one single at the bow thwart. The standing room should be, from sternpost becket to after thwart, about 8'. Much depends upon the steering oar, and it is important to have plenty of room to operate it freely. [Allowing 3' between centers of 12" thwarts, this would give foresheets of 4' at Nantucket, and elsewhere, 7' 6".] Cork fenders along sides below oars, yet clear of water; on each side, under thwarts and close out to side, a cork float 12' long, 12" to 15" in diameter, to assist in trimming and floating the boat if stove or filled; in the ends, air tanks or a mass of cork to *assist* in righting the boat. If the attempt is made to add any more extra buoyancy, the boat will be too heavy to launch and transport. Everything should be made detachable, so that if the boat be found too heavy, a part or all can be temporarily removed."

A sufficient abstract has been given to show that the *basis* of a combination of qualities desirable in a ship's boat has long been in existence.

Before attempting any original design it will be well to analyze these valuable qualities in the light of naval requirements and experience. It appears that all that has been written with regard to life-boats, except the circulars of a few interested inventors, assumes the shore to be the base of operations. Writers on seamanship apparently consider their task limited to instructions how to avoid being wrecked, for none of them attempt to discuss that last terrible evolution.

Other points being the same, a boat's seaworthiness increases with its size; however much, therefore, that of the ship may be deranged by considerations of speed and battery, the boat already has the odds against it, and if it is intended as a last refuge of safety, the seaworthiness should be considered above all other qualities.

From the report of the Life-Boat Committee appointed by the Duke of Northumberland, it appears that the most desirable proportions of a coast life-boat are:

"Length, 100; beam, 30, or for tideways, 25; sheer, 8¾, *i. e.* 1″ for each 1′ of length; weight, 1 cwt. for each 1′ of length; extra buoyancy, 1 ton for each 10′ of length. The internal capacity for water up to level of thwarts to equal the extra buoyancy.

Area of delivery valves to be one square inch to each cubic foot of internal capacity."

From other sources it is learned that the best stability of a parallelopiped is obtained when the *draft equals one-fifth the beam. The height of sides above water should be one-fourth the beam.* The best combination, therefore, for stability would be: *draft = three-twentieths beam; freeboard = four-twentieths beam.* That cross section of the immersed body which offers the least resistance to side motion, and this is very important when broached to, is *elliptical.*

These general features of satisfactory life-boats are interesting, if for no other reason than to show how much they differ from navy boats.

Navy boat-builders, well informed of the prejudices of the service, have generally preferred, in their so-called life and surf boats, to graft a few qualities on the service boat rather than make a radical change. This position seems to be taken in deference to the vanity that places smooth-water speed and trim appearance above seaworthiness.

Fault is not found with the quality of speed, but the method of obtaining it which limits it to smooth water. A good life-boat, but little slower than a whale-boat in the harbor, would easily pass the latter outside. The reason is obvious : speed is improved by rounding off all angles under water. Cambering the keel and raking the stem increase the speed directly. Of all midship sections the semicircle is the fastest; the sharper the bows the better; for the same beam, equal increments of length nearly equally increase the speed.

If these features are developed beyond a certain point, the boat will of necessity have very little stability, being very fine, round-bottomed, and so lively that the weight of one man can dip the gunwale. All the weights must then be carried low ; so low that less freeboard can be allowed than is usual in any but light harbor boats. For fear of top-heaviness, the sheer is made less than one-half what it should be, or is a counterfeit, a mere extension of the sternpost and quick upward turn of the gunwale, with very little benefit to buoyancy. On the other hand, the very feature of a low rail, however dangerous at sea, is of advantage to speed in smooth water, for long-bladed oars, with little slip, can then be used. These, lying low, permit the oarsmen to exert all their strength in *propelling* rather than in *lifting* the boat.

To obtain speed, the lines of flotation of navy boats are made fine, and thus in an instant much buoyancy is lost. The area of the water section being small, a comparatively slight addition of weight, of either cargo or water, largely diminishes the freeboard and renders the boat more liable to ship water than before.

Of course all navy boats must be expected to work in strong currents, and this requirement should be met, in some at least, by *increased length*, to permit finer lines and more oars, rather than by *decreased beam* and consequent loss of power. The size of such a boat, however, would frighten the builder, who would plead for a reduction of beam ; and, if beam is reduced, away go depth and sheer, for the boat, standing high out of water, with cambered keel and flaring bow and stern, would be crank.

This then is the trouble with navy boats ; they will all stand some increase of freeboard, but if given enough, and in addition the proposed allowance of gunwale sheer, their stability is gone, unless the beam is also increased and the bottom flattened. Such a change, however, would indicate a desertion of traditional lines and loss of speed, especially in currents.

As a rule, while life-boats resemble cutters amidships, their bottoms are flatter, and the midship section carried over a larger body. Heavy life-boats almost always have the keel slightly cambered ; if too much so, the boat becomes too lively, pitches, loses speed, and is less steady under sail.

Speed is absolutely essential for surf boats ; without it all the other good qualities of the most perfect boat would be as useless as Saul's armor was to David. It can be obtained in them, without too great loss of stability, by sharp curves at the bow, shallow bearing, and moderate beam. Weight is the only constant element in the persistence of the force that drives the boat against wind and sea. As the resistance increases, the velocity decreases ; weight is, therefore, to a limited extent, an advantage. The bows should not oppose too bluff a surface to the sea, and the end air-cases, if high, should be rounded to offer the least direct resistance to the wind. The crew, too, should be one of power, firmly seated, with sufficient room between the thwarts to permit the independence of action made necessary by the impossibility of precision of stroke.

Speed under sail.— Most navy boats will sail well only with a number of the crew on the weather gunwale and steady bailing by the rest. They *should* sail well even in rough weather without going to this hazardous extreme. One difficulty generally encountered is that the authorities, aware of the crankiness of the boat, cut the sails down until they are "safe" for the most inexperienced coxswain. This, of course, destroys in a great measure the usefulness of the boats under sail, neutralizes available skill and experience, and is a great bar to improvement. It may be truly held by those who, with hand on tiller, feel the boat spring and worm her way onward under all the sail she can stand, now courtesying to a sea, now coquettishly flinging back a veil of spray, that here is opportunity for much careful concentration of mind and sympathy of eye and touch. Cultivate this, and keep full sail power.

Wherever the bearing of the boat can be increased, whether by a wider beam or flatter floor, without much decreasing speed under oars, it should be done; stability will be increased, and greater stability means power to carry more sail.

Stability varies nearly as the cube of the beam, and, less accurately, directly as the length. The greater the beam, the better chance a boat has to ride a breaker safely when broached to, as it is not only more buoyant, but the couple to resist capsizing is greater. Stability is therefore more necessary to life-boats than any other quality. It is better obtained by increased beam, especially in sailing vessels, than by lowering the center of gravity. "The beam of some sea fishing boats is as great as $\frac{11}{28}$ the length. The greatest beam in birds is $\frac{2}{5}$ the length."

Stability is also helped by ballast. "The boats of the National Life-Saving Institution of Great Britain get it with the aid of an iron keel, flatness, and length of floor, and only moderate beam." With regard to a ship's boat, there should be either a heavy keel, sideboard or centerboard, or considerable ballast. The keel would make it difficult to land on a flat beach, or to turn quickly to meet a breaking sea. The centerboard, if used, must be considered in all operations, but its value in connection with the associated qualities of flat bottom, great beam, and light draft makes it seem the best choice. The sideboard is best in that type in which it is desired to keep the weight of a boat down to the lowest point, to permit a small crew to handle it on the beach. As for ballast, the kind having in this case the least objectionable features is water. Its weight need not be transported, and, if confined securely to a reservoir, fitted to fill or be pumped out at will, it has many advantages. The ballast tank should have several compartments to prevent the water shifting in case some leaks out.

"Some of the English sailing life-boats are too heavy to be handled by oars alone in bad weather, on account of the large amount of ballast and draft of water required to enable them to stand up under sail. The ballast is generally water filling every unoccupied space up to the water-line. At first this shifting water seems dangerous, but there is so much that the boat cannot be lively, especially as it is cut off (by the air tanks) from access to the ends or sides of the boat. If these boats were more lightly ballasted, they would answer every impulse of the waves, and the water would continually emphasize this by running toward the lowest point. But heavily weighted and propelled by powerful sails, they cut through every sea instead of rising to it. They are frequently completely submerged and the crew hurled about, but they are stable. . . . One was once upset running for the shore; most of the water having been pumped out, she ran her bow into the hollow, which tilted the remainder of the water to that end, when she broached to, turned over, and broke off her masts, remaining keel up, not being self-righting. Had she been properly ballasted, the sea would have broken over her harmlessly. Since that time she has been ballasted with wood."—(Captain Richard Lewis, "Life Boats.")

It is thus seen how the important quality of speed can be assured only by stability, giving certainty to and room for oar action, and ability to carry considerable sail. Further, it has been seen that stability in the lighter designs is obtained by large shallow bearing surface and increased beam, the latter undoubtedly somewhat at the expense of speed, through increased resistance to wind and sea. In the heavier design, the beam is about one-fourth the length, the lost stability being made up by the addition of ballast in the shape of a metal keel, water, or cork. This increases the whole weight—within limits an advantage—and is also a step towards self-righting.

Buoyancy.—While ships are divided into numerous compart-ments provided with double bottoms and the most powerful pumps, the only boats having extra buoyancy are the steam launches, which cannot be lowered or managed at sea in bad weather. Extra buoyancy has been defined as that buoyancy which the boat possesses after the materials composing it are floated. The more that can be obtained of it, after providing space for the crew and passengers, the better. Not only is the supporting power of the boat increased so much, but its capacity for water is diminished. The manner of its creation and its location are the most difficult problems in designing a life-boat. If too much is placed in the end air tanks, their weight makes the boat unsteady, a large surface is exposed to the wind, the ends of the boat cannot be approached without climbing on them, and finally this buoyancy does not act until the boat is very deep in the water.

If placed on the sides, it becomes almost impossible to capsize the boat, but should this occur, it would be almost as difficult to right her. If placed on the bottom, there is a danger of raising crew and cargo so high as to make the boat unstable. In the Beeching and other life-boats, types of which are in use in many countries, the distribution and amount of this buoyancy is such that while the boat can be capsized only with difficulty by means of a parbuckle, it will quickly right itself when released. This is owing to the turning action of end air tanks, ballast, and iron keel. The false bottom of these boats is placed above the water-line, to afford drainage of water shipped. The space beneath is occupied by ballast and buoyancy tanks, usually filled with cork. A reserve of buoyancy is obtained from a large cork fender running all around the boat just above the load water-line. This adds to the weight, but helps materially when the boat is temporarily swamped and until the water is drained off.

One point must be remembered—with every increase of buoyancy, once having settled its disposition and nature, the weight is increased. Generally speaking, buoyancy is best obtained by detachable air tanks in the upper and cork ballast in the lower parts, thus avoiding the danger of a loss of buoyancy or shifting ballast from a leaky bottom. Some of these tanks, when filling in the space between the under side of the thwarts and the false bottom, are doubly valuable, as they serve to break up and distribute the weight of a sea that may be shipped.

The detachable air tank is of course far heavier than the one built in and forming a part of the skin of the boat. The latter is objectionable—indeed, as bad as none as soon as the boat is stove or becomes leaky; but, until that time, it has all the advantages of the other with the additional one of lightness. Several of the boats, including the first two, most highly spoken of by the committee quoted above, had built-in air cases. They should not, therefore, be rejected without consideration, if it should appear that in no other way can the weight be kept down to that allowed a ship's boat. Most of the life-boats of ocean steamers have built-in end tanks; a few have, in addition, detachable cylinders about 7 inches in diameter under the thwarts near the sides. Built-in cases should be made very strong and with subdivisions, as a rent in the outside skin from rocks or jagged wreckage, or even a leak permitting the compartment to gradually fill, brings as much pressure to bear on its walls as if it were the outside of the boat itself. "At Winterton, Norfolk, 1829, a life-boat, having great camber of keel, was sent off to a wreck. When filled with people, their weight sunk the boat so low that the pressure of the water caused the false bottom to 'blow up.' Whence we may suppose that the outer bottom was leaky and the inner one flimsy yet air-tight.

Self-bailing.—If the boat is not made self-bailing, then precautions to keep the water out are all the more important; these will be discussed under the head of *Freeboard* and *Sheer*.

Self-bailing is commonly effected by closing off all the submerged part of the boat into air or ballast spaces by a false bottom. Above this bottom, ports or scuppers are cut into the side, letting the water run off as fast as it is shipped. Another and more convenient way, as the boat may be kept dry in ordinary rolling and pitching, is to run drainage tubes down through both bottoms between every pair of thwarts. These tubes are furnished with automatic valves opening downward with the weight of water above. They may also be locked or plugged, should it be desirable to increase the ballast when under sail. Some designs employ both methods. With regard to the second method, there are instances noted of the loss of boats through having grounded and choked the drain tubes on shoals and flat beaches.

Self-bailing should be very rapid in order to permit the boat to free herself of one sea and get way on before being struck by another. This can be accomplished only by keeping down the internal capacity of the boat for water, making the freeing ports or valves large, and by raising the floor as high as stability permits, so as to increase the head of flow.

The buoyancy of the submerged portion of the boat must be very great for little draft, in order to hold up boat, crew, and additional weight of water without decreasing too much the head of outflowing water. The required buoyancy can best be obtained without increase of draft by wide beam and long, "flat-round" floor with successive breakers had tumbled considerable water into the boat. Then, when the water was nearly up to the thwarts, the bow dipped under every swell, and the boat seemed viciously undecided as to the side over which it would roll. Had these boats been self-bailing there would have been no difficulty; though, for that matter, that change would have necessitated an entire change of design.

Self-righting.—This quality requires the center of gravity to be kept low, and that of buoyancy high, so that when capsized, the couple formed by gravity and buoyancy will be sufficient to right the boat.

The length of the arm of this couple is increased by concentrating ballast in an iron keel, sometimes weighing over a ton, and raising the end tanks up to and even above a well-sheered gunwale. Opposed to this couple will be two moments, that of a part of the half beam into the buoyancy of the side that is to go under in righting, and that of a part of the half beam into the weight of the side that is to go up. These are reduced to insure self-righting, by narrowing the beam and diminishing the side buoyancy.

The heavy keel neutralizes some of the buoyancy, but it gives back the stability jeopardized by a sacrifice of almost all side buoyancy. The increased weight is certainly of advantage wherever momentum is desirable. Its inertia, however, renders the boat more sluggish with the same power; more liable to be caught in a dangerous position; less likely to benefit by a sudden smooth time in the breakers.

The narrowed beam makes the iron keel a necessity; in fact, makes self-righting a more important quality; but it permits greater speed, and with it, the heavy self-righter holds the palm of power. It is faster at all times in currents. ·

To equal it, the lighter, shallower boat must put more men on its oars; however, the advantages of quickness and handiness go far towards equalizing its efficiency in a seaway through other means.

The loss of side buoyancy is a great blow to sail power; it can be made up only by such a great addition of ballast as to eliminate at once these boats from consideration as ship's boats.

Raised end air-tanks are of use, not only when the boat is capsized, but when it penetrates a breaker, for then its whole buoyancy is felt in lifting the boat; water ordinarily will not break over the bow or stern into the boat. They are also valuable in keeping the weight of crew and water out of the extremities when the boat is swept by a sea. These high ends of course affect the weatherliness of the boat in a strong wind, but is is claimed that this defect has never, of itself alone, prevented the accomplishment of any undertaking of the British life-boats. "Again, it was foretold that, once thrown out of them, the crew could not regain them, since, being so high out of water, they would be speedily carried to leeward out of their reach. In only one of thirty-five upsets has this been the case. Even when under sail, with sheets fast, they have righted and the crew regained them without loss. The one exception was when a boat was turned end-over-end and carried off by the same wave. In other instances, when broached to, one or more men always stayed by the boat, the others being tumbled out to leeward."—(Capt. Lewis, "Life Boats.")

In spite of these strong statements, it would seem that there is not much hope for the success of *the particular trip or undertaking* if a capsize occurs in rough weather outside. The certain demoralization, however slight, of the crew must leave the boat in a less manageable condition after they have climbed back than it was before the capsize. The advantage would therefore seem to be all in favor of the boat least likely to capsize in the first place, unless the circumstances causing the accident are exceptional and not likely to recur.

It is clear that if the boat is not made self-righting, every precaution taken to prevent capsizing becomes more important.

Depth, Freeboard, and Sheer.—The real advantage of stability of form over that of a low center of gravity is, that with a light draft it is possible to carry plenty of freeboard and sheer. By these the water is kept out, the end buoyancy is increased, and space given for clean pulling, or considerable heel when under sail.

The relative advantages of the same freeboard and sheer are much greater in a light boat than they would be in a heavy one.

Waves vary greatly in size and abruptness; a boat will be moved sometimes *on* a large wave, while a smaller one would tumble over the rail. In other words, the greater the inertia of a boat of constant freeboard, the more certain it is that the same wave will come aboard.

With regard to *sheer*, if the rail is straight, there is less surface at the bow for the buoyant action of the coming wave. The rail is low, and before the weight of the forward half of the boat can be transferred and lifted, the crest falls into the boat. The more abrupt the transition from hollow to wave, the more certain this is to occur, making great sheer very valuable in surf boats. For the same waves, the greater the sheer, the more certain it is that the boat will rise to the sea before it tumbles in. If, then, the weights be spaced about the center, the gunwale well sheered, and the bottom slightly cambered, she will lift grandly to almost any sea. The stern must also be sheered and its upper strakes flared, or it will be driven under water.

Sheer is best obtained in combination with rail and end tanks. If the latter are raised above the rail and pointed forward like a whaleback, the same effect is produced as if a much greater rail sheer had been given without these tanks. This method is better for stability and weatherliness under sail; by its adoption, the principle of the conversion of a deep-sea sailing life-boat into a surf boat becomes apparent, the whale-backs being removed in the former case.

Attention is called to the difference in actual benefit derived from the sheers of different boats, all apparently the same on the sheer plan. On the body plan of a lean-bowed boat like all navy whaleboats, the sheer is projected as a small isosceles triangle above the dead flat; in another, as the cutter, the triangle is lower but much broader. The comparative areas of these triangles represent very well the resistance to a topping breaker, and what may be called the *emergency lifting power*—mechanical and hydrostatic when the bows flare—hydrostatic only when they do not flare.

Thus, suppose a heavily laden cutter and whale-boat side by side. The sheer of the latter is somewhat the higher, and, for light surface waves, not sufficiently deep to lift the boat until well under it, might keep out more water. It is so fine and short, however, that it has but little lifting power, and therefore might take aboard a wave that lifts the well-flared, slower rising bow of the cutter. This explains why African surfmen, when employed aboard ship, occasionally prefer cutters for their work in *narrow* surf.

Fittings.—It is customary for life-boats to have a norman post in the bow for towing, and one on each quarter for an after-line when lying alongside. A life line with floats is stopped up every 18″ around the boat, except the two bights in the waist, which are narrow and long, to serve as stirrups in getting into the boat from the water. In some cases, lines are stopped along the bilges to give the crew a hold should the boat capsize. These are always a drag, liable to damage when landing on the beach, and are not required on self-righting boats. A much better scheme is to fit the boat with wrought-iron bilge guards, set up an inch and a half from the bottom, to give a grip. Their ends should be bedded in malleable-iron plates, riveted on the bottom in such a manner as to avoid catching lines, kelp, etc. These bars if made 1⅛″ thick might save the boat from considerable injury when landing on a rocky beach.

All ship's boats intended for use in rough weather should have a *fender*. It should be of cork sewed in canvas or leather, and securely fastened every few inches in a slightly hollowed batten running all around the boat. The fender should vary in size and firmness with the weight of the boat; for the heaviest boats, oak must be substituted for the cork; the latter should ordinarily be 4″ x 7″, and lie just above and touching the deep-load water-line amidships; the former may be 2″ x 6″ wide in the same location.

The *stem* and *stern posts* should have rollers at the bottom of grooves 4″ deep, and wide enough to pay out a good-sized hawser in warping.

Boat-hooks should always have a point, not very sharp, but sufficiently so to prevent slipping and scratching the paint-work. The hook should be very open and rather sharp, so that it can be made to *engage* in the clothes of people in the water, or by a blow be sent into an object that otherwise presents no holding surface. If bowmen are accustomed to splinter up the gangway and scratch the ship's sides, they should be watched, and even made to use the handle of the boat-hook instead of the point. Their carelessness should not be countenanced by destroying the efficiency of so important an implement. The ball at the point of service boat-hooks frequently toggles itself in gratings or guard-rails. Boat officers have sometimes been obliged to shove off quickly from a gangway, leaving, to their chagrin, one or two boat-hooks hanging to the ship's side.

The oars should be light, reasonably stiff, and not too long; balanced on the rail, which tumbles home slightly amidships, for

greater strength.   Oars, like masts, should be strong enough for all
ordinary strains; beyond that, strength is injurious, as it might
occasion the capsize of the boat.   The oarlocks should be swivel, or
tholepin and grommet, high enough above water and thwarts to
permit a high rolling stroke.   In most life-boats and New Bedford
whale-boats the thwarts are eight inches below the rail.   Dovetailed
swivel oarlocks are nearly as bad as inserted rowlocks in case the
oar is jammed.   They should never be used in the surf or at sea.

There is much diversity of opinion about elastic oars.   Doubtless
this quality causes the oar to act as an accumulator of force, which it
imparts to the water uniformly, producing an even pressure throughout the stroke.   It will return all the force put on it when the stroke
is not broken by the motion of the boat or sinking away of the
water.   The bending should not, however, be so great that power is
lost by the oar kicking in the air at the end of the stroke.   A limited
amount of elasticity eases both oar and man, but for quick work, oars
should be rather stiff.   The blades should be nearly six inches wide ;
otherwise, with boats moved slowly against a head sea, there will be
much loss from slip.   It may also be noted that in pulling against
strong currents long oars should be used ; the speed of the blade
being greater, the stroke need not be so fast to keep the strain on
the oar.   There is a narrow limit to these considerations, however,
depending on the ability of the crew to handle heavy oars, and the
roughness of the sea.   The readiness of handling, the smaller
exposure to baffling wave points, and the decrease of work in swinging them, makes the short oar preferable whenever broken water is
encountered.   There should be two steering oars in the life-boat,
one pointed forward, the other aft, so as to be ready to pull either
way without shifting either oar or coxswain.   It is apparent that
with a heavy turning boat, the best result will be obtained from the
coxswain's strength when the oar is dipped closer than in the case
of an easy turning boat.   The loom of the steering oar should be
oval, with the larger axis at right angles to the blade, so that it will
naturally lie, in a wide crutch, in the position for immediate action—
an important quality at times.   There should be a steering crutch
socket in each bow and quarter, for the convenience of a single
steersman coming up on either side of the ship; also a crutch in
both stem and stern posts, for use when two men are at the oar in
heavy work.

Trailing lines should have a close-worked grommet or metal ring sliding on the loom between neat stops so spaced as to permit the oars to be trailed, boated, and handled without repeated adjustment.

The rudder, for sailing and ordinary work, should be of a design easily fitted when in motion. As it is heavy and there is space for but one man to work, this can never be accomplished if the nice adjustment of several points at the same time is necessary. It is better to have a single very stout pintle rove through the several gudgeons and then hasped in at top. The rudder is certain to be exposed to much rough usage in landing, crossing shoals, etc. White yoke ropes, touched up with a rosette or two, will not soil the gloves and will look very pretty on dress occasions. For rougher work, however, a curved metal tiller is preferable. It should have a long bearing in a metal socket firmly stayed by straps to the rudder.

The *stretchers* should be at least six inches wide and canted to give a solid bearing to the sole of the foot. Narrow stretchers, whether they cramp the foot or not after long pulling, are certain to get men in the habit of pulling from their hips rather than from the point of support. Their position depends on that of the thwart with regard to the gunwale, the freeboard, length of oars, and size of oarsman. The best results, therefore, will only be obtained by having a separate stretcher for each oarsman. Once adjusted, it should be securely lashed or buttoned down.

Thwarts.—For communication with the shore through moderately heavy surf and with vessels at sea, for picking up a man overboard, and for ordinary harbor work when it is not desirable to employ a steam launch, a light single-banked surf-boat may be employed. The thwarts in most single-banked navy boats are 32″ from center to center. In most *double-banked* life-boats this distance is 36″; the thwarts being so spaced that, with the crew alone in the boat, the weight and stowage room are nearly equally distributed about the center of buoyancy. The center of gravity should be very slightly in rear of the center of length, to facilitate quick rise to a sea. All navy boats should have the thwarts arranged in this manner at sea; in port it would be very easy to shift them into any position required by the duty, the rail being furnished with additional sockets for oarlocks.

So long as the sternsheets form the only stowage space of navy boats, it will be impossible to adjust their lines to all circumstances of cargo and weather. Many a young boat officer in coming along-

side light has been surprised and mortified to find that his rudder is
all in the air, and, unless his efforts are ably seconded by a sympathetic boat's crew, may carry away the gangway.   This need happen
to him but once, but the same boat used at sea to communicate with
other vessels, to save life, or to pass the surf, is always jeopardized
and apt to broach to, through being down at the head.

The collection of gratings and bottom boards that lumber up
navy boats should never be left in them at sea.   They conceal and
form receptacles for the dirt.   They float about when water is
shipped, and bruise and incommode the crew; or if lashed down,
they must be frequently raised in order to bail.   Navy boats have
so little buoyancy that a great difference is felt in their management, even before the water is up to the sternsheet gratings.   It
is no exaggeration to say that there are boats in the service containing a weight of unnecessary rubbish, which, if turned into
metallic end and bottom tanks, would make them non-sinkable and
perhaps self-bailing.

The weight of footlings and risers, floor pieces and gratings, in bow
and stern would be no more than that of a false bottom of wood or
sheet copper, which would really make a much stronger boat.   The
weight of the keel and unnecessary dead wood at forefoot and run
could be worked into a centerboard well large enough to serve both
as a ballast tank and subdivision of the bottom tank.   With the
weight of the stanchions, boat box, and boat stove case, at least three
water-tight lockers may be built under the midship thwarts, thus
increasing both buoyancy and stowage space without changing the
weight.   Finally, if the boat, instead of being carvel-built, had been
metallic, or, better still, diagonal-built, sufficient weight would have
been saved to increase the freeboard amidships 6", the sheer 12", fit
a proper ballast tank with automatic valves, fit draining valves, and
increase the subdivision of the buoyancy tanks.

These statements are acknowledged to be founded on very rough
calculations; nevertheless it must be apparent that if proper precautions are taken to keep down the weight, a good self-bailing life-boat
need be but very little, if any, heavier than the average navy boat of
the same size.

Material and Construction.—The first boats of the Royal Life-
Saving Institution were clinker-built of oak.   The new ones are
diagonal built of mahogany, which gives great strength and elasticity.
Tredgold, comparing various woods with cast-iron as unity, found for

Oak, good English, sp. gr., .83, strength, 0.25; extensibility, 2.8; stiffness, .093.

Mahogany, Honduras, sp. gr., .56; strength, 0.24; extensibility, 2.9; stiffness, .487.

Thus it would appear that a boat built of mahogany, weighing one-third less than a similar one of oak, and possessing practically the same strength and extensibility of parts, would be actually much stronger, through the mahogany being five times stiffer than the oak. In addition there would be less working, etc., so detrimental to the tightness of built-in air and water tanks. Haswell makes the sp. gr. of *cedar*, .561; of *white pine*, .554; of mahogany, .56.

A few boat-builders will declare the carvel-built boat to be the strongest, possibly because they put in great frames that will stand rough usage. But then the whole strength of the boat depends upon these frames; if one is cracked, as frequently happens in landing in the surf, it is a serious matter, at once throwing additional strain on those adjacent.

In the clinker-built boat this is also true, but to a much less extent; the planks depend on the frames, such as they are, for their shape, but on each other for support.

The diagonal-built boat, however, cannot be injured except locally; it is the strongest, most elastic, and lightest. Its chief objection appears to be its cost.

With regard to metallic boats. It is claimed for the Francis metallic corrugated boats that they are stronger and lighter than any others; that they cannot become "nail-sick," worm-eaten, or water-soaked; they are unaffected by heat of sun or fire; there is no shrinkage to open seams, and they are not liable to be broken in landing; they are always tight and not liable to damage by concussion or blast of cannon in the vicinity; there is no danger to the crew from flying splinters, nuts, or rivets in action. Their first cost is greater than that of wooden boats, but afterward it costs nothing to keep them in order.

There is undoubtedly an aversion in the service to metallic boats. How much the talent on hand in our navy yards for building wooden boats has to do with this can only be surmised.

If a metallic boat is worn out or rusted through, it will go to pieces when least expected, like the famous one-horse shay; some day a member of the crew will step down off the rail through both bottoms. The experience of many officers in the service with worn-out iron steam launches would furnish an interesting corroboration. If this is the only objection, however, it is not sufficient. Pulling boats are out of the water four-fifths of the time and their condition can be easily examined. Their fitness for a cruise must be determined at the navy yard before being assigned to the ship, just as in the case of wooden boats. Like all other parts of the ship's equipment, they have a life, the length of which depends almost entirely upon the way in which they are treated; and their failure need not occur in service unless they have been employed long after their condition indicates the advisability of retirement.

Sails.—Almost all writers on life-boats and deep-sea fishing boats praise the split lug sail above all others. It is not so pretty as the "sliding gunter," but it is less complicated, and certain to set better, especially if the latter has been lowered by cutting off the foot to make it "safe."

The lower lug sail, when reefed down, may become becalmed in the trough of the sea, but it is more stable and easier to handle. Of course a boat may be endangered by carrying too little sail to be easily managed.

In any weather, such extreme caution is but one form of bad seamanship. A coxswain may sometimes be censured for carrying too much sail in rough weather when the very safety of the boat requires her to have way enough to cross the hollows; if necessary, to luff promptly, to fill away as quickly, and to keep up headway with wind abeam when the sheets have been eased freely or let go altogether. The sail that a good coxswain would carry in a steady strong breeze *might* be censurable if carried in a lighter but squally or shifting wind. In fact, a precaution may not be needed sometimes, which, with less vigilance, would fail to save the boat.

Navy boats, to be properly handled under sail, require the concerted action of several people. Neither good results nor mishaps can be ascribed wholly to the skill or ignorance of one man, be he coxswain or boat officer. It is hardly a fair thing to select the time to brail up foresail and jib, and then blame the coxswain for failing to reach the gangway with rudder and mainsail.

Boat drill under sail in a light sea would be excellent training in nerve and seamanship. So long as the present type of navy boats are employed, the crew will be obliged to get up on the weather rail and bail continually. Every man in the boat should wear a life-belt. Many will object, through a foolish pride in their strength, but as soon as the novelty of the drill wears off, they will take a practical common-sense view of the matter and put the belts on without a murmur. It is desirable to have a steam launch cruising near at hand, until the crew has obtained some skill and confidence, and a knowledge of what the boat will stand.

Whatever gear is put in a boat intended to risk a capsize, to accomplish something where the most sanguine can hardly hope for success, should not only be really necessary, but it should be of the simplest description. Tackles, with their "dips" and "thoroughfoots," are an abomination where it is possible to accomplish as much by "swinging off." Wire shrouds and stays, if used at all, should have a rope pendant worked in at a point accessible when the boat is capsized.

In case of a capsize, the boat should be at once cleared of its masts and sails, cutting freely to hasten matters if the water is at all cold; then, riding by the wreckage, right the boat, and rock it on the wave crests until considerable water has been thrown out. After this, with one or two men bailing, she will soon become habitable. Everything must be done with the utmost quickness, for the buffeting of the waves, and especially the chill of the water, make the men lose strength rapidly. Once they feel themselves at all weakened, few will have nerve enough to do more than to hold on.

Crews should be taught to regard a capsize as a not improbable or even rare occurrence ; the resulting condition of things should be pointed out, and the men encouraged to think of the most expeditious way of clearing and righting the boat. While not making the crew in any way reckless, or permitting any laxity of discipline, it may teach it to look on a capsize, not with horror and despair, but as a difficulty, great indeed, but still one that can be surmounted.

Weight.—It will have been seen that in the discussion of the various contradictory qualities of life-boats, the *pros* and *cons* crystallize most thickly on two types as different in design as the American centerboard schooner and the English cutter.

Where weight need not be limited, or the boat landed on a shoal beach, all arguments must narrow down to a decided approval of the heavy life-boat. Its weight and narrow beam form the powerful combination of great momentum and small resistance, enabling it to penetrate and struggle through a sea that would hurl back a lighter boat. The completeness and value of its automatic functions are proven by the fact that a much smaller percentage of people have been lost from capsized self-righters than from non-righting boats.* Yet these boats appear to the seaman very artificial—their power to *resist* capsizing is kept down to permit them to right when capsized; for the same reason stability is impaired by great sheer and high end tanks. All this is adjusted and ameliorated by a massive iron keel and cork or water ballast. These boats are frequently swept by seas, and yet the crews lash themselves to their thwarts with full confidence that the boat will shake herself free. And if it fails—if it is driven back and capsized—still with its wonderful powers it takes better care of its fettered crew than they could themselves if free.

There must be strong reasons for selecting a less perfect boat when the importance of the stake is considered. The main reason is the necessity of limiting the boat's weight.

Aboard ship the greatest difficulty in using the boats in rough weather, or when wrecked, will be in getting away from the ship. This difficulty increases many-fold with the boat's weight, both on account of its requiring more men to handle it, and because the shock of its weight on water or ship is more destructive. Aboard ship heavy boats are a nuisance, however buoyant and valuable when once clear of her; every precaution must be taken to secure them rigidly, and it is a rare thing to find a boat so secured that the lashings may be eased at sea without danger. Heavy boats require a large crew to act as one man; the weight on the davits year in and out is not only a source of deterioration, but increases the ship's top weight and renders her more vulnerable to the onslaughts of wind and sea. To hoist them every member of the ship's company is frequently called to man the boat's falls, and so impossible is it to handle this scattered strength with the required nicety that often the effort dare not be made at all.

Occasion may arise when it is necessary to land in terrible surf, passable by none but the best and largest boats, but ordinarily the choice of the boat will be determined by the ability of the crew to return unaided. In addition, few ships will be able to preserve intact

---

* "35 self-righters have upset, on board which were 401 men. Only 25 men were lost; of other life-boats, 8 capsized with 140 men, 87 of whom were drowned."—(Captain Lewis, " The Life Boat.")

It is no more than an indication that the conditions are different on the Atlantic Coast of the United States that better results are obtained with the light non-righting boat. See abstract from report of General Superintendent Kimball, U. S. Life-Saving Service, page 308.

the trained crew required for a large life-boat. It is customary in very heavy weather to *double bank each oar*, thus making, with a hand at the bow heaving line, another at the quarter, and two men on the steering oar, a crew of twenty-four men for a ten-oared life-boat. What a magnificent barge such a boat would make!

The advantages and disadvantages of the less· perfect though more natural type, which has outlined itself in contrast to the forced design of the heavy boat, may now be considered.

The stability necessary for freeboard, sheer, and clean pulling back stroke is obtained by a small increase of beam and a flat floor, no metal keel being employed. By this change, self-righting is given up (with a saving of one-third the weight), momentum is lost, and speed slightly diminished against head-seas or currents. On the other hand, the lightness and shallow draft makes the boat lively, and it evades or mounts over seas into which the heavier boat would burrow. Recognizing the fact that the proportions of the latter are fixed within certain narrow limits by the variety and contradictory nature of its requirements, while there is great range allowed in the former, it would seem possible to design a faster boat of the lighter pattern, better for every purpose except when the sea is extremely high and rough. To investigate: height of freeboard and sheer need not be so great, as there is less chance of the water coming in or a sea swamping the lighter boat. Also, as the end tanks are not built especially for· self-righting, they may be made much smaller and lower. In this way stability is improved, and the internal capacity for water being diminished, a much smaller demand will be made on it when the boat is swept by a sea, especially as the side-fenders increase the actual submerged beam as soon as water is shipped. It is thus apparent that the stability of the light, unballasted, flat-bottomed boat *may* be just as great as that of the heavy-keel boat of the same beam, under like circumstances. It is only because the non-righting boat *must not capsize* that the beam is at all increased.

Again, with a large, shallow bearing surface, small internal capacity, and great buoyancy, the light boat will keep herself dry, while the heavy boat, with finer lines, will ship much more water, and sinking deeper, drain sluggishly. A condition of the waves may therefore be imagined in which the light boat is always light, dry, and fast, while the heavy boat, ramming every sea and burdened by the constant shipping of water, labors along in the rear.

Let the force of wind and sea increase. The heavy boat crawls on, more slowly but as persistently as ever, for the force of its crew, stored in its weight, is sending it ahead despite the impact of the waves. The light boat, however, is plainly beginning to be distressed; the crew spends more and more power in *rowing up the moving seas,* only to be checked by the driving gale when balanced on their crests.

It has been shown that with almost every desirable quality of a good sea life-boat there are corresponding disadvantages. Experience has taught surfmen what qualities of boats are most important in their locality, and how much other features may be sacrificed to the advantage of these. In the numberless ways in which disaster or need may come to those at sea, still other qualities take precedence. It is more important that a ship's company should be able to get away from a sinking ship and keep afloat, than it is necessary that they should have at their command the ultimate excellence in boats for landing. Yet one and the other may be identical in case of a wreck on the coast, and such is our respect for even a moderate surf that we cannot permit the sacrifice of any qualities essential there. It is certain that if a boat can live in the surf it can in the sea outside, and so the question narrows down to limiting the weight of a surf boat to that which can be handled and carried aboard ship.

It is confessed that the mere presentation of a design, believed to be the type of a man-of-war life-boat, but states the problem; the solution is in the hands of the service. If such a boat is wanted, mature deliberation must remodel all that criticism demolishes.

Detaching Apparatus.—It must be said that the service exhibits but languid interest in the many designs of detaching apparatus, and the reason is simply that no such device can in itself ensure the safe departure of the boat. There are numbers of bad ones, liable to all sorts of eccentricities, and a few really good ones, with all their parts in sight.

All, however, halt at a half-way point; they cause the boat to part company with the falls, but not with the rolling, wave-beaten ship. A few designs recognize this difficulty, and attempt to overcome it with longer, or extensible, davit heads; one device has bearing-off levers hinged at the water's edge. It is surprising that so much attention has been paid to *detaching apparatus* and so little to *means for hoisting.* Generally, the anxious moment with heavy boats comes when the falls are first hooked. With care and plenty of tackle, or better still, as it is unlikely to foul or twist, a long pendant and whip, the falls may be permitted to unreeve; or hooking on may be easily accomplished at an oar's length from the ship's side. The seas are indifferent to swiftering lines; they batter the boat against the side, lift it, and, despite quick work on the falls, let it drop. A misgiving may sometimes cross the mind of the supervising officer as to whether it would not be considered better seamanship by some to let the boat go adrift or swamp astern than lose it in hoisting.

In no boat is simultaneous detaching more necessary, in any weather, than in the heavy ones carrying large crews. Yet it is an exceedingly rare thing to find one of these boats fitted with detaching apparatus. It seems to be decided on shore that these boats shall not be lowered except in smooth water, though they are the only ones that can be expected to make headway against a wind and sea.

On modern ships, the yards, less used for sail now, are still used as derricks in lifting boom-boats and other heavy weights. The mainyard in particular, smoke-begrimed, collecting soot during a cruise to distribute it at leisure until the awful cleaning day comes, must still be straddled every day to rig a "water-whip" or "yard and stay." Many boats now-a-days must still be hoisted out by the ponderous "triatic," the rigging of which requires generally a comprehensive derangement of crew and tackle. Matters are improved on some ships by the lighter boats being stowed amidships, and the heavy ones in cradles immediately under their swinging davits. Still it requires considerable time and method to get all the boats out. Some of them, though the most important in case of shipwreck, are so stowed as to be necessarily the last lifted.

With regard to davits, there appears to have been no advance over those on the first ships built, other than the graceful curves and smaller thickness permitted by the substitution of iron for wood. They still project in an ungainly manner, offering greater resistance to wind and sea than is at all necessary, while their rigidity and permanence compels a general dismantling before coming alongside a dock or other steamer. In fact, people have become so accustomed to this type that many consider them as natural as trees, or, for that matter, think that by their removal men-of-war would be robbed of a most striking characteristic. Yet they are an unnecessary and unsightly encumbrance. Most of the difficulties with boats in rough weather are due to their inefficiency. They are in the way of the battery and a clear all-around view; they increase the top-hamper and gear; invite the destruction of the boats by every heavy sea, and may make ramming seem even more desperate and suicidal than it is.

On several of the new ships a number of the boats are stowed in two rows athwartships, lying in cradles on a superstructure just abaft the forward bridge. On some armor-clad cruising battle-ships they are placed well back from the rail, on a superstructure.* These are excellent places for them, as they are out of the way, easily secured, and permit vessels to go alongside of each other, or the wharf, with nothing to look out for but the guns. The old davits are still retained, however, only, as in the case of monitors, they present new complications and are less efficient than ever.

Undoubtedly the simplest method of all for hoisting in and out these boats or any other heavy weights, and of avoiding the use of the mainyard to get aboard every barrel of provisions, is by light cranes on both sides of the ship, worked by hand and steam.

These cranes should be capable of lifting the steam launch from the water to the ship's deck, of transferring boats from one side of the ship to the other, and, at least one on each side, of swinging a life-boat and crew at least 12′ clear of the ship's side. Each boat should be fitted with its own slings and balanced with the crews on their thwarts. In the case of heavy boats, a two-legged sling and supporting band around the midship section may be used. Davits should not be used at all unless they swing inboard and land the boats, a condition that will usually interfere with the needs of the battery. The crane should be capable of high speed for rough weather work; in addition, to expedite matters, one of the parts of the hoisting whip may be run in by hand.

It may be said that a seaman should not thus be permitted to lose the special skill required in purchasing heavy weights. The answer is that there is no danger of seamanship falling into disuse through lack of difficulties requiring its special knowledge. The navy has every day more and more the need of *thinking* seamen. There are

*The boats of the Maine are stowed in this manner, there being a light swinging crane on each side of the superstructure, by which they are handled. As the arms of these cranes are just long enough to plumb the boats when lying alongside, they will be no better than the old-time davits for hoisting or lowering in rough weather. They are, however, light, neat, out of the way, and useful for many other purposes. It would undoubtedly be an advantage to have the boats of all the cruisers stowed in this manner.

too many actual problems to solve in the new state of affairs to permit the continuance of any drill or work which is no longer necessary.

In justice to the writer, it must be stated that his conclusions with regard to the best type of ship's life-boat were reached and the preceding pages written before he had learned the character of the light self-bailing surf-boats recently tried in the U. S. Life Saving Service. The latest design, the Beebe-McLellan Self-Bailing Surf-Boat, has been in use but two years. It is believed by General Superintendent Kimball to be adapted to ship's use. Through his kindness it is possible to give the plans and specifications in the appendix. It will be noticed that the beam is 28 per cent of the length, the depth, 29 per cent of the beam, the sheer, 1″ to each 1′ of length, and the draft, about one-half the freeboard. The end air tanks do not rise above the thwart line. Some have been fitted with centerboards and a sail. Although only 25′ long, their nominal load is 26 people, including crew. Of course, for ship's use larger boats would be built.

The capacity of a ship's life-boat of this description, not necessarily self-bailing, may be found by the " rules made by the [British] Board of Trade, under the 'Merchant Shipping (Life Saving Appliances) Act, 1888,' to come into force on the 31st March, 1890." Ten cubic feet of boat capacity are allowed for each adult person in a life-boat fitted with air tanks; in others without air tanks, through having more space, eight. The capacity is calculated from this formula: Length × breadth × depth × .6.*

The capacity of a Beebe-McLellan boat 30′ long would therefore be 30 × 8.4 × 2.516 × .6 ÷ 10 = 38. If such a boat was provided with a centerboard, and a tank on each side that might be pumped full of water for ballast when under sail, it would make a fine 12-oared ship's life-boat.

*Life Rafts.—" For every person so carried, there shall be at least three cubic feet of strong and serviceable inclosed air-tight compartments, such that water cannot find its way into them."

Buoyant Apparatus.—"Approved buoyant apparatus shall be deemed sufficient for a number of persons, to be ascertained by dividing the number of pounds of iron which it is capable of supporting in fresh water by 32. Such buoyant apparatus shall not be required to be inflated before use . . . ."

Life-belts.—"An improved life-belt shall mean a belt which does not require to be inflated before use, and which is capable at least of floating in the water for 24 hours with 10 pounds of iron suspended from it."

In order to give an idea of the comparative value of the U. S. Life
Saving Surf-Boat and the heavier types of English life-boats, the
following extracts are made from the statement of General Superintendent Kimball, of the "Organization and methods of the United
States Life Saving Service.   Read to the Committee on Life-Saving
Systems and Devices, International Marine Conference, November
22, 1889."

"The type of boat in most general use in our service, although
properly entitled to be called a life-boat, is distinctively known as the
surf-boat, and this term will be applied to it in the remarks which
follow. . . . Three varieties, respectively designated the Beebe, the
Higgins & Gifford, and the Beebe-McLellan surf-boat, from the names
of the persons who devised the modifications which characterize them,
are the only ones furnished to the stations in recent years.   They are all
constructed of white cedar, with white-oak frames, and their dimensions are from 25 to 27 feet in length, 6½ to 7 feet beam, 2 feet 3 inches
to 2 feet 6 inches depth amidships, and 1 foot 7 inches to 2 feet 1 inch
sheer of gunwale.   Their bottoms are flat, with little or no keel, and
have a camber of an inch and a half or two inches in 8 feet each side
of the midship section.   They draw 6 or 7 inches of water, light, and
weigh from 700 to 1000 pounds.   They are propelled with 6 oars
without sails, and are expected to carry, besides their crews, from 10
to 12 persons, although as many as 15 have been landed at a time in
a bad sea.   Their cost ranges from $210 to $275.   There is no great
difference between the Beebe and the Higgins & Gifford boat, except
that the former has more sheer and is clinker-built, while the latter is
of carvel construction.   The Beebe-McLellan boat is the Beebe boat
with the self-bailing quality incorporated.   This feature has been
added within the past two years, and but few of them have yet been
put in service. . . . Even at those stations where the most approved
self-righting and self-bailing boats are furnished, the surf-boats are
generally preferred by the life-saving crews for short distances and
when the number of imperilled people is not large. . . . As respects
safety they will compare favorably with any other boats.   In 18
years they have been launched 6730 times on actual service, and
landed 6735 people from wrecked vessels.   They have been capsized
but 14 times.   Six of these instances were attended with loss of life,
of whom 27 belonged to the service and 14 were shipwrecked
people. . . . I learn from the annual reports of the Institution
[Royal National Life-Boat Institution of Great Britain] that during

the same period of 18 years, her boats (self-righting and self-bailing) capsized 21 times, attended by loss of life, 68 of the lost being life-boatmen and 7 shipwrecked people. I find by the official report . . . in December, 1866, to Board of Trade (British) . . . that during the previous 32 years, the self-righting boats of the Institution had been launched in actual service 5000 times, . . . 'the 76 life-boatmen lost represented about 1 in 850 of the men afloat in life-boats on service, and the capsizes, 1 out of each 120 launches on service.' . . . In the case of our capsized surf-boats, the 27 men lost represented 1 in 1744 of the men afloat in surf-boats on service, and the capsizes, 1 out of each 480 launches on service. . . . We find, then, 1 capsize of the surf-boat to every 464 persons saved ; a difference in its favor of 172. The self-righting boat lost 1 life to every 136 saved ; the surf-boat, 1 to every 158 saved ; a difference of 22 in its favor. Of the life-boatmen afloat, 1 to 850 were lost by the self-righting boat, 1 to 1109 by the surf-boat ; a difference of 259 in favor of the latter. In the life-boat, one man of the crew is lost to every 157 lives saved ; in the surf-boat, 1 for every 240 saved ; a difference in favor of the surf-boat of 83. . . . Since 1876 there have been put into the United States service 37 self-righting and self-bailing life-boats of the model of a boat received from the Royal National Life-Boat Institution. . . . They are 29 feet 3 inches in length, 7 feet 7 inches beam, 3 feet 1½ inches deep amidships, 1 foot 10 inches sheer of gun-wale, straight-bottomed, pull eight oars, and weigh about 4000 pounds each . . . a heavy iron keel . . . securing the self-righting quality. . . . The boats have capsized once in each 118 trips, and once in rescuing every 146 persons, and one life has been lost from the boats to every 117 saved. There are two other varieties of self-righting and self-bailing boats in the service—the Richardson and the Dobbins. They are modifications of the life-boat just described, though considerably lighter. . . .

"Among the boats at present employed in life-saving institutions, I know of none that can be justly denominated the best life-boat. The type that is best for one locality may be ill adapted or entirely unfitted for another, and a boat that would be serviceable at one time might be worse than useless at another in the same locality. . . . The keeper steers with a long steering oar, and with the aid of his trained surfmen, intent upon his every look and command, manœuvres his buoyant craft through the surf with masterly skill. He is usually able to avoid a direct encounter with the heaviest breakers, but if he is obliged to receive their onset, meets them directly ' head-on.' His practiced hand immediately perceives any excess of weight thrown against either bow, and instantly counteracts its force with his oar as instinctively and unerringly as the skilled musician presses the proper key of his instrument. He thus keeps his boat from broaching-to and avoids a threatened capsize. The self-righting boat is more unwieldy and not so quickly responsive to the coxswain's tactics, and is therefore not so well adapted to our general work. . . . I have given these results in comparison . . . to show that the United States has provided quite as effective means for dealing with the conditions presented to it as the most eminent organization of other countries has for its conditions. . . . May it not be a question whether these properties (self-bailing and self-righting) and the means of propulsion by sails cannot be advantageously incorporated into the surf-boat without materially increasing its weight and draft, and whether such a boat would not be found to be better adapted to perform the general services of life-boats than those which sit deeply in the water, and which on that account, and because of their great weight, are less agile in action and more difficult to transport and launch? And why, since it has been found that the self-bailing principle can be applied to a model thoroughly convenient to be carried on shipboard, may not these vessels even now be supplied with self-bailing boats, in which the liability to capsize is greatly diminished by reason of their ability to immediately free themselves of any water they may ship? "

It is believed that the best boat for ordinary ship's use would be a light whale-boat built on the lines of the Beebe-McLellan boat. If 29' long, 8' beam; depth, 2' 4"; sheer, 2' 5"; draft, 10"; freeboard, 18"; fitted with side air tanks and false bottom; draining through centerboard well, towards which bottom slopes, and carrying two lug sails, it will be a vast improvement on the 29' whale-boat now furnished the new cruisers, especially if a portion of the bottom permits water to be let in as ballast. It may be said that the 32' launches are heavy enough now. It is certain, however, that a boat of that size once shipping a sea would remain unmanageable for so long a time before being bailed out as to run the chance of being repeatedly swept and finally swamped. By adopting a less heavy construction in these boats, it may be possible to put in a false bottom for draining without increasing the weight more than five per cent. The 24' cutters should be either metallic or increased to 26', with 6'

9" beam, flatter bottom, air tanks under bow platform and thwart, around stern sheets, and in half-round tanks under the thwarts out against the side. A ballast tank, if needed, may extend from the centerboard well out to the sides.

The 28' cutters could be better replaced by very light non-sinkable surf-boats. If these are made as light as the combined strength of the diagonal system and mahogany wood will permit, they may be made self-bailing. In all of these boats, air tanks for stowage of boat still, provisions, and repair articles should be built between the bottom and thwarts; limber holes 4"x 18" may be left for passage of water.

It will be noticed that the Beebe-McLellan boat is essentially a compromise; none of the air tanks are detachable; it is clinker-built; there are no air tanks under the thwarts or above the seats in the ends. In consequence, a sea shipped over the bow will run under the thwarts and out over the stern; passengers may be taken in end on. The draining valves are large but simple rubber flaps; they are placed farther from the center than usual in order to act when the boat is rolling.

PART II.

Duties.—The requirements of the coast life-boat have been investigated with the aid of such authorities as are within reach. That phase of the problem is now encountered which can only be solved by the naval officer.

The question is, Do the special duties of navy boats prohibit a radical change if a gain be made in seaworthiness and the weight increased not more than 5 per cent. or 10 per cent.?

Much light will be thrown on the problem by a comparison of the crews and their duties in the past with the changed conditions of the present.

In the modern man-of-war we are at once struck by the simplicity or even absence of sails and spars, and the much greater complication of battery and adjuncts, than was the case in the ships of 1860. In some an attempt is made to retain the great sail power of thirty years ago, but generally the prime object of the ship, that of fighting under steam, has come to the front. The old-time battery was large but very simple, so simple, in fact, that detailed instructions for every move of hand in its service, ordinary and extraordinary, could be gradually mastered by the most ignorant man in a series of easy drills. The handling of sails and spars, however, was a different thing; it was the profession of a lifetime, and many a man efficient enough in gun drill never could become a *real* seaman, no matter what he might be rated. With all the infinite variations of sea, wind, sails, and course, peculiarities of the ship, etc., the progress of a watch was a panoramic problem which the true seaman was constantly solving. No matter how terse the orders of the watch officer, or well drilled the crew, the method of execution was as varied as the circumstances that compelled the change. In fact, a change of any sort *compelled every man involved to think*. This is the grand secret of the adaptability of the sailor. It is not a difficult thing to follow a rope with the eye until the cause of its jamming is determined.

When, however, this is done instinctively, or mentally, as at night; the reflective principle, so weakly illustrated, pervading the whole life, dealing with the irresistible sea, the invisible wind, gaining, in spite of them, an indirect advance where the direct is impossible; when men every hour of the day and night trust unflinchingly to the skill of their own hands and brains in matters where failure would be comprehensive ruin,—then a swiftly decisive but philosophical character is formed, and men are made proud of their profession and ships, when the praise might well be given to themselves. Amid the changing circumstances of every evolution certain necessities rose paramount, and these were recognized and provided for in station-bills of an exceedingly accurate and detailed description. It is the boast even now of many an executive officer that when his crew first came aboard each man was prepared to stow his hammock in the proper netting, take his proper place at his gun, eat at his own mess table, and at sundown evolution tail on to his own boat-fall or yard-rope.

In time of storm and emergency no *seaman* ever failed to find his proper billet, though it was not always the one assigned him on the station bill. Despite the careful forethought and subdivision of labor, in time of trouble the artificiality of the system at once became apparent.

Seamanship was a work of precision only in the final execution; in the preparation, individual intelligence and adaptability had full sway. In the quarter-bill, a large number of men were assigned to each one of twelve or fifteen billets requiring performance of simple and exact duties. On deck, the man made his own billet,—within limits. If he was absent, his work was divided; it did not fail to be done. Hence, in spite of the apparent rigidity of the station-bill, its details were frequently ignored. From this it follows that there were a large number of men whose duties were nominal, or easily transferred and distributed. Thus, having the use of numbers of men without interfering with the routine, a pardonable pride in good appearance originated the present style of navy boat with its large crew and small carrying capacity.

In the future, however, the larger proportion of skilled and rated men, and the very great variety of duties required of men of ordinary capacity, will make it necessary to assign at once all intended substitutes in the battery to regular duty. The neglect of certain of these new duties would have serious consequences, though not at once apparent, as would be the case in sail and spar evolutions. It is all the more necessary that there should always be some one at hand to perform them ; thus leaving but few men whose absence in boats would not be felt. At present, the difficulty is met by detailing a single boat's crew, usually that of the steam launch, to do all regular boat work while in port, in the meantime excusing them from all other duty. It is clear in consequence that little boat duty, other than that of smooth-water drills, may fall to a seaman not in the running boat's crew.

As for the operations of "cutting out" and "landing where likely to be opposed." It is not probable that a battle-ship will be permitted to risk her crew, trained with patience and care, in a landing where likely to be opposed by modern firearms. It is difficult to call to mind any place in the world where a commander would dare to risk his pulling boats in hostile rivers or harbors. On the open coast, the ship's secondary battery can drive the enemy back so that a peaceful landing can be effected, were the boats not certain to be swamped in their attempt to carry a load of passengers through the surf. In fact, operations by pulling boats in closed harbors, unsupported by the ship's powerful battery, are not to be thought of, unless boat guns are employed in sufficient force to overcome all resistance at artillery range before the attempt to land is made.

The terrible rapidity of fire, the range, and consequent power of concentration of modern firearms have compelled many changes in the attack on shore. The loss is minimized by scattering the men, at the same time permitting the largest possible number to keep up the return fire. Many tacticians claim that so long as the ammunition of the attacked holds out and they are not shattered by artillery fire, their front is invulnerable. Front attacks are generally disastrous, and rarely employed except to force the enemy to adopt or keep a certain line of battle.

In boat attacks, catastrophes must occur. Only a very small part of the force can be put in action, while the whole of it forms a target for concentrated fire. It is true that formerly in landing in smooth water—the only place where it was possible—the swifter, low-lying navy boat *had* the advantage of speed while exposed to the enemy's fire. This fire, however, became effective only when the boat was less than two hundred yards from the beach, while now it may be able to stop the boat at ten times that distance. Under such circumstances, the relative speeds of pulling boats are not of much importance. At any rate, the words "likely to be opposed" should be stricken out, with their cheerful indifference and lack of knowledge as to the kind of reception to be expected. The amount of preparation for a likely opposition that could be made in the boats would hardly make any difference. It should be the ship's duty to cover the boats, which in themselves would be helpless, and should be fitted out to land observation party, police, garrison, naval brigade of occupation, or naval brigade prepared to march, etc.

The day of boat operations is by no means past; rather more important now that it is proposed that the defense shall place all its eggs in one basket, all its warlike preparations in harbor-defense vessels and fortifications. The few cruisers that may escape capture for a time will be able to communicate with the beleaguered land only through boats landing on the open coast. The duty of these cruisers is undoubtedly to destroy commerce, but as we have no coaling stations or harbors of refuge, their career will be quickly terminated. Unfortunately, our own commerce is the only one likely to suffer on the seas if such a measure prevails. A cruising fleet of battle-ships would be a moving place of refuge for hunted cruisers.

In siege and blockading operations it will be found that the range of the shore artillery is so great that close observation is rendered too dangerous to be reliable. Occasional landings for the sake of gaining or imparting information will be more valuable than ever. In larger operations it is possible for the boats of a fleet to land on an open coast, fortify their position, and, with the aid of a few ships, hold it until it is made impregnable. There are such points, which, if occupied by an enemy in time, could render the vast fortifications of a port of little value. For, to dislodge the invaders, emplacements must be built for guns heavy enough to drive off the protecting ships. Otherwise the monitors and harbor-defense vessels must come out and fight at sea—if they can. The subject can only be touched upon to show the advantage that the fleet must have whose boats can select some place for landing on the enemy's shore other than that especially prepared for their repulse. In this connection it must be borne in mind that whereas old-time ships were able to run up fortified rivers and clear the woods with shells from a few slow-firing, smoothbore guns, the modern battle-ship may be said to be equivalent in rapidity and weight of fire to a whole fleet of its predecessors, while in accuracy and range it far exceeds any one of them. The U. S. S. Philadelphia or Baltimore, with less than fifty men on deck to handle the secondary battery, could keep up a fire, which, for range, weight of metal and exact pointing, would be superior to that of ten field batteries of six-pounders.

The best preparation for landing where *likely* to be opposed would be to put the boats in a condition to get over the dangerous space as rapidly as possible. A light inclined shield in the bow would be worth three times its weight of rapid-fire guns, although for this special purpose a Gatling will generally be preferred, on account of its light weight, searching qualities, and the absence of cover on the beach. A field- or boat-gun loses its value on approaching the enemy; it incommodes the crew, and makes it impossible to land except in very quiet water, or some distance from the beach.

In landing on the coast, field-guns should be sent in on balsas or catamarans, when they can be floated safely up to dry land. If placed in the boats, they may swamp or stave them in on striking bottom some distance from the water's edge.

Apart from the *method*, the soundness of a policy permitting the hazardous landing of the crew of a modern cruiser may be questioned. Not only must a larger part of the crew than formerly remain aboard to assist in the care-taking and manipulation of the many apparatus, but the loss of a sensible part of such a crew, though at once replaced by new men, would impair the efficiency of the ship for months to come. This is especially important when it must be acknowledged that, so long as the navy is growing and changing, it can *never* possess a full complement of thoroughly trained men. Few landing operations can be carried out except from the fleet, or rather from its auxiliaries; the lightly armed transports and hospital ships carrying extra men, which should always accompany it. These transports could be manned by the Naval Reserve, in which manner that valuable branch of the service would gain actual cruising experience and drill without impairing the efficiency of the more important ships through their short terms of service.

With regard to "cutting out," secrecy and despatch are the most important qualities; in case of discovery and failure, the advantage in speed that one pulling boat has over another is immaterial; the result will be equally disastrous. In case, however, the vessel is boarded, one of the worst defects of navy boats makes itself felt; the boat cannot land more than a "corporal's guard" without weakening its crew beyond a safe limit. In landing, even in smooth harbors, few whale-boats could put more than fifteen men on the beach and leave enough to manage the boat properly.

In any case it must be seen that the demand on battle-ships and cruisers for boats able to land men and stores on the open coast will very probably be one of the features of the next war, no matter where the men are obtained or how remote the occasion. For such a purpose the present type of navy boat is useless, its chief defects being lack of seaworthiness and carrying capacity.

While power and seaworthiness are of the greatest importance, exceptional *speed* may in only one rare instance be more important; that of the *transmission of intelligence*, when neither steam launches, sails, nor signals can be used. Even then the boat must be different from the present type; it must have power, freeboard, draft, and narrow beam; something that could be best illustrated by a double-banked gig whale-boat.

For *laying out anchors* in shallow water or breaking seas, two life-rafts may be employed by a variation of Captain Craven's method; or, for that matter, any pair of broad-beamed, light-draft life-boats would serve better than cutters. A single life-boat may be used in the same manner as the old sailing launch, the draining-valve holes or centerboard well taking the place of the bottom funnel.

In *towing*, *warping*, and *kedging*, the large number of men who can be put on a life-boat's oars ·in any kind of weather without making her loggy or heavy is most important, and may at times save the ship from serious disaster.

The occasion may arise on modern men-of-war in which it will be preferable to *water by boats* day after day for a considerable time, rather than expend coal for distilling. No better boat could be carried for this purpose than a life-boat fitted with end air tanks, and especially is this true when the choice of a landing place cannot be made.

Boats are also used for carrying stores in tow of a steam launch, lightening ship, and making trips when the launch is not lowered. For these purposes, handiness, large carrying capacity, and ability to keep afloat are the vital characteristics; and yet there is hardly a pulling boat in the navy that has an air tank.

Abandoning ship.—It is openly acknowledged that a man-of-war cannot carry a sufficient number of boats to take off her crew in anything but smooth water, and that a certain proportion must be detailed to life-rafts and balsas. It is also admitted that this last nondescript contingent goes far towards spoiling the *drill* of abandon ship, as an indication of the state of discipline and preparedness for emergency aboard ship. It is a feature of the "piping times of peace" that the *outward appearance* is made more important than the actual spirit and purpose of the drill. Divisional competition is a very good thing if *time* is sent to the rear and *thoroughness* is made to take its place.

The question might be asked: Which method would most benefit the actual purpose of this drill? 1st, assign only such a number of men to each boat as would not unduly increase the danger of its loss, and make temporary provision by means of bridges, deck-houses, and life-rafts for the remainder; 2d, do away with the least efficient boat and distribute its weight among all the rest in the shape of increased beam, depth, and sheer, thus increasing the total *safe* carrying capacity. By the substitution of either of these methods the present drill would lose much of its smoothness and precision, but it would be more practical and efficient. Many officers may have noticed that the *operation* of hoisting out boom-boats sometimes litters the deck up more than the *drill* "clear ship for action."

On noting the variety of disasters to which a ship is liable, it is evident that the drill "abandon ship" provides only for that rarest and most merciful of all shipwrecks—that of a ship gradually going down in a calm and placid sea. Instead of doubting the ability of the boats to float their quota, and making the attempt to improve their buoyancy with canvas upper works, they are still further weighted down with heavy cases of *cooked* meats, a large square case, said to contain a boat *stove*, and a Crusoe outfit in a heavy boat box that takes up the space of two or three men. This state of affairs is the accumulation of years of theoretical preparation for a variety of shipwreck that has never come. Every change in drill has introduced additional safeguards and regulations of precaution without weeding out any that may be obsolete.

There are advantages in being able to stow boat boxes and stoves in the hold, and in borrowing the provisions for the occasion. It is certain, however, that the time spent in procuring these articles, should the actual need ever arise, might cost many lives. Within twenty-four hours after the wreck, the amount of water in the breaker will be the uppermost thought in every man's mind, and if the boat stove is fitted to distill it would be regarded as a priceless treasure. As sailors, with the scuttle-butt always at hand, have no reason to deny themselves in the use of water, as is frequently the case with soldiers, they begin to suffer from thirst, or imagine that they suffer, very quickly.

It would seem advisable when at sea to clear out of the boats all unnecessary rubbish, fit a tank under one thwart as a boat-box, one under another for a boat-still, and a third for provisions. All hands could then fight a fire or leak with good heart to the bitter end. There always seems to be a chance left to call in question the judgment of a commanding officer who has to decide the moment *when preparations for leaving the ship become more important than the attempt to save her.* The number of abandoned vessels found at sea emphasizes this feature of a shipwreck.

With regard to life-rafts. Many are highly recommended, but few seamen have exercised with them, and their peculiarities at sea are not well understood. Their chief recommendations appear to be buoyancy and non-liability to damage from staving or capsizing alongside. An excellent feature of some is that the rubber air-cylinders are encased in canvas ones. The inner one being larger than the envelop—like a football bladder—the outside takes the shock, and if injured, the folds permit the inner case to evade puncture.

On the other hand, almost all descriptions of life-raft require time and careful preparation when all hands are working under great excitement. They are charged with being hard to pull and steer against a wind and sea. Having no freeboard, they will be constantly swept by the waves in breaking seas. It seems certain that if the small balsa furnished the navy is weighted down it will capsize like a barrel in a seaway. Nevertheless, in the majority of ships, the salvation of a portion of the crew can be assured only by a liberal supply of *large* life-rafts; especially since the clearing away of the largest boats will sometimes be impossible.

The *catamaran*, as issued at present, is a sheer waste of weight. Though designed to be very buoyant, it is made of heavy wood, of liberal scantling; it frequently leaks, and has not sufficient freeboard with the crew aboard for pulling in rough water. In a seaway it is certain to be unmanageable and repeatedly swept by the waves. A copper catamaran, with wooden fenders, would be more serviceable, more durable, and less expensive to keep in order.

The *dinghy* is intended to be a handy, safe, and economical boat; it, however, frequently fails in all these qualities. It seems fated to be a source of worry and trouble through its being often unfit for the duty on which it is sent. Being a miniature cutter with the scantling nearly that of the larger boat, its weight is all out of proportion to its freeboard and buoyancy, while its low rails make it a difficult boat to pull even in smooth water. In consequence, the dinghy boys almost invariably form a very unhappy quartette.

Seeing the great growth of the metal-working industries in our navy yards, it would seem not impossible to replace the old type dinghys by others of galvanized iron, with end air-tanks and 15″ freeboard. An efficient 19′ dinghy of this description need not weigh 350 pounds.

In connection with these opinions, it is well to consider in detail the demands likely to be made on a ship's boats for life-saving.

These may be divided into two classes, with subdivisions, as follows :

I. In case of the loss of the ship, when provision must be made for the rescue of all hands.

1. Loss by fire, when all hands fight the fire to the last. The loss of the Bombay is an instance where the fire was fought too long, it being impossible to get all the people off the ship.

2. Founders slowly, from collision with other vessels, ice, injuries received in storm or battle, etc.

3. Founders quickly, from various unexpected causes.

4. Wrecked on the coast.

5. Wrecked on outlying shoals or reefs.

And all these disasters, and others unforeseen, take on different shapes and require to be met by different treatment according as they occur by night or day, in rough or smooth weather, in the presence of friends and succor, or alone. The last two subdivisions have in addition as many peculiarities as the coasts on which ships are wrecked.

II. Saving life from the ship.

1. Man overboard under way or in port.

2. Assistance to a vessel in distress.

3. Taking people off a wreck at sea, on a shoal, or on an island.

4. Communication with the shore.

Again, all these important conditions may vary in many ways.

It thus appears that there is really a greater variety of imperative duties demanded of ship's boats for saving life than for all other duties combined. That such demands are made infrequently is hardly the case, considering the long list of minor casualties during the past few years. Even if they were rare, that is no reason why they should be ignored. It would be as sensible for a ship to go to sea with sub-caliber practice-barrels instead of her heavy guns; a much finer drill may be had with the former, perhaps, than with the latter. It is a noble thing to save life, and the qualities of men and boats necessary to success in that would compel success in all other duties. Let seaworthiness be carried to the highest point of excellence in the *navy boat*. Casual routine duties should have no influence in diminishing the peculiar efficiency of such a boat. Comfort will certainly be improved, but the convenience in petty things, and pride in appearance, should no more be allowed to interfere with the grand essentials of a life-boat than they do with the art of war itself.

It is asked that the presence of objections to the proposed change shall not condemn it before due weight is assigned to the related advantages, nor that an advantage shall be rejected as valueless because it is only partially obtained.

Objections to the proposed boats may be *functional* or *materialistic, active* or *latent*. The first at once limits efficiency; the second imposes conditions independent of use; as of cost, care, size, weight, etc.

All that is suggested is that it would be advantageous to exchange a few functional and active objections, of which navy boats have plenty, for a few materialistic and latent ones; the only ones in fact that appear in a life-boat.

There is one last view of the subject, that is, the effect of the possession of good, seaworthy boats, on the tone of the navy at large. The most stirring professional experiences of most officers, calling for all their energies of mind and heart, may be unhesitatingly declared to have occurred in their encounters with the forces of nature. It can almost be said that the exigencies of a battle cannot make as great a demand on the nerve and courage of officers and men as a great storm. Between the old sailing-master and the modern commander there is much difference. The former depended solely upon the power of his crew and his own knowledge to prevail against the storm. A life of this sort made seamen self-reliant, courageous, and persevering; sea fights were fought out to the bitter end, until one of the combatants no longer had the power to strike a blow. In modern times, with the introduction of steam and various labor-saving mechanisms, it is apparent that much of the success of a naval battle must depend upon forces to which no appeal can be made, that the slightest neglect of some principle or precaution in the engine room may throw the fate of the battle adversely, or at least seriously jeopardize success. But this is not all; under the circumstances, it is evident that with the increased range of possibilities, an increased power of adaptation is necessary in the men who fight. Evidently, therefore, it will be detrimental rather than beneficial to make many of these men specialists, and thus, by absorbing the powers of the seaman in any one direction, limit their adaptability to others. In fact, the power of adaptation of the men on deck never was more important than it is at present.

There are many points of isolated and varied responsibility on any one of the new cruisers. The enormous demand for ammunition of all calibers and description, each variety requiring its own precautions and treatment, necessitates a very large number of well trained men possessing greater intelligence and breadth of knowledge than in the past. Considering how much greater the cruiser's powers of offense are than those of defense, it cannot be hoped that an action between them will be even initiated without serious loss. For this reason the battery must be served by men of sufficient intelligence to master the duties and stations of a very flexible quarter bill—a bill that will permit either concentration or scattering of forces, to continue the fire of important guns to the last, or to distribute the crews of those disabled. In fact, every minute of action may annihilate whole gun's crews or mechanisms, and such catastrophes should be provided for in the bill. It may at any time be deemed advisable to withdraw the crews of exposed guns, or guns difficult to supply with ammunition, in order to secure a more rapid fire from others less exposed. These reasons alone seem sufficient to indicate how necessary it is that the fighting portion of the crew should be *homogeneous* and instructed in the essentials of the service of *every* part of her battery. It would also seem criminal to expose on the deck of a cruiser in action a single man more than necessary to make the best use of her fighting powers. The act of calling away from each shielded gun's crew one man for duty in connection with boats or spars must be not only at the expense of rapidity of fire but at the expense of life.

The real problems of naval warfare lie in the proper development and combination of a ship's powers of offense and defense as a whole, and not in the details of the manipulation of any of her separate mechanisms, however intricate they may be. A gun's crew may drill well in a month, yet it may be three years before the ship as a whole can make the best use of her powers.

A modern ship can give a good account of herself only when fought by an intelligent, plastic crew that will not be at a loss to know how to fight when the exigencies of battle have rendered their paper billets obsolete. To attain this state of training with young Americans, it would seem a good plan to form small groups with general instructions. As soon as battle had broken up the finished drill organizations, the groups would take their place. In this way individual experience and adaptability would not be paralyzed by too narrow restrictions. The fact is, the modern battery and its fire tactics must rely upon the *unbidden help* of every one on deck, in the same manner that the old sail and spar evolutions did.

With regard to the seaman. The modern battery, though requiring more precision and permitting many more combinations than that of thirty years ago, is a bad exchange, as an educator, for sails and spars. It is more intricate at first, but once mastered, the treatment of its parts is governed by simple and invariable laws. On the other hand, Nature has a myriad of phases; she is always grand and courage-provoking. A mechanism, to be satisfactory, must always be the same. At times it may be utterly unreliable, or unintelligible, until the cause of derangement is discovered. Were the navy obliged to depend on battery drills alone, or machine drills of any sort, for the training of its *personnel*, the result could not help but be deterioration.

By deterioration is meant simply that the mind, which in the most ordinary duties of the sailor has considerable range of judgment, is limited in the management of machines by a few narrow thumb-rules. The underlying principles will appear to the most intelligent, but beyond and back of it all there must always remain much that cannot be understood or appreciated by any but the specially educated. In fact, however complicated the machinery and ingenious the mind that devised it, only those features which are concerned in its care and manipulation will, with rare exceptions, enter the life of the enlisted man. These principles are naturally made as simple and distinct as possible, in order that they may not require more of the seaman's attention than their relative importance demands. Continual gun or machine drill for an entire cruise can do no more than turn the thinking seaman into an obedient puppet.

To many, it would seem that the days of the seaman are gone; that he should be replaced by a certain proportion of skilled mechanicians and common laborers. But with such a change the very vitality of the sea-fight is lost. A sailor may become as much of a machinist as is required, but as soon as this work takes the nature of a steady vocation, he seems to become less a seaman.*

Seamanship is by no means dependent upon sails and spars; these gave only the best means of its exhibition. All that has been and always will be the best part of the seaman—that readiness of resource and instant adaptation; above all, the quick comprehension of the most pressing need—is of more value now than ever. The question is, how can it best be fostered and developed?

In no way could these qualities be better studied than in boat work—not the boat work of smooth water and a gentle breeze, when half the crew are gradually lulled asleep, but the boat work that makes each man alert for his own safety's sake; the boat work that speedily puts in the coxswain's box a man who can handle the crew as well as the boat; that makes each man there respect himself and his neighbor because their skill and nerve are so essential to the general welfare.

The writer has no thought of encouraging foolhardiness, or of stimulating men to sacrifice themselves, however heroically, where nothing can be gained. On the other hand, he would attempt to cultivate the judgment that discerns the impossible, however much it may be regretted. It is no impeachment of the courage of seamen to say that the more they know of the difficulties of a rescue at sea, the more commendable will be the act of volunteering.

* Machinery covers the earth and is as varied as the requirements of mankind. A machinist must know and do many things, but he knows that, in so broad a field, he owns his title only so far as his knowledge is *exact*. Hence most of his life's energy is expended within narrow limits; sometimes he merely works for the sake of the work without a thought of its ultimate purpose. His vocation is a valuable one wherever exact results are required; but for fighting a very different character is wanted—a character that knows no rule of spirit or material to prevent success.

When to lower a boat in rough weather at sea, just as when to
attempt the passage of surf, is a question dependent upon many
ever differing conditions.   Among these, any one of the three most
important—experience of crew, qualities of boat, and state of sea—
must outweigh all the rest.   To the commanding officer, one man's
life should mean no more than another's, and the staking of six or
eight for one could be justified only when success proved that the
circumstances made the odds more favorable.   For, however much
sentiment may pervade the action and influence the decision, far
away in cool and quiet judgment the deed will be discussed and its
advisability admitted or denied in a very different spirit.

Individual heroism or *esprit de corps* may lead to a contempt of
the judgment that would weigh the chances of success, so that it is no
pleasant task to restrain them.   Yet a barren sacrifice of brave men
should never be permitted.

APPENDIX.

"JAMES BEECHING, BOAT-BUILDER, GREAT YARMOUTH.

Description.—The body of this boat is of the form usually given
to a whale-boat—a slightly rounded floor, sides round in the fore and
aft direction, upright stem and stern post, clench built, of wainscot
oak, and iron fastened.

Length extreme, 36′; of keel, 31′; breadth of beam, 9½′; depth,
3½′; sheer of gunwale, 36″; rake of stem and stern post, 5″; straight
keel, 8″ deep.   The boat has seven thwarts 27″ apart, 7″ below the
gunwale and 18″ above the floor; pulls 12 oars, double banked, with
pins and grommets.   A cork fender 6″ wide by 8″ deep runs around
outside at 7″ below the gunwale.

Extra buoyancy is given by air cases 20″ high in the bottom of
the boat under the flat; round part of the sides, 24″ wide by 18″
deep, up to the level of the thwarts, leaving 10′ free amidships; and
in the head and stern sheets, for a length of 8½′, to the height of the
gunwale; the whole divided into compartments and built into the
boat; also by the cork fenders.   Effective extra buoyancy 300 cubic
feet, equal to 8½ tons.   For ballast a water-tank divided into compartments, placed in the bottom amidships, 14′ long by 5′ wide and
15″ high, containing 77 cubic feet, equal to 2½ tons when full, and an
iron keel of 10 cwt.   Internal capacity of boat under the level of the
thwarts, 176 cubic feet, equal to 5 tons.   Means of freeing the boat of

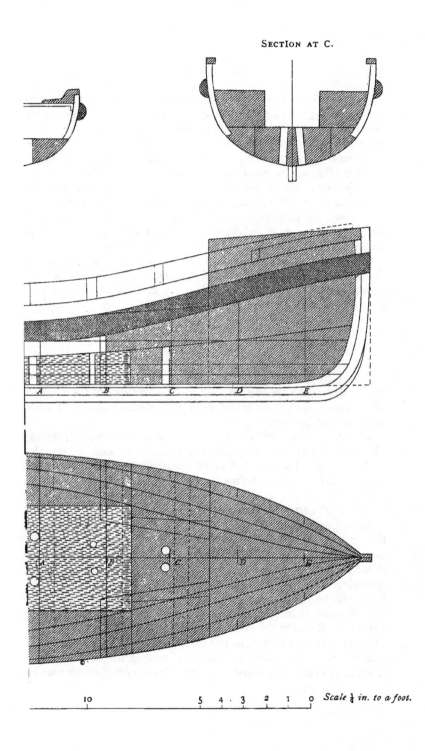

SECTION AT C.

Scale ¼ in. to a foot.

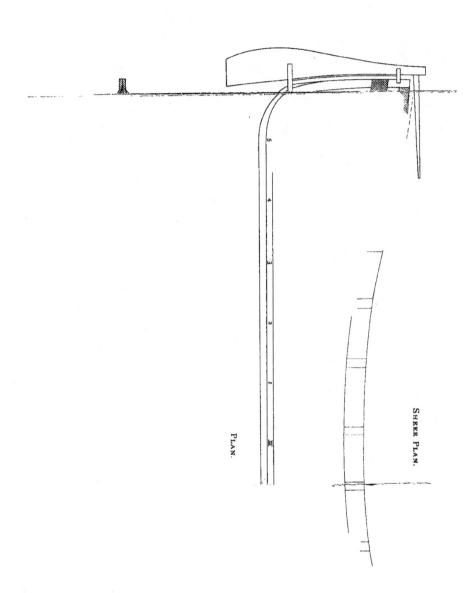

PLAN.

SHEER PLAN.

water, tubes through the bottom, 8 of 6″ in diameter and 4 of 4″ in diameter; total area, 276 square inches, which is to the capacity in the proportion of 276 to 176, or as 1 to .64. Provision for righting the boat if upset, 2¼ tons of water ballast, an iron keel, and raised air cases in the head and stern sheets. Rig, lug foresail and mizzen; to be steered by a rudder; no timber heads for securing a warp to. Draft of water with 30 persons on board, 26″. Weight of boat, 50 cwt., of gear, 17 cwt.; total, 67 cwt. Would carry 70 persons. Cost, with gear, £250.

Remarks.—The form given to this boat would make her efficient either for pulling or sailing in all weathers; she would prove a good sea boat, and in places such as Yarmouth, where there are always plenty of hands to launch a boat, her weight would cause no difficulty. By means of the raised air cases placed at the extremes, the absence of side air-cases for a length of 10′ amidships, the introduction of 2¼ tons of water ballast into her bottom when afloat, and her iron keel, this boat would right herself in the event of being capsized; although from the form given to her it is highly improbable that such an accident should occur.

A passage should be left in the air cases to approach the stem and stern, for on many occasions the only way in which a life-boat can go near a wreck is end on, when the crew of it must be received either over the stem or the stern. The keel, 8″ deep, however favorable for sailing, for steadying her in a seaway, and for aiding her in righting, would be a disadvantage in beaching and would render the boat more difficult to turn in case of wishing to place her end on to a heavy roller coming in. The area of the delivering valves is large in proportion to the internal capacity, and would rapidly free the boat of water, down to the level of her draft, which with her crew on board would not be to less than a depth of some inches above the flow. The air cases are built into the boat, which renders them liable to accidents; if this were remedied and her internal capacity reduced, a 30′ or 32′ boat built on similar lines, with her internal fittings slightly modified, would make an efficient life-boat, adapted for many parts of the coast."

Order of merit, 1. Mark, 84.

"GEORGE PALMER, NAZING PARK, WALTHAM ABBEY, ESSEX.

Description.—The form of the midship body of this boat is that of a whale-boat, with upright stem and stern, rising floor, sides straight in a fore and aft direction, clench built, of elm and fir, and copper fastened.

Length, extreme, 26'; length of keel, 24'; breadth, 6¾'; depth, 3¼'; sheer of gunwale, 20"; straight keel, 3" deep. The boat has 5 thwarts, 24" apart, 9" below gunwale, and 22.5" above floor; pulls 5 oars, single banked, with tholepins. A cork fender 4" wide and 4" deep extends along both sides close up to the gunwale, but does not reach within 18" of the stem or stern.

Extra buoyancy is obtained by detached air cases of wood, 18" square, along the sides up to the level of the thwarts, and in the bow and stern sheets up to the height of the gunwale, the whole divided into 12 compartments; also by the cork fenders. Effective extra buoyancy, 82 cubic feet, equal to 2⅜ tons. No ballast. Internal capacity up to the level of the thwarts, 62 cubic feet, equivalent to 1¾ tons. No means of freeing the boat of water except by bailing. Provision for righting the boat if capsized, raised air vessels at each extremity. To be steered by a rudder. There are two timber heads for warps, one on each bow; wash-strakes around the head and stern sheets, and life-lines along the gunwale. Draft of water with 22 persons on board, 15". Weight of boat, 10 cwt., of gear, 5 cwt.; total, 15 cwt. Would carry 20 persons. Cost £75.

Remarks.—This boat would pull well, is light for transporting along a beach, could readily be manned at any station on the coast, has light draft of water, small internal capacity for holding water up to the level of the thwarts, and detached air cases, which would preserve them from the risk of being stove alongside a wreck; all of which are good points in her favor. She has timber heads forward for securing a warp to, but none on the quarter, and plenty of life-lines.

The boat would not right herself if capsized, and the raised air-cases at each end would prevent any approach to either extremity within 4½', which might prove extremely inconvenient in case of having to receive a wrecked crew on board at the head or stern of the boat, which not uncommonly occurs. The boat is narrow for her length, and her rising floor is not favorable for taking a beach.

This model has been generally adopted by the Royal National Institution for the preservation of lives from shipwreck, and several similar boats are, it is believed, placed around the coasts, as in the Isle of Anglesey and elsewhere in Wales, which, it is said, have been the means of saving many lives."

Order of merit, 8. Mark, 70.

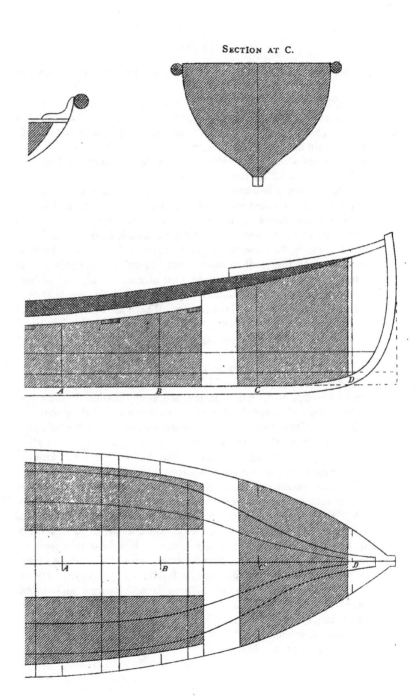

SECTION AT C.

in a fore and aft direction, clench built, of elm and fir, and copper
fastened.

Length, extreme, 26'; length of keel, 24'; breadth, 6¾'; depth,
3¼'; sheer of gunwale, 20"; straight keel, 3" deep. The boat has
5 thwarts, 24" apart, 9" below gunwale, and 22.5" above floor; pulls
5 oars, single banked, with tholepins. A cork fender 4" wide and 4"
deep extends along both sides close up to the gunwale, but does
not reach within 18" of the stem or stern.

Extra buoyancy is obtained by detached air cases of wood, 18"
square, along the sides up to the level of the thwarts, and in the
bow and stern sheets up to the height of the gunwale, the whole
divided into 12 compartments; also by the cork fenders. Effective
extra buoyancy, 82 cubic feet, equal to 2⅜ tons. No ballast. Internal
capacity up to the level of the thwarts, 62 cubic feet, equivalent to
1¾ tons. No means of freeing the boat of water except by bailing.
Provision for righting the boat if capsized, raised air vessels at each
extremity. To be steered by a rudder. There are two timber heads
for warps, one on each bow; wash-strakes around the head and
stern sheets, and life-lines along the gunwale. Draft of water with
22 persons on board, 15". Weight of boat, 10 cwt., of gear, 5 cwt.;
total, 15 cwt. Would carry 20 persons. Cost £75.

Remarks.—This boat would pull well, is light for transporting
along a beach, could readily be manned at any station on the coast,
has light draft of water, small internal capacity for holding water up
to the level of the thwarts, and detached air cases, which would preserve them from the risk of being stove alongside a wreck; all of
which are good points in her favor. She has timber heads forward
for securing a warp to, but none on the quarter, and plenty of life-lines.

The boat would not right herself if capsized, and the raised air-
cases at each end would prevent any approach to either extremity
within 4½', which might prove extremely inconvenient in case of
having to receive a wrecked crew on board at the head or stern of
the boat, which not uncommonly occurs. The boat is narrow for
her length, and her rising floor is not favorable for taking a beach.

This model has been generally adopted by the Royal National
Institution for the preservation of lives from shipwreck, and several
similar boats are, it is believed, placed around the coasts, as in the
Isle of Anglesey and elsewhere in Wales, which, it is said, have been
the means of saving many lives."

Order of merit, 8. Mark, 70.

Scale ⅛ in. to a foot

To face page 375.

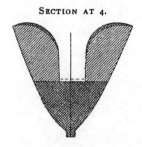

SECTION AT 4.

AIR.

CORK.

BODY PLAN.

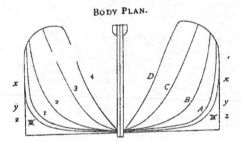

PRINCIPAL DIMENSIONS.

Length extreme . . . . 30' 0"
Length of keel . . . . 24' 0"
Breadth of beam . . . . 8' 9"
Depth to underside of keel . . . . 3' 6"
Sheer of gunwale . . . . 2' 1"
Extra buoyancy, cork and air, equal to . . . 3 tons.
Internal capacity up to level of thwarts . . . 4
Area of delivering valves . . . . 300 sq. ins.
Proportion of delivering area to capacity . . . 1 to .5
Weight of boat and fittings or displacement . . 38 cwt.
Ballast iron band, equal to. 3
Draft of water with 30 men on board . . . 16"
No. of oars double banked 10
Diagonal-built of rock elm and copper-fastened.
Rig of boat, a fore and mizzen lug sail.
Cost complete . . . .

"JAMES PEAKE, ASSISTANT MASTER SHIPWRIGHT IN H.M.'S DOCKYARD, WOOLWICH.

Description.—The form of this boat is that usually given to a whale-boat, having a long flat floor amidships, sides straight in a fore and aft direction, raking stem and stern post, diagonally built of two thicknesses of rock elm, and copper fastened.

Length extreme, 30′; length of keel, 24′; breadth of beam, 8¾′; depth, 3½′; rake of stem and stern posts, 6½″ in a foot; straight keel 4″ deep, and bilge pieces with holes in them to lay hold of on each side on the bottom. The boat has five thwarts, 7″ wide, 28″ apart, 7″ below the gunwale and 15″ above the floors, pulls 10 oars, double banked, with pins and grommets. A fender of cork, 4″ wide by 2½″ deep, extends fore and aft at 4″ below the gunwale.

Extra buoyancy is obtained by cork placed the whole length of the boat under the flooring to a height of 12″ above the keelson, and by light cork or detached air-cases in the head and stern sheets up to gunwale height. Effective extra buoyancy, 105 cubic feet, equal to three tons. A light water-tight deck will be placed on the cork to protect it, and above that a light grating. For ballast, the weight of the cork in the bottom and an iron keel of 5 cwt. Internal capacity for holding water up to the level of the thwarts, 140 cubic feet, equivalent to four tons. The means of freeing the boat of water are by eight tubes of 6-inch diameter through the bottom, and six scuppers through the sides at the height of the flooring, giving a total delivering area of 300 square inches, which is to capacity as 1 to .5. The provision made for righting the boat consists in the sheer given to the gunwales, raised air vessels or cork in the head and stern sheets, and the ballast arising from the weight of cork in the bottom, and the small iron keel. A passage 18″ wide up to within 2′ of the stem and stern is left between the raised air cases in the extremes, and the top of the cases is protected by a layer of cork. Rig, fore and mizzen lug-sail. To be steered by a sweep oar at either end. Timber heads for warps are placed at each bow and quarter, and a roller for the cable in the stem and stern post head. A locker under the flooring amidships for the anchor and cable to be secured down to the keelson, and covered with a water-tight scuttle.

A life-line fore and aft at a foot below the gunwale, and short knotted life-lines to be hung over the sides at each thwart. Draft of water with 30 men aboard, 16″.

Weight of boat and fittings, 38 cwt. Would carry 60 persons.

PLAN

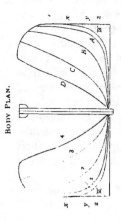

BODY PLAN.

PRINCIPAL DIMENSIONS.

Length extreme . . . . 30' 0"
Length of keel . . . . 24' 0"
Breadth of beam . . . . 8' 9"
Depth to underside of keel . . . . . 3' 6"
Sheer of gunwale . . . . 2' 1"
Extra buoyancy, cork and air, equal to . . 3 tons.
Internal capacity up to level of thwarts . 4
Area of delivering valves . . . . 300 sq. ins.
Proportion of delivering area to capacity . . 1 to 5
Weight of boat and fittings or displacement . 38 cwt.
Ballast iron band, equal to. 3
Draft of water with 30 men on board . . 16"
No. of oars double banked 10
Diagonal-built of rock elm and copper-fastened.
Rig of boat, a fore and mizzen lug sail.
Cost complete . . . .

"JAMES PEAKE, ASSISTANT MASTER SHIPWRIGHT IN H. M.'S DOCKYARD, WOOLWICH.

Description.—The form of this boat is that usually given to a whale-boat, having a long flat floor amidships, sides straight in a fore and aft direction, raking stem and stern post, diagonally built of two thicknesses of rock elm, and copper fastened.

Length extreme, 30'; length of keel, 24'; breadth of beam, 8¾'; depth, 3½'; rake of stem and stern posts, 6½" in a foot; straight keel 4" deep, and bilge pieces with holes in them to lay hold of on each side on the bottom. The boat has five thwarts, 7" wide, 28" apart, 7" below the gunwale and 15" above the floors, pulls 10 oars, double banked, with pins and grommets. A fender of cork, 4" wide by 2½" deep, extends fore and aft at 4" below the gunwale.

Extra buoyancy is obtained by cork placed the whole length of the boat under the flooring to a height of 12" above the keelson, and by light cork or detached air-cases in the head and stern sheets up to gunwale height. Effective extra buoyancy, 105 cubic feet, equal to three tons. A light water-tight deck will be placed on the cork to protect it, and above that a light grating. For ballast, the weight of the cork in the bottom and an iron keel of 5 cwt. Internal capacity for holding water up to the level of the thwarts, 140 cubic feet, equivalent to four tons. The means of freeing the boat of water are by eight tubes of 6-inch diameter through the bottom, and six scuppers through the sides at the height of the flooring, giving a total delivering area of 300 square inches, which is to capacity as 1 to .5. The provision made for righting the boat consists in the sheer given to the gunwales, raised air vessels or cork in the head and stern sheets, and the ballast arising from the weight of cork in the bottom, and the small iron keel. A passage 18" wide up to within 2' of the stem and stern is left between the raised air cases in the extremes, and the top of the cases is protected by a layer of cork. Rig, fore and mizzen lug-sail. To be steered by a sweep oar at either end. Timber heads for warps are placed at each bow and quarter, and a roller for the cable in the stem and stern post head. A locker under the flooring amidships for the anchor and cable to be secured down to the keelson, and covered with a water-tight scuttle.

A life-line fore and aft at a foot below the gunwale, and short knotted life-lines to be hung over the sides at each thwart. Draft of water with 30 men aboard, 16".

Weight of boat and fittings, 38 cwt. Would carry 60 persons.

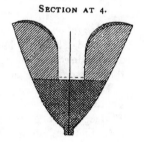

SECTION AT 4.

AIR.

CORK.

BODY PLAN.

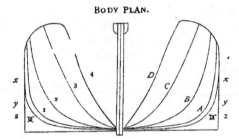

PRINCIPAL DIMENSIONS.

Length extreme . . . . 30' 0"
Length of keel . . . . 24' 0"
Breadth of beam . . . . 8' 9"
Depth to underside of keel . . . . 3' 6"
Sheer of gunwale . . . . 2' 1"
Extra buoyancy, cork and air, equal to . . . 3 *tons.*
Internal capacity up to level of thwarts . . . 4
Area of delivering valves . . . . 300 *sq. ins.*
Proportion of delivering area to capacity . . . 1 *to* 5
Weight of boat and fittings or displacement . . 38 *cwt.*
Ballast iron band, equal to. 3
Draft of water with 30 *men on board* . . . 16"
No. of oars double banked 10
Diagonal-built of rock elm and copper-fastened.
Rig of boat, a fore and mizzen lug sail.
Cost complete . . . .

"JAMES PEAKE, ASSISTANT MASTER SHIPWRIGHT IN H. M.'S
DOCKYARD, WOOLWICH.

Description.—The form of this boat is that usually given to a
whale-boat, having a long flat floor amidships, sides straight in a fore
and aft direction, raking stem and stern post, diagonally built of two
thicknesses of rock elm, and copper fastened.

Length extreme, 30′; length of keel, 24′; breadth of beam, 8¾′;
depth, 3½′; rake of stem and stern posts, 6½″ in a foot; straight keel
4″ deep, and bilge pieces with holes in them to lay hold of on each
side on the bottom. The boat has five thwarts, 7″ wide, 28″ apart,
7″ below the gunwale and 15″ above the floors, pulls 10 oars, double
banked, with pins and grommets. A fender of cork, 4″ wide by
2½″ deep, extends fore and aft at 4″ below the gunwale.

Extra buoyancy is obtained by cork placed the whole length of
the boat under the flooring to a height of 12″ above the keelson, and
by light cork or detached air-cases in the head and stern sheets up
to gunwale height. Effective extra buoyancy, 105 cubic feet, equal
to three tons. A light water-tight deck will be placed on the cork
to protect it, and above that a light grating. For ballast, the weight
of the cork in the bottom and an iron keel of 5 cwt. Internal
capacity for holding water up to the level of the thwarts, 140 cubic
feet, equivalent to four tons. The means of freeing the boat of water
are by eight tubes of 6-inch diameter through the bottom, and six
scuppers through the sides at the height of the flooring, giving a
total delivering area of 300 square inches, which is to capacity as 1
to .5. The provision made for righting the boat consists in the
sheer given to the gunwales, raised air vessels or cork in the head
and stern sheets, and the ballast arising from the weight of cork in
the bottom, and the small iron keel. A passage 18″ wide up to
within 2′ of the stem and stern is left between the raised air cases in
the extremes, and the top of the cases is protected by a layer of cork.
Rig, fore and mizzen lug-sail. To be steered by a sweep oar at
either end. Timber heads for warps are placed at each bow and
quarter, and a roller for the cable in the stem and stern post head.
A locker under the flooring amidships for the anchor and cable to be
secured down to the keelson, and covered with a water-tight scuttle.

A life-line fore and aft at a foot below the gunwale, and short
knotted life-lines to be hung over the sides at each thwart. Draft of
water with 30 men aboard, 16″.

Weight of boat and fittings, 38 cwt. Would carry 60 persons.

Actual cost if built in one of H. M.'s dockyards: Materials, £40;
labor, £45; total, £85.

Remarks.—It is anticipated that this boat, from her form, will pull
fast in all weathers, and be fully able to contend against a head sea.
She would sail well, and from her flat and long floor and straight
sides would have great stability and prove a good sea boat.   From
a slight fulness in her entrance and flaring bow aloft, she will beach
well, and leave the shore through breakers in safety.   It is not probable that a boat of this form could be readily upset; but should such
an accident occur, the sheer of gunwale, raised air-cases in the
extremes, weight of cork in the bottom, and iron keel would cause
her to right herself.   The area of the delivering valves is ample, and
the boat would readily free herself of all water above the flooring
when she has 30 persons on board.   In the possible case of the tubes
through the bottom being choked by the boat grounding on a bar
or bank, there are sufficient scuppers provided in the side to free the
boat of water.

As the greater part of the buoyancy is obtained by cork, all liability
to accident is avoided.   It is proposed to use light fisherman's cork
(of about 12 pounds weight to the cubic foot) for the righting power
in the head and stern sheets, if not found too heavy; otherwise
detached air-cases, divided into compartments to be formed of a
layer of gutta-percha between two thin boards.   The diagonal mode
employed in building gives great strength; the planks are of one
length from gunwale to gunwale, and the keel brought on the bottom,
so that if it were knocked away it would not damage the boat.

The builder claims no merit for his design beyond that of having
selected the best points from the several models selected for competition, and having combined them in a form as shown in Plate III.,
which appears to him better adapted to the general purposes of a
life-boat than any he has hitherto seen.   It is proposed to place the
boat, as soon as completed, at Cullercouts, on the coast of Northumberland, two miles north of the entrance of the Tyne, as a station
well adapted for testing her capabilities."

Attention is called to the way in which the gunwale sheer is discontinued at bow and stern as soon as the ends become too fine for
such sheer to be any use for buoyancy.   Another interesting point
is the extreme cheapness with which the boat was built at a government dockyard.   In 1852, this boat was completed.   It is stated in
an instructive treatise on saving life at sea, issued by the Secretary of

Actual cost if built in one of H. M.'s dockyards: Materials, £40;
labor, £45; total, £85.

Remarks.—It is anticipated that this boat, from her form, will pull
fast in all weathers, and be fully able to contend against a head sea.
She would sail well, and from her flat and long floor and straight
sides would have great stability and prove a good sea boat.   From
a slight fulness in her entrance and flaring bow aloft, she will beach
well, and leave the shore through breakers in safety.   It is not probable that a boat of this form could be readily upset; but should such
an accident occur, the sheer of gunwale, raised air-cases in the
extremes, weight of cork in the bottom, and iron keel would cause
her to right herself.   The area of the delivering valves is ample, and
the boat would readily free herself of all water above the flooring
when she has 30 persons on board.   In the possible case of the tubes
through the bottom being choked by the boat grounding on a bar
or bank, there are sufficient scuppers provided in the side to free the
boat of water.

As the greater part of the buoyancy is obtained by cork, all liability
to accident is avoided.   It is proposed to use light fisherman's cork
(of about 12 pounds weight to the cubic foot) for the righting power
in the head and stern sheets, if not found too heavy; otherwise
detached air-cases, divided into compartments to be formed of a
layer of gutta-percha between two thin boards.   The diagonal mode
employed in building gives great strength; the planks are of one
length from gunwale to gunwale, and the keel brought on the bottom,
so that if it were knocked away it would not damage the boat.

The builder claims no merit for his design beyond that of having
selected the best points from the several models selected for competition, and having combined them in a form as shown in Plate III.,
which appears to him better adapted to the general purposes of a
life-boat than any he has hitherto seen.   It is proposed to place the
boat, as soon as completed, at Cullercouts, on the coast of Northumberland, two miles north of the entrance of the Tyne, as a station
well adapted for testing her capabilities."

Attention is called to the way in which the gunwale sheer is discontinued at bow and stern as soon as the ends become too fine for
such sheer to be any use for buoyancy.   Another interesting point
is the extreme cheapness with which the boat was built at a government dockyard.   In 1852, this boat was completed.   It is stated in
an instructive treatise on saving life at sea, issued by the Secretary of

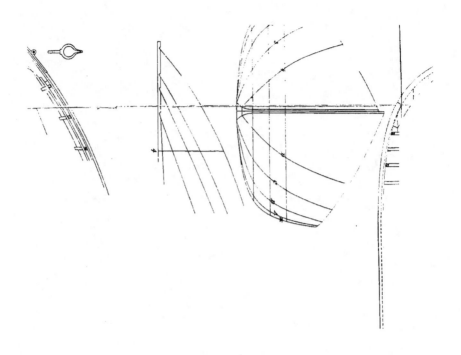

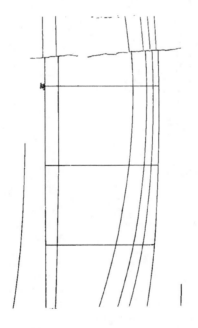

the National Life Boat Institution, that its beam was 8′ instead of
8¾′, that it was furnished with side air-tanks, and air-tanks under the
thwarts; draft, 15″; with crew, 18″; iron keel of 7 cwt. (instead of
5 cwt.); and weighed 46 cwt. instead of 38 cwt. "When she was
hove keel up by a crane, she righted in 5 seconds. When light, she
entirely freed herself of water in 55 seconds. When running for the
beach through heavy rollers, she showed great buoyancy and stability,
without shipping water. She could carry 30 persons beside her crew,
or 42 in all. The present boat in use has been developed from this
one."

SPECIFICATIONS OF THE BEEBE-McLELLAN SELF-BAILING SURF-BOAT.

Dimensions.—Length over all, 25 feet 4 inches; length between
rabbets at sheer-line, 25 feet; greatest breadth outside of planks, 7
feet; depth amidships above keel, 2 feet 5 inches; sheer of gunwale,
2 feet 1 inch.

Keel or Bottom Piece.—The keel is to be of white oak, 2 inches
thick, 9¾ inches wide amidships, and rabbeted ⅞ inch in width for
planks. The ends of the keel are to be steamed and bent up to
shape.

Stem and Stern Posts.—They are to be of white oak, sided 2
inches, molded 4 inches at gunwale, and 5 inches at scarf of keel,
overlapping the keel 8 inches, each lap to be fastened by two galvanized ⅜-inch iron bolts driven and riveted over burrs. These stems
are to be cut out of timber with natural crook to suit the shape.

Frames.—The frames are to be bent out of white oak, siding 1 inch
and molding 1⅜ inches, then cut to fit to the planking. The floor-
timbers running across the keel are to be at least 4 feet 9 inches long
amidships, and to be formed of knees at ends of boat; they are to
be placed 12¾ inches from center to center amidships and closer
at the ends, as per plan. Futtocks are to be placed alongside of
each floor-timber, also between each of the floor-timbers amidships.
Towards either end they are to be spaced, as shown on plan.

Outside Planks.—The outside planks are to be ten in number on
each side of the boat, made of ½-inch thick white cedar, with ⅞-inch
laps, and fastened with four-penny copper nails riveted over burrs,
one through each frame, one between amidships frames, and two
where the distance between futtocks increases. The hood ends are
to be fastened with galvanized iron nails; the bevel-scarfs are to lap

2 inches, covered on the inside by 1½ by ⅜ inch white oak strips with six rivets passing through both, secured over burrs. At the bottom of each plank are battens of ½ by ⅞-inch white oak extending to within 4 feet from the ends of the boat; the one under the top-strake is to be ⅞ inch square, rounded on the outside and extend-ing the whole length of the boat. An outside molding batten at height of sheer is to be of white oak, 2 inches deep, ⅞ inch thick, and rounded on outside; the fastenings to pass through the frames and inner gunwale.

Gunwales and Breast-Hooks.—Inner gunwales to be of white oak, 2 by ⅞ inch, fastened at ends into white oak breast-hooks made of knees with 5-inch throat and 14-inch arms. The breast-hooks are to be riveted through the stems.

Thwarts.—The thwarts, four in number, 7½ inches below the top of gunwale, are to be of 8 by 1¼ inch spruce with two white oak knees on each end on top, siding 1 inch, and with arms on thwart 13½ inches long; each thwart to have a ⅞-inch support underneath with cleats under thwarts and steps on deck. Other supports to be placed in the hold underneath upper supports.

Seats.—Seats on ends to be of ⅝-inch white cedar, supported by 1 by 2-inch cedar carlings; part of the forward seat to be removable; the tops or covering boards of the side air-tanks are continuations of the after-seat. All seats to be supported by ⅝-inch cedar planks. All as shown on plan.

Deck.—The deck is to be of ⅝-inch white cedar, about 5 inches wide, resting on cedar carlings 2 by 1 inch, which are let into risers of 2 by ⅝-inch cedar, all as shown on plan. The deck-planks are to be fastened by five-penny galvanized-iron nails set in and bees-waxed.

Air-Tanks.—The side-boards of the air-tanks are to be of ⅝-inch cedar, to reach from their covering-boards to the carlings of the deck, to which they are to be fastened. They are also to be nailed to the deck-planks in the spaces between the carlings, and bracing-strips of ⅝-inch cedar are to be fastened on top of each carling between the side-boards and the outside planking.

Delivery-Tubes.—The frames of the delivery-tubes are to be made of ⅞-inch white pine, jogged over frames, fastened to them and the outside planking, and provided on each side above the deck with 2½ by 4⅝-inch openings. Inside of these openings are to be valves of rubber packing fastened to the inside of tubes on top by two brass screws passing through a protecting strip ; on the bottom they have three to four ounces of lead clinched on by copper tacks. Around the opening on the inside of the tubes are strips of cedar running from nothing at the top to ¼-inch at bottom, put on with galvanized brads, and with paint between them and the boxes. Tops of boxes to be of ⅝-inch cedar. The tubes to taper in both directions, as shown on plan.

Bitts and Cleats.—Bitts and cleats of white oak forward and aft, as per plan.

Ring-Bolts.—Ring-bolts through both stems of ⅝-inch galvanized iron, rings of 3-inch opening—all well clinched over rings on inside of stems.

Row-Lock Beds.—The beds are to be of white oak, 15 inches long and 1 inch thick, fitted for tholepins and fastened to outer planks and inside gunwale by galvanized iron screws ; they are to be partly let into the inner gunwale. The center of rowlocks to be about 10 inches abaft the after edge of thwarts, and the centers of tholepins to be 5½ inches apart.

Steering Oarlock.—The steering oarlock is to be of composition, shaped as marked in plan, and passing through an opening in a galvanized iron band. This band to have arms 13 inches long, 1½ inches deep, and ⅝ inch thick, and to be fitted around stern-post and to outer molding-batten, and to be fastened by ¼-inch iron rivets, clinched over burrs on inside of breast-hook.

Man-Hole.—The man-hole is to be 17½ inches in diameter, and have a flanged brass ring fitted around it ; the ring to be 1 inch deep and the flange, 1 inch wide ; the flange to be on top of the deck, and secured to it by brass screws.

The cover is to be of cedar in one or two thicknesses, the top being ¼ inch above the deck and extending ½ inch beyond the man-hole, so as to form a shoulder ; a brass flanged ring is also to be fitted to the cover, the ring to be ⅞ inch deep and the flange, ⅞ inch wide ; the flange to be underneath the cover and secured to it by brass screws.

The two rings are to screw into each other, and to be with their flanges ¼ inch thick.

On top of the cover there are to be screwed two oak battens slightly out of parallel, and to the space between them is to be fitted a tapered wrench of oak.

Pump.—There is to be in the hold of the boat a brass pump of 3-inch inside diameter, with rubber boxes and a cover screwed into the top, as shown on plan.

Ventilators.—In each of the side air-tanks is to be a brass ventilator of 1½ inches opening.

Calking.—The deck and air-tanks are to be carefully calked with cotton and white-leaded.

Paint.—The outside of the boat is to be painted with two coats of good oil paint, green to water-line and white from there to molding-batten. The inside of boat is to be either painted or receive two coats of spar-varnish, as the superintendent may direct.

Materials.—All the materials to be of first quality ; all the wood-work to be of perfectly seasoned stuff, clear and free from sap and bad or large knots, and the workmanship and finish to be first-class in every respect, all subject to the approval of the superintending officer.

To face page 333.

U. S. NAVAL INSTITUTE, ANNAPOLIS, MD.

June 3, 1890.

Rear-Admiral L. A. Kimberly, U. S. N., in the Chair.

THE HOWELL AUTOMOBILE TORPEDO.

By E. W. Very, *of the Hotchkiss Ordnance Company.*

PART I.

THE INTRINSIC VALUE OF THE TORPEDO AS A WEAPON OF OFFENSE.

In popular discussions of the value of the torpedo as a weapon of offense, it very frequently occurs that the objective feature of the weapon is lost sight of, or becomes confused with the condition of its development, so that, because with any particular type the results hitherto obtained have been but moderately successful, doubt is expressed as to the intrinsic value of the weapon, and it is even a commonly expressed opinion that sooner or later some form of ordnance capable of discharging large quantities of a high explosive will entirely replace all torpedoes except perhaps fixed mines for the defense of channels.

In naval warfare the torpedo is as distinct a class of weapon and is as independent of all ordnance development as is the "arme blanche" (such as the bayonet, sword, etc.) from the fire-arm in military warfare. The invention of fire-arms created no new objective feature of warfare, but simply provided a means of meeting a necessity that had always existed. During the first period of the introduction of fire-arms as infantry weapons, the line of battle consisted of ranks of arquebus men and ranks of pikemen. The distinct objective features of the two weapons were confided to different corps. The invention of the bayonet enabled one corps to fulfill both features, but neither this consolidation nor any improvement, from the matchlock to the magazine rifle, or from the pike to the sword-bayonet, has in the least degree affected the objective features themselves, nor have improvements in one weapon lessened the necessity for the possession of the other.

Precisely the same conditions exist in naval warfare between ordnance and torpedoes. The former class of weapons is and can only be devoted to the attack of the above-water portions of a vessel, whilst the latter is as exclusively devoted to the under-water body. As ordnance has developed in power, the engines, boilers, steering-gear, magazines and other vital elements of fighting power have been driven for safety to the under-water body, for it is there that the defense is the strongest and the offense the weakest. To leave the attack of the under-water body out of consideration simply because hitherto the development of the torpedo has been but extremely limited, is to omit a fundamental principle of offensive tactics. It has been only one hundred years since Bushnell invented the torpedo. He created no new objective feature any more than did the inventor of the arquebus. It required Cushing's attack on the Albemarle to convince the world that the feature not only existed, but was of prime importance, and from that day the development of the weapon that shall most effectively fulfill that feature has been unceasing, and in the very nature of naval warfare will continue to be.

The torpedo, as a naval weapon of offense, is a permanent weapon, and no ordnance development, no matter what be its nature, can in the slightest degree affect its existence.

CLASSIFICATION OF PURELY NAVAL TORPEDOES.

As in the general discussion of the value of torpedoes, the absolute necessity for the existence of the weapon loses appreciation from confusing it with considerations of the state of development, so in the discussion of development, a failure to properly distinguish the natural classification leads to confusion in the attempt to compare different types.

It would be manifestly absurd to attempt to compare directly and generally a magazine rifle with a field gun, or a field gun with a mortar. Quite as distinct a classification exists amongst torpedoes. Fixed mines have a certain special field of action within which they are undoubtedly superior to torpedoes designed for other fields. Torpedoes that are propelled and guided from a fixed point, with which they are in some manner constantly connected, have also their distinct field. Torpedoes that may be classed under the head of purely naval torpedoes occupy a field of their own, for they are required to be effective in waters of an indefinite depth and under all conditions of movement of the point of discharge.

Considering only this general classification of the naval torpedo, it will be found that, as represented by development, the different types come naturally under one of the three following subdivisions:

1st. The fixed torpedo, carried and used by a vessel at a fixed and very limited distance from it, such as the Spar and the Towing torpedoes.

2d. The semi-automobile torpedo, which upon discharge is independent of the vessel, but which for range is dependent upon the force of projection, and which lacks a complete development of directive force, such as the projectile from the submarine gun.

3d. The automobile torpedo, which, independently of the vessel and irrespective of the means of discharge, maintains its speed by self-contained motive power, and its depth and direction by self-contained directive power.

These subdivisions are developments the one from the other in the order above given, and that the higher development has not rendered the lower one obsolete is due entirely to incompleteness of development. The " bag of powder on the end of the pole " so ably handled by Cushing is clearly but of a most limited efficiency, since it can scarcely be used beyond a distance of fifty feet ; yet it is not entirely obsolete, because certain features which it possesses within that distance are as yet not as certainly assured in the higher developments. The projectile ejected from the submarine gun possesses elements of great simplicity, but it needs but a cursory examination to show that it is an inferior development to the automobile torpedo, for it depends upon the force of ejection entirely for its speed and range; but precisely the same force may be applied in precisely the same way to the automobile torpedo, producing exactly the same result. The automobile torpedo, however, possesses, *in addition* and entirely independent of this force of ejection, an *inherent* propulsive and directive force, which is so much clear gain over the submarine projectile, and any attempt whatever to increase range and accuracy of the latter can only be obtained through an approach to the automobile condition. Therefore the automobile torpedo is the highest (not necessarily the best perfected) development of the naval torpedo.

THE PRESENT STATUS OF THE AUTOMOBILE TORPEDO.

The velocity, range and accuracy of all automobile torpedoes thus far developed are almost incomparably inferior to those of ordnance projectiles, and on this account it is frequently argued that under present fighting conditions it would be very exceptional that the condition would occur where, in action, the torpedo could be used with a reasonable chance of success. This argument is upheld by the evidence (undoubtedly true) that in the wars which have occurred since 1875, at which date the automobile torpedo became a practicable weapon, the results obtained have but a negative value. That is, whilst they may have shown possibilities of future development, there has been no instance of positive success.

That this argument is erroneous is readily susceptible of proof.

1st. We have the undoubted fact that, in spite of all failures and in the face of immense expenditures, every navy in the world not only has made strenuous efforts to develop the automobile torpedo, but the demand for this weapon and the efforts to improve it have steadily increased since the commencement, now nearly twenty years ago. Precisely as with the change from the smooth-bore to the rifled gun, from the muzzle to the breech-loader, from the cast-iron to the built-up steel gun, there is more than a simple novelty involved. The development of the automobile torpedo must go forward, for it is universally and truthfully regarded as an accessory of armament absolutely necessary. It is only necessary to go back thirty years to find the same argument used against iron-clads, or twenty-five years to find it against breech-loading ordnance, or ten years to find it against magazine rifles. The automobile torpedo is as irresistible a development as either of these.

2d. It is conceded by naval tacticians that the general fighting range in naval action will be within seven hundred yards. The automobile torpedo has, within the past five years, been so developed as to become more than a fairly efficient weapon at that range.

3d. It is the automobile torpedo, and that alone, that has forced into existence the worst hamper to effective naval fighting yet known. The net. There is, perhaps, no argument so common against the automobile torpedo as that it is useless because the net will keep it clear of the ship, and yet none is more fallacious. With precisely the same truth might it be said that guns were useless, because armor would keep out projectiles. All naval sea-going vessels now carry nets as a part of their defensive equipment. No commander would use his net in action, knowing that his opponent did not carry automobile torpedoes, and every commander will use his net, knowing that the enemy carries them. A commander without an automobile torpedo, no matter how crude, must hamper himself with a net, whilst he leaves his opponent not only unencumbered, but with a weapon capable on a chance of deciding an action at a single shot.

The conclusion is inevitable. No weapon can replace the torpedo or affect its existence and development, because to the torpedo alone belongs the attack of the under-water body. Of the naval torpedoes as above defined, the automobile is the highest class of development. Types may be more or less efficient, and the sole reason for the Whitehead torpedo having been universally accepted with all its faults is because it has hitherto been the only practicable type. Valid arguments may be made against the Whitehead, or any other type of automobile torpedo. But none whatever can be made against the automobile torpedo " sui generis."

TYPE DIFFERENCES BETWEEN THE WHITEHEAD AND THE HOWELL AUTOMOBILE TORPEDOES.

The first successful type of automobile torpedo developed was the Whitehead, whose only competitor heretofore has been the Schwartz-kopf, which really possesses no type difference. A distinct rival in type to the Whitehead is the Howell, and in following the discussion of the merits of type differences between these rivals, it is necessary to keep constantly in mind the fundamental definition of an automobile torpedo, which is: one that, independently of the vessel and irrespective of the means of discharge, maintains its speed by self-contained motive power, and its depth and direction by self-contained directive power.

In both the Whitehead and the Howell types the speed is maintained by the action of screw propellers. In the Whitehead the screws are actuated by an engine, whose motive power is compressed air, which is carried in a tank in the torpedo, feeding the air to the engine as a boiler feeds steam to an ordinary engine. In the Howell the screws are actuated by a heavy fly-wheel, without the interposition of an engine, the energy of the fly-wheel being the motive force. In this difference of application of motive power appears the first marked contrast of type.

The Howell contains no engine, and thus gains a very important advantage in simplicity and economy over the Whitehead. On the other hand, the motive force of the Whitehead may be kept stored in the torpedo for a certain length of time, so that, in so far as this element is concerned, when it is inserted into its launching tube, it is always ready for instantaneous discharge. The Howell, however, can only have its motive force applied or stored in it after it is inserted in its tube, so that, in so far as fighting conditions are concerned, only the first torpedo inserted can be kept in readiness for instantaneous discharge. This drawback to discharge, however, is less serious than may at first appear, for in action the first torpedo can be discharged at will at full speed. As at present developed, the launching tube may be reloaded, the fly-wheel be spun to full speed, and the torpedo be discharged as quickly as a twelve-inch gun can be reloaded and fired. At half speed of torpedo the rapidity of fire can be nearly doubled. Finally, rapidity of fire is in a very great measure due to the power available for spinning the wheel, and it is beyond all question that this power may be readily and securely applied in much less than half the time required under present conditions, so that the case will be very exceptional, indeed, where in action a torpedo will not be ready at the instant that it is wanted.

In both types the depth of immersion is maintained by a horizontal rudder, actuated by a hydrostatic piston and a pendulum. The arrangement of the combination in the two types is quite dissimilar, but there is no absolute type difference. The discussion of this element belongs, therefore, to a detailed description of the mechanism itself.

In the Whitehead, constancy of direction is maintained by means of vertical rudders, which are adjusted to suit the conditions known or estimated to exist throughout the trajectory at the moment of firing. These rudders must be set just previous to discharge. The Howell has no directive rudders or mechanism whatever, but maintains a rigid direction through the gyroscopic force of its fly-wheel.

In this respect not only does the Howell gain in simplicity in having no vertical rudders or attachments, but, what is of far greater importance, it requires no adjustment for direction before firing. In a very great measure what is gained by the Whitehead in having its motive force stored in it, so that it is ready for discharge when put into the tube, is counterbalanced by the condition that any change in speed of ship, or keel angle of fire or heel of ship after insertion of the torpedo, demands a readjustment of the vertical rudders.

With this drawback to the Whitehead it is almost absolutely necessary, for full efficiency in battle, that the keel angle of fire of any single launching tube should be permanent, whilst with the Howell, since no adjustment is necessary, the launching tube may be pivoted and brought to bear on the enemy at will, with security as to accuracy in flight.

It will be readily understood that an automobile torpedo, from its general spindle shape and complete submersion, is extremely sensitive to the action of disturbing forces, either interior or exterior. The weight of compressed air carried by the Whitehead is with the smallest torpedo made more than twenty-five lbs., and during flight this weight is being constantly decreased. It is therefore quite impossible to maintain the relative positions of the center of gravity and center of buoyancy so that, no matter how exact may be the original adjustment, it will not hold throughout flight, and any variation leads to error either in constancy of depth or direction. In the Howell, the amount and distribution of weight is constant, so that this cause of error is entirely avoided.

In the Howell, the gyroscopic force gives to the torpedo great rigidity against rolling on its longitudinal axis. The Whitehead is most sensitive of all to such a disturbance, and it will be readily seen that the action of rolling directly affects the functions of the rudders, giving to the horizontal rudder a horizontal steering power. It is found necessary with the Whitehead to carry the propellers the one behind the other, mounted on middle line shafts, and it is found that as the speed of these screws exceeds or falls short of a certain normal, either the rear or the front screw tends to rotate the torpedo on its longitudinal axis. For this reason it is necessary to so regulate the supply of condensed air to the engines that the speed of the screws (and consequently the speed of the torpedo) shall be nearly constant throughout its run. This leads to a very appreciable waste of power. On the other hand, with the Howell, the twin screws are mounted on parallel shafts in the normal manner, and no regard need be paid to any difference in rotating power that one screw may have over the other; they balance themselves, and, even if from any cause they did not, the torpedo is held rigid against rotation by the gyroscopic force of the fly-wheel. This being the case, the full power of the fly-wheel is applied to driving the torpedo throughout the run, and as a consequence the speed developed over the first parts of the run may be higher than with corresponding Whiteheads, whilst at an extreme range, accuracy in direction is maintained at a low speed, which is impossible with the Whitehead.

It is considered necessary with any automobile torpedo that in case it should fail in action to strike the enemy, it should be capable of sinking automatically, to avoid the chance of being overrun and exploded by a friend. In a general action also, it is of great importance that a torpedo once discharged should not sheer wide enough from its direction of projection to endanger striking a friend. With the Whitehead, assuming that on discharge its buoyancy is negative, so that it would sink when uncontrolled by its horizontal rudder (which would be the case at the end of its run), it loses so much weight of compressed air as to give it a positive buoyancy, and therefore a special mechanism has to be introduced to allow water to enter the body of the torpedo at the end of its run and sink it. With the Howell this is not necessary, as its weights do not change, and, being started with negative buoyancy, it will sink of its own accord at the end of its run.

It may seem from a cursory examination that with the Whitehead such a disposition could be made that the loss of weight of air could be compensated by admitting water. This, however, is impossible unless the water could replace the air *in position*. Any attempt to do it results in aggravating the variation of the center of gravity with regard to the center of buoyancy, and with water ballast thus taken in, if the torpedo tended to roll in either direction, the water would aggravate the tendency.

As has been explained, when the speed of the screws falls an appreciable amount below the normal, a tendency to heel the Whitehead torpedo is created, and the heeling in turn sheers it broadly out of its course. Therefore, in order to avoid the danger of an erratic course, it is necessary to introduce a mechanism to cut off the air and stop the torpedo when it has run to the limit of range permissible under its constant speed. This difficulty also is entirely avoided with the Howell.

Speed is undoubtedly a most important element of efficiency in automobile torpedoes. With regard to the comparative speeds of the two types, although the Whitehead has records of speeds much higher than any attained by the Howell, it must be borne in mind that with all automobile torpedoes, irrespective of type, speed is in a great measure a function of displacement. The largest Howell torpedo yet built is smaller than the smallest Whitehead used.

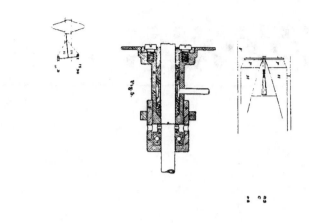

Fig. 4.

Fig. 3.

Fig. 2.

Fig. 1.

There is no valid reason why, for equality of displacement, the speed of the Howell should not be fully as great as that of the Whitehead, and also capable of as much future development.

In the compartment division, section-fastening, bulk-heading, bracing, arrangement of charge and position of details, there is scarcely a single point of resemblance between the Whitehead and the Howell. These details, however, can scarcely be classed as typical, nor can they be compared and their merits be discussed except in the course of detailed description.

PART II.

GENERAL DESCRIPTION OF THE HOWELL AUTOMOBILE TORPEDO.

Plate I.

The general profile of the Howell torpedo is that of a spindle of revolution, the after-body being a true spindle, the middle body a cylinder, and the fore-body an approach to an ogive. There are four distinct and detachable sections: the nose, *A. A.*, which carries the firing-pin and its mechanism; the head, *B. B.*, which carries the explosive charge and detonator; the main section, *C. C.*, which carries the fly-wheel and screw gears; the stern section, *D. D.*, which carries the diving mechanism.

The principal weights and dimensions of the General Service torpedo are as follows:

Diameter of midship section,	. . .	14.2 inches.
Length of body of torpedo,	. . .	116.0 "
Length from tip to tip,	. . . .	128.0 "
Length of nose section,	. . . .	6.0 "
Length of head section,	. . . .	20.0 "
Length of main section,	. . . .	67.0 "
Length of stern section,	. . . .	32.5 "
Weight of fly-wheel, .	. . . .	131.4 lbs.
Weight of explosive charge,	. . .	100.0 "
Total launching weight,	. . . .	475.0 "
Reserve buoyancy with full charge, .	.	3.0 "
Reserve buoyancy with dummy head,	.	13.0 "

THE NOSE.

Plate I.—Figs. 2, 3 and 4.

In order to guard as completely as possible against a premature discharge of the torpedo in handling, the percussion firing-pin is so arranged as to be completely removable, and also to be quickly attached at the last moment before inserting the torpedo in the launching tube.   If the pin alone were made removable, as it is but a small accessory, it might be overlooked in the heat of action.   Also, if any preliminary manipulation of the mechanism were necessary to get it into proper condition for firing, an error or omission might be made.   The entire firing-pin mechanism is therefore permanently fixed in a single hollow bronze casting 17, 17, which is attached to a projecting lip at the front end of the head by a simple bayonet joint, so that a few seconds only are necessary to attach and detach it.

The United States naval specifications for firing-pins of torpedoes demand more functions and greater security than have heretofore been required.   When attached to the torpedo it must be locked safe from discharge whilst handling the torpedo and previous to launching; it must arm itself for firing automatically after launching; it must act if striking an opposing body at an angle of fifteen degrees from its axis; it must automatically lock itself securely at the end of the run of the torpedo.   These functions are performed by the following arrangement of mechanism : a stout steel pin, 18, travels in guides formed in the nose casting, and is actuated by a strong spiral spring, 19.   It is held back in the armed position by a soft metal pin, 20, which seats in a slot cut through the pin, and bears against the outside of the nose.   The outer end of the firing-pin is provided with fan-shaped corrugated horns, 21, which receive the impact blow, and are so shaped and arranged as to prevent glancing or sliding along the object struck when the impact is sharply angular.   The force of the blow is intended to shear the soft metal stop-pin, and thus permit the firing-pin to be driven violently down on the detonator by the spring.

Two small cams, 22, 22, are so pivoted and maintained by the small flat springs 23, 23, that normally they rest against the body of the firing-pin just under a shoulder, so that if from any accident the pin after cocking should be so struck as to shear the soft metal stop-pin, it could not drive down and explode the detonator.   Just in front of the cams is a cross-head, 24, having projections which rest

THE NOSE.

Plate I.—Figs. 2, 3 and 4.

In order to guard as completely as possible against a premature discharge of the torpedo in handling, the percussion firing-pin is so arranged as to be completely removable, and also to be quickly attached at the last moment before inserting the torpedo in the launching tube.   If the pin alone were made removable, as it is but a small accessory, it might be overlooked in the heat of action.   Also, if any preliminary manipulation of the mechanism were necessary to get it into proper condition for firing, an error or omission might be made.   The entire firing-pin mechanism is therefore permanently fixed in a single hollow bronze casting 17, 17, which is attached to a projecting lip at the front end of the head by a simple bayonet joint, so that a few seconds only are necessary to attach and detach it.

The United States naval specifications for firing-pins of torpedoes demand more functions and greater security than have heretofore been required.   When attached to the torpedo it must be locked safe from discharge whilst handling the torpedo and previous to launching; it must arm itself for firing automatically after launching; it must act if striking an opposing body at an angle of fifteen degrees from its axis; it must automatically lock itself securely at the end of the run of the torpedo.   These functions are performed by the following arrangement of mechanism : a stout steel pin, 18, travels in guides formed in the nose casting, and is actuated by a strong spiral spring, 19.   It is held back in the armed position by a soft metal pin, 20, which seats in a slot cut through the pin, and bears against the outside of the nose.   The outer end of the firing-pin is provided with fan-shaped corrugated horns, 21, which receive the impact blow, and are so shaped and arranged as to prevent glancing or sliding along the object struck when the impact is sharply angular.   The force of the blow is intended to shear the soft metal stop-pin, and thus permit the firing-pin to be driven violently down on the detonator by the spring.

Two small cams, 22, 22, are so pivoted and maintained by the small flat springs 23, 23, that normally they rest against the body of the firing-pin just under a shoulder, so that if from any accident the pin after cocking should be so struck as to shear the soft metal stop-pin, it could not drive down and explode the detonator.   Just in front of the cams is a cross-head, 24, having projections which rest

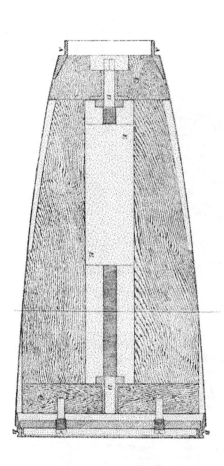

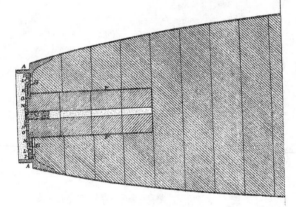

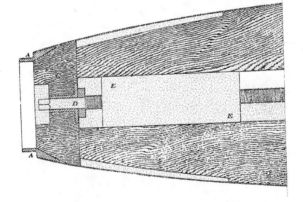

on the cams so that as the cross-head is pushed to the rear the cams are turned out clear of the shoulder, leaving the firing-pin clear. The cross-head is in turn in connection with two small pistons, 25, 25, which are held forward by the cam springs. The front ends of these pistons come out flush with the outer surface of the nose and are entirely open. When, after launching, the torpedo strikes and rushes through the water, the direct pressure on these pistons forces them back against their springs, in turn pressing on the cams and turning them back clear of the firing-pin, which is then completely armed for action. When the speed of the torpedo becomes so reduced as to permit the piston springs to overcome the pressure of water on the pistons, they come forward, the cams turn in under the shoulder, and the firing-pin is again locked. The condition of the firing-pin is at all times plainly visible. The soft metal stop-pin is on the outside, where its condition is always visible. The length of the firing-pin projecting beyond the nose shows whether it is cocked or not. The piston heads being plainly visible, show at all times whether the cams lock the pin or not. Except the fingers be deliberately used to push the pistons back, the firing-pin cannot be unlocked in handling the torpedo. Finally, all parts are secured in the single nose, and in such a way that absolutely no preliminary work is required. At the last moment the nose-piece is secured to the head by a simple twist of the bayonet joint.

It is to be remarked that when the nose is off the head the front of interior of the head is laid bare, so that the detonator itself may be kept out of the torpedo until the last moment. Small holes, 26, 26, inclining strongly backward are pierced through the nose and are left open. These holes leak water into the hollow nose-chamber when the torpedo is stationary or at low speed, which overcomes the reserve buoyancy of the torpedo, sinking it, and finally attacking and drowning the dry gun-cotton detonator, so that if the torpedo fails to make a hit, it locks its firing-pin, sinks and drowns its detonator, being thus rendered completely innocuous.

THE HEAD.

Plate II.

In order that the explosive charge of the torpedo may be readily attached and detached, be stowed compactly and safely in a magazine, and at the same time that the torpedo may be used with perfect

safety in exercise, it is provided with two distinct heads that correspond exactly in profile ; the nose being interchangeable with either
head, so that the firing-pin may be used both in exercise and in battle,
and the connection between the head and the main section being
precisely the same with both.   These heads are distinguished as the
dummy head and the fighting head.

The shells of both heads are made of a single sheet of brass
brazed and spun to shape and braced by bronze rings, $A. A. B. B.$,
at the front and rear ends.   Each of these rings is prolonged slightly
beyond the end of the shell to form bayonet joint locks, the front one
for holding the Nose, the rear one for securing to the Main Section.

The dummy head carries a heavy wooden block, $C. C.$, quite filling
the interior space.   This block has a square hole cut through its
axis and carrying an iron threaded bar $D. D.$ Upon this bar is a
square lead block $E. E.$ The inner end of the bar being squared, it
is readily seen that by turning it the block is traversed back and
forth.   By this means the torpedo is balanced longitudinally.   It is
necessary here to explain that when the torpedo is launched, no
matter whether it has buoyancy or not, its diving mechanism will
keep it at its proper depth.   It is desirable that the greatest weight
of explosive possible should be carried, and also, as has been heretofore explained, that it should sink at the end of its run if it fails to
make a hit.   Therefore, with the fighting head the buoyancy is prac-
cally nil, and the torpedo is permanently balanced in its entirety for this
condition.   For exercise, however, the torpedo must not be allowed to
sink, as it would be lost.   The dummy head, therefore, is lighter than
the fighting head, so as to give about thirteen pounds buoyancy, and
the lead block is introduced in order to give a long arm to a compact
heavy weight in the section that is not water borne, so as to keep the
center of gravity of the torpedo in the same position relatively to the
center of buoyancy as that occupied with the fighting head.   Otherwise there would be a difference of leverage between the two conditions that would alter the steering adjustments.   A single shop
adjustment of this block is sufficient, and it maintains its place unless
moved by screwing the bar.

A complete bulkhead plate screws water-tight into the rear end
of both heads.

In the fighting head, the main part is completely filled with wet
gun-cotton, a small water-tight chamber, $F. F.$, formed of a single
piece of drawn copper being reserved for the dry gun-cotton primer.

This chamber is removable, having a flange *G. G.* at its mouth, which seats on a diaphragm *H. H.* screwed across the mouth of the front casting by means of a ring *I. I.* screwing down on a rubber gasket. A cap *K. K.* covers the primer chamber, being held in place by spring catches *L. L.*, and forming also the seat for the detonator *M.*, which is held in place by the spring catches *N. N.*

By this arrangement the detonator may be removed, and, as the dry primer is contained in a thin tin case, it also is readily removable, leaving the head with only the wet gun-cotton charge, which itself is at all times hermetically sealed in proper shape for stowage in a magazine, the primer and the detonater also stowing separately in their own magazines.

Two small holes *O. O.* are drilled through the cap of the primer compartment, and are filled with a substance that is soluble after long contact with water. As above explained, the holes in the nose admit water at the end of a run. The water attacks the composition, filling the holes in the cover, and after a number of hours dissolves it out, and so drowns the dry gun-cotton primer.

THE MAIN SECTION.

Plate I.

This section comprises the entire cylindrical body of the torpedo and portions of the curved part at either end. The shell consists of three sections of brass plate corresponding to the cylindrical and curved portions, brazed and spun to shape. The section is closed water-tight at both ends, and contains the fly-wheel with its frame, the propeller gears and forward sections of shafting, and the thrust bearings.

The shell is braced against deformation or crushing by six rib rings. A bulkhead ring, 1. 1., which is flanged, and in which are worked the sockets of a bayonet joint, by which the head is secured to the main section. The flange of this ring is threaded, and receives a complete water-tight bulkhead plate, 6. In all automobile torpedoes heretofore constructed, great difficulty has been encountered in making section joints water-tight to resist the great water pressure due to depth of immersion, and in order to secure this very important feature, it has been found necessary to make a junction so complicated as to make it not only a matter of difficulty to separate the sections, but, on account of the wear from frequent joining and disjoining, the torpedo could not be taken apart and put together as desired. This feature has been overcome in the junction of the Howell section in a very simple manner, so that at this important joint the dummy and the fighting head may be quickly exchanged and without danger of compromising water-tightness. As has been described, there is a complete bulkhead plate at the rear of the head and another at the front of the main section. These plates naturally lie on either side of the joint, and when the sections are connected they are only about a hundredth of an inch apart. The space is so small that any leak that may occur through the joint does not admit water enough to be of any consequence, and so long as the joint is reasonably tight, the pressure due to immersion is relieved from any water that may lie between the bulkheads, so that it will not be forced past their joints. The torpedo has been sunk to a depth of forty feet without showing any signs whatever of leakage.

Two intermediate rings, 2. 2. 2. 2., are inserted under the brazed joints of the shell. These are simply plain, flanged, bronze rings. Two rings, 3. 3. 3. 3., support the midship section and at the same time form a part of the assemblage of the wheel frame, the other members of the frame being two plate castings, 4. 4. (shown in Section Plate III., *F. F. F. F.*), forming the bearings for the fly-wheel, which are bolted to the rings. Finally, a rear bulkhead ring, 5. 5., which, like the front one, holds a complete bulkhead, 7. 7., and also forms the seats for the thrust bearings, 8. It is impossible to connect the main and stern sections by a bayonet joint on account of the screw shafts, which prevent the twisting necessary to lock the joint. The strengthening ring of the stern section, therefore, has a lip which fits into an undercut in the flange of the main section ring and is held by screws, *S. S. S.* As this joint comes in the compartment containing the diving mechanism, which must be free to water access, it is not necessary that it should be water-tight.

THE FLY-WHEEL AND ITS CONNECTIONS.

Plate III.

The fly-wheel, *A. A. A. A.*, is of gun steel, drop-forged and treated similarly to the tube and jacket forgings of guns. It has a heavy rim with a solid web connecting to the hub. Secured to the hub and symmetrically placed on each side of the web are two steel miter-wheels, *B. B. B. B.*, which gear into similar wheels, 9. 9., Plate I.,

secured to the inner ends of the screw shafts, the proportion of gearing being as five to four, so that each screw makes 800 revolutions to every 1000 of the fly-wheel. The axle of the fly-wheel, *C. C. C.*, is a single solid steel axle, permanently secured in its seat in the wheel, its bearing ends resting on hard steel rollers, *D. D. D. D.*, in hard steel bearings, *E. E. E. E.*, which themselves seat in sockets cast in one with the frame plates, *F. F. F. F.* . The inner ends of the bearings, *E. E.*, and the bodies of the miter-wheels facing them are grooved and hold steel balls, *G. G. G. G.*, forming ball bearings to take the end thrust of the fly-wheel. Thus the wheel is provided with frictionless bearings, no matter what be the plane of the axle when rotating.

The connection between the fly-wheel and its motor, which forms part of the launching gear, is made through the starboard side of the torpedo by means of clutch couplings to the end of the axle. The right-hand end of the axle is squared, and carries pinned on it a steel end clutch, *H.* A loose clutch, *I. I.*, is held in a stuffing-box, *K. K.*, seated in a prolongation of the frame plate which bears against the shell of the torpedo, a through-hole being cut in the shell and the joint being closed water-tight. This loose clutch, *I. I.*, is so made in order to free the fly-wheel from the friction of the clutch in the stuffing-box. After spinning up the wheel, the moment that the motor is unclutched this loose clutch commences to hang back from its friction in the stuffing-box. This brings the rear sides of the clutch studs in bearing, and as they are cut with a steep slope, the clutch is instantaneously driven out free of the wheel.

In order to preserve the balance or symmetry of the torpedo, the left-hand frame plate is carried out to the shell in the same way as the right-hand one. The interior of this projection is threaded, and a lead disc, *L. L.*, is screwed in to counterbalance the clutch and stuffing-box on the right-hand side. By means of this lead disc alone the entire torpedo is balanced transversely, for a small hole is tapped through the shell, through which a key may be inserted and the disc may be screwed in or out to make the necessary adjustment. Once made, it remains of itself.

Plate I.

The screw shafts proper end at the bearing 10, being secured to the axles of the miter-wheels, 9. 9., by a mortise and tennon connection. This is done with a double object—first, to prevent any skew tendency in the miter-wheel being transmitted to the shaft, and second, to enable the shafts to be entirely disconnected from the contents of wheel frame. In order to neutralize the skew tendency of the miter-wheels, their short axles are held in close bearings in front of and behind the wheels, these bearings forming a part of the wheel frame castings, so as to remain constantly true. The screw shafts are carried straight to the rear through the box 8, forming a part of the rear bulkhead ring, and within which are thrust bearings and a stuffing-box, made necessary by the proximity of the free water compartment. The thrust bearings being placed here have a double advantage. They relieve the miter-wheels of all thrust, and more room is allowed to make stout bearings than if they were placed farther aft.

A broad, stout plate, 11. 11., is soldered to the bottom center of the shell, to which, on the outside of the shell, is bolted a long stud, 12. (S. Plate IV.). The function of this stud is to center and guide the torpedo in the launching tube.

Plate I.—Fig. 5.

The composition of the thrust-bearing and stuffing-box is as follows: Long seats, $a. a.$, are cast in one with the rear bulkhead ring, over which screw caps, $b. b.$ The shaft is slightly increased in diameter at the point $c.$, forming a seat for the steel bearing ring $d. d.$, which has a companion bearing ring, $e. e.$, seated against the sleeve. Steel balls lie between these rings, thus forming a ball bearing. A bronze spanner clasps each of the caps and prevents them from unscrewing, while at the same time it resists any tendency to flexure or spreading of the shafts.

A small bronze loose sleeve, $f. f.$, is slipped on the shaft and lies in the stuffing-box section. This sleeve is pierced with holes and its ends are packed. In this way the stuffing-box is formed, and at the same time provision is made for oiling the bearings, for the oil coming down on the sleeve passes through the holes and is absorbed and distributed by the packing.

THE STERN SECTION.
Plate I.

The stern section is divided by a water-tight bulkhead, 13. 13. 13, into two compartments, the forward one containing the diving mechanism and being open to the free access of water through the inlet holes, *E. E. E.*, pierced through the shell, whilst the rear compartment is closed water-tight and is empty, save the sleeves passing through it, within which are the screw shafts and tiller rods. The rear end of this section is closed by a casting called the tail-piece, 14, which forms in one the butt of the tail and the screw shaft tubes with their cross support *F. F. F. F.* The screw shafts are taken in bearings in the tubes *F.*, and the screw propellers, which are right and left-handed, are screwed to the ends of the shafts, being held fast by end nuts, *G. G.*, which are shaped off in long cones to give a fair run to the water passing the hubs. The triangular spaces, *H. H.*, between the tail body and the screw shaft tubes are covered with plates in order to give a fair flow of water to the rudder and screws. The small chambers thus formed give additional buoyancy also.

The rudder, *I. I.*, is a steel rectangular plate completely filling the space between the outer ends of the screw shaft tubes. In this position it is secure against damage in handling the torpedo and fouling in running. A stout web, 15. 15, stands at right angles to the plane of the rudder, forming a steering yoke, to the ends of which are pivoted the tiller rods, 16. 16, which in turn are directly connected with the diving mechanism.

THE DIVING MECHANISM.

Plate IV.

The bulkhead, *A. A. A.*, separates the rear and water-tight compartment from the diving compartment, both being in the stern section. It is a single casting so shaped as to reduce the water space to the least possible dimensions consistent with the working of the mechanism, and has a broad flange seating on the shell to form a stout stiffening member of this part. The bulkhead, *B. B. B.*, which is the rear bulkhead of the main section, forms with *A. A. A.* a complete water chamber. The tiller rods, *C. C. C. C.*, are pivoted to the rudder yoke, *D. D.*, and pass, inside of sleeves, through the water-tight compartment and bulkhead, their inner ends pivoting directly to their respective parts of the diving mechanism, the upper rod being attached to the hydrostatic piston, *E. E.*, and the lower one to the compound lever of the pendulum, *F. G. H.* These tiller rods are provided with screw junctions, *I. I.*, for taking up lost motion and regulating the angle of the rudder.

The forward compartment being in free connection with the exterior water, the pressure due to depth of immersion is fully borne on the piston, *E. E.* This piston fits loosely in its cylinder, *K. K.*, which is secured to the bulkhead by the posts and nuts, *L. L.* The posts are made hollow and connect with the interior of the cylinder, so that there is free air connection between the space in the cylinder behind the piston and the whole air space in the rear compartment, so as to prevent any back pressure on the piston. A rubber disc covers the piston, and is held water-tight about its edges so as to prevent water getting into the cylinder and at the same time offer no opposition to the free movement of the piston.

Near the front end of the lower tiller rod a seat, *M. M.*, is fastened to it, against which abuts the forward end of a powerful spring whose rear end seats against a movable sleeve, *N. N.* This sleeve screws into the rigid main sleeve of the rod, and a key may be used on the end outside the torpedo to screw it in or out and so alter the tension of the spring which alters the depth of immersion.

Assume that the depth at which it is desired to run the torpedo is ten feet, and that at that depth the total pressure on the hydrostatic piston due to the head of water is one hundred pounds. The rudder being held amidships, let the spring be adjusted to a tension of one hundred pounds. Since the tillers are directly connected, the one to the piston and the other to the spring, it follows that if a pressure of one hundred pounds be brought on the piston, the tension of the spring will be balanced and the rudder will lie amidships. This will occur at the assumed depth of ten feet. If the immersion be less, there will be less pressure on the piston, and the spring will hold the rudder partially down and so steer the torpedo down to its proper depth, and *vice versa.* It should be noticed that the tension of the spring varies inversely as its length, whilst the pressure on the piston varies directly with the depth. Therefore the helm is not thrown hard up and hard down as the torpedo departs from her proper depth, but it is eased over the proper amount to bring her easily to her proper depth. The point at which the helm is thrown hard over depends upon the length and strength of the spring, and as by specification requirements the torpedo must run within two feet of her set depth, the hard-over point is made slightly greater.

As a matter of course the torpedo will always move in the direction of its longitudinal axis. Whilst therefore through the action of the hydrostatic piston the rudder will be brought to a neutral position at the proper depth, the torpedo must be horizontal at that depth or it will continue to go down or up, according to the direction in which it points.   The piston cannot correct the direction of the axis of the torpedo except secondarily.   It is therefore necessary to introduce an element that shall counteract every tendency of the longitudinal axis to leave the horizontal, and this element is the pendulum.

A heavy pendulum, *H.*, is suspended so as to swing in the fore and aft-line of the torpedo.   We may say, therefore, that whenever the axis of the torpedo dips down or up, the pendulum swings forward or aft.   The bob of the pendulum, which is very heavy, is mounted on springs, *O. O.*, on its suspension rods, so that when the torpedo strikes the water in falling from a height, the shock on the suspension points will not be too severe.   The pendulum is connected with the front end of the lower tiller rod by a compound lever, *F. F.*, so as to increase its power.   Assume that the torpedo is at its required depth, its axis horizontal, and its rudder amidships.   Leave, for the moment, out of consideration the action of the hydrostatic piston, and assume that from any cause the bow of the torpedo is tilted down. The pendulum bob at once swings forward, and in so doing pushes the tiller rod back and forces the rudder up, thus tending to bring the torpedo to the horizontal again.

An examination of the combined action of the piston and pendulum shows a valuable feature.   If the torpedo be pointed away from her proper depth line, and so long as she is leaving it, both piston and pendulum work the same way on the helm and combine their efforts to turn her back, but when she turns back, they commence to work against each other so as to ease her gently to her proper line, thus preventing violent-oscillation.

This type of mechanism possesses several features of undoubted superiority over other automatic types, the most important of which are :

1st.   In other types, whilst the mechanism controls the movement, the power necessary to operate the rudder and overcome friction of the working parts is taken from the motive power of the torpedo. All this power is a direct loss to driving power, reducing either speed, or range, or both.   In this type the entire steering power is within the mechanism itself, releasing just that much of motive power to be used in driving the torpedo.

2d.   When a torpedo strikes the water, no matter what the position of the rudder may be, it will receive a violent shock from the impinging water. The pendulum, also, will receive violent impulses both from the shock of discharge of the torpedo and the shock of taking the water. In this type it will be noticed that when the torpedo is ready for discharge, since there is no pressure on the hydrostatic piston, the pendulum is drawn by the spring hard back against a small stop, *P*. It there has a proper and secure bearing against discharge shock. Upon striking the water the rudder is driven up and the pendulum is driven forward, but it will be noticed that the spring receives and eases both shocks, while, moreover, both act in the same direction on the tiller rod. That is, the upper movement of the rudder corresponds with the forward movement of the pendulum. No part is therefore submitted to an undue strain.

3d. The entire mechanism from its front end to the rudder yoke is within a length of two feet and is direct-acting. The force applied to the rudder is always a balanced one of give and take on equal arms each side of the rudder, instead of being all on one side, as in other systems. There is, therefore, less lost motion and a more direct application of power. The importance of this feature will be appreciated when it is considered that in any torpedo the extreme tiller-rod movement from hard up to hard down is never as much as half an inch.

4th. Owing to the compactness of the mechanism and its situation in a section of very small diameter, the water space, which represents a clear loss of buoyancy or carrying power of the torpedo, is reduced to a minimum.

5th. All adjustments are simple, direct, and are made from the outside. The depth of immersion may be altered, lost motion be taken up, and the throw of the rudder be adjusted without the necessity of touching anything inside of the shell.

Finally, by removing the stern section the entire mechanism is laid bare and can be removed or altered without trouble or interfering with any other part of the torpedo.

THE DOW STEAM TURBINE MOTOR.
Plate III.

Rotation is communicated to the fly-wheel by means of a steam motor which is a permanent attachment of the launching tube. The body of the Dow motor is a small cylindrical box about 8.75 inches in diameter by 4.5 inches in depth. The shell consists of a bronze casting, *a. a. a. a.*, having covers, *b. b. b. b.*, through-bolted

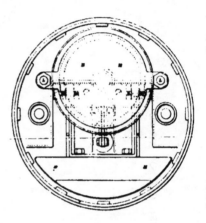

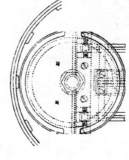

PLATE IV.

to it, which have projections cast in one with them to form bearings for the main shaft. Two discs, *c. c. c. c.*, screw permanently into the wall of the shell, having in turn two smaller discs, *d. d. d. d.*, screwed into and forming a part of them. The interior of the motor is thus divided into three chambers, of which the central one, *e. e.*, receives the live steam direct through the steam pipe (not shown in the figure), and the two outer ones, *f. f. f. f.*, take the exhaust steam which passes out of the motor through the exhaust pipe *g.* The main shaft, *h. h. h.*, is journaled in the bearings formed in the covers and is given a longitudinal play, so as to permit clutching with and unclutching from the machine required to be driven. A sleeve, *i. i.*, covers the central part of the shaft, being keyed to it, but having a slight independent longitudinal play, and to this sleeve is secured a steel disc, *k. k.*, which partially divides the live steam chamber. Two bronze discs, *l. l. l. l.*, are also secured to the sleeve, and it is these discs that are revolved by the action of the steam, transmitting rotation to the main shaft.

Concentric ribs are cut on the opposing faces of the pairs of discs, *c. c. c. c.* and *l. l. l. l.*, which intermesh, and through these ribs a number of angular slotways are cut, those on the stationary discs being at an opposite angle from those on the revolving ones. The live steam entering the steam space *e. e.* passes into the space *m. m. m. m.*, and thence outward between the pairs of discs through their slotways, communicating rapid rotation to the revolving discs and shaft by expansion. After thus performing work, it passes into the exhaust chambers, *f. f. f. f.*, and out through the exhaust pipe. In passing outward through the slotways the steam undergoes seven expansions.

The function of the steel disc *k. k.* is to balance the work done by the two pairs of discs, since practically there are two motors or drivers mounted on a single shaft. Assume that for some reason the right-hand disc is driven harder than the left-hand one. The over-pressure will force the right disc, and with it the sleeve and other discs, to the right, and by this movement the steel disc partially closes the right-hand steam entrance to the chamber *m. m.*, opens and gives more steam to the left-hand one, and thus automatically equalizes the driving force on the two revolving discs.

The left-hand end of the main shaft ends in a clutch, *n.*, and its journal, *o. o.*, is free to move longitudinally, carrying the shaft with it. The longitudinal clutching movement is communicated to the shaft by the stud, *p.*, which works in a guide-slot cut in the starting gear. (See Plate V.) The right-hand end of the shaft projects slightly beyond the end of the cover of the motor, and is hollowed to receive the squared end of an auxiliary shaft, which forms part of the tachometer.

THE ELWELL TACHOMETER.

In such high speeds of rotation as are given by turbine motors, it is of great importance that a reliable speed indicator should always be attached. The Elwell tachometer is the most exact one known. Its indications depend upon the pressure of a column of oil acting upon an ordinary steam gauge, the dial plate of the gauge being marked in thousands of revolutions per minute. The pressure created on the column of oil is due to centrifugal force applied to the oil by the rotation of a small centrifugal pump. The body of the support of the gauge is a brass casting, *r. r. r.*, whose inner end forms a collar clasping the end of the outer shaft bearing and tightened by a screw-bolt at *s.* The small hand-wheel *t. t.* attached to the auxiliary shaft does not belong to the tachometer, but is used when clutching up to engage the clutches. The cap *u. u.* screwing over the outer end of the support casting forms a small oil chamber connected with the tachometer gauge by the pipe *w.* Within this chamber and secured to the auxiliary shaft is a small cylinder having radial slits cut through it similar to the radial guides of a turbine wheel. As this cylinder is rapidly rotated, these slits force the oil out against the sides of the chamber with a pressure proportional to the centrifugal force developed, which itself is proportional to the speed of rotation. The pressure is communicated through the pipe *w.* to the gauge and acts on the pointer. The small conduit *x. x.* leads to an oil reservoir, *y. y.*, which keeps the pump chamber constantly full of oil.

z. z. are oil cups for supplying oil to the main shaft bearings of the motor.

PART III.

LAUNCHING GEARS.

Although, in judging of the perfection of development of any type of automobile torpedo, its means of discharge is left out of account, this latter is so important an accessory that much of the efficiency of the complete weapon depends upon its own completeness.

In the first attempt at perfecting the torpedo the delicacy of its mechanism forbade the employment of any means of discharge by which a sudden shock should be given. Nor could a torpedo be successfully launched from a height above water owing to the shock of impact and the tendency to dive very deeply. Compressed air therefore was resorted to as the expulsive force, and the point of discharge was brought as near to the proper depth line as possible. Both of these elements involved great difficulties of practical application. Compressed air discharge required complications of accessories and of details of the launching gears themselves. As the mechanism of the torpedo became simplified it was made less sensitive to shock, until finally gunpowder has been introduced for direct discharge; but hitherto this has possessed many drawbacks of its own which are very difficult of modification, amongst which the principal are: liability to derangement of the working parts of the torpedo from the impact of unburned grains of powder or the penetration of the powder fluids into its interior; fouling of the bore of the tube; danger of abnormal pressures in some part of the tube, due to a defect in the cartridge or the powder, as happens at times in guns.

In so far as the conditions of discharge themselves are concerned, it must be borne in mind that no matter how efficient a discharge may be, so long as the vessel from which it is made is stationary, such efficiency is no criterion whatever of results to be obtained with a vessel moving. Under-water discharge presents no difficulties whatever when made from a stationary point, but when made from a vessel at speed the drawbacks are very great. In beam fire from a vessel in motion it is apparent that the moment the nose of the torpedo shows beyond the end of the discharge tube, it will be swept violently aft, thus destroying the aimed direction entirely, and endangering the rear end of the torpedo, which may become jammed in the mouth of the tube. To counteract this defect it is necessary to project the guide-bar beyond the tube which will hold the torpedo in line until it is clear of the ship, and then let it go square or even. In straight-ahead fire a difficulty of another kind is met with that, it may be said in passing, is fatal to the success of the submarine gun.

Consider a torpedo discharge tube mounted in the keel line of a ship, under water, its front end open and the ship going ahead at speed. The water inside of the tube has a pressure upon it due to the speed of the ship, but it has no flow, and if a body be placed within the tube it will move with the ship, for the water in the tube has an absolute velocity equal to that of ship. The moment, however, that this body is moved forward and leaves the tube, it enters water having no motion with the ship, and at that moment the body becomes deprived of the driving force due to the speed of the vessel. Whatever be the absolute speed of discharge of the body, if it have no inherent propulsive force it will at once commence to lose speed from the resistance of the water, whilst the ship maintains her speed of advance, with the result that in the case of a submarine projectile, the ship will certainly overrun it. The same thing will happen with an automobile torpedo, unless its screws get a full driving power very quickly and give the torpedo a continuing speed, at least as great as that of the ship.

The moment that the keel angle of the discharge tube is altered to bow or beam fire, the difficulty explained as pertaining to beam fire begins to show itself. Here, then, are under-water difficulties of a most serious nature, entirely caused by motion of the point of discharge, and entirely absent when that point is stationary.

In so far as concerns the necessity for using a guide-bar to keep the front end of the torpedo from being violently swept aft in bow or beam fire, the Howell torpedo requires the application fully as much as the Whitehead. For straight-ahead fire, however, the Howell has a decided advantage over the Whitehead in that, previous to discharge, its screws are already working at full speed and commence to drive the torpedo from the instant it starts forward, so that in clearing the tube there is no check to the speed as occurs to the Whitehead, and consequently no danger of overrunning it.

In above-water discharge, as the Whitehead has no inherent directive force, every effort possible must be made to neutralize tendency to sheer. In beam or bow fire, if either the bow or stern strikes the water first, that end will be swept aft, or, what amounts to the same thing, its forward movement is checked, and consequently the torpedo is given a broad sheer. To counteract this, three precautions have to be taken: It is necessary to attach a long bar or spoon to the mouth of the discharge tube, to which the torpedo hangs itself in coming out and so is made to drop horizontally. If the ship happens to be rolling, the torpedo should be discharged at the moment when she is vertical, but even this will be ineffective if the torpedo has to strike into a wave. The angle of aim cannot be direct on the object to be hit, but allowance must be made for the inevitable sheer of the torpedo, which is always an uncertain amount. Finally the vertical rudder must be set to partially correct a rank sheer. This correction, however, is only of value to a certain point, as beyond it the torpedo will sheer in the other direction.

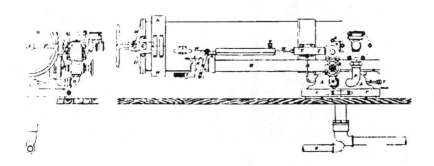

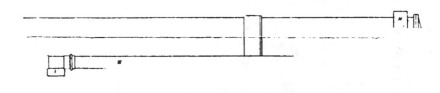

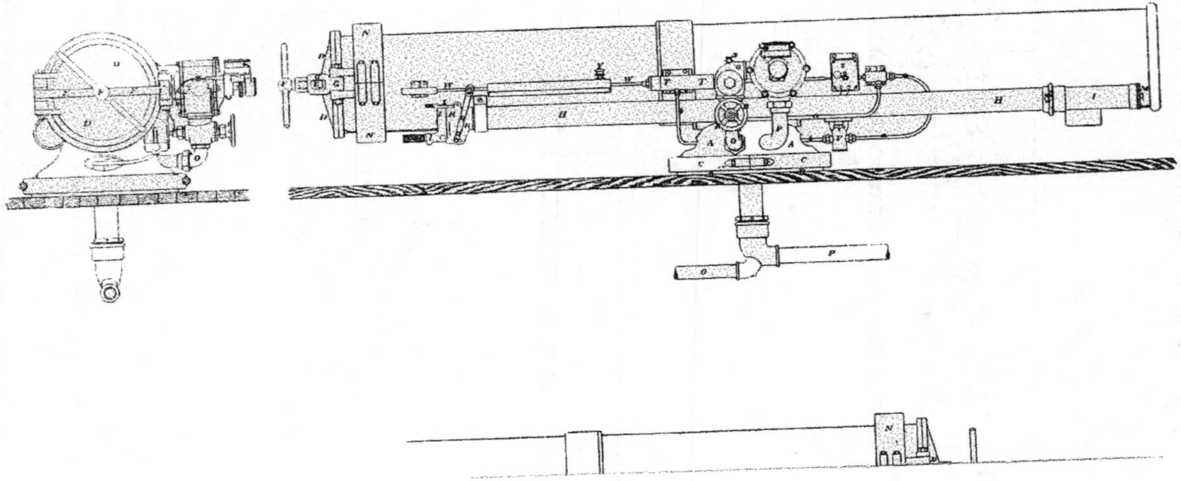

The inherent directive force of the Howell torpedo quite obviates all these difficulties. No projecting spoon is required on the discharge tube, the aim is direct, and no heed need be taken of rolling, heeling or wave surface.

For straight-ahead fire above water, the chances of being overrun are the same in both the Whitehead and the Howell, as in both the driving power of the screws takes effect at the same time. It is an odd accident that in under-water fire the Howell has the advantage of the Whitehead in straight-ahead fire, and the two are equally handicapped in beam fire, whilst above water the conditions of advantage and equality of the Howell are reversed.

THE ELWELL LAUNCHING GEAR.

The launching gear designed by Mr. Elwell, the Superintending Engineer of the American Branch of the Hotchkiss Ordnance Company, possesses a striking feature in the application of the expulsive force. The medium of discharge is gunpowder, but it is so applied that the explosive shock is cushioned against the torpedo, no fouling or solid particles enter the discharge tube, and although ordinary black powder is used, the discharge is practically noiseless and smokeless. The cartridge itself is also in the simple form of ordinary metallic cartridge case ammunition.

Discharge tubes differ in general arrangement, depending upon their emplacement; that is, whether under or over water, and whether fixed or pivoting. With the exception of the accessory features, however, a description of one apparatus will serve for all, as with all gears used for the Howell torpedo every effort is made to maintain as close a similitude as possible.

THE CENTER PIVOT LAUNCHING TUBE.

Plate V.

This gear is designed for open-deck emplacements, where all-around fire is permissible.

The discharge tube is of bronze, bored to a diameter five hundredths of an inch greater than the midship diameter of the torpedo (14.25 inches for the general service torpedo). It is mounted upon a low, broad cone, $A.A.$, whose base rests on a bed-plate, $B.B.$, bolted to the deck, the two being held together by a stout clip-ring, $C.C.$, so that the cone is free to revolve. The interior of this cone may be fitted with a rack and gears, so that the tube may be aimed from a conning tower if desirable. A shallow groove, *s.*, Plate III., is cut the full length of the tube along the bottom of the bore to carry the guide-stud of the torpedo. The rear end of this tube is closed by a door, *D. D.*, hinged to swing laterally, its inside edge being coned and ground to close air-tight. A steel cross-bar, *E. E.*, with a tightening screw, *F.*, through its center, is carried by the same hinges as the door, the free end of the bar being held by a stout bronze loop, *G. G.*, when the door is closed. To lock the door, it is closed, the loop is swung over the end of the bar, and a few turns are given to the tightening screw.

Two brass air tubes, *H. H. H. H.*, are secured to the main tube underneath, one on each side, being connected together at the front end by a cross-pipe, *I.* The tube on the right-hand side, called the firing tube, has screwed to its rear end a small bronze breech-piece, *K. K.*, which is chambered to carry an ordinary metallic cartridge case, and has a simple breech-block, *L. L.*, in which is fitted a hammer, sear and main-spring. The weight of powder used is less than half a pound, with which a discharge speed of over thirty-five knots can be obtained for a torpedo weighing nearly five hundred pounds. The front ends of both air tubes are closed by screw-caps, *M.*, that may be removed whenever necessary to sweep out the tubes. It will be noticed that the forward end of the firing tube is extended well beyond the cross-duct, *I.* This is done in order to form a lodgment for bits of wad or unburned grains of powder that by the explosion will be driven past the duct and be caught and held in this space, precisely as is the case with the cinder-trap of a locomotive engine.

The rear end of the left-hand pipe, called the compression pipe, connects by an elbow with the main tube. Around the rear of the main tube is secured a hollow strap, *N. N.*, into which the elbow of the compression tube opens. The wall of the tube underneath this strap is pierced all around with small square ports cut at an angle, such that the blast of air created by the explosion of the charge will be directed against the door of the tube first, instead of being taken directly on the tail of the torpedo. The air pressure thus created in the main tube drives the torpedo out. At a speed of ejection of about thirty-five knots, a torpedo discharged at a height of about five feet will take the water fully thirty feet from the ship's side.

It is well at this point to call attention to a feature of torpedo use of great importance as affecting over and under-water discharge, that has hitherto not received noteworthy attention. A ship going into action probably will find it necessary to use her net, and if she is only provided with under-water tubes, all her torpedo fire is cut off by the net, whilst if she has tubes above water, she may discharge clear over the net.

The Dow motor is attached to the main tube on its right side, a hole being pierced through in the clutch-line. The steam pipe, O. O., and the exhaust pipe, P., to and from the motor, are carried down into the supporting cone, where a junction box is made, so that the steam pipe goes through the deck inside of the exhaust. This junction is swiveled to permit the system to revolve.

A throttle valve, with a hand-wheel, R. R., gives steam, which is controlled by a regulator valve, S., and there is also connected to the throttle an automatic cut-off.

It is of great importance that, once the torpedo is in the tube in place, the work of clutching, spinning up, unclutching, cutting off steam, freeing the torpedo and discharging it, should all be done in a simple manner, quickly, and with absolute certainty of the proper succession of movements. This work is almost entirely automatic, and is done in the following manner : The small box, V., is a steam cylinder whose piston projects up through the main tube into and filling the slotway for the torpedo guide-stud and forming a stop. This piston is held up by a spiral spring underneath it. To load the torpedo into the tube it is simply necessary to push it in until its guide-stud brings up against this stop, and then close and fasten the door. The clutch hole in the torpedo is then directly in line with the motor clutch, and the moment that these clutches are thrown in action, the torpedo is held firmly against all movement.

A long rod, W. W., performs the work of clutching, disconnecting and firing. The torpedo being in its tube, the powder charge may be inserted. Lift the small spring latch, X., open the breech, and insert the cartridge. It is to be remarked that unless the torpedo is clutched up ready for spinning, it is impossible to cock the hammer, and unless the torpedo is entirely free to leave the tube, it is impossible to fire. The action of firing itself is automatic and is controlled by the lever, Q. By pulling back on the handle, Y., the long rod, W. W., is drawn to the rear, clutching the motor to the torpedo, and bringing the lever, Q., into position, so that the movement of closing the little breech cocks the hammer. If the throttle valve be now opened, steam is given to the motor, and the fly-wheel will be spun up, it being possible to so set the regulator valve that the wheel will run at any desired speed of revolution.

Discharge is operated in the following manner: A small box, Z., contains an arrangement by which a small steam valve may be operated either electrically or by a firing laniard.   The valve works instantaneously, and admits steam into the small pipe, $a. a.$, communicating with the stop-pin.   Steam coming on the upper side of this little piston forces it down so that the pin comes clear of the guide-stud on the torpedo, leaving it clear to leave the tube.   As this piston descends, *and after withdrawing the stop*, a port is unmasked, admitting steam to the pipe, $b. b. b.$, passing to the cylinder, $T. T.$, whose piston is attached to the long rod, $W. W.$, driving it forward. As the rod moves forward it first unclutches the motor, then cuts the steam off from the motor, and finally, at the end of its course, trips the hammer and fires the cartridge.   Thus all the movements are performed automatically, and they can only occur in their proper succession.   The entire time from pulling the firing laniard until the torpedo leaves its tube is but little over one second, most of this time being taken by the torpedo itself gathering movement.

U. S. NAVAL INSTITUTE, ANNAPOLIS, M D.

MARCH 1, 1890.

· DESERTION, AND THE BERTILLON SYSTEM FOR THE IDENTIFICATION OF PERSONS.

By LIEUTENANT ALEXANDER McCRACKIN, U. S. NAVY.

The subject of desertion from the army and navy has been discussed by many persons, and the causes assigned for desertion, with the methods proposed for its prevention and punishment, have been almost as numerous as the number of writers.

It may be taken as an axiom that desertions will always occur. Cases are on record of officers deserting, and the writer has questioned an enlisted man apprehended for desertion, who admitted that he had been well fed and clothed, well treated, had received plenty of liberty, that he had no real cause of discontent, and that he had previously deserted from the Marine Corps! No doubt there will always be such men.

It is idle to write of the evils of desertion and its punishment when wholesale amnesties to deserters are made, and when such principles are promulgated as these by a Senator who lately wrote of desertion as being "an offense which evidences no lack of patriotism and involves the least possible moral turpitude"!

If desertion ended a man's connection with the navy the latter would be the gainer, but unfortunately the deserter thinks it is quite the proper thing to go to another ship, re-enlist under a different *alias*, stay in the new ship long enough to once more become a factor of discontent and disorganization, and then desert again, and so on. Men who have been dishonorably discharged adopt the same course.

As, probably, no humanitarian would desire to thus burden the service, the only question now left of the whole subject of desertion is, how shall the Government be protected from such frauds?

The plan which has been repeatedly recommended officially in the army as the best remedial measure is prohibited in the navy by the 49th Article for the Government of the Navy, which forbids the " branding, marking or tattooing on the body."

There exists, however, a method to which the most fastidious humanitarian cannot possibly raise any objection, viz. the use of photography, and a more exact system of measurements and personal description in the enlistment records than that now in vogue.

There is one system for identifying persons that was inaugurated in Paris in 1882, and which has given such excellent results that it is now used throughout France and is being generally adopted in Europe and the United States, viz. the *Anthropometrical System* of M. Alphonse Bertillon.

The Navy " Descriptive List" gives a man's name, age, place of birth (the foregoing being furnished by the man himself), height, weight, color of eyes and hair, complexion, and permanent marks or scars. The defectiveness of such a description is shown from the personal measurements and observations made in Paris by M. Bertillon with over 10,000 subjects; he found that among a hundred persons of the *same height*, 87 had what is commonly called " brown hair," 10 had blonde hair, 2.7 black, and 0.3 had red hair; one-third had hazel eyes, one-fourth gray, one-seventh blue, and one-fourth of indistinct color.

The identification of a person by the Bertillon system rests on the following measurements :

1. The length and width of the head.
2. The length of the left middle and little fingers.
3. The length of the left foot.
4. The length of the left forearm.
5. The length of the right ear.
6. The height of the figure.
7. The length of the outstretched arms.
8. The length of the trunk.

Perhaps the best concise description of the Bertillon system is that which first appeared in the *Pall Mall Gazette*, and is by the inventor himself. During the last Paris Exhibition the *Gazette* correspondent went to the measurement and identification department in Paris, which is in the Palais de Justice, where he found M. Bertillon " operating in a large square room. There were shelves on one side containing thousands of cards, with the photographs and records of

criminals. Several assistants were busy taking measurements to
add to the collection.

" My system for identifying criminals, said M. Bertillon . . . . is
now in operation throughout the whole of France. It was found that
many old offenders escaped detection. The classification of criminals under their names was not satisfactory. It is in their interest
to keep their antecedents hidden and to give false names. The
Paris police had amassed in ten years 100,000 photographs, but it
was impossible to search this collection every time an arrest was
made. Now, the search can be made in a few minutes. We now
classify our photographs and cards giving the antecedents of prisoners,
under measurements based on a system of anthropometrical descriptions. This system is simple and certain. Identification does not
depend on the uncertainty of a name or the doubtfulness of a photograph. We take no notice of names, and photographs might be dispensed with. Criminals are classified under measurements of
certain bony parts of the human frame. We have here 60,000
photographs and cards with the record of adult male prisoners who
have passed through the hands of the police. We begin our classification with the measurement of the length of the head. We found
by experience that it was better to begin with the head than the
stature. The size of the skull cannot be changed, but prisoners
refuse to stand up straight when their height is being taken. The
exact height could not be obtained except to within three centi-
meters, while the length of the head can be measured to a millimeter.
We divide the length of the head into three classes—short, medium,
and long—which reduces our collection to 20,000. Then we take
the width of the head, and making three subdivisions—of narrow,
medium, and wide—we have 6000 left. Next, we take the length of
the middle finger, and again making the three classes, as we do
with all our measurements, we have 2000 left. We continue on the
same system with the measurement of the foot, the forearm, the
height, etc., until we reduce our collection of 60,000 photographs to 6.
But you will see the system in operation.

" Call in that man who was arrested on the race-course yesterday,
said M. Bertillon to an assistant. The charge against the man was
watch-stealing, but he swore it was all a mistake.

" Have you ever been here before ?   *Mon Dieu*, no.

" Never been measured here ?   Certainly not.

" Where did you come from ?   Geneva.

" Where did you reside last ?   Brussels.

"A peculiarity of the international thief, remarked M. Bertillon. The prisoner submitted calmly to be measured. He was placed on a stool and the length of his head taken with a special compass made for the purpose. One leg of it was placed in the hollow above the bridge of the nose and the other moved round to find the greatest length behind. The compass shows the length to a millimeter. Next, the breadth of the head was taken from one parietal bone to the other. Another instrument was used for taking the length of the middle finger. These three are the surest measurements, said M. Bertillon, and give the best results. The measurement of the left foot was next taken. The prisoner was barefooted, and was made to stand on the left foot when its measurement was being taken. The process continued with the length of the ear, of the forearm, length of the arms extended, the height, and the color of the eyes. The color of the eye is registered according to the intensity of the pigmentation of the iris, but it requires some experience to record this description accurately. It is used most in classification of young persons who have not reached maturity. After the man was measured, a search of five minutes showed that he was an old thief who had been expelled from France and was now liable to a very heavy punishment."

It will be observed that the primary object of the measures taken is for the purpose of classification—*small, medium*, and *large*, or their equivalents, being the divisions used throughout a classification or file—and their principal benefit is in eliminating from a collection of descriptions all except what are in the division of the description sought for; the minute description of features, color of eyes and hair, permanent marks and scars, and photographs, being relied upon to establish the *identity* of the subject in hand, *after* the number of descriptions to be considered has been reduced to a minimum by the previous eliminations.

The sketches show the instruments used, also the manner of taking the measurements and recording them. For the purpose of uniformity the metric system is used, and for the same reason the small book of simple instructions should be followed strictly.

The following measurements of Mr. Geo. M. Portius, of the *American Bertillon Prison Bureau*, Chicago, Ill., taken at different times and places, give an idea of the extreme accuracy of the system :

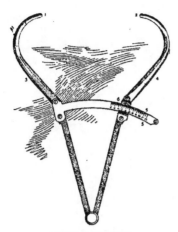

CALIPER COMPASSES.

For measuring the head.

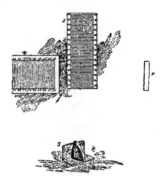

VERTICAL, HORIZONTAL, AND ·SQUARE MEASURES.

For measuring heights and outstretched arms.

SLIDING COMPASSES.

For measuring the ear.

For measuring the foot, forearm, and middle and little fingers.

To face page 364.

To face page 365.

WHERE TAKEN.	OPERATOR.	Height. Metre. Cent.	Out-Stretc'd Arms. Metre. Cent.	Trunk. Cent. Milli.	HEAD. Length. Cent. Milli.	HEAD. Width. Cent. Milli.	Right Ear. Cent. Milli.	Left Foot. Cent. Milli.	Left middle finger. Cent. Milli.	Left little finger. Cent. Milli.	Left fore-arm. Cent. Milli.
Joliet, Ill.	M. H. Luke.	1.763	1.79	93.7	19.5	15.6	6.7	27.2	11.7	9.3	46.2
Jackson, Mich.	A. C. Northrup.	1.755	1.79	93.6	19.7	15.6	6.6	27.3	11.8	9.4	46.1
Detroit, "	H. Wolfer.	1.766	1.79	93.7	19.5	15.6	6.6	*26.8	11.7	9.3	46.1
Chicago, Ill.	G. Bingley	1.76	1.79	93.4	19.6	15.7	6.7	27.2	11.8	9.4	46.4
" "	F. Ferrier.	1.759	1.79	93.4	19.6	15.7	6.6	27.3	11.8	9.5	46.3
Allegheny City, Pa.	A. F. Sawhill.	1.753	1.79	93.3	19.5	15.6	6.7	27.2	11.8	9.4	46.4
Chicago, Ill.	W. McClaughry.	1.766	1.79	937	19.5	15.7	6.6	27.2	11.8	9.4	46.1
" "	Operator No. 1.	1.766	1.79	940	19.5	15.7	6.7	27.2	11.7	9.4	46.2
" "	" 2.	1.765	1.79	940	19.5	15.7	6.7	27.2	11.7	9.4	46.2
Paris, France.	Alphonse Bertillon.	1.76	1.79	940	19.6	15.6	6.6	27.2	11.8	9.4	46.4

*"The *foot* measurement is the *only* one not within the permitted limit of variation. Caused by too great a pressure upon the instrument used."

Slight changes, in the headings only, need be made to adapt the description cards to the Navy Enlistment Record. The system can be used on board vessels which are not supplied with a photographic outfit; for while the photograph is an additional aid in identification, it is not a necessity.

All descriptions being sent to the Navy Department in Washington, the person in charge of the Enlistment Records would soon become very familiar with them and their use.

A bill has lately been introduced into Congress to establish a "Prison Bureau," with "a central office for the identification of criminals." If a *Tascott* be wanted by the "Commissioner of Prisons," he can inspect the files in the War and Navy Departments, and ascertain with certainty whether or not his man has sought security by enlisting in the army or navy,—provided the Bertillon or some similar system be generally adopted.

Not only would the adoption of the Bertillon method of measurement and registration, or such other method as shall minutely describe enlisted men in the army and navy, lead incidentally to the immediate detection of deserters and the dishonorably discharged men who have re-enlisted, but it would further protect the Government by furnishing a sure identification of the holders of honorable discharges, and also of applicants for pension.

			Rem.	
Height,	1 m	67		Head
Stoop,		2,		"
Outs. A,	1 m	75		Ear lg
Trunk,		92		Rgt. Ear w

Remarks incident to Measurments,	Two phot R. l. 8.

(1

	Incl.	recedg		Profile.
Forehead	Hght.	m,	Nose.	
	Width.	m,		L
	Pecul.			w
				Pe

Measured at Joliet,

Remeasured,
When
and Where,

Slight changes, in the headings only, need be made to adapt the description cards to the Navy Enlistment Record. The system can be used on board vessels which are not supplied with a photographic outfit; for while the photograph is an additional aid in identification, it is not a necessity.

All descriptions being sent to the Navy Department in Washington, the person in charge of the Enlistment Records would soon become very familiar with them and their use.

A bill has lately been introduced into Congress to establish a "Prison Bureau," with "a central office for the identification of criminals." If a *Tascott* be wanted by the "Commissioner of Prisons," he can inspect the files in the War and Navy Departments, and ascertain with certainty whether or not his man has sought security by enlisting in the army or navy,—provided the Bertillon or some similar system be generally adopted.

Not only would the adoption of the Bertillon method of measurement and registration, or such other method as shall minutely describe enlisted men in the army and navy, lead incidentally to the immediate detection of deserters and the dishonorably discharged men who have re-enlisted, but it would further protect the Government by furnishing a sure identification of the holders of honorable discharges, and also of applicants for pension.

(Profile —) DESCRIPTIVE. (Full —)

Incl. *recedg.*	(Ridge, *concave*		Beard, *sandy* Hair, *l. chest*	
Forehead Hght. *lw.*	Profile. { Base, *elev* Root, *sw.*		upper, *thin* indented. Complexion, *fair.*	
Width, *m.*	Nose. Length, *m.*	DIMENSIONS. { Projection, Breadth.	Weight, *160 lbs.*	
Pecul.	Pecul. *brown narrow*	lower, *squared.*		
	Pecul. *twisted to left*	Right Ear. Chin. *pointed*	Build *medium.*	

Measured at *Joliet, March 19th,* 1888, by *M. H. Luke.*

Remeasured,
When,
and Where, {

Institution, _Illino_

NAME, _John Smith_

Alias, _Jack Sn_

County, _Cook_

Occupation, _Tailor_

Known or Admitted Former)

Columbus, O

NUM'R'L ORDER	
1.	Hor. rect. sc. 4 ct 10
1.	Mole, 5 c. bel
1.	Anchor tattoe
	above wris
11.	Split Nail M
111	Sc obl. form.
	3 ct. l. temp
111.	Sc obl. backw
	15 ct. r. eye
111	Wart 2 c. bel

The Ro

SPECIMEN OF DESCRIPTION CARD. FILLED OUT.

(Reverse.)

Institution, *Illinois State Penitentiary.* Reg. No. 642.

NAME, *John Smith.*

Alias, *Jack Smith, alias "Slick Jack."*

County, *Cook.* Crime, *Burglary.* Sentence, *3 Years.*

Occupation, *Tailor.* Descent, *Irish.*

Known or Admitted Former Imprisonment, *Done time in Sing-Sing, 2 years;*

Columbus, O., 1 year for Burglary.

U. S. NAVAL INSTITUTE, ANNAPOLIS, MD.

MARCH 1, 1890.

NAVAL TRAINING.

By Rear-Admiral S. B. Luce, U. S. N.

President of the U. S. Naval Institute.

The naval policy of the United States, confirmed by many years of practice, is to maintain a very small floating force. That a similar policy will be continued in the future there can be no reasonable doubt. Such is the will of the people, and naval officers, while accepting the popular decree, must endeavor in good faith to make up for numerical inferiority by excellence of organization and thoroughness of drill. Military and naval histories are not wanting in great examples of discipline more than compensating for disparity of numbers.

Hence, as a supplement to the policy of keeping up a comparatively small navy, must come the policy of greatly increased attention to the question of reserves and to all the exercises which can best prepare the personnel for the great object to which it owes its existence —war. To furnish the best results, these exercises should be carried on systematically and progressively from early youth to mature age.

War is the best school of war; but if we can substitute some other and less expensive method of instruction, we go just so far towards attaining that state of preparation which is the surest guarantee against war. For if war is the great object in maintaining a military and naval establishment, the prevention of war is a still greater object. That fact cannot be too strongly emphasized. The prophylactic system is the wisest in this as in other cases, and it is just here that we find ourselves in touch with the Universal Peace Societies.

It is the trained athlete who enjoys the greatest immunity from aggression. The cloud of war which has so long been hovering over Europe has not burst, simply because all the great powers are prepared for it.

The ideal system of training comprehends mental, moral, and physical culture.   One of the deepest thinkers and closest reasoners of modern times remarks that the man who, in danger or upon the approach of death, preserves his tranquillity unaltered, and suffers no word, no gesture to escape him which does not perfectly accord with the feelings of the most indifferent spectator, necessarily commands a very high degree of admiration.

The heroes of ancient and modern history, he continues, who are remembered with the most peculiar favor and affection, are, many of them, those who in the cause of truth, liberty, and justice have perished upon the scaffold, and who behaved there with that ease and dignity which became them.

War, he explains, is the great school for acquiring and exercising this kind of magnanimity.   Death, as we say, is the king of terrors, and the man who has conquered the fear of death is not likely to lose his presence of mind at the approach of any other natural evil. In war men become familiar with death, and are thereby necessarily cured of that superstitious horror with which it is viewed by the weak and inexperienced.   They consider it merely as the loss of life, and as no further the object of aversion than as life may happen to be that of desire; they learn from experience, too, that many seemingly great dangers are not so great as they appear, and that with courage, activity, and presence of mind, there is often a good probability of extricating themselves with honor from situations where at first they could see no hope.   The dread of death is thus greatly diminished, and the confidence or hope of escaping it augmented.   They learn to expose themselves to danger with less reluctance; they are less anxious to get out of it, and less apt to lose their presence of mind while they are in it.   It is this habitual contempt of danger and death which ennobles the profession of the soldier, and bestows upon it, in the natural apprehension of mankind, a rank and dignity superior to that of any other profession.   The skillful and successful exercise of this profession in the service of their country seems to have constituted the most distinguishing feature in the character of the favorite heroes of all ages.

The word "soldier" is used here in its generic sense, and applies equally to those who fight under the flag of their country, whether at home or abroad, whether on land or at sea.   But if we make a distinction between those who embrace the profession of arms on land, or the "soldier" proper, and those who pass their lives at sea in the service of their country, then are the observations in respect to the profession of the soldier far more applicable to that of the sailor.

During long years of peace the life of a soldier is one of comparative inactivity, and unattended by those dangers that "try men's souls." It is not so with the sailor. For although his country may enjoy continuous peace, yet he himself is constantly battling with the elements. His whole life may be said to be passed in confronting danger.

"They that go down to the sea in ships, that do business in great waters, see the works of the Lord and his wonders of the deep; their soul is melted because of trouble, and they are at their wits ends," was not written by the Psalmist of men who pass their lives in ease.

The constant contending with winds and seas develops those rare qualities in the sailor, the exercise of which is demanded in a like degree by no other calling. Of this familiarity with difficulty and danger comes a contempt for it, as was remarked of the soldier inured to war. Hence that quality of reckless daring so characteristic of the sailor—a quality of inestimable value in war.

Formerly, the best and most thorough school of training for the young seaman was on board the merchant sailing ship engaged in foreign trade. This was due to the smaller crew of a merchant ship as compared to a ship-of-war, and, as a consequence, the greater and more constant demand for personal exertion; to the more economical fitting of rigging, sails and spars, which increased the chances of casualties; and to the urgent necessity for making quick passages in the interests of the owners, which required the "carrying on" of sail. A youngster's first tussle with a royal in a fresh breeze will long live in the memory of the man. He has no time to feel dizzy or sea-sick, nor does it matter whether he hangs on by his teeth or his toes. He must, unaided, roll up that royal and pass the sea-gasket, and do it, too, in a reasonable time, or somebody would know the reason why. From that he goes to wrestle with an obdurate topgallant-sail in a stiffish blow; to stowing a jib when every 'scend of the ship would seem to plunge him in the angry seas; to taking in a close reef. Rain and sleet and snow and ice and dark nights, and gleams of lightning giving a transient view of the wild seas, and deafening thunder, are accompaniments which he must take as they come.

" Though the tall mast should quiver as a reed,
And the rent canvas, fluttering, strew the gale,
Still must he on ——— "

The wonder often is that he can do any work at all aloft under
such circumstances ; that he has any mind or strength beyond that
which is absolutely necessary to self-preservation.

THE SCHOOL OF THE TOPMAN.

No one can deny that what may be called the school of the topman
on board a man-of-war is, or at least at one time was, one of the
most difficult and perilous that could be undertaken by men in time
of peace.   Seafaring people whose duties rarely, if ever, carry them
above the vessel's rail, are often appalled by the dangers of the sea.
What, then, shall be said of those whose habitual duty is high up on
the " giddy mast "—far above the rail?   The duty of the young sailor
on board a square-rigged sailing vessel is principally aloft, where it
would seem to require all his care and strength to keep from falling
to certain death.   This is true even under the most favorable circumstances, as, for example, when loosing, furling, bending or unbending sails while at anchor and in pleasant weather.   The slipping
of the foot, the missing of the hold, or the parting of a rope may
readily send one to inevitable destruction.   How much, then, are
those risks increased when the ship is in heavy weather at sea !   One
would suppose that at sea the sailor aloft had as much as he could
do to hold on for dear life, to say nothing of doing anything like
real work.   But sail must be reduced as the storm comes on, and
the " common sailor," as he is called, must reef and furl, as the case
may be, let appearances be never so uninviting ; for the safety of
sails and spars, nay, of the ship itself and all her crew is often dependent upon his exertions.   He is cool and self-possessed in the
presence of this imminent danger, and works as only brave and
hardy men long accustomed to danger can work, and the sails are
reefed or furled and safety insured through his courage and endurance.

Heavy rain-squalls may make the canvas as hard and stiff as boards,
and blinding snow-storms benumb his limbs, and night add its terrors
to the scene, till the heart of the boldest may well quail with fear,
and yet the true sailor, noways daunted, speeds him to his task.   It
is not uncommon to find " ordinary " seamen, not wanting in manli-

ness, but whose previous experience has been confined to "fore-and afters," absolutely refuse under such conditions to lay out on the yards. They would, as far as they were concerned, allow the sails to thrash themselves to ribbons first, and no coaxing or driving could induce them to leave the security of the tops.

Thorough seamen, on the other hand, would lay out on the yards, not only without fear, but with a certain degree of cheerfulness, let it blow high or blow low, and having accomplished their end, would come down from their perilous labors with the utmost unconcern, just as if their conduct were not beyond all praise. This total uncon- sciousness of his own high merit is certainly not the least admirable trait of the sailor's character.

The two representative seamen just referred to, the one who braves danger and he who shuns it, form an interesting and instructive study. On deck, in pleasant weather, there is to the casual observer little to distinguish one from the other in their capacity of sailor; but viewed under the conditions just depicted there is a marked contrast.

In the one, the entire mental, nervous, and muscular systems are under the absolute dominion of the instinct of self-preservation. To grasp and retain hold of the nearest object which promises adequate support engages his whole thought. He is breathless through his exertions in going aloft, bewildered by the warring of the elements and a general sense of insecurity, and the sum of all the work of which he is capable is to save himself from being blown out of the rigging into the raging sea.

In the other we find total self-abnegation. The rolling and pitch- ing of the ship, the loud flapping of the canvas heard above the howling gale, the surging of the yard, the driving rain, the blackness of the night, all combined have no terrors for him. With no thought of himself, his energies are all centered in the work before him. How he manages to get from the rigging to the yard and thence to the yard-arm, and why when there he is not shaken off into the sea, seems little less than a miracle. But he is there for a purpose, and that purpose he accomplishes, and accomplishes well. For the time and place, he is possessed of all the attributes of heroism.

Says a recent writer on physiology: "The man accustomed to use his muscles seems to obtain from them, *without effort*, a much more considerable amount of work, and this without an increase in the muscular fibers sufficient to account for the greater ease with which they contract. The nerve seems to transform a moderate stimulus, which passes along it, into an energetic one, and a *man accustomed to work performs, without effort of will,* movements which would formerly have caused him excessive voluntary strain."

"*The power of automatism acquired by daily practice comes to our aid constantly in the performance of difficult and rapid movements.*"

"It is incontestable that certain faculties of the soul come into play in bodily exercise to excite the contraction of muscles and to co-ordinate movements; it is also incontestable that these faculties are improved and developed by exercise."

"The faculties which preside over the co-ordination of movements are developed by the performance of difficult exercises, and their improvement endows a man with the quality we call skill."*

It is impossible for any young man to go through the school of the topman and become an able seaman, referring always to the sailing ship, without having his moral being permanently affected by it. Indeed, it is well known that such an experience does affect character, and has endowed the sailor with those high qualities of self-reliance, endurance, courage, and patience under difficulties which have always characterized him.

We venture here to quote Dr. Lagrange once more :

"The faculty which orders a muscle to act and which gives it the stimulus necessary for its contraction is called the *Will :* it also is developed and improved by the repeated use made of it. It shows its acquired superiority in the sphere of movement by a greater persistence of effort, by a greater tenacity in muscular action. The person who every day, *in spite of the different pains of fatigue, sustains energetic and prolonged muscular efforts,* acquires a greater power of *Willing,* and from this acquisition result certain very striking changes in his moral disposition. The habituation to work gives to a man greater energy of will considered as a motor force, and from this change of a moral order, as much as from that of a material order, results a particular form of courage which we may call *Physical Courage.*"

"Physical courage is manifestly increased by the practice of muscular exercises. It is almost exclusively in men whose daily work is

* Physiology of Bodily Exercise. By Fernand Lagrange, M. D. The International Scientific Series. D. Appleton & Co., New York. This work is commended to all interested in education in general and physical culture in particular.—S. B. L.

laborious, or who are given to violent exercises, that we see bold and energetic actions. The practice of muscular work and the habituation to bodily exercise dispose a man to brave all forms of material danger."

" In difficult exercises [such as our topmen were formerly subjected to], all the *psychical* faculties associate in the work of the muscles; hence arise the most characteristic conditions of difficult exercises: they need brain-work. Judgment, memory, comparison, will, such are the *psychical* factors which preside over their performance. The cerebrum, the cerebellum, the sensory nerves, are organs whose very active concurrence is indispensable.'

Seamen accustomed to running up the rigging four and five times a day, sometimes when at sea oftener—itself one of the severest of physical exercises,—to constant rowing in boats, often for long distances, and the handling of heavy guns, put forth "energetic and prolonged muscular efforts" to an extent hardly contemplated by the author.

The abolition of sails, the use of the hydraulic or pneumatic gun-carriage, and the indiscriminate use of the steam-launch has put an end, in a very great measure, to such severe physical training. This by way of parenthesis. To continue:

The school of the topman was, literally, the school of danger, as we have seen, and it prepared the man-of-wars-man, as no other school could, for his duties at the gun. The gun-captain, above all, must be possessed of those very qualities which were the product of the severe training of topmen—physical courage, self-possession, endurance, and the automatic movements due to habituation.

There is, or at least there was, not a little in the training of the young sailor that reminds one of the Spartan youth as moulded by Lycurgus after the Dorian model. The young Spartan was reared to a life of hardship, privation, and discipline. He was trained to courage, and taught acuteness, promptness, and discernment—qualities essential to a soldier. Such an education, the historian tells us, produced an athletic frame, simple and hardy habits, an indomitable patience and quick sagacity. " Their bodies," says Gillies, " were early familiarized to fatigue, hunger and watching ; their minds were early accustomed to difficulty and danger."

Somewhat analogous to the training of the sailor as a preparation for war on the ocean, was the chase as a preparation for war on land. The chase was not inappropriately termed " mimic war."

Wellington was always ready to praise the dash, courage, and endurance of his regimental officers, the sons of English squires and the younger sons of the aristocracy. They had been brought up to ride to the hounds, and to acquire, in early youth, a quick eye, a ready hand, and a good digestion. With these came the moral qualities, courage and endurance. The history of our own wars furnishes similar examples.

THE DECK-HAND.

In attempting to present a picture of the sailor "handing and reefing" in bad weather, there is no wish to convey the idea that bad weather is the rule at sea, or that the sailor was always to be found at the weather-earring. On the contrary, during a three years' cruise on some stations, the gale may prove a rare exception. The point is simply that when the danger does come, the sailor, by virtue of long training, is prepared to meet it, not only on the topsail yard, but wherever it may overtake him or in what form soever it may come. Battle is only another form of danger, and as he has but one life, it makes little difference to him whether he loses it at the hands of an enemy or through the perils of the sea.

It is an interesting question as to how far the character of the sailor, as heretofore known, has been and is yet to be modified by the use of steam as a means of propulsion.

By the multiplication of steam machinery the artisan is saved much labor, and the sailor has come in for his full share of the manifold benefits conferred by steam.

It is no longer by his constancy and skill that his sails are trimmed to the ever varying winds and his vessel turned again and again to her course. The powers of a thousand horses are now harnessed in his service, and his vessel is driven on in spite of winds and seas. What labor he is saved! What days he is spared of baffled skill and deferred hopes and all their chastening influences! What dangers he escapes! What exemption from a thousand perils! To-day the sailor may enjoy his repose on deck even during the heaviest gale at sea, albeit the firemen and coal-heavers are sweating out their lives feeding the insatiable maw that creates the motive power.

Formerly to get a frigate under way and stand out to sea required the presence on deck of the entire crew. It was an operation that called for the exercise of skill on the part of the officers, and the utmost promptness and alacrity on the part of the sailors. The same was true of entering port and coming to an anchor. How is it now? The complement of the Chicago is, let us say, three hundred and eighty-odd men. Of these there will be about one hundred belonging to the engineer's force. Throwing out the latter and the special class of petty officers, messmen, band, etc., there will be left about two hundred and ten or fifteen petty officers, seamen, landsmen, and boys. Of these two hundred and ten or fifteen it requires just two to take the ship to sea—one at the steam steering wheel, the other at the lead. We are not considering the engineer's force. That force does all the work in getting up steam, firing, running the engines, etc. Our present inquiries lead us to the deck. There we find, in round numbers, some two hundred seamen of various classes. What part do they take in getting the ship to sea or on entering port? No part whatever. It is proper on such occasions to station hands at the life-buoys, at the engine-room bell, on the lookout, at the signals, and the life-boat's crew at their boat ready for service. These are all positions of inactivity, calling for no mental or muscular exertion. On going to sea, the anchor once catted and fished, which may be done chiefly by means of a steam winch, the man at the wheel and the man at the lead are the only two of the two hundred who are called upon to take an active part in the operation. The one hundred and ninety-eight might as well be asleep in their hammocks for all the use they are in the management of the ship. So far from being of any use, the question is what to do with the crowd of men who are lounging about the decks. And it is this same question of what to do with the crew, how profitably to employ them, that frequently comes up on board the sailless steamer.

In bad weather at sea the steamship sailor will naturally "get in out of the rain." There is nothing in particular for him to do, and he does nothing but make himself as comfortable as circumstances admit. The man at the wheel is sheltered by the pilot-house; the men on the lookout alone have to expose themselves.

It is to the mercantile marine that navies in general look for their reserves of seamen in time of war. Let us consider then, for a moment, what kind of sailor the mercantile marine is producing. The great majority of tonnage in that service is found in steam vessels. The crews are made up of the engineer's force and the deck-hands. The latter have comparatively little work to do. The merchant steamer, when loaded by stevedores, casts off her lines, steams out from her dock and proceeds to sea. There is one man at the wheel. As the pilot takes charge, there is no one at the lead. The crew clear up the decks and make everything secure for sea. Thanks to steam, there is little or nothing for the deck-hand to do while at sea. On reaching her port of destination the vessel goes to her dock, throws out her lines and makes fast. Voyage after voyage is made without even so much as lowering a boat.

With such slight demands on the muscular and mental activities, the deck-hand could not exist on the food once thought good enough for a sailor. Fortunately, with the change in the mode of life comes a greatly improved dietary system. But even with the latter improvement, it does not seem possible that the deck-hand can get enough bodily exercise to keep him in the best condition. Says Lagrange, to quote that authority once more, " Muscular exercise has a considerable influence on the process of nutrition, and it is to this influence that are due the changes which occur in the conformation of a person whose muscles are habitually in action." What comes of no muscular exercise ?

There are numerous examples of individual heroism among deckhands, but it can scarcely be said of them, as a class, that their daily labors are such as to contribute to a high physical development and the physical courage which accompanies it.

Another quality found in the sailor was his contentment with ship life. He was, to borrow a phrase from the scientist, in perfect correspondence with his environment. The long sea-voyages in sailing ships during his minority habituated him to the life and its attendant hardships. He soon learned to content himself with an environment which would be intolerable to a landsman. But, owing to the quick passages made by steamers and the comparatively short time actually spent at sea, the deck-hand does not become habituated to ship-life—his preference is for the shore.

In the event of our becoming engaged in a maritime war, the deckhands of our merchant steamers would soon find their way into the navy and form parts of the complements of our crews.

This would not be so deplorable as at first glance might seem, for the tendency in the navy itself is to man our ships with deckhands. We say the tendency is in that direction, how far off soever may be the end to be accomplished.

Let us begin, as has been suggested, by putting the naval apprentice in barracks on shore, where he may be perfected in infantry drill, with a little marline-spike seamanship thrown in. After a course of soldiering on shore he will be put on board a steam practice ship, as has also been suggested, to steam in and out of port when the weather is pleasant, and to play at making sail when the wind is fair. From this he goes to a schooner-rigged steamship, or one with military masts, in the general service, where his instruction as an infantryman and marine artillerist is continued *ad infinitum*.

The difference between the man so trained and the deck-hand of a merchant steamer is one of degree only. Neither one is a sailor, nor possessed of the admirable characteristics of the sailor, while the sphere of action of both is confined to the deck. The trained man is, of course, the superior in general intelligence, orderly habits, discipline, and knowledge of arms; but he is not a sailor. And herein we find another of the many examples of history repeating itself; for we would, under such a system, have much the same general division of labor that prevailed on board the Roman war-galley; that is to say, the rowers who furnished the motive power of the galley would be represented by the engineer's force, and the *classarii milites*, or fighting men, would be represented by the marine artilleryman. There was a third class mentioned in the accounts of the Greek and Roman navies of the ancient civilization, but it seems to have excited little attention at the hands of the historian; this class was called *mariners*. They attended to the work about the vessel and had the care of such sails as were used.

It is generally conceded that if sails are to be abolished, sailors will no longer be needed. But *seamen the navy can never dispense with*.

If, on the other hand, sails are to be retained, then the sailor class must continue to exist; and it will be for the naval administrator to determine how best to shape the course of naval training so as to combine all the good qualities which have heretofore characterized the sailor, with those which go towards the "make-up" of the sea-man-gunner. Now, have we advanced sufficiently far in our knowledge of steam and electricity to warrant dispensing with the wind as a motive power?

SAILS.

The first essential in the discussion of questions affecting naval policy is to state the end proposed: whether we are always to act on the strictly defensive, or whether, in case of being forced into war, we are to be prepared to assume the offensive whenever such a course seems wisest.

A naval policy is largely the result of racial characteristics. The instincts of one race may lead to one line of policy; those of another race to quite another line. Thus one race may be brave and chivalric and excel in individual genius, and yet have no aptitude for foreign trade, or taste for maritime affairs of any description. Having no love for the sea, they would keep within their own borders. "We are a self-supporting people," they would say; "why concern ourselves with ancient laws of exchange of commodities with the effete monarchies to the east and to the west of us? We are a law unto ourselves."

Under such a régime our national policy would scarcely reach beyond our own boundaries. Motives of economy, rather than public utility, would control the development of the navy, and defensive tactics govern its operations.

As the navy would be for harbor defense only, the original monitor type would serve the purpose. The coal-yard and the dry-dock would always be within easy reach, hence there would be no need of masts nor sails, nor yet sailors, nor would the vessel's bottom require copper-sheathing. The time of such vessels, when in commission, would be spent at a wharf, for convenience of landing the crew for infantry drill. Meanwhile our fisheries would die out, and the American flag would disappear from the seas.

Under the dominion of another race, one, let us say, with an hereditary genius for maritime affairs, a different policy would obtain. If, for example, the Anglo-Saxon race continues to shape the destinies of these United States, then it is probable that we shall see a gradual revival of American shipping, the relegation of our foreign trade to American bottoms, and a return to our former naval policy. For among the characteristics of the Anglo-Saxon race are to be noted successful enterprise in foreign trade and a love for the sea.

This is no place to discuss the relative merits of the two systems. Each one is consistent in its several parts, and has its claims to consideration. But a line of naval development should proceed on one theory or the other.

The naval policy entered upon shortly after the close of the war of 1812 was that of a maritime people noted for enterprise and sagacity in mercantile affairs. Our "white-winged" commerce spread over the most distant seas, and thither our war-ships followed to give it moral and material support, whether the trader itself at sea, the merchant in a distant land, or our representatives accredited to foreign governments. Our ships, though few in number, were the best of their kind, and represented every class. We had fast frigates to overhaul merchantmen, and noble battle-ships that could stand up to any foreign battle-ships they were likely to fall in with abroad. The battle-ships were constructed on the theory, that while they furnished the best system for the outer line of coast defense, they could, if required, be sent on long cruises in distant seas. They were used as flag-ships on foreign stations, a certain number being held in reserve. They were "unwieldy" only in the imaginations of those who had never sailed in them.

A revival of our early naval policy would call for a battle-ship on each foreign station, with a proportionate number of cruisers in addition.

Formerly our ships were in a large measure self-sustaining. They are not so now. Having no outlying stations for coaling, docking, or repairing, and being entirely dependent on the good-will of those who, in a maritime war, might find it to their interests to throw impediments in the way of our receiving aid, the rig of our ships and the preservation of their bottoms from fouling become questions of the highest importance. As the character, moreover, of the average ship will qualify the character of the average crew, it is germane to our subject to bestow a passing glance on the rig of war-ships in general. This we may do without venturing to argue the questions involved.

According to the prevailing systems of naval tactics, the battle-ship finds her principal rôle in the line of battle. As her province is to fight in concert with other battle-ships, she should excel in manœuvring qualities, hence she is provided with twin screws. If, during battle, one screw becomes disabled, she is not thereby rendered helpless, for she can keep moving by means of the duplicate screw. To navigate under sail, she would require such enormous spread of canvas that, considering her peculiar character as a fighting machine, it has been thought better to give her none at all. Instead of sails, therefore, she is provided with two heavy upright spars called "military masts," from the tops of which sharpshooters, screened by steel shields, may fire down upon the decks of opposing battleships as they crash by each other in the shock of battle. The military mast will be also used as a place of observation during battle. In the latest additions to the French ironclad navy they are so used. The French battle-ships Le Hoche and l'Amiral Baudin have on their military masts places especially designed for the commanding officer and the torpedo officer, from whence, over-looking the field of battle, they can direct the movements of their ship, distinguish in the mêlée friend from foe, and send out their torpedoes intelligently. Farragut first suggested and practiced this plan. The battle-ship, moreover, does not let go of her base of sup-plies, unless indeed she can fight her way to the enemy's coal-pile.

The rôle of the cruiser is altogether different. She has no place in the line of battle. Her manœuvring qualities need not be so great as to sacrifice for them certain other desirable qualities, hence she does not require twin screws. She will, as a rule, cruise singly, and fight single-handed, and, presumably, not at very short range. As a "commerce destroyer," which is, according to our authorities, the principal office of the steel cruiser, she will roam the seas in quest of prey, and remain out during long periods. She should therefore have full sail power, and her bottom should be protected from fouling. She severs herself completely from her base; hence the more self-supporting she can be made, the more efficient she will prove in war. If it is established, as now seems likely, that twin screws give better general results than single screws, then they should be made to uncouple.

The new steel cruiser is known familiarly as a *paper-sided* ship, to express the ease with which her sides may be penetrated and her vitals reached by guns of very small caliber, such as are assigned to secondary batteries. To give the "paper-sided" commerce-destroyer the characteristics of an armor-clad battle-ship proves one of two things: either that the prevailing systems of naval tactics are fundamentally wrong, or that—but this is wandering from our subject.

The first Naval Advisory Board, assembled under the Depart-ment's order of June 29, 1881, was composed of the best representa-tive talent of the navy. Rear-Admiral John Rodgers, one of the most distinguished officers our navy has produced, was presi-dent of the Board. Chief Engineer B. F. Isherwood represented the Engineer Corps with marked ability, and John Lenthall, one of the ablest naval architects of his day in this or in any country, stood for the Corps of Naval Constructors. *The Board was unanimous in recommending single screws and full sail-power for cruisers.*

For convenience of reference we subjoin a table of sail areas, giving the proportion of sail spread by one of the old sailing ships, the Constellation; by the steam sloops-of-war, and by the steel cruisers, with a view to establishing a comparison between the several classes.

SAIL AREAS.

"As a rule, there should be given about 36 square feet of plain sail to every square foot of midship section."—(Theo. D. Wilson, Chief Constructor, U. S. Navy.)

"The Board is of the opinion that all classes [of ships recommended] should have full sail-power, the amount of sail surface not to be less than twenty-five times the area of the immersed midship section."—(Report of Advisory Board of June 29, 1881.)

"In place of the second-rate 14 knots vessels of the spar-deck type recommended by the majority, we propose to substitute ship-rigged first-rates of the single-deck type ... of 4354 tons displacement and 20,000 square feet of sail surface."—(Minority report of Advisory Board of 1881.)

	Area of Midship Section.	Area of Plain Sail.	Square Feet of Sail to 1 Foot of Midship Section.	
Constellation,	545	18,000	33	sailing ship.
Brooklyn,	628	19,327	30.77	wood, single screw.
Richmond,	560	16,600	29.64	" "
Lancaster,	721	20,700	28.71	" "
Hartford,	641	17,877	27.90	" "
Chicago,	790.24	14,880	18.83	steel, twin screws.
Petrel,	289.5	4,850	16.78	" single "
Boston and Atlanta,	675	10,400	15.40	" twin "
Newark,	807.23	11,932	14.78	" " "
Yorktown,	435.3	6,352	14.58	" " "
Baltimore and Charleston,		0.000	0	military masts.

From this table it appears that had the Chicago been sparred in accordance with the recommendations of the first Advisory Board, she would have had a spread of canvas equal to that which she now carries plus the amount carried by the Yorktown. Even then she would have been under-sparred.

The lessons furnished by the recent experiences of the Brooklyn, the Iroquois, and the Inman steamer City of Paris, should not be thrown away upon us. Let us take a glance at the performance of the former, premising that it was an exceptional year for trade winds, which were "conspicuous by their absence" both in the Pacific and the Atlantic. Where the maximum strength of the trade winds should have been found, only light airs and calms prevailed. Furthermore, she fell in with bad weather immediately after starting on her homeward passage, and for eight days after sailing still had the coast of Japan in sight.

The Brooklyn, Captain Byron Wilson, having broken her shaft, left Nagasaki, Japan, on the 5th of September, 1888, under sail, reaching Honolulu, Sandwich Islands, distant 5000, in 40 days; from Honolulu to St. Thomas, W. I., distant 16,000, she made the run in 141 days; from St. Thomas to New York, distant 1700, in 17 days; arriving in New York April 25th, having spent 198 days at sea in traversing 22,700 miles. She passed 33 days in port, viz. 26 days in Honolulu and 7 days at St. Thomas. While rounding the Horn, under whole topsails and topgallant-sails, there was a period of two weeks during which she averaged 200 miles a day. It is proper to state here that her propeller had been hoisted in on deck, and the aperture between the stern-posts planked up, giving her a run as clean as that of a sailing ship. This was an altogether exceptional advantage.

The steel cruisers have twin screws, which reduces the chances of similar accidents; for with one screw disabled they can steam very well with the other. Their use obviates, moreover (at least so it is claimed), the necessity of sails.

Let us now place the Baltimore at Nagasaki with one screw disabled and no sails, and ask to which ship one would confide his fortunes in a passage to Honolulu of 5000 miles: the Brooklyn or the Baltimore? The old or the new? The former, though dull, can sail, and could manage to fetch some place, let the winds be never so adverse. The Baltimore also may manage to get in somewhere. But if her one screw, doing double duty, *should* break down in mid-ocean, the ship would be rendered absolutely helpless. She could do nothing.* Speaking for ourselves alone, we would prefer to take our chances in the old Brooklyn.

---

* The following notice, taken from a recent paper, is slightly suggestive of the condition of a sailless cruiser on breaking down in mid-ocean: " Halifax, April 4th, 1890. Captain Hire, of the brigantine Alejo I. from Barbadoes, reports that on Monday last, while south of the Georges, he sighted a steamer flying the signals, ' *Want immediate assistance.*' ' *Starving.*' He bore down for her, and found her to be the steamer Southgate, from Placentia, N. F., for New York. Her shaft was broken and her provisions nearly all gone. She had been in a helpless condition for a number of days, *not having sufficient sails with which to make headway.*"

What if the accident to the City of Paris had occurred when only half-way across?

· To provide for such a contingency as the disabling of the motive power, though one of the conditions of the problem we are called upon to solve, is not the most important one. The main question is one of logistics, and as such is to be dealt with by the naval strategist and tactician. In the case under consideration the great desideratum is the coal supply, or, what amounts to the same thing, the husbanding of that already on hand for the hour of need.* The steel cruiser, with fine lines, twin' screws, triple expansion engines in separate water-tight compartments, and great coal capacity, renders possible long passages at sea under steam. For, with increased speed, we now have comparatively less expenditure of coal. This leaves little to be desired if eternal peace could be assured. But the naval tactician demands that there shall be a still greater saving of coal, a still greater economy of the vital principle to which he owes his very existence on the field of active operations in war.

The transatlantic steamer, with a fixed route and short stoppages, may make the round trip without coaling away from a home port; and the English ship-of-war is pretty sure to find an English coal depot in almost any part of the world. With the United States the case is different. Having no coal depots beyond our own ports, we must calculate upon a cruiser carrying fuel enough during peace for long passages, with a liberal margin for such contingencies as are likely to arise at sea—bad weather, delays, casualties, etc. During a maritime war, when coal may be reasonably refused by neutrals, the difficulty of obtaining supplies abroad would be greatly aggravated, hence the necessity of greatly increased coal endurance, beyond that already given.

As a "commerce destroyer," the steel cruiser would have to remain out for long periods, watching for her prey and chasing suspicious craft. Stationed on one of the great highways of ocean commerce, she would lie in wait for traders, who would be warned of her presence and give her a wide berth. She would have to seek them. Under similar conditions, a cruiser with full sail-power could reach her destination with her bunkers full of coal, and in many cases could do a great deal of cruising under sail, thereby holding on to her coal for battle or for chase. The underlying principle is that logistics should conform to strategy, not strategy to logistics.

---

* *Logistics* is the branch of the military art which has to do with the details of moving and supplying armies or fleets. The logistics of this case is the coal supply, which is of the very first importance. Under certain conditions coal may rank above ammunition in the scale of military values.

The Newark carries 850 tons of coal. Should disturbances in Rio place in jeopardy the interests of American citizens engaged in busi- ness in Brazil, it might be desirable to send her there. She could steam the entire distance, 4733 miles, in twenty days at her most economical rate, 10 knots per hour, with an expenditure of 500 tons of coal, or say 25 tons per day.

The Brooklyn, with her propeller in place, could make the passage under sail. It would take her three times as long, perhaps, but she would arrive at her destination with a comparatively small expendi- ture of coal.

After being in Rio for some time, the coal question forces itself, let us suppose, upon the attention of the captain of the Newark. Distilling and cooking and running the dynamo must go on, and the coal supply gradually but surely melts away. Strained relations with the authorities on shore render it impossible to get coal there ; and as the Riachuelo is close at hand, the American admiral is loath to let the Newark go. But go she must, there is no help for it, how- ever urgent the necessity for her presence, for she must have coal at any cost. So with due regard to the margin of safety, there comes a a day when she must steam down to Monte Video, 1200 miles further south, to fill up her bunkers.

The Newark is no sooner at sea than it is discovered that her speed has been diminished fully 25 per cent, owing to a foul bottom. Under such conditions the Riachuelo, with her clean copper-sheathed bottom, could easily overhaul her in the event of actual hostilities. Knowing this fatal weakness of our cruisers, it would be strange indeed if the authorities on shore did not resort to those dilatory tactics by which negotiations would be prolonged and a rupture forced only after coal and patience had been exhausted.

Let us suppose a similar case in the Pacific, where the necessity might arise for sending the Charleston, with her two military masts, to Valparaiso, a distance of about 5000 miles. The Charleston's coal capacity is 800 tons, giving her a steaming radius of 6450 miles at 10 knots an hour. The passage at the most economical rate of speed would require 650 tons of coal, leaving but 150 tons for " coal pro- tection " in the event of a battle. The Hartford could have made the passage under sail.

Or, let us place the Charleston at Honolulu, and require her to go thence to Valparaiso. The distance is say 5900 miles, approaching the limit of her steaming radius. On reaching Valparaiso, coal becomes an absolute necessity. There is plenty of it on shore, but, unfortunately, the Blanco Encalada, an armor-clad, copper-sheathed vessel of war, lies between the Charleston and the coal-pile, and the gravity of the questions at issue between Chili and the United States forbids the former giving supplies to a possible enemy. The Hartford could have made the run under canvas, and reached her destination with enough coal for emergencies.

Let it not be supposed that we advocate a return to the Brooklyn class. That class is cited merely to show that we have, in times past, built steam cruisers that would do well under canvas, and thereby enabled to husband their coal supply for emergencies. Our contention is that the steel cruiser should be able to do the same. And it certainly seems only reasonable to ask that the cruiser of our future navy should be furnished with the means of reaching a distant point where belligerent operations may ensue, with her bunkers practically full, for two reasons: first, that she may be prepared for battle, chase, or a protracted blockade; and, secondly, that her full bunkers may afford the coal protection they were designed for. Full sail-power would furnish the means. We are not wholly ignorant of the objections urged against putting masts and yards on modern cruisers. But if the naval tactician declares that the cruiser must have full sail-power, we venture to assert that the naval architect will be able to overcome the objections and solve the problem.

If the naval architect declares that the combining of all the essential features of a fast steam cruiser with efficiency under sail is not an insoluble problem, then the naval tactician would be enabled to avail himself of one of nature's most potent forces, the ceaseless winds, while he held his principal motive power in reserve for the day of battle. His position might then be likened to that of the general of an army who should hoard his own supplies while living on the enemy. The architect having so decided and furnished a sail plan, the seaman then steps in to arrange the details, so that a large spread of canvas may be rapidly reduced, and sails and spars so disposed of as to present the minimum of resistance when steaming head to wind.

So far, our highest authorities (those already quoted) have declared for full sail-power, and such we believe will be the final verdict for war-ships designed for service in distant seas. As a corollary to the proposition of full sail-power comes the demand for copper-sheathing on the bottom of our cruisers. As those already built have neither, it is very evident they are not intended for foreign service during war.

The question of copper-sheathing has been so thoroughly discussed by Mr. Philip Hichborn, U. S. Navy, in his exhaustive article on " Sheathed or Unsheathed Ships,"* and by Lieut. S. Schroeder, U. S. N., in " The Preservation of Iron Ships' Bottoms,"† that there is little if anything left to be said on the subject. It is only referred to here as belonging to the " tactics of the ship," which has to do with the question of the training of the personnel of the ship.

The history of the Civil War furnishes a practical illustration of the soundness of our views on the question of sails. Captain Raphael Semmes, of the Alabama, writes :‡ " I was much gratified to find that my new ship (the Alabama) proved to be a good sailer under canvas. This quality was of inestimable advantage to me, as it enabled me to do most of my work under sail. She carried but an eighteen days' supply of fuel; and if I had been obliged, because of her dull sailing qualities, to chase everything under steam, the reader can see how I should have been hampered in my movements. I should have been half my time running into port for fuel. This would have disclosed my whereabouts so frequently to the enemy that I should have been constantly in danger of capture ; whereas I could now stretch into the most distant seas and chase, capture, and destroy, perfectly independent of steam. I adopted the plan, therefore, of working under sail, in the very beginning of my cruise, and practiced it unto the end. With the exception of half a dozen prizes, all my captures were made with my screw hoisted and my ship under sail; and, with but one exception, . . . I never had occasion to use steam to escape from an enemy."

In another place he says : " She (the Alabama) was a perfect steamer and a perfect sailing ship at the same time." Her propeller was constructed to hoist clear of the water. The reverse of this picture is also shown : " The reader has seen that the Sumter, when her fuel was exhausted, was little better than a log on the water, because of her inability to hoist her propeller." Will the steel cruiser do any better at sea "when her fuel is exhausted"? The Sumter had a good supply of sail and one screw to drag; the steel cruiser has no sail and two screws to drag.

* Vol. XV, No. 1, 1889, Proceedings U. S. N. Institute.
† General Information Series, No. VII, 1888.
‡ Memoirs of Service Afloat.

THE TRAINING SQUADRON.

Assuming that the naval strategist will decide that, whatever practice may obtain abroad, the naval policy of the United States requires a full-rigged, copper-sheathed steam cruiser of high speed, then the demand for sailors will continue. It was to supply that demand that the Training Squadron was put in operation in 1881 (see General Order No. 271 of 1881).

The first essential in organizing a system of training is to know exactly what is to be produced. " Going into training " means the preparing of the brain, the nerves, and the muscles for the accomplishment of a certain specified end, whether it be for championship in the ring, success in the boat-race, or generally for success in any undertaking requiring strength and skill. Mental stimulus is necessary to complete muscular development. It matters not what may be the exciting cause—whether ambition, the pursuit of pleasure, or the necessity of providing one's daily bread—the result is much the same. A boy gladly expends in a game of baseball as much heat and energy as would be required to saw a cord of wood ; the latter would disgust him. Men whose sole object is the increase of muscle tissue soon turn with loathing from the tread-mill system of the gymnasium ; adequate mental stimulus is wanting.

The object of naval training such as we are now to consider is to produce a man who, to the fine qualities of the sailor as we have endeavored to portray them, adds skill in the use of the high-power guns of the present day. That is the object of our naval training system as we understand it. It is to produce men like those who handled the lock-strings of our 9 and 11-inch guns and Parrott rifles during the late Civil War—prime seamen, cool in the face of danger, courageous, and, withal, capital shots—men like the one who fired the Cumberland's last gun.

It is important that we should have armorers in the navy, and gunsmiths, and bookbinders and typesetters and the rest. But, before all else, we must have seamen and good gun-captains, with an abundant surplus of those who are to fill their places when they fall in battle. In other words, our guns' crews are to be made up of trained men. The highest expression of the trained man is to be found in the able seaman and expert gunner combined. It was for him that the rating of *seaman-gunner* was established. It was expected that seaman-gunners would be assigned as gun-captains of the guns of our primary batteries, and that in time every gun afloat would have a seaman-gunner for its captain. It was with that view that a school of practical gunnery was established on board the Minnesota in 1881 and General Order No. 272 of that year issued. The plan failed then through causes not necessary to specify, only to be renewed now under conditions likely to ensure its permanency. We may therefore safely assume that a gunnery-ship will hereafter form part of our naval training system, and that the seaman-gunner will no longer be degraded by assigning him to such duties as ship's lamplighter, typesetter and the like.

One of the most important truths to be understood by those having to do with naval training is, that unless pupils under training are early habituated to life on the water, much of the time, labor, and expense devoted to the work will be thrown away. Why train the naval apprentice for civil life?

It should be made an offense for any officer to moor a school-ship, or a training ship, or any vessel used for the education of young seamen, to a wharf or dock unless actually necessary when the vessel is under repairs. Under the specious plea of convenience of landing for infantry drill, the pernicious system of tying up the school-ship to a dock has wrought its full measure of evil upon the service. It is to be hoped, in the interests of the profession, that the practice is now at an end forever. Let the young sailor understand from the beginning that his only way of getting from his ship to the shore is by means of a boat, and let the boat and its management become thoroughly familiar to him, from the very outset of his career.

Infantry drill necessarily forms a part of the programme of exercises of the stationary school-ship; but it is just as necessary that the apprentice should learn how to get into a boat with his arms and accoutrements on, as it is that he should learn to march with them on. The few months spent on board the stationary school-ship are devoted to the "breaking in" process, to prepare the apprentice for the next stage of development, the sea-going training ship. With that view he is taught to take care of himself and his belongings, taught to keep his person and his clothes and bedding clean and in good order, to conform his words and deeds to the rules of the ship. In short, he is given his first lessons in naval discipline, which, if looked into, will be found to mean a great deal of self-discipline. We cannot begin too early with that.

English grammar, and arithmetic, and geography, and infantry drill are excellent subjects for the occupation of the mind of the young seaman; but there is an occult process of absorption going on without cessation, by which the mind of the naval apprentice becomes embued with a taste for the service which ends only with life itself. That part of the education is worth all the rest put together; it cannot be imparted, it comes of itself to the boy who is endowed by nature with an aptitude for the sea. He soon finds himself in perfect correspondence with his environment; the ship itself and all it contains; the sailors and their peculiarities; the guns and arms, suggestive of battle; the system by which the crowded decks seem never in confusion; nay even the very smell peculiar to ships. All at first have a fascination for him, and then become a part of his life. Thus he acquires from the very beginning something of the ways of sailors, their modes of expression, habits of thought, manners, customs—the first insight into the *technique* of his calling. He also learns their stories, their songs, and their traditions. The attraction of his surroundings *draws out* his powers. The inward being is in sympathy with the external life. *He is in correspondence with his environment.* This is education; it comes not of books. *Educare* means to "draw out," not to "put in" or cram.

There are, it is true, boys who, having no aptitude for sea life, find all this positively distasteful. When this is due to racial characteristics, or, what amounts to the same thing, the law of heredity, it is difficult to overcome. But as it is very common to find among boys cases of retarded development, both mental and physical, we should not, in the earlier stages of training, be hasty in determining a boy's fitness for the service.

The routine of drills and exercises on board the stationary school-ship prepares the boys for the next step in the process—the cruising ship. The readiness with which the boy falls into place on board the latter is the measure of the success of the former.

It must be confessed that officers are sometimes unreasonable in asking too much of the boy who has been but six months on board the school-ship. They expect him to take the wheel, or heave the lead, or run up aloft like an able seaman. But the school-ship does its part if it gives the apprentice a good grounding in the merest rudiments of his calling.

It is to the cruising ship, however, that we must look for the most important results, for it is here that the apprentice is expected to discover if he is made of the right sort of stuff. The cruises of the training ships should always be to foreign ports. We want blue-water sailors, and they can be made on blue water only. Hanging around the coasts and inland waters and anchoring every evening or two, so that all hands may enjoy a night in, is very pleasant but it is not business. The argument in favor of home waters is that by being frequently at anchor the boys may be taught more. If that means that they can be taught more out of books, it is probably true. But we have started out with the distinct understanding that the object of the training ship is to produce sailors—deep-water sailors; and the only way to do that is to make deep-water cruises. The young sailor may then gain that experience at sea which goes so far toward forming his character. No artificial method can take the place of actual experience at sea. The West Indies furnishes an admirable cruising ground for training ships, particularly during the winter months.

By letting the ships cruise singly, with a view to their meeting at a certain time, under squadron organization, for competitive exercises, preceding the annual examination and dispersion of the advanced classes, the very best results follow. The spirit of rivalry stimulates each one to do his best.

Each ship's company of boys should be divided into two classes. At the end of the winter's cruise the advanced class only should be transferred to the general service, leaving the remainder to take another cruise, the complement being filled up from the stationary school-ship. This ensures having always a fair proportion of the boys who are familiar with the ship. The second class of one cruise becomes the first class of the cruise following.

The permanent crew of the practice ship (the petty officers and seamen) should be carefully selected from general service men. It is from them that the boys get their earliest and most lasting impressions. The longer the boy can be kept on board the sailing practice ship, provided she be kept cruising, the better.

The fact must not be lost sight of that the practice ships, though intended primarily for the training of the naval apprentice, are at the same time an admirable school of practical seamanship for the young officers who go out in them.

It has been objected that service on board an obsolete sailing ship is no preparation for duty on board a modern steam cruiser. This is a great mistake. The handling of a sailing ship at sea, under all the varying conditions of wind and weather, stimulates the faculties as no other experience can.

The broadsword and the rapier (or foil), as weapons, are obsolete, and yet constant and intelligent practice with them, and with the " gloves," trains one into better command of the temper and of the limbs. Such exercise, properly conducted, gives a quick eye, a good wind, and an active liver, and enables one to keep cool and self-possessed in the presence of a certain class of dangers. The weapons themselves are not subsequently brought into requisition, they are simply used as a means to an ulterior end. It is the same with the sailing ship. Service in one teaches an officer what he cannot so readily learn on board a steamer. Among other things it teaches him to control men.

Of two officers, in other respects equal, that one will make the best steamer officer who has had the most experience in sailing ships.

From the training squadron the apprentice is transferred to the general service, where he remains till he comes of age.

It is much to be regretted that a more uniform system of treatment could not be accorded the apprentice in the service at large. Perhaps such a thing is too much to ask of human nature.

During one period it was the practice, when a ship was paid off, after a cruise, to give the apprentices a leave of a month or so, with orders to report on the termination of their leaves on board the Minnesota, where they were to serve out their unexpired terms. This gave a good opportunity to judge how the boys had been cared for by the officers during the cruise. The greatest difference was observable in the apprentices from different ships. One set of young men (for such was the rapidity of physical development that the majority presented anything but a boyish appearance) would have plenty of money due them, each a complete outfit of clothing, clean and in good order ; be manly in bearing, respectful in demeanor, and thoroughly well up in the details of their duties, whether aloft or at the guns. Of such a set, nearly all would be as far advanced in rating as the regulations would permit. Another draft of apprentices, who had returned in another ship, would be the reverse of all this. It is scarcely necessary to add that the former set would, as a rule, express a partiality for the service, while the latter would not. The differences were so marked, it was suggested that on a ship being inspected at the expiration of her commission, a special report should be made of the apprentices and the progress they had made during the cruise.

This is a very important phase of the subject, and has a direct bearing on the question, Will the apprentice, after attaining his majority, re-enlist for the general service?

It is for the apprentice who, on his discharge, re-enlists within a stated interval that the gunnery-ship is intended. Here he goes through a course in practical gunnery and cognate branches, is instructed in practical electrical work and torpedo service, and qualifies for the rating of seaman-gunner. The course on board the gunnery-ship will probably include instruction with the diving apparatus. This completes the system of special training.

WAR PROBLEMS.

We have spoken of a Training Squadron: there are two of them; one for boys, the other for officers, seamen and marines. The latter is composed of the ships of the North Atlantic Station. That is the high school, so to speak, of our naval training. Assembled annually for exercises on an extended scale, it becomes the best school of application we could have. Acting in conjunction with the War College, the operations of the squadron occupy at once the highest plane attainable during a time of peace. The following plan of operations has been attended with success:

Two or three representative officers of the squadron meet a like number of officers of the college, the whole occupying the position of a general staff to the commander-in-chief. The latter indicates the broad lines of operations to be undertaken, the staff arranging the details. If the commander-in-chief decides, for example, to attempt to force an entrance into Narragansett Bay, the attack and defense will be arranged with that view. If it be to attack New York by entering Long Island Sound at its eastern end, then the defense may be designed to prevent his passing the Race or getting by Plum Island; or the problem may be to effect a landing on Long Island, and to oppose such landing.

Whatever the military and naval problem may be, it is carefully worked out in detail by the staff, and maps and written instructions prepared for the information and guidance of the principal officers in the same manner as if it were an operation in war. When this has been done, the officers are assembled and the entire plan is explained by one of the staff. The object is generally to illustrate some military principle or to execute some strategic movement, or perhaps to show the practical working of some tactical manoeuvre.

At this meeting it is expected that every officer, particularly such as are entrusted with important commands, will interrogate the lecturer until the part each one has to perform is thoroughly understood. Umpires are appointed who are to take notes as to the carrying out of the plan in practice. At the appointed time the projected movements are carried out as nearly according to the programme as possible. In due season, after the conclusion of the operations, the officers are again assembled and the criticisms of the umpires read. A general discussion then takes place. By this intelligent method of conducting exercises, officers and men are brought to a realizing sense of what hostile operations might lead to, and, next to war itself, learn from them more of their business than they possibly can by any other method. Not only are such exercises highly instructive, but they lend a pleasing variety to the monotony of a cruise, and interest, in an unusual degree, all who take part in them. They are to be commended, if only on hygienic principles.

In addition to the war problems, there may be projected a number of exercises in which all the ships of the squadron take part; such as distant expeditions in boats necessitating an absence of four or five days, encampments on shore of a week's duration, night attacks by the boats or the ships, etc., etc.

RESERVES.

If a settled policy calls for a small navy, it should also call for a large reserve of force, wherewith to expand that navy in time of need. The fisheries and the mercantile marine generally supply the reserve. As early as 1778, it was declared in Congress that "the fisheries of Newfoundland were justly considered the basis of a good marine." The great benefit of having such a class of hardy seamen to draw upon was felt in the war of 1812; and in 1813 Congress passed the act granting a bounty to fishermen engaged in the bank (Newfoundland) and other cod fisheries, with a view to encouraging that industry and nursery of seamen. The act of July 28, 1866, however, abolished the granting of bounties. The greatest number of seamen in former years were produced in the sailing ships engaged in foreign trade. That source has almost ceased to exist—almost, but not entirely. In some lines of trade, sailing ships are still profitable. As long as they exist, Congress should oblige or induce the masters or owners, by bounty or otherwise, to take boys out in numbers proportioned to the ship's tonnage.* By this means our merchant vessels will be training up seamen that would be a very material aid in manning our ships in time of war. Fifty years ago it was estimated that about ninety thousand seamen were engaged in our foreign and coasting trades and fisheries. It was further estimated that of that number, owing to the interruptions of trade by a maritime war, about thirty thousand would be available for the navy. How many seamen could we count upon to-day ?

THE YACHT SERVICE.

When vessels of war were built of wood, the gun was required to send a projectile through about 30 inches of tough live oak. As a consequence, guns of heavy caliber were in demand. In the war of 1812, much of our success was due to the fact of having guns of larger caliber than our opponents. From that day the caliber of guns increased till it got up to our IX and XI-inch smooth-bore guns of the Dahlgren pattern. Small vessels, or such as had light scantling, could not carry these guns, and small smooth-bore guns were of little use. But owing to the great changes brought about in late years, a moderate-sized yacht can carry a gun that could place the thin-sided Chicago, of 4000 tons, *hors de combat* at a distance of a mile and a quarter. The Hotchkiss 9-pounder, or the 6 or even the 3-pounder, which could be carried by many of our yachts, will penetrate the side of the new steel cruiser at 2250 yards, and a chance shot might reach her boilers. Some of our steam yachts could do better. The 4-inch breech-loading rifle, weighing 3200 pounds, throwing a 30-pound projectile, will penetrate 6 5 inches of iron at 1000 yards. This gun mounted on a pivot, on board a fast steam yacht, would prove very effective.

It may be seen from these changes in the character of the offensive and defensive powers of war vessels how a flotilla of steam yachts might thus be made a valuable auxiliary to a fleet of battle-

---

*All foreign-going sailing ships should take out, as part of their crews, boys who are not able seamen, to do nautical duty, as follows:

Vessels from 100 to 300 tons, at least 2 boys.
"   "   300 " 500 "   "   4 "
"   "   500 " 700 "   "   6 "
"   "   700 " 1000 "   "   8 "
"   "   1000 " 1400 "   "   10 "
"   "   1400 and over,   "   12 "

ships, assigned to the duty of guarding our coasts in time of war, as ocean scouts, despatch vessels, or for torpedo service.

Statistics show that the majority of the crews of our yachts are of Scandinavian origin; they come of a great race of seamen.

"When the Romans invaded Britain, the Brits had no fleet to oppose them. We do not, until a later period, meet with that love of the sea which is so characteristically English; not before the gradual absorption of the earlier inhabitants by a blue-eyed and yellow-haired seafaring people, who succeeded in planting themselves and their language in the country."

"To the numerous warlike and ocean-loving tribes of the north, the ancestors of the English-speaking people, we must look for the transformation that took place in Britain. In their descendants we recognize to this day many of the very same traits of character which these old Northmen possessed."

"Britain finally became the most powerful colony of the northern tribes. At last the land of the emigrants waxed more powerful than the mother-country, and asserted her independence; and to-day the people of England, as they look over the broad Atlantic, may discern a similar process taking place in the New World."—DU CHAILLU, *The Viking Age.*

If, with due respect to the distinguished author, instead of "emigrants" in the last clause of the above quotation, we should read *conquerors*, the statement would be nearer to historical truth; and nearer still if it be made to refer to an earlier period of our own history rather than "to-day." The passage is transcribed for the reason that it gives a picturesque view of the influence of race upon the progress and direction of national development. The lesson is full of meaning to us. It is for the naval administrator to trace along the streams of immigration the strain of a race whose "school of war was on the sea"; it is for him to ascertain to what extent it exists in this country now, and how far it may be made available in the event of war.

Boys of a sea-loving race will naturally seek the sea; it is the part of wisdom, therefore, for the Government to make the paths straight which lead to the gratification of those tastes, that the training of young seamen may be carried on unceasingly, and to a far greater extent than will ever be possible, even in a navy many times greater than our own. Indeed, from motives of the highest patriotism, as well as from professional considerations, the naval administrator is directly concerned in what is popularly termed the rehabilitation of our mercantile marine; he is interested in the establishment of an executive department of the Government having for its object the promotion of ocean traffic, and the supervision of all public business relating to the same; in the building of merchant sailing ships and fast steamers; in the registration of merchant seamen, and, in general, in whatever tends to foster and increase foreign commerce under the American flag.

Senate bill 1628, reported January 6, 1890, for the "encouragement of commerce, the protection of navigation, and the improvement of the merchant marine in the foreign trade," proposes a measure of great national importance and of incalculable benefit to the navy.

A broad and comprehensive scheme for giving the utmost freedom of scope to our maritime interests is, indeed, but part of the policy of keeping the navy down to the merest nucleus of a floating force.

DISCUSSION.

NEWPORT BRANCH.

NEWPORT, R. I., *June* 26, 1890.

The meeting was called to order by Rear-Admiral Luce, President of the Institute, and the members were requested to nominate those they desired to fill the offices of the Branch, which resulted in the election of Commander T. F. Jewell as Vice-President and Professor Charles E. Munroe as Corresponding Secretary.

Commander Jewell was then called to the chair and made the following opening remarks:

Gentlemen:—We have assembled this evening for the purpose of discussing a paper on "Naval Training," a subject of vital importance to the service in these days when new appliances and new methods are replacing the old. We are fortunate in having with us the distinguished author of the paper, an officer whose experience in the service comprises nearly half a century, who has ever been foremost in advancing the interests of the Navy, whether it be in the training of men, the education of officers, the discussion of the type of vessels-of-war, or the methods of handling them; and who, although he has passed the period of his active service in the Navy, is still as alive to its needs and interests as any of those present. He kindly offers to give us a résumé of his paper, and I now have the pleasure of introducing Admiral Luce.

Rear-Admiral LUCE, U. S. N.—*Mr. President and Gentlemen:*—As some of those present may not have seen the paper about to be discussed, or may not have had time to read it, a few explanatory remarks may not be out of place.

I am glad to notice among those present a member of the medical profession, for one of the principal questions under consideration touches upon matters in which they are directly concerned. The first question, briefly stated, is, "How far has the introduction of steam, as a motor power, modified the character of the sailor?"

That is the first question treated of in a general sort of way by the paper. Or, to put it in another form and judging from current events, "How far *will* the character of the sailor be modified by the exclusive use of steam as a motor power?"

The sailor exists under two characters. First, in the popular estimation, which pictures him as a man given over to drunkenness and debauchery, and addicted to profanity. The second is his real character, given by those who best know him : owning all the weaknesses of humanity, indeed, in common with all men, but possessed, more than is common, of certain qualifications which are the product of his arduous life—such as physical courage, muscular power, self-reliance and endurance—qualifications of the highest value in war, for that is, after all, what everything about us means—war, or the preparation for war. Now, it is owing to the very fact that sailors have such a superabundance of physical force—such an excess of vitality—that, when released from the restraints of discipline, those forces seek a vent ; and when the controlling influence of a strong will becomes clouded by drink, riotous conduct is generally the result. But it is a base aspersion on the class to say that sailors, as a rule, are given to intemperance. My own experience goes to show that such is very far from being the case. What are called North country sailors are noted for their sobriety ; and there are many like them.

Now, admitting that the exclusive use of steam has a tendency, and a very decided one, to change the physical and moral qualities of the sailor, the next question is, "Is it possible to devise a system of training, by means of which all the good qualities ascribed to the sailor of the past may be preserved and all the objectionable ones eliminated?" For my part, I believe it is not only possible, but I believe we are doing it now. I have great faith in our naval training system. It has its "ups" and its "downs," of course, in common with the rest of the world. With the exception of an occasional set-back, I believe the progress made by our system of naval training has been fair. If there be any fault to find, it is due to our peculiar system of naval administration, and not to a want of personal interest in the apprentice system. To avoid the chance of being considered "personal" in anything I may say, it may be as well to state that no reference is made to the present time or to the present administration, but rather to an indefinite period in the past.

With the idea of beginning a course of training which was, in the end, to produce a man-of-wars-man, an effort was made some time ago to cultivate in the young apprentice a fondness for the water. With that view the New Hampshire's boats were to be seen every afternoon under sail and filled with boys. Boat sailing was a recreation the boys looked forward to on Saturdays with great pleasure. But far better than the boats was the schooner Wave. The best boys were selected for her crew, and, when she went out, one boy was placed at the wheel, and another at the lead, the others alternating with them. In this way, under the guise of recreation, the greenest boys were enabled to pick up something of sailorizing ; but the important part of it was that any taste they might have for the water was brought out by this species of amusement. Going out in the Wave and in the cutters, under sail, was held out as a reward to the boys for good behavior. Besides the schooner Wave there was a large working model of a full-rigged brig, built up on the New Hampshire's first launch. By the use of that model the boys were enabled to learn the names and the lead of all the rigging, instead of getting it by heart out of a book. In the little brig they had all the spars and sails and rigging of such a ship as the Boston, so that the boys not only saw the lead of the ropes, but the effect of hauling on them, the manner of making and taking in sail, etc., etc.; in short, it was such a working model as enabled the instructors to take the boys through all that branch of practical seamanship in an intelligent manner, by object-lessons. The dismantled hulks of the Wave and the little brig now lie in the cove at the training station in an advanced state of decay, and scarcely ever does one see a boat belonging to the training squadron under sail. Base-ball and field sports generally completely usurped the place of sailing and fishing parties.

Now, it may be said in answer to all this that boat sailing is out of date ; that the steam launch has rendered sails for man-of-war boats unnecessary. But it sometimes occurs that a ship-of-war has but one steam launch, and that it may be necessary to send out several boats at the same time, and not all of them in the same direction. Moreover, it may, and sometimes does, occur that the distances the boats may have to go are great. Let it be supposed that a boat expedition has to be sent out with the prospect of a week's absence from the ship. In that case the boats must be self-reliant. For such contingencies our boats should be constructed and rigged with a special view to carrying capacity and for sailing, and our boats' crews should be good boat sailors. I have seen our training cruisers come in here and their boats constantly go to and from the shore under oars when they had a fair wind and could sail. I may say that such is the rule. I will go further and say such is the rule not only on board our training ships, but on board our ships generally throughout the service. Furthermore, the very reverse of this is the rule on board of foreign ships-of-war.

It will be observed that when foreign ships-of-war arrive here their boats will, as a rule, go to and from the shore under sail whenever the wind permits. It is so all over the world. I have often and often seen foreign ships-of-war, particularly the English and French, send their boats on shore in bad weather, with their sails reefed down, when we did not dare to lower our boats.

It may and probably will be answered that, in this age of progress, with steam, electricity and big guns and torpedoes and the rest, boat sailing is a very insignificant affair, and really beneath the notice of men of science. That may be true. But a taste and an aptitude for boat sailing are indications—simply *indications*, but altogether wanting with us—of a taste and capacity for sailorizing in general, if the word "sailorizing" may be permitted. Boat sailing is alluded to here, I may say by way of apology, because it belongs to the alphabet—the *alpha beta* of naval training, which is the subject under discussion.

It is commonly said that our earlier impressions are the more lasting. Now, the first and earliest impressions of the naval apprentice are all connected with fresh milk and green grass, with base-ball and foot-ball. He sees the officers deeply interested in homing-pigeons, and expending their energies on lawn tennis, while their principal instructors—the school-masters—as soon as the toils of the day are done, seek their homes on bicycles. Now, there is something commendable in training carrier pigeons, especially for war purposes, and bicycles are fast becoming a military necessity ; but to familiarize naval apprentices, to whom we are to look as the forming of the very bone and sinew of the Navy, with such matters, can only result in one way, and that is to unfit them for life on board ship. As a question of naval training, then, we may reduce the proposition to its simplest terms, and ask the direct question, Which is better as a recreation for the naval apprentice, base-ball or boat sailing? One cultivates muscle, lungs and liver, it is true ; but the other cultivates a taste which it is desirable above all things to cultivate—a taste for a life on the water; and the cultivating of the physical powers will come in the natural order of events on board the training cruiser. One encourages a taste for the land, the other for the sea. Now, we have plenty of landsmen in this country, but we have very few sailors ; and we need to begin in early youth to cultivate them. Let us remember that it is necessary not only to train young men for the Navy, but to induce them to stay in the service after they are trained.

The subject of recreation has been dwelt upon because I hold it to be a very important factor in education. Boys from agricultural districts who love country life find ship life intolerable. They may be attracted for awhile by the novelty of the change, but as soon as that wears off they become unhappy, and, failing to procure a discharge, they will desert. The same may be predicated of the city boy who loves city life.

On the other hand, there are boys, whether from the rural districts or the city, who have a natural taste for the sea. Between these two lies a third class, who evince no decided partiality for one or the other—for the land or the water—but who very naturally will be attracted by what they find most pleasant.

As many of the boys who enter the service do not know their own minds, do not know what their tastes are, do not suspect what is in them, it is manifestly impossible for the closest reader of character to discover their aptitude for the service. Time alone can unfold their characters, their proclivities and their abilities. Taking the average boy, therefore, with a very dim perception of the possibilities within him, and knowing perfectly well that to fit him for ship life is one of the conditions of the problem of naval training, it seems to be the dictate of reason that the process of weaning the "raw recruit" from the land and fostering a preference for the water cannot begin too soon. Hence "sailorizing" and aquatic sports should have the preference on board the stationary school-ship over soldiering and field sports.

The result of my observation is that the officers of the training station have done the very best possible with the means furnished them. If the proper tools have been withheld, they cannot be justly blamed for indifferent work.

One word as to the New Hampshire. She was brought here in 1881 and moored in the stream with four line-of-battle-ship's anchors and 90 fathoms on each chain. It was intended that she should remain there permanently, but, owing to a misapprehension on the part of those concerned, she was unmoored and hauled in the mud on the south end of Coasters' Harbor Island. It was a mistake; but the mistake will prove a fortunate one if it has taught us the evil effects of too close a connection between the school-ship and the shore.

Surgeon J. C. WISE, U. S. N.—It has been said that, after all, the greatest wealth of a nation consists in the possession of a vigorous population, and that it cannot be maintained that money is the sinews of war, if the sinews of men's arms are failing (Lord Bacon).

Able military critics assert that high courage is the concomitant of bodily soundness, and that the power of armies is dependent upon the capabilities of the individual soldier. Comparative study justifies the statement that the strongest animal is the most enterprising and courageous.

Of all the factors entering into the acquirement and maintenance of manly strength and development, that of occupation is second to no other in importance, carrying with it another equally so—that of nutrition, man's occupation having direct relation to his ability to procure his food supply.

The results of occupation as affecting growth and development have been recently illustrated by Mr. Walter Galtho, in a study of "The Physique of European Armies." It appears from the writer that one of the effects of the operation of the corn laws in England has been to increase the urban population at the expense of the rural districts, which has had a very depreciating influence on the recruiting material of the country by a marked decrease in the vigor of both men and women.

In France this change of the population from country to city life is exciting the serious attention of government, while in our own country the deserted farm land in one State alone is enormous. It cannot be questioned but that in a few years this change will materially involve the physique and character of the people, wherever it operates.

Omitting for the present the consideration of occupation on the moral and psychical characteristics, we will note here its bearing in a physical way.

With the advance of civilization and the increasing divergence of mankind from primitive methods of life, occupations have become specialized, resulting in asymmetrical growth; that is, the growth of one part in a greater ratio, and often at the expense of another, frequently resulting in a positive deformity. As illustrative of this we will cite the tables of Drs. Chassagne and Dally, entitled "Influence Précise de la Gymnastique," being the result of an examination of the pupils of the military school at Joinville, France. The pupils were divided into five (5) classes : 1. Men who had followed rural occupations; 2. masons, builders, sawyers, wheelwrights; 3. smiths, farriers, mechanics, bakers, carpenters; 4. shoemakers, tailors, watchmakers; 5. students, clerks, architects. The growth and development of the first class was found superior in almost every particular; the upper extremities of the smith and baker were found unduly developed, while the lower were lacking in normal growth. The class leading a sedentary life was found deficient in muscle and overburdened with fat, while the fourth class, using a stooping posture, showed the evils of contracted chest and stunted growth. Other evidence is equally strong. Dr. Beddoes found the citizen of Glasgow and Edinburgh, on an average, from one to two inches shorter and from 15 to 20 pounds lighter than his country neighbor, and so it was with Sheffield, Exeter, and other cities in England.

It might seem that to state these facts was but to establish a truism; unfortunately, the *truth* is the phase of a question which we most reluctantly concede. Reflection on what has been said will, in most respects, answer the question suggested by Admiral Luce's paper, "What will be the effect, physically considered, on the sailor, by the substitution of steam for sail power?" The change will be for the worse; it will be closely akin to that following the surrender of country life for town life; it will be the result which will follow sending most of the crew below decks, in place of life on the upper decks. And what of those who remain? We might cite as the finest specimen of manhood, in all the attributes of courage and conscious power, the Norsemen and Vikings, exulting in muscles of iron, and as fearless as the very seas which they conquered—indeed, it is difficult to conceive of any avocation in life which more fully calls for the putting forth and exercise of all the powers of the body, and many of the finest qualities of the mind, than that of a seaman; the multifarious duties of ship's life diversify and equalize the efforts of the system, occupy and recreate the mental faculties as do no other. As to the amount of labor of which the seaman is capable and accomplishes, let us reflect that in going aloft he raises the weight of his body to a corresponding height, and that this is done often under the greatest difficulties, incident to wind and sea. If it be true, as good authority states, that a day's work for a strong, healthy adult is equal to 300 tons lifted one foot high, then we can easily understand how much in excess of these figures must be the work of a man who labors night and day, for the sailor off watch is actually at rest but a small part of the time. Then as to the character of the work, how wide is its range! Take, for example, an able seaman, whose muscles stand out in such relief as to be easily traced from origin to insertion, trained in many a contest, bending over the yards and snatching the sails from the very teeth of the storm. Contrast the man in this duty with him who works his watch-mark or embroiders his cap with a skill the daintiest fingers might envy. As a matter of fact, he has educated the "*fidicinales*"; a series of delicate muscles known only, as a rule, to fiddlers in the manipulation of the bow. The entire muscular system is complete in growth, and normal in development—using these terms in their scientific sense—the first being a mere increase in anatomical elements, the latter expressing capability to assume higher duties and functions—the man has had, in plain words, a muscular education in all that relates to variety and association of complex and simple movement. Consider the picture which Admiral Luce draws of the topman aloft. His intelligence keenly alive, his special senses equally on the alert, his wiry and elastic muscles acting in exact co-ordination, so as to bring the various forces in the line of the desired resultant. And what can we say of this splendid animal in other respects? Is he really the "*bête noir*" that we do not wish to occupy the sidewalk with? We all know his vices. Do we remember his virtues? England owes a lasting debt to an old surgeon of the fleet, Dr. Thomas Potter, who in those days, when scurvy claimed more victims than the sea and the enemy combined, evinced an intelligence a hundred years ahead of his time, and a philanthropy unsurpassed, kept the scourge of ships in abeyance and the fleets well manned. He has left us this picture of a blue-jacket: " In the hour of battle he has never left his officer to fight alone, and it remains a solitary fact in the history of war. If in his amours he is fickle, it is because he has no settled home to fix domestic attachments; in his friendship he is warm, sincere, and untinctured with selfish love. The heaviest of metals, as Sterne calls it, becomes light as a feather in his hands, when he meets an acquaintance or old shipmate in distress. His charity makes no preliminary condition to its object, but yields to the faithful impulses of an honest heart. His bounty is not prefaced by a common though affected harangue, of assuring his friend that he will divide his last guinea: he gives the whole, requires no security, and cheerfully returns to a laborious and hazardous employment for his own support. Was I ever to be reduced to the utmost poverty, I would shun the cold thresholds of fashionable charity to beg among seamen, where my afflictions would never be insulted by being asked through what follies or misfortunes I was reduced to penury."

It is well said that the whole life of a sailor is one of confronting danger, and it is this which has indelibly impressed his character; it is the constant association with danger which gives the sailor less dread of it, and makes him in this respect, as in most others, "*sui generis.*"

Such is the man as evolved under the old conditions : what will he be in the new ?

Let us state the proposition as Admiral Luce puts it : " The complement of the Chicago is, let us say, three hundred and eighty-odd men. Of these there will be about one hundred belonging to the engineer's force. Throwing out the latter and the special class of petty officers, messmen, band, etc., there will be left about two hundred and ten or fifteen petty officers, seamen, landsmen and boys. Of these two hundred and ten or fifteen, it requires just two to take the ship to sea—one at the steam steering wheel, the other at the lead. We are not considering the engineer's force. That force does all the work in getting up steam, firing, running the engines, etc. Our present inquiry leads us to the deck. There we find in round numbers some two hundred seamen of the various classes. What part do they take in getting the ship to sea or in entering port ? No part whatever."

"On going to sea, the anchor once catted and fished, which may be done chiefly by means of a steam winch, the man at the wheel and the man at the lead are the only ones who are called upon to take an active part in the opera-tion." "So far from being of any use, the question is what to do with the crowd of men who are lounging about the deck." That the new life is in sharp contrast with the old we will all admit.

The Chicago has masts and sails, but we will leave them out of the count, as it seems the time must come when they will be as difficult to trace as the lost link in the process of evolution.

Let us not lose sight of the fact that while the sailor has been losing the characteristics of the old régime, he has been steadily acquiring those of his land congener—the soldier. What then of the change? Can great-gun drill, small arms, broadswords and similar exercises take the places of those which have become obsolete? It is impossible: the sailor, as we understand him, will go with the sails, and we shall see his like no more.

The difference in the two conditions is immense and various in its operation; it is not unlike that existing between a heavy Percheron and a Kentucky thoroughbred—it is the difference between a sham and a reality. The one condition brings into action the ambition, the courage, all the power, the very soul of the man; the other is one of a purely mechanical character, a perfunc-tory performance, an automatic process, and to many a drudgery. In the one the end is the safety of the ship, the life of the crew and the applause of heroic action; in the other it is a question of the liberty-list, a step to a higher rating, a bronze medal perhaps. The process will not be unlike that of the Scotch Highlander who gives up his life as crofter, a life where he has been face to face with Nature, and enters the factory at Glasgow or Edinburgh. One state imbued him with power and independence, the other makes of him a spiritless creature, who is conscious that a horse might do most of his task as well. In no way can the change be for the physical welfare of the man; in most relations there will be agencies for positive deterioration. It may be pertinent at this time to allude to the means by which we may in a measure counteract this result and preserve for our seamen at least a fair amount of health and something of the old spirit which animated them.

Primarily, we need a higher standard of recruiting both for the general ser-vice and apprentice system, and far more care in the selection of cadets at the Naval Academy. Let those who are to deal with this problem in future beware how they accept the too common statement that gunpowder has made any two given men equal. The time will never come when individual prowess will be below par in great physical contests: science is mighty, but science and human power are mightier. Dr. Parkes says on this point: "After the invention of gunpowder the qualities of strength and agility became of less importance, and athletic training was discontinued everywhere. But within the last few years the changing conditions of modern warfare have again demanded from the soldier a degree of endurance and a rapidity of movement which the wars of the eighteenth century did not require." We are told that in the athletic games of the Romans, officers and men exercised alike. Marius, it is said, never missed a day at the Campus Martius, and Pompey was able at 53 years of age to run and jump with any soldier in the army. The sum of a man's ignorance was expressed by saying " he knew neither how to read nor swim." In feudal times, as we well know, skill in arms and strength of body decided the contest. At this time all European nations teach gymnastics systematically in their armies.

Our late war is replete with incidents where but for the individual strength of the promoter, many important enterprises must have been defeated. The courage, tenacity and endurance of Cushing and his crew are qualities as valuable now as when he destroyed the Albemarle. As before said, the best authority and ablest experience go to prove that physical strength, courage and daring go hand in hand. We must have good physiques to commence with. At the Naval Academy there is undue pressure in the matter of physical examination, brought about by position and influence, resulting in the acceptance of candidates destined to a short career, preparatory to the retired list. This statement is substantiated by the not uncommon retirements in the lowest grades of the Navy. A standard of physical strength for recruits is a purely arbitrary one, fixed by the State and based essentially on the demand for material, the scale varying with the necessities. Napoleon, in revoking his orders to enlist the youth of France, added that they but "strewed the wayside and filled the hospitals." For general service the best material of our Navy is from the merchant marine, and the worst the landsman. In the training service too great dependence is put on a lad's filling certain requirements in mensuration, and too little attention to heredity, temperament and general adaptability. He should be not only organically sound, but, to meet the life and hardships of a sailor, he should be robust. At one time the proportion of apprentices found physically disqualified was appalling. With a vigorous constitution to commence the race, the physical necessities for a good " man-of-wars-man " is more than half met. The introduction of seamen into the service through the training system is all-important : here the habits so necessary in after-life are engendered and fostered, especially a growing love for gymnastics and athletics generally. Let them continue to learn their seamanship as now, in *cruising sailing* ships. It would seem an imperative necessity to increase this arm of the service and offer greater inducement to make it a life-occupation, as it is estimated that now not over ten per cent continue in the Navy.

Other most important adjuncts toward securing the end in view will be gained in close attention to discipline which permits no idleness, but, as Admiral Luce suggests, it is difficult to know what to do with the crowd of loungers on deck. Great-gun drill, small arms, broadswords at sea ; battalion drill and field sports in port. Let the liberty, as a rule, be in the daylight, and all hands on board to supper. (And it surely behooves officers to heed this injunction also.) A due regard to those most important functions of sleeping and eating must be had, and all provisions for a vessel leaving the country should be inspected by a board of officers who will form part of her complement. Errors in regard to the age and other qualities of provisions are generally detected when it is too late to find any remedy, and as a rule the alternatives are that the man shall sacrifice his stomach or the government sacrifice the provision—the former too often prevails.

In the economic processes of nutrition there is a condition known to physiologists as metabolism, the normal requirements of which are the supply of foods qualitatively and quantitatively of the proper kind, the laying up of this food in the body, a regular chemical transformation of the tissues, the formation of excretory products, and their timely and complete elimination from the system. The maintenance of this condition in proper adjustment means health and strength, depending closely upon a just relation between food, rest, recreation and labor. To preserve this happy state amid the increasing and perplexing problems which now environ us, demands all the care and consid- ,eration that we can possibly bestow, and when we have produced a sailor in good physical condition, experience shows that, as a rule, he is among the willing and intelligent members of the ship's company—apathy and inertia being often the result of obscure disease.

A conscientious attention to detail can alone keep our naval force in a state of fair physical efficiency, without which there can be no other.

Surgeon C. A. SIEGFRIED, U. S. N.—Not being a member of the Institute, and taking part in this discussion by invitation, I must apologize for giving you my views on this important matter of the Navy *personnel.* As Admiral Luce truly says, the object is to obtain discussion, to unify, if possible, our views, seeking only the improvement of the service. To much that is written by Admiral Luce there can be no objection. It is beyond argument, while, besides, I do not pretend to know much of the subject beyond the parts connected with the *personnel,* the health and strength of the men, their status and general condition for usefulness.

According to my understanding of the *present* needs and conditions of the service, the papers of Admiral Luce and Surgeon Wise on *personnel* are not very pertinent to the subject in hand, since we must take things not as we might wish them to be, but as they really are. None of us can deny the changed conditions, the dissatisfactions of men, the coming into use of inventions and practical use of forces in our Navy unthought of in the good old days of sailing ships and long cruises. I need not describe the ships in use by great naval powers and those we are putting in commission, and the prevalent idea that "cruising" will be given up. There are no governments extant building sail-power ships. Hence we must train our men to use these new tools. If men universally are being taught more, paid more for their labor, their social and ambitious cravings stimulated and fostered by our modern American life, we cannot do otherwise in the Navy. Why do men flock to cities, leaving the finer, more healthful country life for the slums and crowded tenements? It is because more mind satisfaction comes from it, they see and hear more; the modern spirit is striving for recognition and to find other things to do in rising to a higher plane. As a rule the American of the present will not easily submit to discipline or regularity of work, nor to force

compelling obedience, and all these are essential to efficiency. Hence our
material comes from in great part the lower order of society—from the vicious,
the unruly, and the foreign-born. Now, to train these for the present naval
uses·is a complex task, and only done partially and with loss and waste. If we
were to go back some years it would be a simple matter to put them in ships,
and by the force of association and attrition they would soon learn the simple
duties that once pertained to our ships. They would in time be moulded into
what is known as the " blue-jacket "—a being unsafe to the last degree in his
tastes, habits and practices, but a good sailor and topman, handy with the plain
old guns, with their creaking trucks and simple paraphernalia. Now, will this
blue-jacket, so well able to pass the weather earing in all weathers, childish
in many things, and not given to learning new tricks, once he is formed by the
force he only respects while on board ship—will this man do for the machines
now building for our Navy? Will he be of much use in the Navy—the mastless hulls, filled with intricate appliances that seem to be formed for purely
destructive purposes only? There will be here and there lonely cruisers of
course, as exceptions. All these craft need mechanicians, marine artillerymen,
electricians, and a few lonely steerers and lookouts. I believe most of you
admit that naval force *now* is only embodied and intrinsically potent in the inventions of recent years; this, in the full sense—ship construction, materials,
guns, explosives, torpedoes, and the use of newly discovered forces and
motors.

If the previous speakers argue in favor of sailing ship training to make men
nervy and courageous in these days, such training is to my mind a great waste
of time. Life is short, and familiarity with danger in any shape breeds contempt for it, and courageous nerve born of the trained mind is far more reliable.
The statement that the blue-jacket of the good old days was a finely developed
man, healthy and capable of any exertion or work, I take exception to.
Neither did his mind control his actions. He had well-developed arms and
chest ·muscles, but from the waist down he was weak in proportion; his legs
were spindles; he was and is always beaten in tugs-of-war contests with the
marines. I need not dwell on the quality of his self-control—his brain and
morals. We must train the mind more; and if men in cities set apart time and
place for health exercise, so must we do, as is even now done in other navies.
The parts of the essays relating to the physiology of exercise, the virtues of
the old sailor, and those matters, I do not argue against; they are matters of
fact, but don't help us much here practically—they belong to books. It is not
in reason to keep apprentices apart from ships; but will many argue successfully that the former or present system is anything but in great part wasteful,
and only too well calculated to inculcate immorality and stunted physique? I
know of reports of commanding officers who speak to this effect. Low, damp
decks, dark corners, crowded living and berthing accommodations, fighting and
oaths, no respect but for force—all this only too common. And the returning
liberty-men! What a commentary on the system and the material of the blue-
jacket of the past! Why stick to the past in *personnel* only? Less men will
be needed for these new ships, and they must be better trained and paid; they

must be skilled of hand and self-respecting members of society, and willing parts of the force. Why must we accept the vicious and disobedient, in fact the worst boys of families, as a rule? What has brought the seaman's profession so low, into such disrepute? Can it be the well-known life of debauchery, the riot and excess upon all occasions? Or have the unhealthful, unnatural conditions of his life changed his nerve cells, so that his brain craves morbid excitement and excess, his brain becomes not a crown of reason over his actions, but simply guides a cunning animal? The average working life of a sailor in the merchant service is given as below ten years, and somewhat higher in navies! I deny that training of boys in sailing ships can be compared in usefulness with the training that can be given in what is known as the "Barrack System." From six months to a year ashore, and then to our ships as they are. Familiarity with modern appliances and dangerous instruments of warfare, mind and hand training, morals, drills, gymnastics, how to live in health by rule, the management of boats and small vessels by short trips—these all at first will go farther towards producing valuable material for our Navy. But the stock must be improved, the age limit changed to from 16 years to 20. The stunted, the vicious, and the whole idea that the Navy is a reformatory school, must be abandoned. As we gradually grow to this, so will our seamen increase in numbers and improve in quality, and *Americans* may be attracted. There are no American seamen now, comparatively speaking, and we must form a body to man our national arm afloat.

I take exception to Surgeon Wise's statement regarding the liberties of men and compelling their return to supper. The uneven and varying treatment of men in ships is one of the chief causes of dissatisfaction, as well as the manner of ratings, pay, etc. So long as the work and discipline permit, the return to supper of liberty-men is not in accordance with life and habits ashore, for, as things are, work ceases at sundown and recreation begins. Men must be treated differently, rated by boards of officers, broken only by courts-martial, better paid, and the service of the Government be made an honor for them as well as for the officers. If Admiral Luce means the smell of tar and hemp I quite agree with him, *that* is agreeable and healthful, but the "smell of a ship" denotes to me only lack of fresh water and honest cleanliness. Shining brass-work and bright upper decks have commonly gone with dirty rags and saliva for the former, white cloths covering long lines of ill-smelling clothes-bags and dark corners below, and a bucket of fresh water to wash the bodies of a gun's crew. Good sound brains and bodies, and with good morale, are produced without sending men to sea in sailing ships, and we need such for our new navy. It is more agreeable for us in some ways, and life at sea is romantic and often poetical; and who can deny the charm of sailing in the trades, the swish of the rushing waters as she heels over, the merry sail, oh!

. Men are not free agents. We use new inventions because of some saving of time and labor in the attainment of an end, though we know the value of many old methods. If one refused to adopt new methods, or to adapt one's self to the new conditions, so much the worse for the man or the nation. There is nothing gained in pointing out the ugliness of things, the noise, ill-health conditions, and brutal matter of fact short cuts to an end. Do men remain contented in quiet country ways, with their green fields and serene atmosphere, because some of us know it is better and more beautiful? Where is there beauty, peace or health in the modern town with its hideous suburbs, railway tracks, garbage heaps, its tenement districts, its crime and genteel beggary, and its wasteful government? These things are with us face to face, as we have the modern navy problem and the new war conditions; the marine arm of the national force and the *personnel* are forming under circumstances different from those preceding our time. If a modern ship-of-war can annihilate a whole fleet of old-time frigates, in just so much must the *personnel* be improved. We must come to this sooner or later. The officers and men must approach nearer each other; the man to know more—a higher type—and the officer to apply his own hand to the plow on occasions, to teach and as an example. There is less room for carelessness and incompetence due to neglect. I take it that we all know the chief characteristics of our pushing, inventive and leading officers, many among the younger men; they do not hesitate to use their hands to set matters going or to show men by example; they are in the work and realize the difficulties of the problem, and are helping on the newer state of things and the organization, finally the effectiveness of the Navy. The whole tendency of modern war forces is to obtain the utmost efficiency by the plain force of hard material agencies. Sentiment or romance don't apply in these times. Nations by diplomacy will avoid war to the utmost, but preponderating force must rule, and we know that they will be consummated by a few swift merciless strokes. Within recent years I have seen the flower of the navies of England, France and Italy, great monsters, mastless, and containing within them the best results of the science and inventive genius of modern times, and I beg to be excused from believing any longer in the training of our seamen in obsolete sailing ships. We now have no colonies, and I may say no merchant marine or sailors, and we must form our *personnel* from American citizens. The problem is hence more difficult for us, as we do not have the conscription, nor have we a surplus population or a great commercial marine, as is the case of England.

Boatswain H. SWEENEY, U. S. N.—*Gentlemen:*—Having been in the seaman-gunner class, I feel that I have been in a position to understand to a considerable degree their grievances. The question has been put, "How are the seaman-gunners to be induced to remain in the service?" This question arises from the fact that the majority of these men, after being trained, find that the Navy does not pay, in the majority of their cases, as much as their services will command elsewhere; therefore it appears to me that the way to retain them is to pay them fully as much as others will. It is, perhaps, true that, having been so well taught, they should feel themselves under some obligation to the Government, and for *that* reason remain; but even so, they are not violating any obligation by remaining out of the service after serving an enlistment and after being discharged. I have said above that in the majority of their cases they are not well paid, and I will try to show that such is the case. We will say that there are six seaman-gunners sent to a vessel; they are all of the same class, possibly. One of this six will, in all probability, be rated electrical machinist, at $70 per month, one will be armorer, at $45, and the four others may get the rates of oilers or water tenders, at $36 and $38 respectively, and these differences of pay create a feeling of discontent. I think that the pay of this class of men should not be less than $45, even though the established pay of the rating which they hold be less, and if they are competent to fill the ratings which command a higher rate of pay they should be given the preference.

Commander C. M. CHESTER, U. S. N.—In reply to Lieutenant McLean's interrogatory, I may say that the men are kept in the Coast Survey from the fact that they are unhampered and can go or stay when and as they see fit. Apropos, however, of this question of increasing the inducements for seamen to remain in the Navy, I can say that officers are keenly alive to the necessity for doing so, and that boards have recommended increased pay and permanent ratings, and that such a scheme is now in the hands of the President. He has the authority to make the necessary changes, but not the means to pay for them, so that it is doubtful when this can be put in operation.

THE CHAIRMAN.—There are few of us, I imagine, who will not agree with Admiral Luce in insisting upon the great importance of the physical development of the men who are to form the crews of our vessels-of-war, and I for one agree with him in believing that the sailing ship is the best possible school in which a man-of-wars-man can be trained. There is no question, I think, that the "reckless daring so characteristic of the sailor," the quick-wittedness, the ready resource of the man who has spent his earlier years on the "giddy mast," are still the most valuable qualities to be sought for in the modern man-of-wars-man. But are we to go back to the sail period to procure these most desirable characteristics. Granting that spars and sails have seen their day and that the crews of our ships are to be deck hands hereafter, is there no way in which we can secure the results aimed at? Are we to look upon our vessels-of-war as gymnasiums merely? Were the Spartan youth not "trained to courage, and taught acuteness, promptness and discernment" without the assistance of spars and sails? Were the hardy fishermen of the Banks, from whom our Navy was so largely recruited in the earlier history of our country, the products of square rigs and towering masts? Were they not rather men who, from their earliest years, had been accustomed to taking their lives in their hands, who had, by hard experience in the face of danger, become self-reliant and quick-witted, and acquired those very qualities that made them so desirable an addition to the crews of vessels-of-war? It seems to me that it will not be necessary to take a step backward and supply our cruisers with full sail power in order to develop the muscles of our men, or to train them to that alertness which we all admire in the true sailor.

It must be for some years yet a matter of opinion as to whether, in order to get the most efficient service from our cruisers, they should carry any sails at all. It may be hoped that the value of sails for cruising purposes in that type of vessels may be determined within a few years, when we consider that of the five first-rate steel cruisers on our navy list, the Chicago and the Newark are barks, the Philadelphia and San Francisco are three-masted schooners, and the Baltimore has military masts. Who can doubt what the answer will be? How much cruising could the Chicago do under sail? Does it not seem probable that she would be a more valuable vessel if the weight of her masts and yards and sails, and the storeroom full of supplies to keep them in repair, were replaced with coal? You cannot put everything into a ship. Sailing power must be sacrificed if we are to carry heavy guns and ammunition, and coal for long passages. The lack of coaling stations alone, it seems to me, is an argument for doing away with useless sails and increasing coal capacity.

Examples drawn from the Brooklyn, the Hartford and the Alabama prove little. The Baltimore could have made that hypothetical voyage from Honolulu to Valparaiso in twenty days and then had coal enough to steam thirty days longer, or perhaps twice as long as the Hartford could steam starting with full bunkers and helped by her sails, and when she had arrived she would at least be able to get away if she found it too hot. The Hartford, even if she had been able to make the passage before the emergency that called her there had passed, would be able neither to fight nor run away.

It seems to me that the Admiral's argument points rather to the great necessity there is that our country should establish coaling stations abroad. Perhaps it may be impossible that this should be done peacefully to any great extent, but it certainly could be done in the Pacific, and not improbably in the West Indies, without resort to force. We are wasting a golden opportunity in the Sandwich Islands at this instant. Wise statesmanship should anticipate the day when our increasing manufactures shall exceed the needs of our own people, and when our commerce shall be reaching out to the markets of the world.

By all means let us have training ships with full sail power in which to create our sailors. But let our cruisers not be hampered with useless material for the sole purpose of keeping the physique of our crews up to a standard which could be otherwise maintained.

OFFICE OF THE ADMIRAL,
WASHINGTON, D. C., *June* 14, 1890.

Dear Sir :—I have read with much pleasure the article of Rear-Admiral Luce on "Naval Training." It is written with the usual ability that marks everything coming from his pen, and it would be well for the Navy Department if it should take to heart some of the principles laid down by the author.

I think Rear-Admiral Luce handles too tenderly those parties in the Navy, whoever they may be, who instituted the system of reducing sail power on ships-of-war and supplying everything with military masts, without considering whether the ships were intended as cruisers or as vessels for coast defense.

Not one of the new vessels hitherto planned or built is fit for war purposes in times of hostilities. "Cruisers" cannot cruise for want of sail power, and so-called "line-of-battle" ships cannot go into battle for want of proper endurance. In time of war the result will be that the present Navy will be laid up on account of too much military mast and the entire lack of sail power, without which a vessel of the Navy is not a perfect machine. This question of sail power is one on which the efficiency of the service hinges.

It is not to be supposed that the series of able articles now appearing in the Proceedings of the U. S. Naval Institute will be without effect upon their readers, the young officers of the Navy, who are coming forward and expect to manage the affairs of the service. Every officer should bring to bear what influence he possesses to cause all cruisers and line-of-battle ships to be fitted with sufficient sail power to enable them to cruise for long periods at sea, if necessary, without entering port except for provisions, so that they may save their coal, as did the rebel steamer Alabama, for an occasion when they may be called into action or to chase an enemy too fast to be overtaken under sail.

This is the battle confronting the young officers of the Navy, and it must be fought *now*. We demand for our sea-going vessels thirty-five feet of canvas for each square foot of midship section. As to the coast defense vessels, they may be fitted with as many military masts as the constructors choose to supply. I think the article under consideration will have the effect of drawing more strongly the attention of the service to this vital question, and the fight should commence at once to ascertain whether the ships of the future are to be fitted out under the same régime of mistakes that has characterized the Navy for the past ten years, which has given us a "squadron of evolution" that cannot *evolute*, and, after training officers and men at the training-school under spars and sails until they become good sailors, has relegated them to ships where the military mast is the order of the day and where hoisting ashes is the nearest approach to seamanship an apprentice boy can learn.

I would encourage the discussion of this subject, and only regret that Rear-Admiral Luce did not say more about the matter, as no one knows better than he the necessity for keeping up that great element of power, the seamanship of the Navy, which has made our naval service particularly distinguished in the several encounters in which it has met foreign foes.

Very respectfully yours,

DAVID D. PORTER, *Admiral U. S. Navy.*

To the SECRETARY U. S. NAVAL INSTITUTE.

WASHINGTON.

Lieutenant E. B. BARRY, U. S. N.—Admiral Luce has given us an admirable essay on matter not, I fear, sufficiently often before the minds of naval officers of to-day.

The main idea of the essay is training for war. I venture to ask, however, if it be true the only preparation for war is to be found in knowing how to "hand, reef and steer," with some infantry drill thrown in? Is the soldier any less a soldier because he carries a magazine rifle and does not fight "shoulder to shoulder," as in days gone by? If the sailor of to-day differs from the sailor of Nelson's time, I take it the reason is because he is a better man—not inferior physically, certainly superior mentally—and in his fighting capacity measured to-day.

The decay (?) of the seaman has been going on for more than a hundred years, but to-day he is as active, as vigorous, and a thousand times better educated than was his great-grandfather. That the school of the topman used to prepare the seaman for his duties at the gun *physically*, no one can deny, but with change of battery and change in the method of working it, his old duties have disappeared. No longer does the handspikeman " heave " while the tacklemen " haul " until the gun points at the huge hull a few hundred feet distant and the gun-captain pulled the lock-string "when she was on." In those days it did not matter about fine shooting, and the vicious custom of making gun-captains survives to-day in our watch, quarter and station bills. Number 301 or Number 702 becomes captain of a gun, not because he can shoot, but because his rate as petty officer puts him there as a captain of top or something else, but not as a gun-captain : we have no such rate.

No wonder the "sailor" petty officers made the best records in the earlier attempts at systematic target practice. They were the only men aboard the ships that had fired great guns.

We are beginning to recognize that a man-of-war is intended to shoot her guns and to shoot them straight, and we are gradually putting gun-captains aboard ship who are meant to be captains of her guns. Let us have the rate next and get rid of top-captains in mastless vessels. The trained seaman-gunner is so vastly superior as a gun-captain to the old top-captain that comparison is useless.

Sails in a war-ship already have sunk to a subordinate position. If it be found the cruiser cannot be made effective in war-time with masts, yards and sails, she will be accompanied, of necessity, by a collier and coaled on her station. I think we all will agree with the Admiral that sails will be used in peace, especially on ships making long passages, but it seems a very doubtful question if a cruiser will not be "cleared for action" when war is imminent, rather than run the risk of being caught on her cruising grounds by a mastless vessel, able, owing to absence of spars, to steam much faster than she can.

The "base of supplies" of a battle-ship is almost anything selected for that base. It seems certain that fleets of modern battle-ships will not cruise in war-time without taking their base of supplies with them. A fleet as now organized contains supply-ships for coal and for ammunition. With their small steaming radius, modern battle-ships could not carry on an aggressive campaign without a movable base. What applies to them will apply to the cruisers.

Twin screws and the hull as built for them have overcome violent opposition, so it is but natural to suppose, were it acting to-day, the Naval Advisory Board of 1881 would favor twin screws for cruisers, whatever might be its opinion of sail power. I think the Admiral's remarks about the Brooklyn and the Baltimore rather favor twin screws.

In case of "strained relations" with Brazil I think the solution of the coal question for vessels like the Newark would be comparatively simple. Far different, however, would be the solution of the questions of fouling and copper sheathing. "As the cruisers already built have neither full sail power nor copper sheathing, it is very evident," says the Admiral, "they are not intended for foreign service during war," and with this he dismisses them. This is not the place to discuss the merits or demerits of copper sheathing, but an adequate coal supply taken to the station will keep one of these cruisers going as long as the state of her bottom will permit, not to mention the strange anomaly called " coal protection," which vanishes through use just when it is wanted the most.

· The Admiral says "the highest expression of the trained man is to be found in the able seaman and expert gunner combined." Now, what is to be the able seaman? Is he to be a man trained in the "school of the topman," when topmen, so called, are not wanted; or is he to be a physically well-developed, healthy man, with a gunnery training added? All things being equal, the trained shot can beat the seaman shooting; the eye is not made quicker nor the brain more active by "laying out and passing the weather earing." If this were true the best target practice would be sail and spar drill. No seaman is a good shot *because he is a seaman*, but because he individually has the aptitude for shooting, and with equal practice he would shoot just as well if he never went above the rail.

I venture also to differ with the Admiral about the typical apprentices returning from two kinds of ships. "It is scarcely necessary to add," he says, "that the former set (the better set) would, as a rule, express a partiality for the service, while the latter would not." When the better boys seem desirous to return to the service the cause is to be sought, not in seamanship, but in crew. Put a lot of apprentices aboard an American ship with an American crew and they usually will be content. But our Navy is not the place for Americans; our officers say openly they prefer Scandinavians, as the Admiral himself hints, and the American boys soon see their country's service is no place for them. As a consequence, few of the best boys re-enlist, except for the benefits of the gunnery class, and many desert disgusted before their term of service has expired. If we want these young men to come back, let us convince them they will serve on American ships. I believe an honest effort to man our ships with our own people, aided by the co-operation of commanding and executive officers, once the ships are in commission, will soon bear fruit; then the Olsens and Nielsons that refuse to become United States citizens, but who are so eagerly sought after by officers, will give place to Americans, who, even if not "so easily governed" as the mass of foreigners forming our ship's companies, will be ready to serve their own country and, if necessary, to die for their own flag.

WASHINGTON.

Ensign A. A. ACKERMAN, U. S. N.—While agreeing in many details with the distinguished essayist, some slight differences at the start lead to widely differing conclusions. It can hardly be otherwise. Youth looks with hope to the future. Age and experience recalls the circumstances which have developed their own powers and would have their successors fare as well. The certain advance of any art or profession arises from the change and readjustment of petty details, these later cause modifications on a higher plane. In this way only can we explain why the views of the Advisory Board of 1881 would hardly be adopted in every respect by a similar board to-day.

The people want an efficient navy, and they will obtain it in time with no great haggling over the price. The prevailing sentiment will declare itself in favor of such types of ships as will best accomplish the most urgent duties of a navy, and the training of the crews will be best accomplished by perfecting them in the performance of their duties on those ships. One of the conclusions derived from the British naval manœuvres of last year shows it to be impossible to work a modern war-ship with a crew which has had only a general training on some totally dissimilar and possibly obsolete vessel. The multitude of changes in appliances since 1864 and lack of the correcting adjustments of war, places the problem on a platform of rational investigation. It should not be solved arbitrarily, and it cannot be entirely covered by experience. Let us consider a few of the changes since the time of the Alabama.

There are at present in existence 2,000,000 miles of land telegraph lines; 120,178 miles of cables, of which 13,178 miles belong to different governments. The whole of this has been built within the last fifty years, *and most of the cable within the last twenty-five years.* It takes twenty minutes to send a message to Egypt from London; less than an hour to Bombay; to China two hours; and less than three hours to Australia. This means of communication is rapidly extending. By the cable, commerce can now be warned when to sail and when to avoid danger; it can also be transferred to a neutral flag in a few hours after the declaration of war, even though it be in a far distant corner of the earth. As for the cruiser, once located, she can be pursued and captured or crippled even though victorious, unless she is able to stand punishment.

From Mulhall's Dictionary of Statistics we find that in 1864 the total registered steam tonnage of the world was 1,843,479; in 1881 it was 4,751,988; in 1888 it had grown, according to Lloyd's Register, to 7,021,000. Mulhall's Dictionary gives the ratio of steam to sailing tonnage in 1860 as 1 to 12. [This in itself is a commentary on Admiral Semmes' boast that, with half a dozen exceptions, all his prizes were taken while under sail and with his screw hoisted.]

In 1888, according to Lloyd's Register, the ratio of steam to sailing tonnage *was* 1 *to* 1.4. This is not a fair comparison of their respective values as carriers. The steamers of 1888 were much faster than those of 1864, they get over half as much more work out of their coal, make quick trips, and probably carry from five to ten times as much freight as an equal tonnage of sailing vessels.

This remarkable change makes even more probable the immediate observation and location of a cruiser, and consequent warning of commerce. Even though the cruiser's depredations were confined to the slow sailing vessels liable to run over time, she would probably be spoken every day by a neutral steamer unless operating in out-of-the-way districts. Even supposing the means of communication to be what they were twenty-five years ago, would the ravages of a dozen Alabamas be as great a national calamity to any European power as it would have been at that time? Certainly not if they took their prizes under sail. A considerable proportion of the commerce of the world passes through our own or contiguous ports. We cannot consider much of this commerce liable to capture by cruisers, unless we declare war with a haste improbable under a government so efficiently checked and counterchecked. Commerce may perhaps be obstructed in the West Indies or with the British Provinces, but this commerce might with but little permanent loss be transferred to a neutral flag or be placed under convoy. It seems, therefore, that if we are not only to inflict damage on our enemy's commerce, but to exhibit such a power for the continued destruction of his property as will make him wish for a speedy though expensive peace, our most important work will be done in waters far from home.

Consider England's commerce in this connection. Her external trade was over £700,000,000 last year. Of this £144,000,000, *over one-fifth*, was with the United States, which has sent to England more than 40,000,000 quarters of wheat and flour in a single year. These stupendous figures only partly indicate the sacrifices these countries must make merely by closing their ports to each other.

The conclusion is irresistible that if "commerce destroyers" are to take any part in forcing an advantageous peace with a country that has already made such sacrifices for its principles, they must be able to accomplish an appalling amount of destruction. Not only this, but a large part of the commerce of England lies along the great trade routes from England to Gibraltar, through the Mediterranean to India and China beyond—routes lying almost wholly in enclosed seas and guarded by stations and battle-ships. It therefore seems evident that the commonly accepted type of "commerce destroyer" must be satisfied with the occasional capture of a slow steamer on the outside trade routes to the Western Hemisphere, South Africa, and perhaps Australasia. As these slow steamers are warned beforehand, and may be transferred, no damage is likely to be inflicted on England's commerce within sufficient time to influence a peace.

Nor would the total stoppage of this outside commerce work as great injury to English ship-owners as in the time of the Alabama. The rates of carriage to-day are but little more than one-twentieth what they were thirty years ago, and although by means of faster steamers the number of trips is increased, these ships are earning much less money than ever before. As for English manufactures, they may be easily transferred on interior lines and shipped in neutral bottoms over the few routes where the insurance would otherwise eat up the profits. All this, of course, is supposing that we had these trackless wastes efficiently patrolled.

It may now be positively affirmed that the roving cruiser can have little or no effect in bringing about a peace or relieving the pressure on our home ports. It would take a long time to seriously weaken such a colossus as England's commerce unless we attacked its main arteries. But to do that we must be able to fight. The attack must be prompt, determined and sustained; all else is mere bushwhacking of no national importance. We must expect to see some of our coast cities bombarded, some of our ports blockaded. Worse than that, our labor market will be glutted; there will be no occasion for immense armies to resist invasion or to invade Canada, the border States can take care of that. But our wheat must rot in the granaries, our freight cars must rust on the sidings. There may be something of a grim satisfaction in the thought that those who most strongly oppose an increase of our Navy will be made to suffer first.

It must be admitted that while nearly all the world admired and comforted the Alabama, most European powers to-day would look with satisfaction on such a humiliation of the United States as would compel a modification of our protective tariff. Neutrals would submit to a great deal in order to bring about so desirable a result. Whatever is done, therefore, our coast must be kept clear of the enemy by a powerful off-shore fleet; while touch should never be wholly lost with a single one of the "commerce destroyers" which may form an integral part of the national defense. It is evident that if our cruisers are not to be sent out into trackless wastes of water, there to carry on the Alabama style of warfare, the arguments in favor of full sail power and sheathed bottoms lose all their force.

The difficulties of coal supply and fouling cannot be ignored; they must be met and overcome. To sacrifice the advantages of a modern cruiser in order to evade the accompanying difficulties, is to make our strategy dependent on inferior logistics. The essayist has given a number of possible examples in which modern cruisers may have been embarrassed by want of coal. Just sufficient data is furnished to demonstrate the peculiar weakness of these ships and the corresponding advantages of sails. Such a comparison is incomplete and, it may be said, unfair to our naval administration, as a notable lack of foresight or strategy is indicated in each example. Our Government sent coal to Samoa in a sailing ship: had the need of it been considered sufficiently urgent, it could have reached there much earlier in a steamer. The fastest steamer on the Greely Relief Expedition was the collier Loch Garry. The difficulties of coaling at sea have never been fairly met; when they are, they will be overcome. It will not be difficult for an eighteen-knot collier to keep in touch with two twenty-knot armored cruisers. She can run away from the battle-ships, and her consorts can take care of all else.

If all our ships were provided with long-armed swinging life-boat derricks, upon which an endless chain could be run, and self-bailing, unsinkable launches, coal could be easily transferred in a seaway from prize or collier to the launch and thence to the ship. After watching the Pacific mail steamers lightering freight in the heavy swell at Champerico and San José de Guatemala, it does not seem impossible to select a time when the same could be done farther out at sea, especially if the colliers were provided with efficient lighters. It will be said that this increase of complication may be avoided by the use of full sail-power cruisers; let us consider whether the accompanying disadvantages are not greater than those intended to be avoided.

In order to derive as much benefit from sail power as possible, the essayist has given the "commerce destroyer" but one hoisting screw. She cannot manœuvre, therefore, as well as the all-steam cruiser with twin screws, and knows that in close fighting the latter will have the advantage; therefore it is presumed she will fight "not at very short range," though it is evident that the range will not be at her selection. Where the means of offense are so much greater than those of defense, the fight cannot be too fast and furious on the ship that wins the victory. The commander having the most numerous battery must feel that at long range he is simply playing with his good fortune and running up his casualty list. The quicker the enemy is made to feel his superiority, the shorter the battle and the less costly in men and equipage. The weaker vessel is the only one to benefit by prolonging the conflict, as a chance shot may cripple her opponent and at least shorten her cruise.

But suppose a full sail-power cruiser goes into action as valiantly as of old, will the gallant master's division be sent from behind their gun-shields out into a hail of iron and fire from Gatlings and rapid-fire guns to clear away the wreckage of sails and spars? Will not the crew suffer from falling spars and rigging and the langridge of outside boats? Will this not impede the service of the guns? Will not the commander, be he ever so gallant a seaman, find it better to steam ahead blindly in seeming rout and confusion, rather than foul his propeller in trailing wreckage? And will not, therefore, the faster all-steam antagonist select that quarter for attack from which it can deliver the most murderous fire with least damage in return?

Suppose the all-steam cruiser considers herself too weak to fight the sailing .cruiser—the latter is shadowed and kept from doing harm. Vessels will be warned off just as the escaping Constitution warned the American merchant-men. The sailing vessel, bigger and stronger, attempts in every way to shake off her companion; she has been observed by neutral steamers, faster steamers of the enemy have run away from her, and soon from every quarter their cruisers will be sweeping down upon her. She might as well be captured as made useless by a crippling engagement, so she seeks to exhaust the steamer's coal. But her own is going fast, she dare not let steam go down, she is hampered by her sails, fearing that·if set her antagonist will select that time for an attack. Her time of usefulness is over. Even though she seeks a fight with her nimble antagonist, she cannot force it.

It seems most certain that our possible enemy will be more inclined to peace by what our cruisers *may do* than by what they *have done*, and if that impression is not produced in less time than it takes to seriously foul a cruiser's bottom, then the attempt to force an advantageous peace by destroying his commerce is a failure, and we had better put the full strength of our Navy into fighting ships. Privateersmen can look after commerce. The best combination of commerce destroyer and fighting-ship capable of running away from battle-ships is the "armored cruiser," or at least a well-protected cruiser. There are many battle-ships such vessels need not fear, especially under certain conditions of weather. Their economical rate of speed should be high, permitting them to throw off clues to their pursuit and to appear unexpectedly in new fields.

They must also have the highest possible speed for the following reasons:

1. As the area in which the cruiser is located varies as the square of the

product of her highest speed into the number of hours since observation, the inverse ratio of this may indicate the comparative chance of finding her.

2. Without a very high speed the most valuable prizes would escape her.

3. Prizes would be made only after an alarming expenditure of coal in chasing.

4. The shorter the chase the less danger of interruption by darkness, fog or other features of weather, proximity of the enemy or neutral waters. The less danger of observation by neutral and others and consequent location.

5. When interrupted or pursued by a stronger but slower vessel, the higher her speed the more certain of escape ; the shorter the dangerous period when accident or breakdown would be fatal ; the less expenditure of coal. Even if the pursuer is faster, the chase will be longer and the opportunities of escape through accident, change of weather, etc., would be greater.

It is certain that if a naval architect can build an all-steam nineteen-knot cruiser in a certain displacement in order to devote 12 to 15 per cent of that displacement and 3 per cent more to masts, spars and rigging, he must rob her engines, her battery and her coal supply. A day or two less at full speed, a fourth less battery power, a knot or two less in the chase.

Training in some respects is a suppression of individuality, a unification of independent members of a society. No man or officer is disciplined or capable of creating discipline in its highest state who cannot suppress his individual. traits for the time being in the humdrum of routine. A military life is full of petty exactions and observances. These expend much time, and by fatiguing otherwise vigorous energies, incentive to competition or. progress is often wholly lost. After considering the changes in commerce and material since the time of the Alabama, it would be well to consider the changes in the people from whom our future man-of-wars-man must be drawn, now that aliens are no longer permitted to enlist.

Our people are better educated, better clothed and better fed than they were , twenty-five years ago. [There are undoubtedly numerous cases of destitution in the land, but in almost every.particular the individuals will be found to be the lowest type of foreign emigrant, and utterly unskilled. If they are skilled workmen they are certain, with the aid of public sentiment, to obtain satisfactory pay.] American workmen work neither as long nor as hard as they did in 1864, but they use their brains as well as their hands. In some respects .old trades, through subdivision, have been swept away. Others may by contrast appear more gloomy than ever, but the elements of improvement are at work among them all. Hardly a day passes but what we hear of some great strike —not for bread, but for a principle—not for more money, but for less work and the same money. The spirit of progress is abroad in the land ; the educated workman respects the law, for he has learned that it protects him as well as his employer. He is independent, as quick to discover and repudiate cajolery as to take a stand against arbitrary misgovernment. Times have changed ; a strike is no longer made by a deluded mob of fanatics beating itself to pieces against the bayonets of soldiers, but a disciplined army, provisioned, paid and

led as such. As such the nation approves of them and values them, the sentiment of the public is with them as long as their actions are lawful. But a single
overt act, and their cause is lost—they are scattered and swept away by a right-
eous public indignation.

Such are our people. This is a very different picture from that only twenty
years ago. The American man or boy who ships to-day wants, should get, and
the people will see that he gets, more consideration than would have been
shown him then. The prospective reward must be greater also in order to
retain valuable men. There is something terribly repugnant in the thought
that we can only retain men in the service through their ignorance and help-
lessness on the beach. The remedy is plain; it will no longer be a harbor
for aliens; it should not be for the weak and criminal, for stepsons and incor-
rigibles. The Navy should offer nearly the same advantages of education,
social equality and home comfort that any plain trade does. Patriotism, a love
of travel and adventure, should not be expected to do too much.

The contrast is not a pleasant one to the man who smarts under restraint
aboard ship and is an uncouth " common sailor " on shore, and his brother
who, with not a particle more of self-denial or self-restraint, reaps all the
benefits that this free land bestows upon an honest workman.

Ordinarily the men supposed to be eligible for naval service in time of war
are engaged in fishing and river work. The fact is, however, a more independent, yet helpless and useless, lot of men than these can hardly be imagined.
Every State in the Union should send its quota to our training school, but
these boys should be guaranteed something in return: 1st, a good common
school education by regular certificated school-teachers holding naval warrants, drawing a pay and occupying the social position to which they would
be entitled in private life. A rigid curriculum should be established and a
military government free from any tinge of paternalism. It is hard to make a
rollicking school-boy take much thought of the future unless the advantages
of position can be brought home to him, so after spending one year at school
they should be distributed among the various ships of the fleet. On these
ships they should be carefully guarded and instructed, an officer being detailed
to the ship for that special purpose. They should form a separate division of
inspection and muster, but at quarters should be distributed as they could be
used to best advantage. On no account should they be advanced, petted, or
granted liberties not allowed the humblest landsman aboard ship. They
should be compelled to obey their petty officers. One year of hard work in
such a position should decide the boy whether the reward is worth the effort.
If he elects to remain in the service he is returned to the training school
barracks, given one year's additional schooling with practical exercises, and
drafted to a ship for a three years' cruise. If at the expiration of that time his
conduct has been good and he be able to pass such examinations as may be
prescribed for him, he shall be admitted to special training courses in electricity, gunnery, torpedo practice, the construction of ships, guns, mounts,
etc., etc. Not too much should be attempted in the one year or eighteen
months' shore duty thus allowed. Whatever is done should be as thorough

and practical as possible. According to his merits, as determined by competitive examination, the seaman apprentice may now be assigned aboard ship to one of the inferior rates pertaining to his specialty, but if not so assigned he should be given a slightly increased pay on account of his examination certificate. Opportunity should be given him of studying both theory and practice of his specialty. On the completion of this cruise he may, on passing a proper examination, receive, with the approval of his late commander, a rate of petty officer of the Navy. As a large increase in number of warrant officers is needed to properly carry on the details of the work of construction of the new Navy, there would quickly be an opportunity of again promoting deserving petty officers. Warrant officers of a certain age and experience who shall have attained practical excellence in any branch may be given a commission and assigned to special duty. In this way an apprentice boy could be led to feel that with hard work he could keep mounting all his life. He would respect his seniors all the more through feeling how difficult it was to attain their position.

The most important step in the whole advance is the creation of petty officers of the Navy. At present we have no petty officers. Every man aboard ship feels that the petty officers get too little pay and consideration to amount to much. There is a very wide gulf between them and the warrant or commissioned officers and hardly any difference between them and their men. This trouble has been due to the fact that in order to educate young officers they have been required to perform the duties of petty officers; this has been excellent practice for the young officer, but it has made the petty officer a man of no responsibility. On modern ships, however, with their large and complicated batteries, there will be room for all, and every petty officer should be made to feel his responsibility and then treated with the consideration that that responsibility merits. He should be educated to command his men;—anything to break the widening gap between officers and men. To this end it is absolutely necessary that petty officers, until they are furnished by the training school, be taken from the best men in the fleet, be given comfortable quarters, and relieved to a certain extent from the many galling surveillances to which enlisted men are subjected.

With regard to the officers and men generally, it may be said that training is not at all progressive. So long as an officer or man performs his special duty satisfactorily, that is all that is required; there seems to be little or no preparation for higher duties. For the sake of its own interior training a ship must go to sea *frequently*, not necessarily continuously. At sea its manœuvring powers should be tested by every officer aboard ship. It is odd that the watch officer is not practiced more in these evolutions. There should be target and torpedo practice, boat expeditions, offensive and defensive fleet tactics off the ports and at sea; then let the ship go into port again, and require every line officer aboard ship to familiarize himself with the harbor plan and sailing directions, so that the entry will be a true lesson in piloting. Let the naval cadets trace the harbor plans and work out with the aid of boats cross bearings of turning points and range lines. Let them study the fortifications, food, water and coal supplies, the language, habits and history of the people. Reduce the volume of their log-book copying and column-ruling, and improve its quality and value by only selecting important days. Let each divisional officer make it possible for his men, through a little judicious information, to spend a profitable and pleasant time ashore; and then to sea again to apply the quickened powers of observation to the ship itself and her functions. Let all drills go on intelligently, progressively; there are some movements of hand in all drills so important that they should be instructive; this faculty can only be obtained by repetition.

In training alternates to the various billets they should not be so hampered and nursed as to fail to form intelligent opinions of their own. This holds even more for the officer than for the seaman. The watch officer is relieved on going into port, and is, perhaps, stationed in the waist, where his duties are purely nominal, and nothing can be learned of the handling of the ship. He may never even learn how to swing his ship around an obstruction, the degree of helm for certain appearances of current, etc. He has his duties to attend to; piloting the ship is the responsibility of some one else. And yet it would be considered a disgrace for any officer to admit that he did not know how to handle his ship, how to compare his chart with the view around, when to start his turns and with what amount of helm. Somehow he is supposed to absorb all this practical information, when every one knows that it can only come with experience and the closest observation of those who are expert, the harbor and the chart, at one and the same time. This is but one phase. Instruct any officer in the duties of the next higher grade and you give him a bird's-eye view of his own. He becomes more competent, more apt to provide for contingencies, and more sympathetic and conscientious in his duty towards his superiors.

Lieutenant A. B. WYCKOFF, U. S. N.—The paper on "Naval Training," by Rear-Admiral S. B. Luce, is most admirable. Mental, moral and physical culture are all absolute essentials in naval training. The kind and degree of the latter is likely to prove a most difficult problem on board our mastless vessels. How shall we obtain the same physical development of our seamen and the moral effect upon character which the work aloft gave in the old Navy? There is great danger that physical development will be neglected in the new Navy, with steam capstans, mechanical gun-carriages, and no sails or spars. The pulling boat should give some legitimate exercise if not replaced entirely by the steam launch, but this is all that remains.

In my opinion, the essayist's solution of this question is the correct one. Our cruisers should retain their spars and sails and be copper-sheathed. The spars should be as light as possible, so that most of them can be stowed on deck when steaming to windward. The twin screws should uncouple, so that fair passages can be made and the coal saved for emergencies. I know there are serious objections to retaining spars on our cruisers because of the difficulties of finding stowage room for the sails, rigging, blocks, etc., and the dangers incident to battle from falling spars and top-hamper fouling the screws. But if the naval architect can solve the problem of stowage room, let us retain our spars in peace times and dismantle the cruiser, if necessary, upon the declaration of war.

The condition of the Dolphin and Yorktown, at the time of this writing, sufficiently proves the necessity of copper sheathing and the great gain in economy and efficiency thereby.

The naval apprentice should be enlisted for eight years. Six months should be spent at the training station at Newport, sleeping in a hammock and instructed as at present, but with more modern appliances. His education should be continued on the training-ship for one year. These vessels should be fully sparred, and have modern batteries and a complete electrical outfit. The apprentice's first regular cruise should be on the light-sparred cruiser, where the work aloft will do so much to advance his physical and moral development. After three months' leave the apprentice could be sent to complete his enlistment on the battle-ship or coast-defense vessel. During both cruises his professional education should be carefully carried forward. The modern cruiser and battle-ship furnish every facility for completing the education of the apprentice, if regular and progressive instruction is given.

This can be done by the watch and division officers, if a little common sense is infused into the discipline of our men-of-war. The present age of the lieutenant's-watch officers of large vessels is forty years. Five years from now it will be forty-five years. Let them stand their watches at night at sea when another officer is needed as an efficient lookout on the forecastle. But in the daytime and in port let the ensign become the regular officer-of-the-deck, and the old lieutenant be relieved from "treading pitch" night and day, which he is physically incapable of doing and at the same time attend efficiently to the hundred other duties now incumbent upon him. The ensign can perform this deck duty just as well, and the lieutenant is left free to carefully carry forward the instructions of his division, man by man, in every necessary detail. In addition, the lieutenant can perform all the extra duties which now devolve upon the regular watch officer, and thus both will be left with some little time for recreation and the study necessary to keep abreast of the advance in their profession. It is an impersonal matter with me, as I hope my watch-standing days are over.

The petty officer of the service must and will be a more intelligent and important personage, and be fully qualified to always take charge of the boats except when cadets are sent in them for practice and experience. The ensign port-watch officer will thus be relieved of the most disagreeable duty which his rank has heretofore performed.

I feel sure the apprentice can be given all necessary knowledge in gunnery and electricity on board ship, and at the same time not be prepared for a lucrative position with some electric-light company, as our seaman-gunners seem to be at present.

WASHINGTON.

Ensign J. B. BERNADOU, U. S. N.—I have read this most valuable paper with care and attention. In a section headed " Sails " the Admiral has set aside for a moment the subject under consideration, with the obvious purpose of throwing light upon the general conditions that are associated with naval training, and has taken up the question of our future naval policy.

Two lines of development, specified as being distinct from one another, are briefly discussed; the first assuming the construction of a fleet of vessels of the monitor type—to be left unsheathed, and kept within easy reach of the coal yard and the drydock; the second looking to the creation of a fleet of battle-ships and cruisers, of which the former may be fighting machines with twin screws and military masts, while the latter are to be provided with single screws and possess full sail power.

The national policy entered into at the close of the War of 1812, and the rapid growth at that time of our merchant marine, are dwelt upon as lessons for the future; the naval organization of the period is explained, and it is stated that "a revival of our early naval policy would call for a battle-ship on each foreign station, with a proportionate number of cruisers in addition."

In relation to this portion of the paper, certain questions suggest themselves. Do the conditions that obtained at the close of the War of 1812 bear close relation to those of the present day? Why are the two systems of defense above mentioned to be taken as entirely separate and distinct from one another? Why should coast defense be coupled with disappearance of merchant marine? And why should not naval development, starting from a defense basis, be extended to the limits of aid to the creation and support of a national commerce?

In 1812 the territory of the United States might have been roughly described as a strip of coast land backed by a wilderness; to-day the dimension of our country from east to west is greater than the sea-coast line from north to south; in 1812 we had but one great coast line, now we have two, and two land frontiers extending from ocean to ocean.

Owing to our fortunate natural position, however, and the character of our frontiers, the fear of a land invasion enters only as a minor factor into plans for the national defense. The points at which direct attacks are to be expected are upon our sea coasts, and these we must defend. Our Navy at present, with all vessels building and appropriated for, is but the nucleus of the armed protection that the interests of our country demand. It is necessary, therefore, to continue the development so auspiciously begun, and in doing this we must proceed with prudence and expedition, setting forth our wants in the order of their importance, and in the way best calculated to appeal to Congress and the people. We must excite in our behalf the sympathies and the interests of the capital, labor, production and manufactures of our country.

It is more than probable that so long as we are utterly without a mercantile marine worthy of our name, so long as we are willing to hire foreigners to make money at our expense by carrying our products over the seas, that our Navy will remain of very moderate proportions. The way to obtain a powerful navy is to build up a merchant service. As soon as our flag floats over American merchantmen upon all seas of the world, the great private interests at stake will develop a national character, and the need of adequate protection for investment will be felt—a protection only to be afforded by numerous and powerful squadrons of vessels of war.

Until such a time comes, however, let us develop our Navy with a view to what it would be most needed for in time of trouble in the near future, and let us keep the heavier ships where they would then do the most good. We cannot get large sums of money to protect interests that do not exist and that, as yet, only exhibit a probability of development.

ANNAPOLIS, MD.

Lieutenant WAINWRIGHT.—In his paper on Naval Training, Admiral Luce brings forward a number of points, all of vital interest to the naval service. To some extent they require separate treatment, and may be discussed more intelligently if placed under their several heads. The first question raised is, What class of cruiser shall we build? This may be considered from its strategic, logistic, and economic side. The second question is, How shall we train our naval seamen? This again may be considered under two heads—one the kind of training necessary to enable them to carry on their duties in a modern vessel of war; and the other, under the supposition that such training is not sufficient to develop a good fighting man, What additional training is necessary, so that the naval seaman shall not be found wanting in time of war?

First as to the class of cruiser. The cruiser has many important duties to fill. First and most important is as an adjunct to, and part of a fleet of battle-ships. Nothing has been shown more conclusively by modern naval manœuvres than the necessity of a number of cruisers to accompany the battle-ships. Without cruisers to act as scouts, the fleet is shorn of much of its strength and importance either for offensive or defensive warfare. For this service they not only do not need sails, but cannot afford to carry the weight. Speed is of prime importance, and the coal endurance must be greater than that of the armored vessel. The service must be performed under steam, and the cruiser must always be ready to do her uttermost and cannot afford to meet the resistance of the spars in contrary winds, or to carry the extra weight of sails and spars. The next most important service is to go to some foreign port, to protect our interests in case of disturbance. Here again speed is of prime importance. True, if she have full sail power she may arrive at the desired port with full bunkers and prepared to make a long stay without coaling. But frequently, if not usually, disturbances come without warning; and it would be a rash commander who dared to make his passage under sail, leaving our interests and the lives of our citizens unprotected for an indefinite time. Does any one remember the amount of time occupied by some of our old vessels in making the passage, with their screws hoisted, from New York to Rio? Some may argue that sail can be used with favorable winds; but this has always proved of doubtful economy in time and in coal. Take Admiral Luce's example of disturbances at Rio—certainly we must proceed under steam at a fair rate of speed. We must provide coal, not diminish the supply carried, by using a portion of the displacement for weight of spars and sails.

We need coaling stations, but without them we certainly can provide for such emergencies as the one above mentioned. Coaling steamers would be sent to meet the cruiser, or to accompany her. There is one thing certain, that if two vessels are built on the same displacement, the one that carries sails and spars must be less efficient than her competitor in some particular. Leaving out the question of stowage room and the possibility of disembarrassing the vessel of these encumbrances in time of actual combat, the weight devoted to sails and spars in the one vessel, can be devoted to armament, protection, machinery or coal in the other; her offensive, defensive or motive power, or endurance must be decreased to enable her to carry top-hamper. This is a law that cannot be overcome by the genius of the Naval Architect. He may be able to build one to carry top-hamper that will be the equal or the superior of his rival's mastless cruiser; but with the same genius he can himself build a mastless cruiser superior to the one he has built to carry sails. The remaining important war functions of the cruiser are as a destroyer or protector of commerce. The importance of the commerce destroyer is greatly exaggerated in the minds of non-professional men; but all naval strategists are agreed that while great damage can be inflicted on an enemy by commerce destroyers, they will have little effect on the ultimate results of the war. Our late war furnishes us with a good example: the Alabama and others destroyed an immense amount of property, almost ruined our ocean carrying trade, and the Alabama's work is frequently cited as an example by the advocates of commerce destroying; but it did not affect the results of the war even in the slightest degree; but few vessels were diverted from their real war duties, and the injury to the pecuniary resources of the country available for waging war was unfelt. And this is true of all like cases. Privateering and commerce destroying are the resort of the weak who desire to inflict some damage on the strong, but they have no lasting effect, and have never been the cause of closing a war—the weakest must go to the wall. Still we must be prepared to do something in the way of commerce destroying. Shall we require special cruisers with good sail power, capable of remaining at sea for a long time? Such vessels would labor under two grave disadvantages; they would be unable to catch merchant vessels of like or greater displacement, and must fall a victim to commerce protectors unhampered with sails. To protect our commerce as it grows and needs protection we must acquire coaling stations, and until we have them, confine our attempts to destroy an enemy's commerce within the limits of the steam cruiser.

The question of economy in time of peace may be raised also. This might be answered by the fact that it would be too expensive to design vessels for peace purposes only; but we can go further and show that even in time of peace sail power is an expensive adjunct to a modern cruiser. Admiral Colomb has calculated the amount of coal saved by the use of sail on a three years' cruise, and compared it with the interest on the cost of sails and spars, together with the expense of repairs, and found it would have been more economical if steam had been the only motive power used. A vessel in time of peace is only of use when in communication with the shore; the time spent going from port to port is wasted except so far as training goes, and for the training of the officer, the vessel should be using the motive power that it would use in time of war.

The question of sheathing is somewhat outside of the subject of naval training; still the paper of Naval Constructor Hitchborn has been mentioned. It is a most able paper, but far from conclusive on the subject. There are many examples that may be quoted against him, and he ignores the fact that even vessels whose bottoms are sheathed with copper will foul in time. It is a question of degree only. All bottoms foul, and many examples of copper-sheathed vessels with very foul bottoms can be cited. The real question is, Is there sufficient difference in amount and time of fouling to over-weigh the known disadvantages of copper sheathing? The weight of expert testimony is against sheathing except for special purposes.

The question now comes to naval training. If sails are to be used no longer on men-of-war, we no longer need sailors. Our seamen must steer and heave the lead, they must be good artillerists, and should make fair infantrymen; and they must be able to handle their boats skillfully, both under oars and under sails. Steering is no longer a fine art, and the helmsman can readily be taught to keep his vessel close to the course. It is far more important to have a good marksman at the lockstring than it was of old. To make a direct hit is of more importance, the opening range is greater, more money is thrown away when shots are wasted, and the guns are instruments of precision, admitting of a higher grade of marksmanship. Our great-guns, small arms, torpedoes and electric lights, all require men of a higher grade of intelligence than was the old sailor. It would seem to be unquestionably a fact that the best training for work on a mastless cruiser would be on a similar vessel, unless such work is of a character to prevent the full development of the mind and body of the men. It seems to be almost universally admitted that the class of work required on a modern man-of-war is not of the kind necessary to keep the men in good physical condition; and, as for their minds, it is apt to turn the seaman into a mere mechanic and to destroy his distinguishing characteristic, his adaptability and readiness in emergencies. Gymnastic exercises have been proposed and even adopted, to some extent, for the improvement of the bodily strength of the seaman, and no doubt, if scientifically carried out, finer physical specimens can be produced than even by the old training with sails. Undoubtedly the strongest claim that can be raised for the retention of sails at the present time is that it was an excellent way of training seamen for their war duties, and the arguments on this head are the most difficult to answer. In questions of tactics, logistics and economics, facts may be brought forward to answer the arguments; but in questions of training, the new methods have not been systematized sufficiently, or long enough in operation, to offer any results as evidence. We know that the old sailor was a fine specimen and answered his purposes admirably;—will the new seaman be as good in his place? A great deal of sentiment is mixed up in this question: much romance hangs over the career of the sailor. Many of us have devoted a large portion of our lives to learning how to handle vessels under sail, and no one with more marked success than the writer of the paper under discussion.

We know that a man who is a good sailor must have a quick eye and be ready in emergencies, and that he can readily learn to handle a vessel under steam properly. But, as far as the men go, have we not exaggerated their capabilities in our minds? A little while ago, if the question arose of introducing a new weapon, the cry was raised, it is too intricate to put into Jack's hands. A good topman or good boatswain's mate was put at the lockstring; occasionally he was a good marksman, but frequently he was a very poor one. He was courageous, possibly more so than the average soldier, owing to the frequent encounter of dangers during a sea life ;—if sails are retained for training solely, will they still have this same effect? I hardly think so ; in time of real hazard, rather than incur real danger, the sails will be furled, and the commander, who would not be justified in risking the lives of his crew and the safety of his ship, would place his reliance in his steam power. Certainly, as a mere exercise, the gymnasium might take the place of the sails and spars. It may be well to retain them on vessels devoted to the training of apprentices ; some little knowledge of sailorizing is still of advantage, and they could be utilized for physical culture. But with the sails there must also be modern equipments and modern weapons. The boys should be made familiar with their real surroundings, and not with some ideal condition pertaining to ancient romance.

The question of retaining the apprentices and seamen in the service is one both of economics and of training, but largely the former. Will any amount of training on board ship, living on board ship, teach a boy to love the sea ? If he have clean, well-lighted barracks when not cruising, in place of being crowded on an old, decaying vessel ; if, when he is at sea, he cruises in a modern ship with modern weapons, is he not more likely to learn to love the service ? True, by training men sufficiently long on board ship, they may be unfitted to earn their living on shore and thus forced to re-enter the service ; but we do not want this kind of man, for surely, when his term expires, he will spend his time on shore in debauchery, and only re-enlist when forced by his necessities, bringing with him a constitution sapped at its foundations, and a temperament dissatisfied with its surroundings. The wages, comforts and certainty of position of the seaman must be such as to induce him to remain in the service. The apprentice should be required to make one cruise after he has reached the age of 21, then he will be better fitted to make a selection of his future life. If he then finds he has good pay with reasonable opportunities of increase, that he is well treated and his comforts fairly well looked out for, and his position certain, he will choose to remain in the service. One thing will greatly add to continuity of service, and that is the abolition of the receiving ships and adoption of barracks in their places. If the men can live in the barracks, with the ordinary privileges of other men of their class, they will find that they are better lodged, clothed and fed, have more to spend than those of their own class, and they will remain contented with the service, while their training can be conducted under more favorable conditions than on a receiving ship.

We will always need seamen; they need not be sailors and they must be more than mere mechanics. A higher type of man is necessary than of old: he must be more highly educated, have better morals, and be more carefully trained. His regular exercises can be supplemented by gymnastics ; and, if to encounter danger is necessary, he may confront it in boats under oars and sail, in bad weather and through surf. The ship-of-war must be allowed to follow out its lines of highest development, and the seaman must be trained to suit the development of the machine and to use it to the best advantage in time of war.

SAN FRANCISCO.

Lieutenant J. C. WILSON, U. S. N.—Having received an advanced copy of a paper on Naval Training, written by Rear-Admiral S. B. Luce, with a request to send opinions, etc., on the same, I feel inclined to comply with the request, not because I believe I can discuss the subject as ably as the writer has done, but because I believe it is desirable to have as many expressions of opinions on this important subject as possible. I regret that absence from the city prevented my receiving the paper in time to study it thoroughly, and that lack of time precludes anything but a brief discussion.

The reasons given by the writer for rigging our cruisers with full sail power are very pertinent, and he presents the subject in a manner not generally considered by the naval expert, viz., the influence of the life of a topman on his character as a man-of-wars-man. There can be no question but that constant necessity for activity of action both of body and mind, exposure to performances of hard and perilous duties, and the sacrifice of all other considerations to those duties, as well as familiarity with danger, cultivates the highest attributes which combine to make the heroic sailor and model man-of-wars-man. This is certainly a point in favor of full sail power on our cruisers worthy of much consideration, but it is after all secondary to the other reason given for advocating sail power, viz., the ability of the cruiser to cruise independently of coaling stations. This in our country becomes a very important consideration, and the writer aptly illustrates how either the Charleston or Newark in time of need might be worse than useless.

It is a question which we must decide for ourselves, as the circumstances governing other nations in their construction of a navy may be entirely different from those which should govern us. We cannot afford to stand by and let the experts of other countries study out questions of construction, equipment, etc., considered best for their naval policy, and then adopt them for ourselves. We must study the conditions under which our ships are to be used, and construct them accordingly.

Practically speaking, we have no coaling stations, and consequently, in time of war, no means of obtaining coal out of our own country. Without coal our cruisers of course would be useless, and they would be under the necessity of always remaining within striking distance of a home port. Under these conditions their sphere of usefulness would be very much reduced, and the boast that we could sweep the high seas of an enemy's commerce could hardly be realized. It is very desirable, if not indispensable, that a cruiser should be able to keep the high seas for months at a time, not only to reduce the probability of capture to a minimum, but to be able to remain out in search of merchant vessels.

The reason assigned for making our cruisers practically mastless is to obtain the maximum of speed and handiness, both very important qualities for a cruiser to possess. She needs speed to be able to choose her distance in case of being brought into action, or to escape from a more powerful foe, if necessary. The experience of the Alabama, as quoted by the writer, illustrates that great speed is not essential to a successful commerce destroyer.

We then have the question before us in the equipment of a cruiser, whether it is better to obtain increased speed and handiness by discarding all sail power, or to sacrifice a moderate percentage of these qualities in order to obtain the advantage of a cruiser's being able to keep the high seas independently of coaling stations. I do not think there can be any question as to which alternative possesses the greatest merit for our cruisers, which cannot depend on coaling stations. I agree with the writer that they should be rigged with sufficient sail power to enable them to cruise under sail alone, thereby sacrificing, if necessary, a moderate percentage of speed and handiness. Accepting this idea as correct, it becomes necessary, as the writer remarks, to so construct and rig our cruisers that the minimum amount of speed and handiness will be sacrificed, and it would seem not a very difficult problem to find a system of telescoping the masts, and stowing the yards, so that these qualities would be but little interfered with. The screw could be disposed of either by hoisting or uncoupling, as deemed most advantageous. As twin screws seem to give the best results and are desirable for many reasons, they might be fitted to uncouple.

The question of sheathing the bottoms of iron and steel ships seems to me to have been pretty generally decided against the practice. I think it is on record in the English service where sheathing has been removed because it was found to be positively dangerous to the safety of the vessel on account of the galvanic action set up between the two metals. Again, it increases the displacement (and, in consequence, reduces the carrying capacity somewhat), besides changing the lines of the immersed section of the ship, which combine to decrease the speed. The only advantage of sheathing is that it renders the ship independent of docks, and this of course is a very great one, but as a ship can now go from six to eight months out of dock without serious consequences, it would seem that the disadvantages of sheathing more than outweigh the advantages. I should conclude, then, to let the cruisers go unsheathed, but by all means give them sufficient sail power for cruising purposes. Speed under sail is not essential, as with our modern high-power guns of great accuracy merchant ships can be brought to at longer distances than they could during the days of the Rebellion, and, in case of necessity, steam can be used.

The advantage to be gained by sail power in ships in the education of the man-of-wars-men, as so clearly set forth by the writer, applies especially to the apprentice system, and there is even more reason for retaining it on "training ships" than on cruisers. We should have at least one modern-built, armed and equipped small-sized cruiser, with full sail power, for the advanced class of apprentices, suggested by the writer. The younger boys could learn the duties of a man-of-war sailor in any ordinary cruiser, and should be kept at sea (or at least cruising) constantly until promoted into the advanced class on the modern ship, where more time could be spent in port and devoted to gunnery and similar drills. The barrack system, which I think should be adopted in place of the receiving ship for the "general service recruit," would not do for the apprentice system.

In this discussion I have confined my remarks to the points brought out by the writer, but believe that the whole system of recruiting and training of men to meet the requirements of the new Navy should be radically changed.

I think the thanks of the service are due to Rear-Admiral Luce for so ably opening the subject of "Naval Training" for discussion, and so clearly setting forth the necessity of sail power in our new cruisers.

PROFESSIONAL NOTES.

DETERMINATION OF THE ACTUAL TRACK OF A VESSEL DURING TRIALS ON MEASURED COURSES.

' By Ivo Chevalier de Benko, Lieutenant, Imperial Austrian Navy.

[Translated by Lieut.-Comdr. E. H. C. Leutzé, U. S. N., from *Mittheilungen aus dem Gebiete des Seewesens,* Vol. XVII, No. 12.]

In the following we have in view only the cases where the greatest exactness is necessary to determine the actual wake of a vessel and the time required to make it, as for instance in trials for speed.

As the contracts with shipbuilders generally call for premiums or penalties for 0.1 of a knot of speed, we will assume that any method which will assure exactness to that limit will be permissible.

Speed trials consist in running the ship over a well known or easily measured course and taking the exact time required to do it.

The exactness of the results will therefore depend, 1st, on the exactness with which the ship is kept on the measured course; 2d, on the exactness with which either the prescribed or the actually made course can be determined; and 3d, the exactness with which the time is taken.

Under 1 we can omit such faults as would arise from the state of the sea and strong winds, as they are difficult of determination, follow no laws, and would generally be avoided by choosing suitable weather. Their effects could be felt in the difficulty of keeping the vessel on her course, she being exposed to forces which may be considered to act in the same direction as the course and also perpendicular thereto. The harmful influence on the machinery caused by the laboring of the vessel would also have to be taken into consideration.

The first two mentioned causes of error are, however, similar to some which are unavoidable during good weather, namely, the errors due to bad steering and the influence of currents.

The error caused by bad steering can be compared with that made by the surveyor when he chains distances which are to be measured. In that case nearly every link of the chain makes a small, unmeasurable angle with the direction of the actual line to be measured.

We may perhaps assume that as a general thing a ship will run in regular succession three minutes on her course, three minutes on one side and for the same space of time on the other side, and will then return again to her course for another three minutes, etc. The time of a period would therefore be nine minutes, during two-thirds of which the vessel would be on her course. It would be easy to determine, arbitrarily, the length of the curves and their number, and also the angle of deviation from the course. After that a formula could be established for this error. For instance, if the deviation from the course should be a ¼ point and the distance 30 nautical miles, we would have the formula $30 \times 1852 (1 - \cos \frac{1}{4}$ point$) = 67$ m. or 0.1206 per cent. If, then, the speed of the vessel had been 20 knots the error would only amount to

0.024 knot (this would be the maximum error no matter how much of the time the ship is off her course, as long as it is never more than ¼ point).

It will therefore be seen that the error caused by bad steering is small, and by close attention to the steering it would be so reduced as not to be large enough to be taken into consideration. If it is deemed advisable, however, to take this error into consideration, it would be necessary to note the length of the times and the amounts the ship deviates from her course, and also where the deviation reaches the maximum; the amount of the latter must, of course, also be noted.

The following will show how a convenient formula for this calculation can be established.

The curve representing the course of the ship is similar, as far as the error in question is concerned, to the motion of the pendulum.

The angular velocity of the sheering vessel is comparable to the velocity of the pendulum. Supposing then that the steering is regular, the error will then be the difference between the curved course $ABCDA$ and its projection AA, Fig. 1.

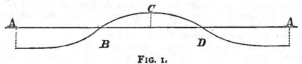

FIG. 1.

It would therefore only be necessary to establish the formula for the curve and calculate its length.

In actual practice, however, the curve is too irregular to conform to a fixed law, and it would be necessary to observe the deviations (a), noting the time in minutes during which each deviation exists, and then calculate the error due to each a and its corresponding t, as follows:

From Fig. 2 we have

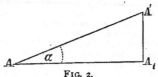

FIG. 2.

$$AA_1 = AA' \cos a \quad \therefore \quad f(\text{error}) = AA' - AA_1$$

$$= AA' (1 - \cos a) = \frac{tv \times 1852}{60} \, 2\sin^2 \frac{a}{2}$$

$$= \frac{1852}{60} \sin 1'' \left(\frac{2\sin^2 \dfrac{a}{2}}{\sin 1''} \right) tv = [6.17510 - 10] \, vt \, \Delta,$$

in which the quantity in parenthesis is the logarithm of the actual coefficient, Δ is the well known number of Delambre of which there are tables, and v the mean velocity obtained by the formula given on the last page of the article. The total error is the sum of the partial errors thus obtained, i. e., $F = [6.17510 - 10] \, v\Sigma t\Delta$, and if t be the same for every observed value of a it may be written before the sign of summation, and the calculations can be made by means of tables of the number Δ which are entered with a to degrees and tenths.

2. If so called "measured miles" are not available, it can be presupposed that it is possible to get accurate data for the position of and distances between prominent points on the coasts from the Coast Survey Office. The probable

errors which are given by this office, in connection with other data, will determine whether the chosen points are sufficiently accurately located.

In choosing a course its length should be the first consideration (if 1, 2 or 3
miles are wanted). The length will be governed by the requirements of the
case. After that the situation and direction can be determined. In regard to
situation, it is necessary to choose a location free from strong winds, heavy seas
and irregular currents (specially those near the mouths of rivers). In regard to
direction, it is desirable that it should be normal to that of the ranges which
are to determine the limits of the run, that 'the marks of these range lines
should be distinctly visible, that the marks of each range are not too far apart,
and finally that the ranges both have the same relative direction to the measured line. As the ranges should be sharply defined, large objects, such as
church steeples, should be chosen ; coast lines, etc., which are changeable,
should be avoided.

It is not essential that the ranges mentioned above should be perpendicular
to the course, but it is necessary to have at each range a third fixed point
upon which angles may be taken for determining the exact points at which the
line actually cuts the limiting ranges. The moment the limiting range AB is
on, the angle between A and C (Fig. 3) should be measured, and it will then
be known, the triangle ABC being known, at which point, E or E', the range
has been crossed. If A and C are not too high it will not be necessary to
reduce the measured angle to the horizon ; it is not difficult to prepare tables
for this purpose, which are entered with the heights of the objects and the
measured angle.

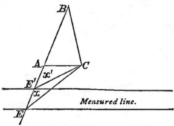

FIG. 3.

The strict formula would be $\cos \zeta' = \dfrac{\cos \zeta - \sin a \sin \beta}{\cos a \cos \beta}$, in which ζ represents the measured angle, a and β the height angles of the objects, and ζ' the
projection of the angle ξ.

As a and β are generally small, Legendre's approximation formula can be
used,

$$x = \left(\frac{a + \beta}{2}\right)^2 \tan \frac{\zeta}{2} - \left(\frac{a - \beta}{2}\right)^2 \cot \frac{\zeta}{2},$$

to calculate a small table for the required correction. The table would be
entered with a and β for height of eye from 3 to 8 meters and with ζ to degrees.

But as (Fig. 3) the angle AEC should be chosen as near to 90° as possible,
this correction of the observed angles becomes of no importance as long as
A and C are not located too high.

I will now show how it is possible to find without much trouble the correction to be applied when the range is crossed at E' instead of E as required. It
has already been mentioned that if the ranges determining the limits of the
course are not perpendicular it will be sufficient to have triangles at each end
whose position and distance from the course are known. For instance, in Fig.
4 AN is the course and OCM and EGD the known triangles ; the angles OAN

and ANE and the distance AN are then also known. The lines OMA and EGN represent the limiting ranges. In order to verify if the ranges are passed at the prescribed points A and N, the angles MAC and GND are measured; they are of course known, as the situation of the triangles and their distance from the measured line are known.

The fact that in Fig. 4 the prolonged course AN goes through the triangle point C is irrelevant to the matter, it is only thus indicated to show how easy the reduction of the course becomes when the ranges are perpendicular to the course and its prolongation cuts through one of the points of the triangle.

Let omc and egd represent the sides of the triangles, s the course, and a the verification angle OAC, and v the angle END. If the angle measured on the range EG be $EN'D = V'$ instead of $END = V$, the vessel will then have

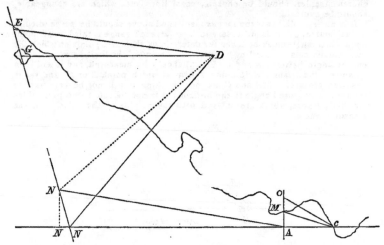

covered the distance $AN' = s'$ meters instead of the distance $AN = s$ meters, and $s' - s = c$ will represent the correction due the difference $v' - v$ in the angles.

As the triangle END can be solved by means of the triangle EGD and the angle V, it is proper to assume that the piece NN' belonging to the triangle ANN' can also be considered as the change in the side EN of the triangle END due to the change in the angle V.

$$NN' = DN \frac{\sin (V' - V)}{\sin V'}.$$

In this formula only V' is changeable; it is therefore easy to prepare a table for NN' whose argument is the corrected angle V' (corrected for instrumental error and reduced to horizon).

$AN' = s'$ can be found without difficulty from the triangle ANN', as $AN = s$ and the angle END is assumed to be known. We would have

$$s' = (s + NN') \sqrt{\frac{1 - 4sNN' \cos^2 \dfrac{N'NA}{2}}{(s + NN')^2}},$$

where the right side of the equation only contains known quantities, and it would be again easy to prepare a table for the different values of NN'. But as NN' only changes with v' it becomes possible to combine the tables for NN' and S' into one table for S' with the argument v'. In order to facilitate the computation of the table an auxiliary angle $(\cos^2)$ is introduced under the sign of the square root.

But it now becomes necessary to mention why the range is crossed at N instead of N'; in other words, which part of the value NN' must be attributed to the engine. It can be considered that NN' is caused by poor steering, incorrect compass error, and finally through the component of the current normal to the course. It has already been shown under 1 how to eliminate the error due to bad steering; the error due to the normal component of the current can be eliminated by the projection of S' on AN_1, or by finding the means of calculating the rapidity, namely,

$$S = s + NN_1 = s + NN' \cos N'NN_1,$$

the error due to the influence of the compass error. There only remains then the error due to faulty compass correction. This can be reduced to a minimum by establishing on the coast a range parallel to AN on which the error can be accurately determined.

It is therefore necessary to make the following calculations for each limit of the measured course:

$$NN' = DN \frac{\sin (V' - V)}{\sin V'}$$

and

$$S = s + NN' \cos N'NN_1.$$

Therefore if you let $NN' \cos N'NN_1 = C_n$ and the analogous quantity for the other limit equal C_a,

$$S = S + C_n + C_a,$$

from which a table with double argument (the verification angles at A and N) can be constructed beforehand and the determination of the speed easily accomplished.

It will be seen that S is sharply defined as the speed maintained by the performance of the engine; it is entirely independent from the " *cross* " component of the current. It furnishes at the same time a good guide for selecting the course; should it, for instance, be a short course that it is repeatedly run over in a comparatively short time, the current can be considered the same in any two consecutive runs, so that the place can be selected without paying any attention to the currents. Should it be a long course, however, then it would be well to select a place where the current sets across, so that it can be accurately determined or eliminated by the method we have above indicated.

If it is certain that the current sets across it is of no importance if its strength is constant or not.

In conclusion I would call attention to the fact that the mean of several runs over the same short course is not always accurately determined; the first and last run should be only given half weight. Under supposition that the current remains constant during the run in one direction and return, we have

Run.	Partial Mean.	Mean.
A		
	$\dfrac{a+b}{2}$	
B		
	$\dfrac{b+c}{2}$	
C		
	$\dfrac{c+d}{2}$	$\frac{1}{10}(a + 2b + 2c + 2d + 2e + f)$
D		
	$\dfrac{d+e}{2}$	Or with only five runs
E		
	$\dfrac{e+f}{2}$	$\dfrac{a + 2b + 2c + 2d + e}{8}$
F		

or in general

$$V_n = \frac{1}{2(n-1)} \; V_1 + 2(V_2 + \ldots + V_{(n-1)}) + V_n).$$

The assumption that the current does not change in a certain time becomes more tenable as this time is reduced. On the contrary it can therefore not be assumed that the current remains the same during all the runs, but it is perfectly correct to assume this to be a fact during two consecutive runs.

3. I have shown under 1 how the error due to bad steering can be calculated, and under 2 I have shown a method to eliminate the compass error and cross current. The time observations remain to be discussed.

It is important the coincidence of the range signals at both ends of the run should be observed by the same person, and it is not immaterial to whom this duty is entrusted. The choice of a proper person becomes more important when the ranges do not cover but pass above and below each other.

It is dangerous to use the sight vane of an azimuth compass, as it may lead to grave errors, as it is very difficult for the observer to judge if the compass is horizontal, that is if the wire of the sight vane is perpendicular. On the other hand there are persons who have peculiar faculty for judging vertical coincidences. During calm weather a plumb-bob may be of service.

The time is generally noted with chronoscopes (Marenzeller watches); this seems to me to be superfluous unless the watches are repeatedly compared with a well regulated chronometer, as the irregularity of the personal error will certainly be considerably larger than the rate of a good chronometer. I would prefer to mark time by a good chronometer which beats to seconds, the seconds to be counted aloud by an assistant.

An error of 1 second in running over a course of 1 mile at the rate of 20 knots amounts to $\frac{1}{180}$ of the observed time or 0.5 per cent, which would give an error of 0.111 in the speed to be determined; it is therefore plain that the times of observations during such run are of the utmost importance. It becomes therefore necessary that the comparison of the chronoscope with the chronometer should be correctly taken, that is, at the same temperature and in the same position.

The valuation of the comparisons is simple.

I will review the methods recommended in the above.

Five persons are required for the observations :

    One person to observe ranges.
    One person for marking time.
    One person for taking the angles.
    One person to observe the steering of the vessel at the compass.
    One person to assist at the compass (should be provided with a watch).

The next step is to establish the course by means of the verification angles reduced to the horizontal plane; the correction for poor steering is then applied. The preliminary data required for the latter, namely the speed, is obtained by dividing the length of the course by the time according to the formula quoted below.

Finally the observed times must be corrected after comparison, and then the equation

$$V \text{ knots} = \frac{S \text{ Meter}}{T \text{ Sec.}} \times \frac{3600}{1852} = (0.28866) \frac{S \text{ Meter}}{T \text{ Sec.}} \quad (1852, \text{ J. P. M.})$$

computed. In this equation $S =$ the exact course in meters, $T =$ the exact time in seconds, and V the speed in knots. The number in parenthesis is a logarithm.

· The above method is particularly useful on long courses, where steering on a range cannot be resorted to.

SPEED TRIALS OF FAST SHIPS.

A REVIEW.

The Journal of the American Society of Naval Engineers, Vol. II, No. 1, contains under the head above, two highly valuable papers on the subject of speed trials, the object of the discussion being to bring out the best method of ascertaining the speed developed by the new protected cruisers, Philadelphia, San Francisco, etc. The idea of the speed trial is to ascertain the exact distance a vessel can cover in four hours, uninfluenced by wind or by current. A scientific standard of comparison being needed, the best speed under the most favorable circumstances is required. The three principal methods proposed are : By a continuous run at sea, the speed being based upon the number of revolutions of the screw found necessary to give a knot in smooth water. By a similar run, the speed being based upon the number of revolutions of a measuring screw, rated in smooth water. By a continuous run at sea past a series of buoys or stations on shore, so arranged as to give the distance that the ship is likely to make, provision being made for accurate determination of the tide and current at frequent intervals along the course.

It is manifestly impracticable to obtain a course free from wind and current, and impossible to steer a vessel without allowing it to deviate from the straight line connecting the terminal points. The last method appears to be the favorite one at present; but each method has its objections, some of which are pointed out clearly in the papers above mentioned. Mr. Isherwood plainly states his doubts as to the accuracy of the measured mile trials, and yet he uses the results of some of these trials to throw doubt upon the method of revolutions of the screw. The paper above, translated by Mr. Leutzé, shows clearly the source of inaccuracy in those trials, and in place of taking the slip of the Boston's screws to discredit the method of revolutions of the screws, the irregular slip, even at such widely varying speeds, should have been sufficient to throw doubt on the distance traveled as determined by the observations. This translation of Mr. Leutzé's shows how the distance traveled should always be increased more or less to allow for inaccurate steering, and also that although the method of running backwards and forwards over the same course, when only short intervals of time elapse, will correct the error due to the component of the tide acting in the line of the vessel's keel, it ignores the other component which increases the distance both in going and in coming.

It would seem as if all of the three methods were correct theoretically, the practicable application of them being the great difficulty. The great objection to the last method proposed is the expense. Large signals must be erected and their positions carefully determined, the course must be well buoyed, and a number of vessels anchored at various points along the course so that the strength and set of both surface and sub-surface currents can be observed; then with a smooth sea and little wind the speed could be determined with a near approach to accuracy. There is one point that will always throw some doubt upon the results reached by running over a buoyed course, and that is the error produced by the current. No doubt, with a number of stations and numerous current meters, the strength of the current at various depths can be ascertained and its full effect upon the vessel be calculated; but the resultant direction will be always somewhat in doubt, particularly when it can only be approximately determined by the compass. This may be a small error, but when taken together with the necessary inaccurate steering of the vessel, the results will be further from the truth than if using the simple and less expensive method of taking the number of revolutions made by the screw in a given time.

The advantages of the measuring screw proposed by Mr. Isherwood are that the screw may be rated by means of a medium sized vessel, over a short course at not above medium speeds.   Once rated, the course may be set in any direction from the land, keeping in deep water and having the wind abeam.   The actual distance over the bottom becomes of no importance ; the current, therefore, does not affect the results, nor does the course steered, so long as the helm is not put over so as to retard the ship.   In other words, if the measuring screw be properly rated, the exact number of revolutions in precisely four hours will give the required speed.   If we can obtain the rate of the vessel's screw we have all the advantages of the measuring screw without the necessity of providing a special instrument, and then there must always be some doubt as to the rigidity of the measuring screw when used on vessels of high speed.   There is one point about the vessel's screws : why rate them at low speeds, when only high speeds will be used ?   It is obvious that if anything happens to the machinery so as to cut down the speed beyond a very slight amount, the trial will be useless for purposes of comparison, and the contractors would certainly demand another trial.   Therefore the screws need be rated only for the best speeds of the vessel and speeds slightly below the best.   The vessel can be sent over the measured mile with good way on and with plenty of steam ; the mean between two runs made at short intervals apart will eliminate the error made by one component of the current, and by carefully determining the terminal positions, not relying on the ranges alone, the error produced by the other component can be removed.   Then by applying the corrections for deviation from course, the exact distance traversed can be obtained.   A break-circuit chronometer would give exact time intervals, and an improved counter would record the number of revolutions.   A few runs at each speed would suffice to rate the screw.   It is possible that the change of load during the four hours' run would affect the slip, but as the high speeds are within the squatting limit, this does not seem probable, but it might be readily tested on the measured mile.

In spite of Mr. Isherwood's argument, the revolutions of the vessel's screws would still appear to be likely to give the nearest approach to accurate results.   In fact, for one usually so accurate, Mr. Isherwood has been quite careless in his presentation of the facts.   His mean of all the slips, in the case of the Boston, should have been 12.75, not 11.75 per cent, and this would have given a smaller variation between the experimental and the calculated speeds.   Again, if he had not been trying to make his facts prove his case, he would hardly have been willing to assert that the mean slip, ascertained from widely varying speeds, could be the true slip, or that experiments made with unsatisfactory precautions were of use in proving his theory.

In all the methods proposed there are chances of error, in fact none are strictly accurate ; but the one that will produce the most regular results, with the nearest approach to accuracy and the least expenditure of money and time, is the one proposed by the Engineer in Chief, if it be so modified as to leave out the lower speeds he proposed, and the contractor be required to run over the measured mile at the highest speed, and if no speed below 18 knots be used to determine the number of revolutions per mile.   The distance through the water is what we want ; we can ignore the current and avoid the effect of the wind, the number of revolutions and the elapsed time can be noted with precision, and the results must be accepted as a very close approximation to the truth.   R. W.

PROPOSED NAVAL MESSENGER PIGEON SERVICE.

Most of the European governments have now fully recognized the practical value of homing pigeons as messengers, and possess a complete system of pigeon stations along their respective coasts, under direct control of the government.

Canada has quite recently followed their example by establishing an organized system of messenger pigeon stations throughout the Dominion, extending from Halifax to Windsor and connecting her principal seaports with the interior.

We advocated in the Proceedings No. 47 (1888) and No. 48 (1889) the establishment of a similar system at our principal seaports of the Atlantic coast and on some men-of-war. So far, no organized service of messenger pigeons has been established in the United States Navy, but some interesting trials have been made at the United States Training Station at Newport. Commander T. J. Higginson, U. S. N., recently in command of the U. S. S. New Hampshire, says, in answer to an inquiry as to advisability of establishing a connected service of messenger-pigeon stations along the Atlantic coast: "In answer to your letter concerning carrier pigeons, I beg leave to state that I would favor a system of stations along the coast, with a central station for breeding, and I think Newport would be the best place for that purpose; breeding the birds here and transferring them while young to the other stations so that their first flight would return them to their permanent homes.

We have at this station erected a pigeon-house and have some choice messenger stock. Although our cote is still in its infancy, we have made some very interesting trials with our birds, and have been much pleased with some of the results. One of the pigeons flew from the Hen and Chicken's Lightship to our cote, a distance of twelve miles, in 16 minutes and 35 seconds. One of our birds has a famous record and is a sister to the famous homer Akron which won the international gold medal in 1887 for best speed. The brother of one of our flock flew from Washington to Fall River, a distance of 365 miles, in 11 hours and 7 minutes. The parents of several of the carrier pigeons belonging to this station have records of over 400 miles, and some of them are from imported pure Belgian breeds. Pedigree is the principal requisite in a homing bird.

Several of our birds were taken to New York on the Juniata last year, with intention of liberating them along the coast. The weather, however, was thick and they were not flown. While at Brooklyn one of the pigeons escaped from the Juniata and it was considered lost, as it had never flown a greater distance than from Point Judith, but great was our surprise when in a few days the bird arrived at his home here safely and in good condition.

I think it would be a good plan for all naval vessels leaving Newport to be supplied with carrier pigeons for the purpose of sending communications ashore and to train them for long distance flights. The use of carrier pigeons as bearers of dispatches would prove of great use to the naval service, and I am heartily in favor of establishing carrier pigeon lofts along the Atlantic coast. I have no doubt of the success of the undertaking if it were organized and thoroughly executed."

This is very encouraging and a strong indorsement in favor of establishing a naval messenger pigeon service. The use of homing pigeons in the merchant marine is quite common. It is a well known fact that many captains have pigeons on board for use in communicating with the vessel from the small boats away from it or from shore. These birds, it is said, never mistake any other vessel for their own when at dock or in the harbor. Among the numerous instances of the use of pigeons for sea service we mention that of a bird which, liberated from the steamer Waesland at one o'clock in the afternoon, when three hundred and fifteen miles from Sandy Hook, was at his loft in the evening. Another let go from the Circassia at nine in the morning, when two hundred and fifty-five miles out, brought a message in the afternoon. Major-General D. R. Cameron, Director of the Messenger Pigeon Association, Canada, says in his letter: "I am of opinion that a most important branch of the pigeon service will be connected with coast service. The evidence that these birds can be relied on to cross 400 miles of the ocean is apparently thoroughly reliable.

. Amongst the many ways in which pigeons capable of doing so much might be used, not the least interesting to a large and very influential part of the public would be the conveyance of intelligence from the passenger vessels crossing the Atlantic. Reducing the belt within which vessels between New-foundland and the coast of Ireland are beyond the reach of telegraph stations by 800 or 900 miles, would be a result of the highest importance." Another writer on the subject says : " It is a wonder to me, after these experiments, that the captain of any vessel should leave the shore without the means of com-municating with it." Among recent experiments in France and Italy the official reports show that during the squadron manœuvres pigeons were freely used as messengers, and often arrived many days before the dispatch vessels sent at the same time.

A system of lofts to be situated at the principal navy yards along the Atlantic coast could be established at a very small expense to the government, as the homing pigeon fanciers throughout the country are anxious to see the govern-ment take hold of the matter and are willing to give their hearty support to the enterprise.

We suggest a connected system of twelve main naval lofts to be situated at the following navy yards and stations :

1. Portsmouth.
2. Boston.
3. Newport. Already established on receiving ship New Hampshire.
4. New London.
5. New York.
6. Philadelphia. Receiving ship St. Louis.

7. Washington. (Central Station.)
8. Annapolis. (On the Santee.)
9. Norfolk.
10. Port Royal.
11. Key West.
12. Pensacola.

The greatest distances being between the last four naval stations, some inter-mediate points would be needed between them to insure a connected service. Beaufort, N. C. (or Fort Macon), Wilmington, Georgetown, Charleston, Savannah, Jacksonville, Tampa, Cedar Keys, and Apalachicola, would be desirable " points de relâche." Several of these stations have already private lofts. Key West has a loft belonging to the Army Signal Service. From these lofts any vessel could be supplied with pigeons and cotes be built on board some men-of-war as in the French Navy. .

Advantages of an Organized Service of Messenger Pigeons.

A service of messenger pigeons for naval purposes could not be improvised at short notice, and the birds would require long and careful training before being of any use as bearers of dispatches.* In war time the occasions are innumerable when serious derangement of plans, loss and discomfiture may be involved by the absence of previously organized provision for the rapid trans-mission of news. The advantage in favor of the side possessing such facilities over an opponent without them is enormous. .

. War vessels defending a coast are frequently without the means to transmit vital intelligence to the mainland. If provided with trained and reliable mes-senger pigeons they could send communications ashore over a distance of several hundred miles, signal the approach of the enemy's fleet and report his every movement.

. In peace, vessels leaving and approaching the coast could report the posi-tion of disabled vessels needing assistance, and signal their own needs and locate wrecks. At places remote from telegraph stations, light-houses, camps and squadron manœuvres, interrupted electric communication and many other

*From the report of Major-General Cameron.

circumstances afford numerous occasions for employing homing pigeons as messengers.

The fact that homing pigeons can fly several hundred miles a day at sea, that birds can be bred and trained on board ship, that they can be accustomed to the noise of the guns, that they can recognize their own ship among others, that they can be relied upon, as proved by numerous experiments, to carry news from the fleet to the shore (and under favorable circumstances from the shore to the fleet and from one vessel to another), when beyond the range of heliograph and electrograph, should suffice to secure the support of the government to this new enterprise, and encourage the speedy establishment of a permanent system of naval messenger pigeon lofts at the principal navy yards and stations along the Atlantic coast.

H. MARION,
Assistant Professor, U. S. Naval Academy.

THE TIME AND DISTANCE REQUIRED TO BRING A SHIP TO A FULL STOP AFTER THE ENGINES ARE STOPPED AND REVERSED.

BY ASSISTANT NAVAL CONSTRUCTOR WILLIAM J. BAXTER, U. S. N.

In handling ships in a crowded harbor during fleet drill, in avoiding a friend, or in ramming an enemy, it is of the highest importance for a naval officer to know the time and distance required to bring the ship to a full stop, when moving at varying speeds, after the order has been given to stop and reverse the engines. The captains of merchant steamers, from their constant practice, would be expected to estimate with more accuracy than naval officers, whose opportunities for knowing their ships are much more limited; but the results of investigations by the Office of Naval Intelligence (General Information Series No. VIII) show that even with these merchant captains the tendency to underestimate both the time and the distance is somewhat general.

It would doubtless be a great convenience if every ship carried the means of accurately determining the time and distance required for these manœuvres, so simply arranged as to be ready for instant use by an officer totally unacquainted with the ship. To determine this data accurately for each ship would require an elaborate series of experiments; but, by the methods suggested below, it can be obtained and put in a shape for ready reference, with but little inconvenience and a small amount of calculation, with so close an approximation to theoretical accuracy as to fulfil all practical requirements.

The following is the principle of the method:

Suppose the curve $OAEB$ to have been constructed, having times, in seconds, as abscissae, and velocities in feet per second as ordinates; at any point B the velocity in feet per second is represented by Bb, and the time in seconds that will elapse from zero velocity is represented by $Ob = T$; if cb represent a unit of time, the distance passed over in this time, dt, will be represented by $vdt = cb \times cd =$ area $cdBb$, and as the same is true of any point in the curve, the whole distance passed over will be represented by $\int^T vdt =$ total area of the curve $OAEBb$. The same is true of other points on the curve; thus, the area OAa represents the distance that would be passed over before the velocity is reduced to zero, and Oa represents the corresponding time. These areas can be taken as the ordinates of a new curve; thus, to any convenient scale, $A'a$ represents the area OAa, $B'b$ represents the area

OAEBb, so that for any velocity represented by *Ee*, *E'e* represents the distance and *Oe* the time. If, then, a set of these curves is provided for a ship,

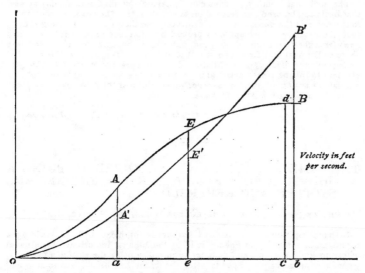

Velocity in feet
per second.

by an instant's inspection, an officer, knowing the ship's velocity, can tell the distance and time required for her velocity to be reduced to zero and the ship come to a full stop.

To find the data necessary for constructing these curves two methods are suggested.

FIRST METHOD.

Let the ship steam at her highest possible speed, having one officer detailed for observing the speeds, and another for observing the corresponding times. At a given signal, whose time is noted, let the order be given to stop and reverse the engines, and as the ship loses speed the observers note the speeds and the corresponding times, continuing these observations as rapidly as possible until the ship comes to rest, and have the data entered on a form, as follows:

U. S. S. OHIO.—*February* 16, 1890.

Knots per hour reduced to feet per second.		Speeds in knots.	Times observed.	Elapsed time in seconds.	Direction and force of wind.	Condition of sea.
20—31.80	10—15.90..........	16	10 48 0	243	Three points abaft the beam; Force, 3.	Slight swell.
19—30.20	9—14.31..........	14				
18—28.62	8—12.72..........	12	49 18	165		
17—27.03	7—11.13..........	10				
16—25.44	6— 9.54..........	8	50 02	121		
15—23.85	5— 7.95..........	6	50 27	96		
14—22.26	4— 6.36..........	4	50 56	67		
13—20.67	3— 4.77..........	2				
12—19.08	2— 3.18..........	0	52 03	0		
11—17.49	1— 1.59..........					

To draw the curves, use zero velocity and the time corresponding for the origin of co-ordinates; convenient scales are: 1 inch equals 10 seconds, and ½ inch equals 10 feet per second. To plot a point on the curve, say that corresponding to 8 knots, from the above table, the ordinate 12.72 is found in column 1, and the abscissa 121 in column 4; other spots are plotted in the same way, and a fair curve is drawn through them. As all subsequent results depend for their accuracy on the accuracy of this curve, it should be checked by an acceleration curve.

To construct this curve, let $OABC$ be a time-velocity curve, Oa, ab, bc being units of time, since the acceleration

$$= \frac{dv}{dr},$$

$$\frac{dv_1}{dr} = \frac{Aa}{Oa} = \frac{Aa}{1},$$

$$\frac{dv_2}{dr} = \frac{Bb_2}{Ab_2} = \frac{Bb_2}{ab} = Bb_2 = B'b,$$

$$\frac{dv_3}{dr} = \frac{Cc_2}{Bc_2} = \frac{Cc_2}{bc} = Cc_2 = C'c,$$

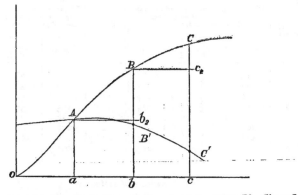

and a curve of acceleration can be drawn having $Aa = Aa$, $B'b = Bb_2$, $C'c = Cc_2$, etc., as ordinates, showing the acceleration at any time and velocity. The relation Force — Resistance $=$ Mass $\times$ Acceleration is a fundamental one, and it is readily seen that when the ship is moving ahead at uniform velocity the acceleration vanishes and Force $=$ Resistance, but after the engines are stopped and reversed this becomes — Force — Resistance $=$ Mass $\times$ Acceleration, or Force $+$ Resistance $=$ Mass $\times$ — Acceleration. This negative acceleration is composed of two parts, that due to the resistance offered by the water to the onward motion of the ship, and which varies with a varying power of the speed, and will thus quite disappear when the ship comes to a full stop; the other part is that due to the force of the engines acting through the propeller and pulling the ship astern. A few seconds will elapse after the signal is given before the engines can be reversed, and after reversal before they are acting efficiently, but after this time their force is practically constant, and a constant acceleration is produced. It should be noted, however, that these accelerations cannot be exactly evaluated, because " Mass " in the equation consists not only of the known mass of the ship, but also the mass of the water which is drawn after the ship, the value of which is unknown.

The acceleration curve, as used as a check on the velocity curve, must be a fair curve, must be nearly parallel to the axis of abscissae near the origin, and contain no re-entrant curves. If it does not fulfil these conditions there are personal errors or errors of observation in the velocity curve, and the latter must be altered until it and its acceleration curve are fair curves.

Having thus obtained an accurate velocity curve, its integral or curve of areas must be found, and this can be done accurately enough by using the trapezoidal rule with an interval of about one-tenth the total time; thus, from full speed to full stop, 250 seconds later, the distance is

$$25(\frac{25.4}{2} + 25.1 + 23.6 + 20.5 + 16.8 + 13.2 + 10.0 + 7.2 + 4.6 + 2.2)$$

$$= 3397.5 \text{ feet} = 1132.5 \text{ yards.}$$

Performing the same operation at each interval the distances are found and plotted to convenient scales, and a fair curve drawn as in the accompanying diagram.

To use this diagram it is only necessary to know the speed at which the ship is moving, and the corresponding distance and time are readily found, and a table similar to the following can be made for ready reference.

Speed knots.	Distance and time required to bring ship to a full stop after engines are stopped and reversed.			Speed knots.	Distance and time required to bring ship to a full stop after engines are stopped and reversed.		
	Yards.	Ship-lengths.	Time.		Yards.	Ship-lengths.	Time.
			Min. Sec.				Min. Sec.
1	14	$\frac{1}{6}$	18$\frac{1}{2}$	9	255	3$\frac{1}{8}$	2 12$\frac{1}{2}$
2	27	$\frac{1}{3}$	35$\frac{1}{2}$	10	318	4	2 22$\frac{1}{2}$
3	42	$\frac{1}{2}$	52$\frac{1}{2}$	11	390	4$\frac{7}{8}$	2 34
4	59	$\frac{3}{4}$	1 07$\frac{1}{4}$	12	465	5$\frac{3}{4}$	2 45
5	80	1	1 22	13	548	6$\frac{3}{4}$	2 56
6	108	1$\frac{1}{3}$	1 36	14	658	8$\frac{1}{4}$	3 11
7	147	1$\frac{3}{4}$	1 49	15	816	10$\frac{1}{8}$	3 30
8	195	2$\frac{3}{8}$	2 01	16	1095	13$\frac{1}{2}$	4 03

SECOND METHOD.

At convenient opportunities, when the ship's speeds are different, let the signal be given to stop and reverse the engines, noting the time required to bring the ship to a full stop, after the signal is given, recording the data, thus:

Date.	Speed of ship before signalling to stop and reverse.	Time required to bring ship to a full stop.	State of Sea.	Remarks.
January 3, 1890....	14 knots.	192	Smooth.	3 boilers.
December 17, 1889.	10 "	143	Smooth.	2 "
February 2, 1890...	2 "	35	Moderate swell.	2 "

The velocity curve can then be plotted from its co-ordinates in columns 2 and 3, and the acceleration and distance curves drawn as in the first method.

In addition to the error due to neglecting the mass of the following water, there is another error due to the assumption that one velocity curve can be made to satisfy all the requirements; thus, let *OABC* be the velocity curve at

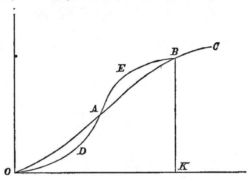

15 knots, and *ODAEB* that at 14 knots, then these methods assume that the area *ODA* = area *AEB*, *i. e.* area *OABK* = area *ODAEBK*, the latter is a little greater, so that ships will slightly overrun the distances as given by these methods. If a complete series of experiments could be carried out on one ship, the probable rate of error could be determined and so tabulated that when applied to the results obtained by these methods, practical accuracy would be obtained.

The data and diagrams described, however, will furnish the necessary information so accurately as to be of great service to officers; and when used in connection with the tactical diameters, precision in fleet manœuvres can be more easily secured.

BIBLIOGRAPHIC NOTES.

AMERICAN CHEMICAL JOURNAL.

VOLUME XI, No. 7, OCTOBER, 1889. Drs. Wolcott Gibbs and H. A. Hare begin the systematic study of the action of definitely related chemical compounds upon animals. H. W. Wiley proposes a method for determining molecular weights, based on the assumption that the rise in the boiling point of a solution multiplied by the molecular weight of the substance dissolved is a constant. R. Hitchcock studies the action of light on silver chloride. Reviews and Reports: Recent researches on cobalt and nickel; Treatise on the Principles of Chemistry (Patterson-Muir); Watt's Dictionary of Chemistry (new edition); A Manual of Assaying (W. L. Brown).

VOLUME XI, No. 8, DECEMBER, 1889. A new method of determining gas densities, by J. P. Cooke. Reviews and Reports: On the nature of tellurium; The condition of matter in the vicinity of the critical point.

VOLUME XII, No. 2, FEBRUARY, 1890. W. M. Burton proposes a method for the detection and estimation of petroleum in oil of turpentine adulterated by the former.

VOLUME XII, No. 3, MARCH, 1890. Gibbs and Hare continue the study of the action of definitely related compounds on animals. The paper on the revision of the atomic weight of gold, by J. W. Mallet, begun in the February number, is concluded. The atomic weight found is 196.91 ($H = 1$).   C. R. S.

BOLETIM DO CLUB NAVAL.

OCTOBER TO DECEMBER, 1889. Naval apprentice school. Theory of the rudder. A new gas-check. On the study of naval tactics. Naval history of Brazil. Modern artillery. Movements of the fleet.
                                                J. B. B.

THE ENGINEER.

FEBRUARY 15. Overheated boiler plates.

An abstract of a series of experiments made by M. Hirsch to determine the effect of rapid evaporation on fast steaming boilers.

Racing of marine engines.

MARCH 1. Roberts water-tube boilers. Lift of safety valves. Boiler design. Facts about aluminum.

MARCH 15. Corrosion of propeller shafts. Cylinder lubrication. Electric welding.

APRIL 12. The use and abuse of forced draught.   J. K. B.

JOURNAL OF THE ASSOCIATION OF ENGINEERING SOCIETIES.

VOLUME IX, No. 2. Some tests and observations on building stones. Notes on the harbor facilities of Cleveland for handling coal and ore. Some points on bridge inspection.   J. K. B.

JOURNAL OF THE AMERICAN SOCIETY OF NAVAL ENGINEERS.

The Marshall valve gear. Tests of Worthington pumping engines. Turkish torpedo-boats.

A description with illustration of the hull and machinery of the torpedo-boats built by the Germania Co., of Berlin and Kiel, for the Turkish Government. On the trial trips these boats repeatedly reached 21.8 knots with an extra weight of 6 tons coal on board. On the three hours' run at sea an average speed of 21.3 knots was obtained.

The second trial of the U. S. S. Baltimore.

After the first trial of this vessel some changes were made in the machinery, the most important being the increase of the pitch of the propeller to about 21.5 feet and an overhauling of the main valves to secure a freer exhaust area, a later cut-off in H. P. cylinders and an overhauling of the air pumps. On the trial a reserve supply of fresh water was carried in the double bottom to make up for losses and to avoid danger due to foaming from a mixture of salt and fresh water. The indicated horse-power for all the machinery (maximum) was 10725.60. The maximum speed obtained was 20.6 knots, and the H. P. developed 9568.47. From an examination of the data in the two trials, the increase in horse-power in the second trial was due to higher steam pressure, greater air pressure and the use of the two auxiliary boilers, coupled with the better use of the steam due to freer passages. As a result of the second trial the contractors receive a premium of $106,442 for excess of power developed.

A description of the hull and machinery of the torpedo-boat Cushing.   J. K. B.

MECHANICS.

FEBRUARY, 1890. Steam boiler design. Air supply to boiler furnaces. Sprague electric locomotive for metal mines and freight tramways. Steam jackets.

A paper by Prof. Dwelshauvers-Derry of Liege, on the mode of action of steam jackets, with experimental data showing the reasons of their economy.

Shaking grates on the Pennsylvania R. R.

A description of the Reagan grate, with the results of a successful experimental test of a continuous run for a week.

MARCH. The properties of aluminum, with some information relating to the metal.

A reprint of a paper read before the American Institute of Mining Engineers, in which the properties of the metal are very fully discussed under the headings of: Purity of aluminum, its properties with reference to specific gravity; the action of heat; corrodability; its mechanical properties; conductivity; the action of impurities; the alloys of aluminum; methods of analysis; methods of preparation of aluminum, and practical hints on the subject of the treatment of the metal.

Aluminum bronze and brass as a suitable material for propellers.

An article by E. H. Cowles, descriptive of the various metals that have been used for screw propellers, a comparison of aluminum bronze with the alloys now used, the advantages claimed, and to which are added tables of compositions and tests.

Electrical accumulators or storage batteries.

APRIL. Steam boiler design. The elements of gropostatics. Theory of chimney draught. A steam-tube cleaner.   J. K. B.

MITTHEILUNGEN AUS DEM GEBIETE DES SEEWESENS.

VOLUME XVII, No. 12.   Treatise on the three arms of modern naval battles and the tactics determined by them. Determination of the actual track of a vessel during trials on measured distances. Facilitation of the "day's work" by means of tables of hour angles, etc. French fleet manœuvres, 1889. Trials of steel plates for boilers of torpedo-boats. English and French naval arsenals. Influence the heating of a vessel exercises on the deviation of the compass. Cellulose on vessels' bottoms. Coast defense. Lacquer for ships' bottoms, prepared by the Japanese Holta. Proof of new armor in the United States. A new shot tray. Installation of torpedo on several battle-ships and cruisers. Large search light by Saulter, Lemonnier & Co. Spectacles for firemen. Completion of the U. S. Monitor Puritan. Annual coal consumption. The largest sailing vessel. Turkish torpedo-boats. Torpedo-chaser Schachin-derga. Turkish corvettes Susot and Beyrut. Literature: The compass on board, a hand-book for commanders of iron vessels.

VOLUME XVIII, No. 1.   Compasses on board modern war vessels. Coal consumption on board French war vessels. Statistics on the management of the Imperial German navy. Estimates for the French navy for 1890. Trials of compound armor-plates of English and French manufacture. A new auxiliary motor. New English cruisers of the first class. New Spanish cruisers. Launch of the French cruiser Jean Bart. New war vessel of the United States. Launch of the U. S. cruiser San Francisco. New navy-yard of the United States. The dynamite cruiser Vesuvius. New German auxiliary cruisers. The Chinese armored cruiser Tschih-Yuen. Spanish torpedo gunboats El Temerario and Nueva España. Faults in the new English gunboats. A new torpedo-boat for the India Government. Trials of French torpedo-boats. Launch of the French torpedo-boat Avant-Garde. French torpedo-boat Agile. Trials of submarine boats. Manœuvres in the cyclones of the South Indian Ocean. Washington International Conference. The highest maritime council in France. Names of Russian vessels in course of construction. Strengthening the defenses of Malta. A new species of armor. A new gun for vessels. New pneumatic gun. Guns for the Chilian navy. The 12 cm. and 15 cm. Armstrong rapid-firing cannon. Speed of English men-of-war. Alison motograph. Greatest depth of the Mediterranean. A new fog-signal apparatus. Literary Notices.

VOLUME XVIII, No. 2. The English fleet manœuvres of 1889. Incrustation in marine boilers. Discoidal projectiles of Captain Chapel. Navy of the United States. Skoda rapid-firing cannon. Trials with the engine regulator of the fish-torpedo. New vessels for the French navy. English protective-deck cruiser of the first class. Rapid-firing cannon for the English armor-clad Trafalgar. Launch of the German dispatch vessel Matsusmia. French armored cruisers of the second class. Navigation by night in the Suez Canal. Two cruisers for the Royal Greek navy. New vessels for the Portuguese navy. Changing of old Spanish armor-clads into cruisers. End of the engine trials of the Agile. Electric workshops in Constantinople. Difference of level of different oceans. Electric motors for the transportation of projectiles. The torpedo-boat Cushing. Literary Notices.

VOLUME XVIII, No. 3. The International Maritime Conference at Washington. Improvements on ships' engines. Management of boats in the surf. The " Canet " rapid-firing gun. Comparative speed of wooden and iron sailing ships. Relative to the building of battle-ships for the U. S. Navy. The new armored cruisers of the U. S. Auxiliary vessel for the U. S. The new French coast defense armored cruiser. The French armor-clad Le Hoche. Change in the masting of the Imperieuse and Warspite. Torpedo-gunboat Almirante Luich for the Chilian navy. New lifeboat. American torpedo-boats. New English regulation for gun salutes by vessels. Launch of Turkish war vessels. New gunboats for Portugal. Literary Notices.   E. H. C. L.

THE RAILROAD AND ENGINEERING JOURNAL.

FEBRUARY, 1890. Counterbalancing the revolving and reciprocating parts of locomotives. Radial valve gears.

A paper read before the Hull Institute of Engineers and Naval Architects, by J. K. Smith, containing a description with sketches of the varied forms of radial gear, from the earlier inventions of Hackworth up to the modern adaptations of Jay, Marshall and Morton, with a discussion on the relative merits claimed for each type.

Water-tube boilers.

An abstract of a paper read before the Institution of Civil Engineers by W. I. Thorneycroft. The author commenced by discussing the kinds of circulation in various forms of tubulous boilers, arriving at the conclusion that circulation in order to be perfect must be systematic. By circulation was meant the motion of water contained in a generator from the upper surface of the liquid down to the lowest part, and returning again to the upper surface, in contradistinction to the motion of water simply from the point where feed enters to a point in the boiler where it becomes steam. Having defined this term, all boilers were divided into classes depending on the method of circulation. The boilers of Perkins, Herreshoff, Temple and Field were compared and discussed. The failure of the boilers of the Propontis was attributed to the fact that the upper ends of the tubes were of such large diameter that instead of steam and water passing over in foam, steam alone left the tubes, and all impurities brought in with the feed gradually accumulated on the upper part of the tubes and ultimately led to their destruction. The lightness of structure and strength to resist internal pressure were next mentioned, and the advantage of the tubulous boiler in this respect fully discussed. The paper concluded with the results of some recent trials by Professor Kennedy on boilers sent out by the author's firm, both under natural and forced draught, with a full description of the method of working. At the most economical rate the evaporation reduced to standard amounted to 13.4 pounds water per pound of fuel, and the following heat table showed the way the heat of combustion was utilized :

Heat expended in heating and evaporating feed water..........86.8 per cent.
" " " raising temperature of furnace gases.........10.8 "
" lost through formation of carbonic oxide.................. 0.5 "
" " by radiation and otherwise unaccounted for.......... 1.9 "
_____
100.0 "

The high percentage of evaporation represented the efficiency of the boiler, and was simply equal to the ratio between actual evaporation and that theoretically due to perfect combustion, or 13.4 : 15.4.

. . Torpedo-boats for the Italian navy. United States naval progress.

Recent circulars from the Navy Department on the construction of machinery and hull of new vessels. Progress in ordnance and armor plates. Notice of competitive trial of armor plates to be held at the Proving Ground.

APRIL. Grüson's quick-fire guns. Boilers with corrugated fire-boxes. The English armored cruiser Impérieuse. United States naval progress.

Description and drawings of boilers and engines of cruisers Nos. 7 and 8.

Interoceanic communication by way of the American isthmus.

Historical account by Lieutenant Barroll, U. S. N., of the attempts that have been made to connect the Atlantic and Pacific oceans by a canal across the American isthmus, and a comparison between the American and the Egyptian isthmuses. · J. K. B.

TRANSACTIONS OF THE AMERICAN INSTITUTE OF MINING ENGINEERS.

The form of crater produced by exploding gunpowder in a homogeneous solid. Notes on fuel-gas. Notes on some coals in Western Canada. Concentration of low grade ores. The Davis-Colby ore-roaster. Electrical accumulators or storage batteries. The peculiar working of a blast furnace. Natural gas explorations in Eastern Ontario. J. K. B.

REVISTA DE LA UNION MILITAR, BUENOS AYRES.

MARCH, 1890. Railways, with regard to their military importance. Our military situation. Foreign notes and reviews. J. B. B.

REVUE MARITIME ET COLONIALE.

JANUARY, 1890. The scientific mission to Cape Horn (continued). The earth magnetism. Meteorologic periods. A study of the preparations and plan of operations against Sontag (Tonquin). The main features of the organization of the coast defenses of Germany. Oceanography: Topography of the sea, etc. (continued). Naval chronicles.

FEBRUARY. Reports of the prize essays of 1889 at the Academy of Sciences. A history of a fleet of the past. Maritime chronicles of Antwerp, 1804–14. Oceanography: Statics (continued). Naval chronicles: The Gibraltar station ; New regulations for firing salutes, etc.

. MARCH. Oceanography: Statics (continued). A history of a fleet of the past (continued). Notes on the diurnal variations in the direction of the trade winds. Movements of the atmosphere. Naval chronicles: The Channel squadron ; The armored ship Victoria and her guns; The armored Trafalgar, etc.   J. L.

REVUE DU CERCLE MILITAIRE.

JANUARY 12, 1890. A German criticism of the French army. Strategic army transportations. Cavalry and the smokeless powder.

JANUARY 19. The soldier's hygiene.

JANUARY 26. A proposed manual of field artillery fire. Supplying ammunition to artillery in the field.

FEBRUARY 2. Night marches and attacks. Infantry advanced posts.

FEBRUARY 9. Supplying ammunition to artillery in the field (continued). Night marches and attacks (continued).

FEBRUARY 16. Renewing the supply of artillery in the field (continued). Night marches and attacks (ended).

MARCH 2. Getting field artillery ready for war. Espionage in war.

MARCH 9. The German navy. New fire regulations in the Russian army.

MARCH 16. Movable strategic bridges.

MARCH 23 and 30, and APRIL 6. Getting field artillery ready for war (continued). The German rifle, model of 1888. The Italian army mobilization.   J. L.

RIVISTA MARITTIMA.

JANUARY, 1890. Steamships and steamship lines—an historical and descriptive sketch, by Salvatore Raineri. A military study on the defense of the maritime frontier, by G. G. New French war ships, by D. G. The compass in modern war vessels (transcription of Proceedings of Royal United Service Inst.). Study on coast-defense tactics (from the same). Description of H. M. S. Blake.

FEBRUARY. Steamships and steamship lines (continued). The Institute of Electrical Technology, by G. Bertolini. The economy of steam navigation. Study on coast-defense tactics (continued).

MARCH. Steamships and steamship lines (concluded). Résumé of Hertz experiments on electrical waves. The economy of steam navigation (continued). German coast fortifications. The submarine boats Gymnote, Goubet and Peral. J. B. B.

RIVISTA DI ARTIGLIERIA E GENIO.

JANUARY, 1890. Repeating arms (with plates), by Lieutenant-Colonel Viglezzi. The difficulties in concentrating fire of field batteries, and how to overcome them, by Captain Leser. Notes: On field shrapnels; Russian 6-inch field mortar; Double-acting fuze for same; New electrical apparatus for exploding mines.

FEBRUARY. The various uses of illuminating gas. Indirect fire of field artillery. Austrian view of the French artillery. Notes: Coefficients of elasticity and resistance of iron, steel, bronze, and copper; The Maxim gun; Penetration of small projectiles; Apparatus for tempering projectiles.

MARCH. Variations of trajectory in elevated regions. The Austro-Hungarian artillery. Russian field operations. Notes: The new Austrian rifle; Experiments with Nobel's smokeless powder; On firing against balloons; Artillery *vs.* cavalry. J. B. B.

REVISTA MARITIMA BRAZILEIRA.

OCTOBER AND NOVEMBER, 1889. The torpedo school of application. On naval manœuvres and working ship. The preservation of torpedoes. The whale fishery. On explosives. New lighthouses of Brazil. Foreign notes and reviews.

JANUARY AND FEBRUARY, 1890. On naval manœuvres and working ship (continued). On explosives. Foreign notes and reviews. J. B. B.

REVISTA MILITAR DE CHILE.

JANUARY, 1890. Armstrong rapid-fire guns, by Don J. C. Salvo. Instructions for target firing (small arms). First assistance to the wounded on the field of battle (trans.). Transportable turrets for field fortification (trans.).

FEBRUARY. Necessity for compulsory military service. Comparative weights of infantry equipments of different nations. Instructions for target firing (continued). First assistance to the wounded on the field of battle (continued).

MARCH. Modern artillery. Battle formations. Instructions for target firing (continued). J. B. B.

SCHOOL OF MINES QUARTERLY.

Irrigation engineering. Patent equivalents. Indian grass oils.

THE STEAMSHIP.

FEBRUARY, 1890. On the stresses produced in circular metal plates by unequal temperature. Curves of statical stability. High-speed engines for cargo boats.

MARCH. Supplementary note on Bremme's valve gear. Radial valve gears.

APRIL. Notes on the application of speed formulæ. On the loss by condensation and re-evaporation in steam cylinders. Protection of propeller shafts. Boiler furnaces.

A paper read before the N. E. Coast Institution for the purpose of eliciting opinions and discussion on the advisability of further increasing the thickness of furnace flues.

Friction of screw propeller engines.

A paper read before the Institute of Marine Engineers, and dealing with the losses due to friction on the various parts of the marine engine, from the pistons to the propeller blades; the question of rough surfaces and lubricants ; the form of steam valves ; the different kinds of metal used in bearing bushes, and the materials used for propeller blades.

The increase in the British navy.   J. K. B.

THE STEVENS INDICATOR.

VOLUME VII, No. 1. The transmission of power by wire rope. On the most economical engine for small power. Testing machines. Notes on Rankine's treatment of chimney draught. Design for a hot-air heating apparatus. Stevens Institute course in experimental mechanics.   •   J. K. B.

TRANSACTIONS OF THE AMERICAN SOCIETY OF CIVIL ENGINEERS.

NOVEMBER, 1889. Experiments relating to the hydraulics of fire streams.

An investigation in detail of the scientific principles governing the customary means of delivering water for extinguishing fires. The experiments described, in completeness and accuracy, are about the most reliable yet undertaken. The subjects investigated were : the coefficients of discharge of nozzles of various forms, loss of pressure by friction in fire hose, effects of curves in line of hose upon loss of pressure, effect of reduction of area at coupling, heights and distance reached by jets of water under various pressures, influence of size of jet upon height attained with a given pressure, and efficiency of different kinds of nozzles.   J. K. B.

TRANSACTIONS OF THE NORTH OF ENGLAND INSTITUTE OF MINING AND MECHANICAL ENGINEERS.

JANUARY, 1890. Coal mining at Warbora, East Indies.
     J. K. B.

INSTITUTION OF MECHANICAL ENGINEERS.

1889, No. 3. Description of the Eiffel tower. The ether-pressure theory of thermo-dynamics applied to steam. The rationalization of Regnault's experiments on steam. On gas engines, with a description of the simpler engines.   J. K. B.

UNITED SERVICES GAZETTE.

FEBRUARY 8, 1890. Range-finding; its destined effect on tactics. Modern military rifles and how to use them.

. FEBRUARY 15. The ship's chronometer; its history and development. The Barracouta explosion. The future American navy.

. FEBRUARY 22. Lessons to be learnt from naval manœuvres. The defense of ports and coaling stations. Coast defense. The health of the navy. The naval uniform question. Malta harbors.

MARCH 1. Lessons to be learnt from the naval manœuvres. Gunnery trials of the Trafalgar. Naval medical service. The health of the navy.

MARCH 8. · The Royal United Service Institution. Exit masts, yards and sails. The health of the navy—III.

MARCH 15. The navy estimates. Lessons from the naval manœuvres. Sea chests. Naval amenities.

MARCH 22. The navy estimates. Report of the Army and Navy Commission. The Coast Guard. The navy estimates.

MARCH 29. Institution of Naval Architects. The navy estimates. The administration of the services. Naval amenities—II.

APRIL 5. The navy estimates—IV. The administration of the services. Great guns. Weight of machinery.

APRIL 12. Arrival of the Calliope. Naval notes. Naval amenities.

APRIL 19. Notes on recent naval manœuvres. Bray's Gunnery Instructor. The Barracouta disaster.

APRIL 26. Notes on the defense of a modern fortress. Messenger swallows. Naval amenities—IV.

MAY 3. The Duke of Edinburgh and the missions to seamen. Naval manœuvres at Hong Kong. The Coast Guard. The defense of Australia—I.

MAY 10. Salve Calliope!—an epic of March 16, 1889. Engineer Corps of the United States navy. The defense of Australia—II. The Coast Guard. Sight tests for volunteers. Actuarial calculations.

MAY 17. Gun trials of the Howe. Stern-wheel gunboats for the Zambesi. The Coast Guard. War cruisers.

MAY 24. Forced draught.   R. W.

LE YACHT.

JANUARY 11. Editorial on the promotion of lieutenants by selections for 1890. The trial of Le Gymnote. The use of life-buoys. The increase in the British navy for 1889–90.

JANUARY 18. Editorial on the Newfoundland and Miquelon

fisheries. The use of life-buoys. The International Marine Conference.

JANUARY 25. Editorial on the large battle-ships. Notes from foreign ship-yards. Review of the merchant marine. Description and cut of the cruiser Le Hoche.

FEBRUARY 1. The International Marine Conference. First-class Yarrow torpedo-boats; cut of the disposition of the ejecting tubes. Review of the merchant marine.

FEBRUARY 8. New vessels appropriated for. Budget of 1891. Rules of the road proposed by the International Marine Conference. Notes from foreign ship-yards. Review of the merchant marine.

FEBRUARY 15. Editorial on the proposed increase of the United States navy. Description and plans of a petroleum torpedo-boat. Notes from foreign ship-yards. Review of the merchant marine.

FEBRUARY 22. Description of the twin-screw steamer Normannia of the Hamburg-American line. The accident in the fire-room of the Barracouta during her trial trip.

MARCH 1. Editorial on deep sea navigation for men-of-war as a school of instruction for officers and men. The use of flying squadrons. The necessity for "station navies" in the colonies. Article on the proposed transfer of the Bureau of Ordnance to the Torpedo Bureau. Review of the merchant marine. Cat-boats of the United States. Notes from foreign ship-yards.

MARCH 8. Editorial on smokeless powder. Description and plan of the San Francisco. The Italian torpedo-ram Fieramosca; description and cut.

MARCH 15. Proof firing of the Trafalgar's guns ; cut of the ship and arrangements of the rapid-firing guns. Review of the merchant marine.

MARCH 22. Editorial on the naval appropriation for 1890-91. Cruisers of Great Britain for the Australian squadron. Review of the merchant marine. Notes from foreign ship-yards.

MARCH 29. Editorial on the appropriations for the British navy for 1890-91. New vessels, types, etc. Notes from foreign ship-yards. Sketch of the Italian torpedo-boat Aguila.

APRIL 1. Editorial on battle-ships. Review of the merchant marine. The accident to the City of Paris.   A. C. B.

THE PROCEEDINGS

OF THE

UNITED STATES NAVAL INSTITUTE.

Vol. XVI., No. 4.   1890.   Whole No. 55.

U. S. NAVAL INSTITUTE, ANNAPOLIS, MD.

THE PROTECTION OF THE HULLS OF VESSELS BY LACQUER.

BY LIEUTENANT J. B. MURDOCK, U. S. N.

Among the many problems arising from the use of steel in naval construction, none is more important than that of protecting the under-water body of a vessel from the corrosive action of sea-water. The problem is a modification of the old one of protecting the bottoms of wooden or iron vessels, as steel is much more liable to corrosion under water than iron, and steel vessels, being generally constructed of much lighter scantling, are proportionally more injured by an equal amount of chemical action. So far as the purposes of construction are concerned, no material equals steel; but if steel ships cannot be protected against the destructive action of the medium in which they must exist, the question assumes another phase, and economical results might be obtained by reverting to iron or wood. As, however, the use of steel enables the naval architect to obtain results in speed and strength and lightness of hull which are otherwise unobtainable, this reversion is practically out of the question. Steel must be used and protected.

In General Information Series No. VII., Lieutenant Schroeder gives a most complete and interesting résumé of the whole question of protection of the hulls of vessels.   He refers on page 277 to the process of lacquering, which has been tried on several of the vessels of the Japanese navy, but without giving any details.   It has been my good fortune while on the Asiatic Station to meet Mr. Hotta, and to receive from him quite full information as to his process of lacquering ; and I have also had an opportunity, through the courtesy of the officials of the Yokosuka dockyard, of inspecting two of the vessels of the Japanese navy while in dock for the purpose of receiving or renewing their lacquer coats.   It was thought that the information thus obtained might be of interest to the members of the Naval Institute, as having an important bearing on one of the most prominent professional problems of the day.

Nearly every book on Japan refers in greater or less detail to the subject of lacquer, but until the substance was analyzed only little was really known of it.   It may not be amiss in the consideration of our subject to examine somewhat in detail into the constitution and method of preparation of lacquer, using for the purpose such published accounts as have appeared in the proceedings of scientific societies, or in the works of residents of Japan who had carefully investigated the subject.   The lacquer tree is indigenous to Japan, and is also found in Corea, China, and the countries of Farther India.   In Japan it is systematically cultivated.   New plantations are being laid out in Japan and will commence to yield in five years, and as the whole coast country of Asia is practically available for the cultivation of the tree, it is safe to say that a great demand, such as might be caused by an extensive application of lacquer to steel or iron ships, would create its own supply.

Lacquer is obtained from the tree by making incisions in the bark.   Usually several cuts are made approximately parallel to one another and at various points on the circumference.   The lacquer exudes from these incisions in the form of a thick grayish juice, and is gathered by a wooden spatula.   As already stated, the trees commence to yield when five years old, and yield for ten or fifteen years.   An inferior lacquer is sometimes obtained by grinding the small branches and twigs of the tree, but nearly all that is used is obtained from incisions in the bark.   It is purified by stirring in a tub with a wooden spade, by which process the excess of moisture is given off and the consistency slightly increased.   The pure commercial lacquer has a specific gravity of about 1.002 at 20° C., is of a grayish white color and dextrinous consistency, and possesses a characteristic sweet odor.   In contact with the air it darkens and hardens, forming a film that protects the lacquer underneath. The ordinary method of application to any surface is by taking a small quantity on a wooden spatula and placing it on the surface to be lacquered, working it down into a thin uniform coat by repeated strokes with a flat camel's-hair brush, the strokes being made in different directions, but in ordinary work most commonly at right angles. The lacquer rapidly darkens to a dark brown, and afterwards dries, forming a lustrous coat. Colors are given to the lacquer coat when desirable by mixing metallic body pigments such as vermilion, cinnabar, ochre, or orpiment.

Lacquer has been frequently analyzed, with somewhat varying results. The analyses given below were made under the direction of Prof. H. Yoshida, Asst. Professor of Chemistry in the Imperial University at Tokio, and recognized as one of the best authorities on lacquer in Japan. They are from a paper contributed by Prof. Yoshida to the Proceedings of the Royal Society of Edinburgh, on the Chemistry of Japanese Lacquer.

The constituents of pure lacquer are found to be a resinous acid, gum arabic, water, and a nitrogenous residue. The method of analysis adopted was to extract the resin from the lacquer by treating it with absolute alcohol, evaporating the solution and drying at 105° to 110° C. The residue was boiled with water and the extract evaporated on a water-bath, giving the amount of gum. The residue from the water solution consists of a coagulated nitrogenous substance and small quantities of coloring matter. The sum of the percentages of the above subtracted from 100 gives the amount of water and volatile matter.

The percentage composition of pure lacquer was obtained from a sample specially collected under official supervision in the province of Yoshino, celebrated for producing the best lacquer in Japan. It yielded the following results:

Soluble in alcohol (urushic acid), . . .	85.15
Gum arabic, ...	3.15
Nitrogenous matter, ...	2.28
Water and volatile matter, ...	9.42
	100.00

Other unadulterated samples gave the following: Number 1 was from Hittachi, 2 from Sagami, 3 from Yechigo, 4 from Sagami, and 5 and 6 commercial lacquers, origin unknown, bought in Tokio.

	1.	2.	3.	4.	5.	6.
Urushic acid, . .	64.62	68.83	66.92	80.00	64.07	58.24
Gum arabic, . .	5.56	5.02	4.75	4.69	6.05	6.32
Nitrogenous matter, .	2.10	2.01	1.72	3.31	3.43	2.27
Oil, ...	0.09	0.06	0.06	?	0.23	?
Water and volatile matter,	27.63	24.08	26.55	12.00	26.22	33.17
	100.00	100.00	100.00	100.00	100.00	100.00

The quality of the lacquer is determined by the amount of urushic acid. 5 and 6 are fair commercial lacquers. The common adulterations are water and oil. An excess of the former can be expelled by stirring the lacquer. Paint oils are frequently added to the lacquer to increase the transparency and covering power, and may be used up to 20 or 25 per cent without impairing the drying power of the lacquer, although the power of resisting chemical agents, the most valuable property of lacquer, is much reduced.

The principal and by far the most important constituent of the juice is the *urushic acid* obtained from the alcoholic solution. It is a pasty substance of a dark brown color, of specific gravity 0.9851 at 23° C., and having the characteristic smell of the juice. It is quite insoluble in water, and is unacted upon either in air or oxygen, but is soluble in benzene, ether, carbon bisulphide, and chloroform. It is permanent up to 160° C., but above 200° decomposes slowly with carbonization. It is very poisonous, and produces most disagreeable itching when applied to the skin, causing what is commonly known as lacquer poisoning. It is probable that this poisoning is also partly due to a volatile substance given off in the drying of the alcohol solution. When produced by the urushic acid, the best remedy is a strong solution of acetate of lead, which forms a urushiate of lead. This is, when formed from solution, a grayish flocculent precipitate.

Urushic acid dried at 100° C. was analyzed by Mr. Hiraga, and the analysis communicated by Prof. Roscoe[1] to the Manchester Literary and Philosophical Society, as follows:

	I.	II.	Mean.	Required by theory for $C_{14}H_{18}O_2$.
Carbon, . .	77.09	77.01	77.05	77.06
Hydrogen, . .	9.28	8.75	9.01	8.28
Oxygen, . .	13.63	14.24	13.94	14.66
	100.00	100.00	100.00	100.00

[1] Memoirs, Manchester Literary and Philosophical Society, 3d Ser., Vol. VII., p. 249.

. Prof. Yoshida examined many of the metallic salts of urushic acid. They are readily obtained from the alcoholic solution. The iron salts are quite complex and numerous. A small quantity of one of these salts added to ordinary lacquer produces the black lacquer of commerce. Long-continued action of strong hydrochloric acid on urushic acid produced an isomeric form, a β-urushic acid. Chlorine and bromine produced series of substitution products, as did also nitric acid. Many other interesting chemical compounds were obtained, but their consideration is foreign to the present subject.

The *gum* in the lacquer juice seems to be in the form of gum arabic. An examination gave the following results:

		Arabic Acid ($C_{12}H_{22}O_{11}$).
Carbon,	42.47	42.11
Hydrogen,	6.40	6.43
Oxygen,	51.13	51.46
	100.00	100.00

The physical properties of the gum are those of gum arabic. It exists in the juice in the form of an aqueous solution, and even in hardened lacquer the gum can be extracted by treating it for a long time with water. As it is not therefore an essential part of the varnish, its function appears to be to keep the different constituents of the juice in a state of uniform distribution, and also to assist the adhering power of the lacquer when applied to a surface.

The most important practical question connected with lacquer is that of the drying. Prof. Yoshida examined this most carefully, and his experiments appear to be convincing. It was found, as already stated, that the urushic acid was permanent in air, nor did a mixture of the acid and the gum in the proportions in which they exist in the juice show any tendency to harden. The hardening was found to cccur only in the presence of the nitrogenous matter that remained after the urushic acid and gum arabic had been extracted from the juice by absolute alcohol and boiling water. The residue thus obtained is a coagulated albuminous substance, having no action whatever on urushic acid. If, however, the gum is extracted by cold water, the residue shows very marked diastatic properties. Its analysis is as follows:

	I.	II.	Mean.
Carbon,	59.52	59.72	59.62
Hydrogen,	7.62	lost	7.62
Nitrogen,	5.48	...	5.48
Oxygen,	26.18		26.08
Ash,	1.20	...	1.20
	100.00	...	100.00

The substance is slightly soluble in water, and in sodium chloride and weak alkaline solutions. It does not act on sugar solution or gelatinized starch.

Previous experiments having shown that lacquer deprived of this nitrogenous matter would not dry, two series of experiments were made to determine the practical conditions under which the diastatic action of the albuminoid was best carried on. In the first, a small quantity of pure lacquer juice was heated on a water-bath, and the heated juice was then laid in a thin layer on a glass plate and put in a box, the air of which was kept moist. The temperature of the box was kept at 20° C.

1. The ordinary juice dried at 20° C. in 3½ hours in air, and in somewhat less than 2 hours in moist oxygen.

2. The juice heated to 30° C. dried in 4 hours.

3.   "   "   "   40°–43° '   4½ "

4.   "   "   "   50° C.   ' time not known.

5.   "   "   "   55°–59° C. dried in 24 hours.

6.   "   "   "   above 60° C. did not dry.

In juice containing a larger proportion of the albuminoid and water the upper limit has gone as high as 73° C., beyond which no drying took place. The lower limit has been found to be 0°–2° C., but this loss of activity is only temporary, the lacquer drying readily as the temperature rises. The upper limit being that of the coagulation of the albuminoid, the diastatic action ceases completely.

The second series of experiments was designed to illustrate the nature of the chemical action involved in drying. Samples of juice were put under bell glasses containing different gases, at the temperature of 13°–15° C., with the following results:

1. In dry air the sample did not dry.

2. In moist air the sample dried in 4 hours.

3. In dry oxygen the sample did not dry.

4. In moist oxygen the sample dried in 2½ hours.

5. In dry hydrogen the sample did not dry.

6. In moist hydrogen the sample dried imperfectly in a day and a half.

7. In dry carbonic acid the sample did not dry.

8. In moist carbonic acid the sample dried in two days.

9. In dry nitrogen the sample did not dry.

10. In moist nitrogen the sample dried imperfectly in a day and a half.

More careful experiments made in hydrogen and carbonic acid in eudiometers standing over mercury showed that the lacquer would not dry.

As the result of these experiments the inference is legitimately drawn that the presence of oxygen and moisture is necessary for the drying of lacquer, and that the action is best carried on at a temperature of about 20° C., ceasing at the limits of 0° and 70°.

In order to determine the exact nature of the chemical action involved in the drying, a specimen of pure juice was taken, rapidly heated over the water-bath so as to coagulate the albuminoid, and the heating continued long enough to expel all water. Analysis then gave the following:

Carbon,	75.54
Hydrogen,	8.97
Nitrogen,	0.11
Ash,	0.21
Oxygen,	15.17
	100.00

Lacquer naturally dried was heated to 100° C. and analyzed:

Carbon,	70.87
Hydrogen,	8.225
Nitrogen,	0.092
Ash,	0.032
Oxygen,	20.781
	100.00

The general action in drying is, therefore, that the lacquer takes up oxygen, and Prof. Yoshida states that the difference between the analyses is accounted for by the supposition that each molecule of urushic acid, $C_{14}H_{18}O_2$, has taken up one atom of oxygen and changed into *oxy-urushic acid*, $C_{14}H_{18}O_3$.   To test further this supposition, urushic acid was repeatedly subjected to the action of strong chromic acid containing an excess of sulphuric acid.   The product was cohobated with alcohol to dissolve the unattacked urushic acid, and dried at 106° C.   The result was a brown powder.   The mean of four analyses of this substance as compared with the theoretical composition of oxy-urushic acid is as follows:

	Mean.	$C_{14}H_{18}O_3$.
Carbon,	71.61	71.79
Hydrogen,	7.94	7.69
Oxygen,	20.45	20.52
	100.00	100.00

This oxy-urushic acid is a very stable substance, being unacted upon by any reagent except hot fuming nitric acid, which gradually changes it into a yellowish spongy substance, $C_9H_{18}(COOH)_4$. It is to the presence of this oxy-urushic acid in hardened lacquer that its strength and durability are due.

Prof. Yoshida sums up the chemistry of lacquer in the following words:

1. "Lacquer juice consists essentially of four substances; viz., urushic acid, gum, water, and nitrogenous matter.

2. "The main constituent, urushic acid, is a stable acid, capable of forming many well-defined salts and derivative compounds.

3. "The gum is probably identical with gum arabic.

4. "The nitrogenous matter has a composition allied to albumen, with much less nitrogen, and has a peculiar diastatic property.

5. "The hardening of the juice is due to the oxidation of urushic acid ($C_{14}H_{18}O_2$), into oxy-urushic acid ($C_{14}H_{18}O_3$), which is effected by the aid of the nitrogenous matter in presence of oxygen and moisture.

6. "The hardening can only take place within narrow limits of temperature, viz., between 0° to 2° C., on one side, and the temperature of the coagulation of the albuminoid (60° to 70° C.) on the other.

7. "The gum is essential in keeping up the emulsion and uniform distribution of the constituents; but in the hardened lacquer it is injurious, causing blisters on newly lacquered ware when treated with water for a length of time.

8. "Any quality can be conveniently given to the juice by the

addition of pure urushic acid, which brings down the relative amounts of gum and nitrogenous matter. The lacquer becomes thus more transparent and gains in strength.

9. "The mixture of 20 to 25 per cent of drying oil with the juice does not much impair the essential quality of the lacquer."

THE LACQUERING OF SHIPS.

The idea of lacquering iron and steel vessels as a protection against the action of sea-water was suggested to Mr. Hotta, a lacquer manufacturer of Tokio, by the observation that pieces of old lacquer recovered from the sea showed but little action, the lacquer being practically unattacked. As the Japanese were then purchasing iron and steel ships from abroad, and were encountering the same difficulties that were met with elsewhere in protecting the metal, experiments were made on special test-plates, which were immersed in sea-water for considerable periods, generally at the Yokosuka dockyard. The first results obtained were not fully satisfactory, but were very encouraging, and the tests were continued, varying slightly the composition of the lacquer, or adding chemicals to assist in obtaining the desired results. In June, 1886, a practical test was made by lacquering about 1200 feet of the bottom of the Fuso-Kan, using the preparation of lacquer that at that time had given the best results. The ship was docked again in September, 1887, and the condition of the lacquered portion was so satisfactory that the Admiralty gave an order to lacquer the whole bottom. In December, 1888, the ship was again docked, but the lacquer coat was found to be so good that no repairs were made. In June, 1889, the ship was again docked, the lacquer being still satisfactory. In each case anti-fouling paint was applied over the lacquer. The Fuso was docked once more in April, 1890, and although the lacquer covering was almost perfect, it was for some unknown reason all removed by scraping, and the bottom was painted.

Many other vessels of the Japanese Navy have since been lacquered, a list being appended. Experimentation has been going on continually. The work is all done by Messrs. Hotta & Co., they holding a monopoly under the laws of Japan, practically the equivalent of an American patent. Not content with merely protecting the metal against corrosion, the contractors have endeavored to meet all the requirements of the case by providing an anti-fouling lacquer preparation, as well as an anti-corrosive, the use of metallic anti-fouling paints over the lacquer having been found to be injurious, the urushic acid of the lacquer sometimes attacking the metallic base of the paint, resulting in the practical destruction of the useful qualities of both. This preparation was developed experimentally, and test-plates coated with both protective and anti-fouling lacquers having given most satisfactory results immersed in sea-water at Yokosuka for eighteen months, the Japanese Admiralty ordered the lacquering of the new despatch-vessel Yaeyama with both kinds of lacquer. The work was performed in July, 1890, and the result will be watched with interest, as the test-plates remained perfectly clean; and if the same protection is afforded to the Yaeyama under the ordinary conditions of service, the anti-fouling lacquer will have vindicated its claim to be the equal if not the superior of any similar composition known.

The protective or anti-corrosive lacquer is mainly lacquer, small quantities of some inert minerals like mica or kaolin being added to increase the covering power and body of the preparation. The composition of the different coats differs somewhat, that applied directly to the skin of the ship containing the largest proportion of lacquer.

In the first experiments special pains were taken to have the plates perfectly clean, the metal being washed off with acid in order to remove all oxides, but this process was soon discontinued, as the lacquer coats when applied to the clean iron were found to blister considerably. In later experiments the steel was merely brushed clean, removing all non-adherent oxides or films, the lacquer being applied over all adherent substances. An important point in which lacquer differs from ordinary protective compounds is in its insulating power against galvanic action. It is well known that if a steel plate having magnetic oxide of iron on its surface be exposed to sea-water, a strong galvanic couple is set up between the oxide and the steel, underneath the protective covering, and pitting of the metal results. With lacquer the case is different. Lacquer has no chemical action on the magnetic oxide; and if the plate is *dry* when it is applied, galvanic action is prevented by the waterproof and insulating properties of the lacquer coat. The exclusion of water prevents any action whatever, and the best results with the test-plates have been obtained on plates on which the presence of the magnetic oxide was ignored. The removal of this oxide, which is ordinarily considered necessary, is therefore avoided.

The method of lacquering is as follows : The ship is docked and the bottom carefully scraped of all yellow rust, old paint, or other matter that is not strongly adherent. If old paint adheres firmly, it is considered unnecessary to remove it. The bottom of the ship is then shut in by screens of old canvas suspended from just above the water-line to the bottom of the dock. In winter, stoves or other heating apparatus are placed inside this screen to raise the temperature and thus facilitate the drying of the lacquer. It has been proposed to allow exhaust steam to escape inside the canvas screen to secure the requisite warmth and moisture for the drying; but although this practice is common in lacquer manufactories, it has not yet been tried on ships. In summer, the screen around the bottom is necessary in order to screen the metal from the direct rays of the sun, which might raise the temperature to such a point as to impair the lacquer by a partial coagulation of the albumen. With the screen in place, the lacquering can be carried on in almost any weather. When everything is in readiness, the first coat of protective lacquer is applied, and worked down into a smooth uniform coating by a soft brush, as already described. One man can lacquer about 500 square feet, one coat, in eight hours. The time of drying of the first coat, which, as already stated, is almost pure lacquer, may vary from three or four hours to one day, according to the temperature and moisture of the air. In cold weather the drying process is tedious.

As soon as the first coat is dry the second is applied. This may contain mica or kaolin in small quantities, as also the outer protective coats. Five protective coats have generally been applied to the Japanese men-of-war, but a smaller number may be used when the anti-fouling lacquer is applied over them. The inner coat of the anti-fouling preparation is principally lacquer, the amount of poisonous mercury salt contained in the preparation increasing with each coat. The only ship that has thus far received both preparations, the Yaeyama, has four coats of the protective and three of the anti-fouling.

The number of coats considered necessary is at least three of the protective and the same number of the anti-fouling lacquer. Under favorable circumstances the ship would not be in dock more than six days, but ten would be more probable. The cost of the lacquering is stated by the contractors as five yens for thirty-six square feet, about thirteen cents U. S. gold per square foot. If the work were to be carried on more generally, the expense could be reduced, as a permanent corps of employes have to be maintained, although frequently out of work. Work has been done at much less than the above rates. The chief, in fact the only serious, objection to the use of lacquer is the expense. If dockage is cheap, this is not excessive, but the long time involved for the proper drying of the lacquer coats renders the operation a very costly one when the dock charges are high. The vital question is whether the protection afforded is worth the expense. If absolutely complete, preventing all deterioration or fouling of the hull, it would be economy to pay the highest charges in order to have the vessel always sound and in condition for service. It is just becoming known outside of professional circles that steel vessels are expensive, and that more money may be spent in excessive coal bills in trying to force a foul hull through the water than would be expended with proper economy in the frequent dockings necessary for keeping the hull clean.

It cannot be said with certainty that the use of lacquer is more expensive than the ordinary methods of protection. This depends very largely, of course, on the dry-dock charges. The contractors claim that one thorough lacquering of a vessel's bottom will keep it clean and protect the metal for three years. The expense of this operation may be figured up as follows:

Taking the Charleston, for instance, we have:
Cost of cleaning and lacquering 20,000 sq. ft. at 13 cents, $2600
Docking and nine "lay days," 4300
$6900

For painting and cleaning we have:
Docking and one "lay day," $1400
Painting, etc., (estimated), ... 600
$2000

Allowing the claim of the contractors that one lacquering is sufficient for three years, and taking the common estimate that a steel ship should be docked every six months at least, the total cost for *three years* is:

Lacquering, ... $6,900
Painting, ... 12,000

These prices are for Yokosuka dockyard. Accepting the data in Naval Constructor Hichborn's article on the "Sheathing of Ships"

in relation to the work on the Chicago in government docks, we
have the cost of one docking and painting at New York as follows:

Docking, ... $400
Painting, ... 1000
$1400

Assuming that the cost of lacquering in the United States would
be twice that in Japan, we have:

Docking, ... $400
Lacquering 21,000 sq. ft. at 26 cents, ... 5460
$5860

Taking the cost as before for three years, we have:

Painting, ... $1400 × 6 = $8400
Lacquering, ... 5860

It may be questioned whether the lacquer will last three years,
and it is also possible that the painted ship would need docking
oftener than once in six months.   In the absence of data a careful
estimate is impossible, but enough has been shown to render the
statement probable that protection by lacquer is not in the *long run*
expensive.

The experience of the Japanese navy must be largely relied on,
and the unanimous testimony of all the naval officers whom I have
met is that lacquer affords excellent protection to the hull, but is
expensive.   It is noticeable that the work is being continued in the
Japanese navy in spite of the expense.

Through the courtesy of the officials of the Yokosuka dockyard I
was allowed to inspect the condition of the lacquered bottom of the
Takatchiho in January, 1890, the ship having been in the water since
May, 1889.   The water-line belt had been lacquered in September,
1886, and repaired in May, 1889, when the rest of the ship was
lacquered, as a result of the good condition in which the belt was
then found.   When the bottom was examined in January, 1890, it was
found that on the bilge and floor plates the lacquer was perfectly
smooth and unbroken and had afforded complete protection to the
metal.   On the sides below the water line there were numerous
small blisters, averaging about a quarter of an inch in diameter; but
these were dry inside, the lacquer coat being unbroken and the

metal underneath was bright and uncorroded. Occasionally larger blisters were found which contained water, the film of lacquer having become broken. Underneath these the metal was dull but uncorroded, and there were no signs of rust. In cases where the lacquer had been scraped off, rust cones had formed, and their position marked the number of breaks that had occurred in nine months. In the entrance, especially in the wake of the anchors and chains, the lacquer was considerably broken and the metal consequently rusted, but in no part of the hull was there any extensive corrosion or pitting, except *underneath* the lacquer, showing that it antedated the application of the lacquer in May, 1889. An interesting feature illustrating the effect of the lacquer in preventing galvanic action existed in the starboard side of the run, where some of the plates showed extensive corrosion under the lacquer, apparently the result of galvanic action between the steel and the propeller and its fittings. Here was not a single rust spot, showing that no corrosion had taken place since the application of the lacquer.

The impressions derived from the appearance of the bottom of the Takatchiho were that lacquer is a perfect protection against the action of sea-water, *so long as the coat remains unbroken.* Although much more elastic and adherent than any kind of paint can be, it is somewhat susceptible to mechanical injury, and especially so forward where the anchors and chains and the impact of floating bodies are liable to break it. Every break becomes a spot of corrosive action or pitting. As it seems impossible to prevent this injury, and as the protection afforded by the lacquer is that of the worst portion, it would seem desirable in practice to dock the ship oftener than once in three years for examination and, if necessary, for repairs to the lacquer coat. If this were done, the metal of the ship would suffer but little deterioration.

Another use of lacquer that has not been tried as yet is as a substitute for cement on the inside of ships and for the protection of the inner skin throughout. There can be no question that its use here would prevent all rusting, as it seems absolutely unalterable in air. It has been used with success as a substitute for galvanizing, and seems to admit of numerous applications in places where metal is to be protected against the chemical action of gases.

Messrs. Hotta & Co. are making preparations for carrying on the lacquering of ships in other countries than Japan, and it is possible that in the near future the process may become widely known.

MEN-OF-WAR LACQUERED BY MESSRS. HOTTA & CO.

Japanese.

Fuso.—June, 1886, 1224 sq. feet lacquered for trial. Sept., 1887, entire bottom lacquered. Dec., 1888, docked but no repairs made. June, 1889, slight repairs made. March, 1890; lacquer scraped off.

Riujo.—April, 1888, armor shelf lacquered. A wooden ship copper sheathed; armor belt much corroded.

Tsukushi.—August, 1887, entire bottom lacquered. June, 1888, docked but no repairs. Feb., 1889, slight repairs.

' Naniwa.—Sept., 1886, 5200 sq. feet (water-line belt) lacquered. May, 1888, additional surface lacquered. Feb., 1889, entire bottom lacquered.

Takatchiho.—Sept., 1886, water-line belt lacquered. May, 1889, entire bottom lacquered. Jan., 1890, docked and slight repairs made.

Atago.—May, 1889, entire bottom lacquered.

Torpedo-boats 1, 2, 3, 4, 5.—April, 1887, entire bottom lacquered. These boats have all been docked since and slight repairs made as necessary.

Kotaka.—Sept., 1888, entire bottom lacquered.

The above were lacquered with the anti-corrosive preparation only, generally five coats. Metallic anti-fouling paint was used over the lacquer, and the vessels had to be docked to renew this paint.

Yaeyama.—July, 1890, entire bottom lacquered with four coats of protective and three of anti-fouling preparations.

Russian.

Battle-ship Dmitri Donskoi.—Nov., 1886, armor belt partly lacquered. Nov., 1887, armor belt wholly lacquered. Oct., 1888, lacquer on steel portions was found to be in very good condition. On the zinc sheathing it had been detached through the action of the urushic acid on the zinc.

Admiral Nakhimoff.—Aug., 1889, steel armor belt found to be very much corroded and was therefore lacquered.

NOTE BY LIEUT.-COMMANDER CLIFFORD H. WEST, U. S. N.

Messrs. Hotta and Company of Tokio, Japan, have sent two plates to the United States for trial by the U. S. Navy Department. One plate is of steel and one of iron, each four feet square, and covered with three coats of anti-corrosive and three coats of anti-fouling lacquers. These plates arrived at New York City in November, 1890. , Chief Constructor T. D. Wilson, U. S. Navy, has directed that the plates be submerged in tide water at the Navy Yard, New York, for a period of three months, when they are to be taken up, and a report made to the Bureau of Construction and Repair as to their condition.

U. S. NAVAL INSTITUTE, ANNAPOLIS, MD.

THE SYSTEM OF NAVAL TRAINING AND DISCIPLINE REQUIRED TO PROMOTE EFFICIENCY AND ATTRACT AMERICANS.

By Lieutenant W. F. Fullam, U. S. Navy.

The abandonment of sail power in ships of war—even in cruising ships; the introduction of many complicated weapons that require great care and skill for their effective use; the probability that naval battles will be shorter in the future than in the past and blunders far more fatal—these and many other facts emphasize the importance of a change in the training of men-of-wars-men. The modern ship has become pre-eminently a floating battery, as seen in the change of type from the Hartford to the Baltimore. It follows, logically, that the men to man such ships should be, pre-eminently, naval gunners—not sailors of the old school. All other than military elements have been reduced to a minimum in naval warfare and in naval architecture, and it is consistent with such a transformation that the training of the personnel should be more military, if efficiency in war is to be the aim of a navy.

Modern ships are building for the United States Navy, to be armed with the most improved weapons. The efficiency of these ships in battle will depend upon the organization and training of the personnel. A gun or a torpedo will do little harm in unskilled and unpatriotic hands. Powerful ships will not win with crews composed largely of badly trained foreigners. It is essential, therefore, that the personnel of the navy should receive as much attention as the material of war during this era of change, and it may be well to consider the service to-day, particularly the enlisted man as we find him on board ship—his habits, character, and training—with a view to improvement in organization and esprit.

It will not be denied that far greater skill and coolness are required to point a gun or a torpedo afloat than from the shore. In fact, all the conditions of battle on board ship call for quicker action, a truer eye, and greater intelligence than is demanded in directing the same weapons on land. *It follows, therefore, that a navy should not be formed of men less skillful and less trustworthy than soldiers, but rather of a higher type if possible.* Let it be considered how far the system now in vogue in the United States Navy is designed to attract and develop such men.

Personal observation and study of the men on board the new cruisers must convince the most conservative officer that the sailor fails to meet modern requirements, and that this failure is to be attributed more to his training than to a lack of intelligence. Officers who are in daily contact with, and who are charged with the military instruction of, the sailor will naturally be the first to discover his faults: they will see that he is not attentive enough; he is too careless in little things, thoughtless in handling new and complicated weapons. It is difficult to fix his attention, to stop his talking, laughing, and trifling at drill. Much time is wasted in his instruction—months are consumed where days should suffice—and the best results are never attained. The sailor will work uncomplainingly for hours scrubbing decks and paintwork; but the divisional officer, in attempting to arouse in him a spirit of pride in military matters, finds himself heavily handicapped at times by the extreme listlessness or weariness of the sailor, who appears to regard military duties as of secondary importance.

With petty officers the faults are still more glaring. They are often appointed with no view to their military capacity nor to their skill as marksmen, but simply for their ability and willingness in scouring and cleaning, and in working with sails. Few of them can drill a squad. When given the command of men, a petty officer will often blush to the roots of his hair. The men grin and think it funny—the idea of giving a petty officer any military duty! The petty officer has been made a nonentity in military matters. He has been given little authority, and is not respected as he should be by the men. He cannot always be depended upon, and will not, as a rule, report infractions of regulations, no matter how serious.

But the men and petty officers are not so much to be blamed as the system under which they have been reared. They have never been trained in a military manner. They have never been given responsibility in matters of discipline. As in the old days of sails, when activity aloft and seamanship pure and simple were the chief requisites, the strictly military duties on board ship are performed by the marines. Here we come to the real cause, direct and indirect, for the absence of military instincts among men-of-wars-men—the fact that a company of soldiers is placed on board ship to do the military duties, to watch and search and discipline the sailor. Quite naturally the latter feels relieved of all responsibility in such matters. Petty officers regard discipline as foreign to their duties, and they are perfectly excusable for taking this view of the case. A general laxity results, as might be expected. To get ahead of the marines, to circumvent them and deceive the officers who direct them, is regarded as perfectly justifiable. A policy of espionage tends to bring out vicious qualities and to stifle the better impulses in all men—even in sailors.

Nothing could be more harmful to the sailor than the presence of the marine guard afloat, because it prevents the development of a military spirit and deprives the sailor of the duties and responsibilities that cultivate the qualities we most require in these days— exactness, care and trustworthiness.

Moreover, the effect upon the sailor is discouraging and debasing. He is searched at the gangway like a pickpocket and shown in a dozen different ways that he is neither respected nor trusted. The good as well as the bad, the petty officers among them, are frequently subjected to humiliating treatment—treatment that has the inevitable tendency to drive good men out of the service and to keep those in who have little self-respect. No man—not even a sailor—is made trustworthy if he is never trusted, nor respectable if he is never respected. A system of discipline ignoring such a cardinal principle is responsible for the low standard in the navy, and for the fact that the tendencies and habits of the men forward of the mast are not what they should be.

In discussing this question of the employment of the marines afloat it is to be understood that the efficiency of the marine as such is not attacked. It is simply a question of the moral effect upon the sailor of placing a guard over him *and teaching the officers not to trust him.* Aside from their duties as police and sentries, there is no necessity whatever for retaining marines afloat. A special corps of sharpshooters is no longer necessary, because all sailors should be taught to shoot as well as marines, and it is reported that the latest target returns at the Navy Department show that the sailor is not inferior to the marine with the rifle, notwithstanding the fact that the marine would naturally strive to excel in this one particular. And considering the type of man that is needed under the conditions of modern warfare afloat, there is no more reason for having a special police force on board ship than there is for a similar force in every regiment of soldiers. The man-of-wars-man of to-day should be as competent to perform all military duties afloat as is the soldier to do the same duties on shore. It was after a careful consideration of modern requirements that the Organization Board, forced logically to the conclusion that marines were no longer needed on board ship, recommended by an almost unanimous vote that they be withdrawn from service afloat.

The presence of the marine on board ship degrades the whole service in that it deprives the sailor of a military training, reduces the petty officer to a nonentity, and thus throws many trifling duties and burdens upon the commissioned officer. The latter must follow up the afterguard-sweeper to see that he gets the sand out of the corners, and superintend everything day and night, trusting nothing to the petty officer. His time and energies are wasted in doing a petty officer's duties. The petty officer, relieved of his proper functions, becomes one among the men, and the latter respect and obey officers and petty officers accordingly.

The complaint is frequently made that the sailor is listless and indifferent and that he shows a deplorable lack of personal interest and pride in the service. The feeling between officers and men is not always what it should be. But it is very easy to account for this condition of things. When a sailor comes on board ship he is practically given to understand that there are certain duties that he is incompetent to perform ; that he cannot be trusted in many respects; that he cannot step into the presence of his commanding officer and carry his orders; that he has nothing to do with the discipline of the ship—such duties belong to the marines. If it is desirable to cultivate manliness among sailors, such a system is sure to fail because it works directly against human nature.

On the other hand, the marine is taught to regard himself as the military factor on board ship; that he is to be trusted at all times ; that he is the medium of communication with the captain; that he must hold himself above and aloof from the sailor; that (as a distinguished officer expresses it) he is "the bulwark between the cabin and the forecastle."

Note the difference: the sailor is told that he is untrustworthy; the marine is told that he must be trustworthy. The sailor is told that he must be watched and searched; the marine is told to do the watching and searching. The sailor is told that military duties and responsibilities are not his; the marine is required to be strictly military at all times, careful, thoughtful, and ever attentive to little details. The sailor is debased, and his interest and pride in the service are lessened by such teaching; the marine is brought under elevating influences, his manliness is appealed to, and his efficiency is obtained at the expense of the sailor. Apply to the marines and their non-commissioned officers the treatment accorded the sailor in matters of discipline, and the corps would be ruined in a month. Here is the explanation of the sailor's attitude, the reason for his lack of pride—there is a "bulwark" between officers and men. They have lost touch with each other, and there cannot be that mutual feeling of confidence and respect that should exist between them. When this "bulwark" is removed, and the sailor is told that there is no duty too good for him on board ship; that he is required to be trustworthy and respectable; that he must do military duties; that he will be supreme in his own sphere, and that promotion will result from the faithful discharge of military obligations—when this is done, the sailor and the petty officer will be as trustworthy as the marines. *A better class of men—more Americans—will enter the service and discipline will be greatly improved.*

As a rule the marines are not better men than the blue-jackets. The majority of them come from the same class. Scan their faces and they are no brighter. There is no element in the corps as valuable and intelligent as the apprentice, if the latter were only retained and developed. It is principally a difference in training and in clothes. A good man is more likely to be recognized in a frock coat than in a blue shirt. Why should better men enlist in the marine corps? Neither the pay nor the opportunities for promotion are so good. If trustworthy men can be found for $13 and the clothing allowance of $3 per month, it is because they are trusted and treated like men. Treat the sailors more like men and a better class will enlist. It is not so much because the peculiar duties required of sailors are objectionable. *Men will accept these conditions if they are paid, promoted, and respected for faithfulness and efficiency.*

Aside from the moral effect of refusing to trust sailors to discipline themselves, there are certain practical considerations calling for the abolition of marines on board ship. The engineers' force in a modern ship is proportionally much larger than in ships of the old type, and there are fewer hands available, as a rule, for the duties of cleaning and caring for a ship. Out of a crew of 180 men how many do we get to scrape the ship's bottom in dry dock? Not more than 50! Modern guns and steel ships require more care for their preservation, and there should be as few passengers and idlers as possible. Every man on board ship should do his fair share of the cleaning and drudgery. It is absurd that there should be forty marines on board the Boston, occupying one large compartment of the berth deck, contributing little to the cleanliness of the ship, with nothing to do but to watch the scuttle-butt, attend the commanding officer, stand by for ceremonies, and guard the men who form the bone and sinew of the service and upon whom the navy must depend for success in battle. All these duties are of a comparatively trivial nature and could be performed by twelve or sixteen men *if petty officers were utilized in matters of discipline*. If there is room for forty marines on board the Boston there is room for forty sailors in their stead, and the latter will be available for general naval duties, and will contribute far more to the efficiency of the ship in time of peace and in time of war.

To cite foreign navies as authority for the retention of marines is a poor argument and its weakness is easily exposed. Some foreign services have marines afloat and some have not; this argument works both ways. It is admitted that much can be learned by the study of foreign naval establishments, particularly in such matters as matériel, organization, and preparation for war. But there are a few problems that each nation must solve for itself, and one of these is the proper system of discipline and treatment for the personnel. A system applicable to one nation may not be to another. Under a republican government, with voluntary enlistment, men must be attracted rather than forced. Under the monarchical forms of Europe, men recognize caste and are more willing to accept subordinate positions, and enforced enlistment keeps the ranks full. But the American is not so willing to acknowledge the superiority of somebody who is NOT his superior. The American man-of-wars-man may not be willing to admit that he is less trustworthy than the marine, *and we ought not to bid for men who are willing to do so.* He will not, as a rule, accept a position offering little in the way of promotion, and we ought not to bid for men who have no ambition. He will not enter a service where, *in the name of discipline*, respect and consideration are withheld from him. Foreign wages and work-ingmen's conditions are not acceptable to Americans, and no more are foreign systems of naval pay and discipline. It is folly to bemoan the fact that the American is unlike the foreigner in this respect—the fact must be recognized. The navy cannot undertake to reform 65,000,000 people. The wise policy for the navy is to conform to the national institutions of the country it is supposed to serve, and naval officers will find in the study of the personal traits of their own people the only sound and practical principles upon which to base a system of discipline for the United States Navy.

The present condition of the personnel of the navy is eloquent proof that the system of pay, discipline, and promotion is not designed to attract and retain a sufficient number of trustworthy Americans. Not half of our sailors are American born. All lan-guages are heard on our decks. It is an international navy. Modern ships are building to be manned by this heterogeneous crowd! What pride can an officer feel in his vain attempts to arouse some national spirit and esprit in such crews? There are some excellent men in the service, to be sure—some Americans who are perfectly trustworthy, though little trusted—a few who have a natural love for the navy and are not driven out by their discouraging surround-ings. Many of the foreigners are good men also; but the navy should not be a training school or an asylum for such. The fact that the service is more pleasing to foreigners than to Americans proves that the treatment of the personnel is already in conformity with foreign rather than with American institutions. To continue such treatment, or to look abroad for a solution of the problem, will simply perpetuate existing evils and give us a *navy largely composed of men who have no pride* in the service, and foreigners, many of whom cannot speak English. An ancient system applied to modern con-ditions, a foreign mold for the native mind, will never attract the type of men needed to make the much vaunted "new navy" an efficient and patriotic service.

It is easy to explain why Americans of the proper class do not flock into the line of the navy, and why so many refuse to stay after serving for a time. It is simply because the profession of a man-of-wars-man is not regarded as being sufficiently respectable. The "Jacky" that we hear so much about is popularly regarded as a harum-scarum

creature, quite unlike other human beings, quite unworthy of trust
and responsibility. Officers have done much to give the sailor this
reputation, by their attitude toward him. If they insist that he
cannot be trusted and must be watched and disciplined by marines,
how can people at large think any better of the sailor? When he is
fortunate enough to get a billet as ship's writer or yeoman, when, in
other words, he ceases to be a sailor and becomes a non-combatant,
he dons a sack-coat, gets good pay, and is regarded as being respectable—he is no longer searched at the gangway. There is not much
inducement for a man of any ambition to remain a blue-jacket. Promotion in the line, to a petty officer's rank, means practically nothing.
Although it has always been recognized that an army cannot be efficient
without good non-commissioned officers, it appears to have been
decided in the United States Navy that a petty officer can have
nothing to do with the military discipline of a ship. In this the man-
of-war element is belittled—it affords no career. The non-combatant
and fancy elements are exalted, although the service is maintained
for purposes of war. A citizen of this country cannot be advised to
put his son into the navy with the idea that he shall follow for life
the profession of a man-of-wars-man. The young man must work
for a clerkship if he wishes an honorable and paying position afloat
in a ship of war.

In view of these facts it is not to be wondered at that 90 per cent
of the apprentices leave the service. As soon as the novelty wears
off they see that they can do better outside. They find little national
pride and spirit in crews composed so largely of foreigners, and the
petty office carries with it little that satisfies ambition. As a rule,
the best among them leave the navy and some of the poorest remain.
The tendency is toward the survival of the least fit. To blame the
apprentice school is folly. That school may not be perfect, to be
sure, but it is far more admirable than the service at large, considering
what the latter fails to do. If the apprentice is not up to the proper
standard when graduated to the cruising ship, it is the duty of the
latter to develop and improve him. But neglect of the apprentice
is not unusual, and the treatment he receives is not always such as
to create in him a feeling of personal interest in the service. Whatever may be said of the apprentices, they form the best and the most
intelligent element in the navy. That such material should not be
turned to account is, in itself, enough to condemn as impracticable a
system of discipline and training which, in its unbending and

un-American requirements, drives back to civil life an element that should be retained at all costs.

The service afloat should be, above all things, a *practical school.* Examine it as such. It has already been shown that the sailor and the petty officer are not properly developed, and that the apprentice —the best American element—is lost. Surely these are not good "practical" results. Next consider the officers. The Naval Cadet is not developed as he should be after he enters this "practical school." The system is rather one of repression as far as he is concerned. Individuality and independence are constantly discouraged. In his relations with his seniors there appears to be a fear lest he may assert himself too much. Many of the duties to which he is assigned belong properly to petty officers. The reason given is that he must be "kept busy." Suppose the graduates of West Point were "kept busy" by assignment to the duties of non-commissioned officers? The latter would be ruined, and the efficiency of the army would be imperiled. The result in the navy is similar.

There are, of course, certain subordinate duties that a cadet must perform. But he should be given as much watch duty as possible, and his manliness should be cultivated from the start. He can be 'kept busy' on board a modern ship and developed at the same time, both mentally and physically. Instead of copying the log it would be better to require him to write up the systems of ventilation and drainage, the machinery, etc. He should be required to study and report upon the watch, quarter, and station bill, which will, of course, embody all the principles of modern organization! Instead of discouraging the formation of an idea or the expression of an opinion by a cadet, it would be wise, rather than dangerous, to invite both, and to point out his errors, as would be the course with any young gentleman in civil life. Such treatment would inspire genuine respect for his seniors. Officers who have passed through the steerage, and whose memory has not failed them, should have learned how respect and hearty support are best secured from cadets, and the treatment that produces the opposite result. The Naval Cadets, as graduated from the Naval Academy to-day, are not less "practical" in all that properly belongs to their station than any other grade in the navy; and a searching examination of all officers would establish this fact, even in the case of many who pride themselves upon being "practical."

The ensign, at times, has been given no better chance to develop than the cadet. At an age that would make him eligible as a Senator or a Representative in Congress—at an age when he is best qualified mentally and physically for watch duty, and most needs contact with, and practical experience in command of men, he is deprived of such duty. It is unnecessary to dwell upon this subject. Every ensign should stand regular watch on board ship, even if he cannot be a wardroom or divisional officer. No harm will be done by increasing the number of watches. Officers will devote more time to the study of tactics and practical naval matters. *Those who have passed through the junior grades during the past decade can bear testimony to the fact that their minds have been subjected to a smothering, stunting process that has not properly prepared them to meet the demands that war would inevitably impose upon the younger officers of the navy.* Nothing could be less "practical" than this.

Nor does the divisional officer escape the professional wet-blanket. His intelligence is not fully utilized afloat. The fatigue of petty duties prevents the proper development of his mind professionally. By reason of his contact with the men on deck and at drill, he may see much that his superiors have no opportunity to observe. He may notice the beginning of bad tendencies, the cause, and the necessity for changes in organization and training. But he is seldom encouraged to report what he sees. He is often considered a growler or a disturbing element merely because he is using his brains. A few vain attempts to improve matters, by suggesting changes or reporting defects, may meet with the cold-water treatment, if not with something worse, and he soon falls back into the old groove—into what has been called the "don't care stage." Zeal is discouraged, ambition is murdered.

It is admitted that a subordinate officer, to be useful, must perform uncomplainingly the proper duties of his grade. He must not seek to tear down without building up. He must not expect that his suggestions will always meet with the approval of his seniors. On the other hand, it has always been recognized that to be successful a commander of men must utilize the intelligence, stimulate the ambition, encourage the zeal, and turn the very vanities and weaknesses of his subordinates to account. Failing in this he cannot inspire genuine respect, and he contributes little to efficiency. Military discipline, and the bearing of officers toward each other, should not be such as to forbid hearty co-operation and enthusiasm in professional matters on board ship. And yet it cannot be denied that, as a rule, officers in the junior grades of the United States Navy find themselves treated with more consideration and their intelligence more fully utilized on shore at the Bureau of Ordnance, at the Naval Academy, at the Intelligence Office, and other stations, than afloat in ships of war. *In this there is a natural tendency for brains to seek the beach and stay there.* The system that produces this result cannot be considered "practical" in a naval sense. Now as to questions of organization and routine in this "practical school" afloat: ingenuity could not well devise a worse system of messing than that now in vogue. The best marksman in a ship may be in the powder or navigator's division, and the poorest at the heaviest gun. Target practice is conducted so carelessly at times as to arouse no enthusiasm or pride among officers and men. There are few instances where a spirit of rivalry has been cultivated among the different ships of a squadron. *Owing to the fact that there is little care shown in selecting gun-captains, and no permanency in the rate, there is no certainty that the men who get the little experience at target practice in time of peace will be at the guns in time of war.* In fact, it is evident that in many cases they will not or ought not to be there. Therefore, as far as it may influence accurate shooting in battle and determine the result of a fight, much of the ammunition expended in target practice might as well be dropped overboard.

There is, or should be, one essential difference between a yacht and a man-of-war. On the latter, men should be trained to take more pride in military duties than in anything else. Routine cleaning is very important, but it should not be kept uppermost in the minds of men-of-wars-men. Suppose a regiment on shore spent most of its time in cleaning its barracks, and suppose non-commissioned officers were appointed for ability in that direction, would efficiency as a military organization be judged by such work—work that women could do as well as men? The same principle applies to the navy.

To sum up, it appears that in the service afloat, sailors, petty officers and officers are not properly developed in a military direction, and that important problems in organization and training are left unsolved in the absorbing work of carrying out a "routine." But what is the "practical" value of a routine if it fails in such particulars? It has come to pass, as a result of such a routine, that the service afloat is less "practical" than any other part of the naval establishment, and that is saying a great deal.

It will, perhaps, be admitted that the navy of each country should be organized and trained with due regard to the peculiar conditions which war would impose upon that country. In the case of the United States the navy is very small, considering the task that would fall to it in the event of war. It is a mere nucleus, and would of necessity be recruited to three or four times its present size. It would fight at great disadvantage along an extensive coast, against heavy odds, with scant material, and with comparatively few ships and guns. Every man and officer would suddenly find himself in a position of responsibility far in advance of that he holds to-day. It follows from this that men and officers should be prepared for quick promotion—ready for the duties of the next higher grade. If possible, every man should be an American, competent to lead the recruits. In short, it is undeniable that this nucleus should be the most perfectly trained of all navies.

But, as it is organized and conducted to-day, these essential conditions and requirements are to a great extent ignored in the United States Navy. Instead of preparing a man or an officer to do the duties of his own grade or the next higher, he is required in many cases to do the duties that properly belong to the next lower. Instead of encouraging officers to study and anticipate the war problems that may at any time fall to them to solve, their minds are kept on petty officers' duties. Instead of developing the petty officer and preparing him for greater responsibilities, he is practically reduced to the ranks. Instead of attracting Americans, we cling to a system that invites foreigners and repels Americans. Instead of establishing a high standard, and developing a trustworthy, self-respecting set of men, we establish a low standard, and neglect to develop the personnel in a military direction; thus in time of peace there is no proper preparation for war. It does not appear that any genius is displayed in directing the minds of men and officers. In fact, it would be difficult to devise a system better designed to unfit men and officers for the ordeal of war—the emergency for which they exist.

REFORMS IN TRAINING AND ORGANIZATION.

To prepare the personnel of the United States Navy for war, three important reforms must be sought:

1st. To attract Americans and create a true national spirit in the service afloat.

2d. To raise the standard and develop the type of man required under modern conditions in a navy which is a mere nucleus in time of peace.

3d. To improve the discipline and military efficiency by inspiring not only the officers, but the men with a feeling of personal pride and interest in the service.

To secure the first result, the passage of the Alien Bill is absolutely necessary. It is reported that there was some opposition to this bill on the ground that its passage would render it difficult to recruit the navy. But the folly of such opposition is apparent. It is important that the sailors, as well as the officers, should be citizens of the United States. If the conditions are such that Americans will not enlist in the navy, the sooner these conditions are changed the better. A system that attracts foreigners should give way immediately to one that will attract Americans. The bill should pass, even if the ships of the navy are laid up for lack of men or manned with half crews for months to come. Proper concessions to national traits and institutions will convert the navy from an international to an American service. Until these concessions are made there can be no American navy, properly speaking. It will continue to be a more or less mercenary service, inefficient in time of peace, and lacking in the loyalty and esprit essential to success in war.

QUARTERS AND COMFORTS FOR THE MEN.

An important concession must be made in the matter of quarters and the conditions of life on board ship. Americans of the proper class will not submit voluntarily to the discomforts that foreigners are compelled to accept under enforced enlistment. This fact may as well be recognized first as last. Ships like the Charleston, Philadelphia, Baltimore, Newark, San Francisco, and Yorktown are not designed to attract Americans, but rather to keep them out of the service. The uncovered gun-decks of these ships are an abomination in more respects than one. The living space for men is insufficient, and in bad weather they will, at times, be compelled to live little better than pigs in a pen. It is an unfortunate fact that some of these cruisers will be far more uncomfortable for their crews than ships of the old type. In this there is a step backward, an utter disregard of the means necessary to attract the men of the proper class for modern men-of-war's-men. Such a lack of consideration for the comfort of men will delay, if it does not prevent, the Americanizing of the navy. No more ships should be built without flush decks, and all those now finished or building should be decked over without delay. The slight additional weight may reduce the speed ⅜ to ¼ of a knot, but the gain in the efficiency of the navy by such a concession to the men will compensate ten-fold for the trifling loss in speed. In the craze for speed all other elements of efficiency should not be forgotten.

The men need more locker room on board ship. As a rule they have no proper place to keep their rain clothes and rubber boots, and they are usually compelled to lash their peacoats in their hammocks or to stow them in various corners, where they soon get dirty, wrinkled and ragged. A ship's company dressed in peacoats is a sight that gives evidence of shameful neglect. More attention must be given to many of the little things affecting the comfort of the men, if it is desirable to make them contented and fond of the service.

PAY AND PROMOTION.

To secure the second reform—"To raise the standard and develop the type of man required under modern conditions"—there must be an increase of pay, and better opportunities for promotion in the service. Proper promotion will be afforded by making the combatant petty officers the most trusted and responsible men in the discipline and organization of a ship. Permanency in ratings, and the option of retirement after a reasonable continuous service, will keep good men in the navy, ensuring them a proper reward for long and faithful service. No doubt the Board of Organization solved this problem, but their report has not been published.

It may not be necessary to materially increase the pay all along the line. If the petty officers are well paid and are not deprived of their rank except by sentence of a general court-martial, the induce-ments for good men to enter and remain in the service may be suffi-cient, because the proportion of petty officers in a ship's company is so large that a man may reasonably hope to get a petty officer's billet after a short term of service as an apprentice, or seaman. It may be well to pay seamen $30 per month; as in the Lighthouse Service. Thus a comparatively small increase in the appropriation for "Pay" may serve to give us an American navy. If the matter is properly represented, Congress will no doubt respond. It is not probable that the people will consider it good economy to spend $7,000,000 annually for an international navy, when an increase of about 10 per cent may secure a body of intelligent' Americans who will give a far better account of themselves when war comes. It is beneath the dignity of a great nation to hire foreigners to defend its flag, in order to save a few dollars. The United States can afford to employ Americans for that purpose, no matter what it costs.

ORGANIZATION AFLOAT.

The third reform—" To improve the discipline and efficiency by inspiring not only the officers, but also the men with a feeling of personal pride and interest in the service "—may be secured by changes in the organization and training on board ship.

Under a republican government, with voluntary enlistment, it is evident that the navy must compete, to a certain extent, with the trades and civil employments if good men are to be induced to enlist. It must hold out equal opportunities for an honorable career —otherwise good men will refuse to stay in the service. Good discipline must be secured by first appealing to the personal pride and interest of men, and not until such a policy fails should there be a resort to severity and force,—otherwise, foreigners, the lower elements and the slums must be depended upon for recruits. *The navy of the United States cannot be formed of good and trustworthy men until the latter, as well as the officers, are permitted to have the true interest of the service at heart, the same feeling of personal pride, the same hope of permanent and lucrative employment, and the same certainty of reward and proper treatment in return for faithful service.* There is no alternative.

To create this feeling of personal pride and interest among men-of-wars-men should be the constant aim of the service afloat. This policy, as well as the necessity for the military training of sailors, requires the immediate withdrawal of the marines from service on board ship, and the recognition of the fact that the sailor of to-day must be the equal, if not the superior, of the soldier in skill, trustworthiness, and respectability.

In briefly outlining a system that shall require, above all things, the military development of the man-of-wars-man, the blunt remark of a former messmate—" It is about time that the battery instead of paintwork and sails should form the basis of modern organization " —may be taken as a text. It is a rock upon which to build.

Imagine a modern ship in battle. Surely the intelligent spectator of a naval fight would consider the rapidity and accuracy with which the great-guns were fired as the governing element of victory. He would look to the patriotism and cool courage of the men who aim the guns as the principal force influencing the result of the fight. Napoleon's maxim—" FIRE IS EVERYTHING, *the rest is of small account*," is as true at sea as it is on shore. *And yet, in the United States Navy, this maxim is utterly ignored.* The men who pull the lockstrings and upon whom the issue of battle depends are appointed, as a rule, with no regard to military fitness. They are given less pay than the ship's bugler, or the cooks and stewards, and the same as the tailor and the barber. They are not trusted to govern themselves or others, and are regarded as inferior in military matters to the landsman who, as a marine, comes on board ship *for the first time in his life.* It is pure sophistry to attempt a defense of a system so sublimely absurd.

As the gun, the important weapon, should form the basis of modern organization, so the gun-captain should be the leading man in the discipline of a ship's company. The one principle naturally follows the other. The leading men in time of war should be the leading men in time of peace. The men who have the greatest responsibility in battle should not be relegated to a position of inferiority when the fight is over. This is the key to the problem. The petty officer must not be robbed of his just reward. *He must have the same duties in the organization and discipline of a ship that the non-commissioned officer has in the army.* This can never be while the marines remain on board ship. They are an ever present obstacle to the development of the petty officer.

In modern organization, a new rate of Gun Captain should be created, with pay of $50 per month in the beginning, and with longevity of $1 per month for each year of service. These men should be required to qualify as marksmen, as well as seamen, before receiving their rates, and they should not be disrated except by sentence of a general court-martial. They should be gun-captains in fact, responsible for the bright-work and the condition of the battery, the men being taught to obey them implicitly. They should be required to know the elements of tactics and to drill and instruct the men frequently, under the supervision of the officers.

In the organization of a ship there should be no such names as forecastlemen, afterguard, etc., but the men of each division of guns should be required to clean a certain part of the ship. " Lay aft the Fourth Division," instead of " Lay aft the Afterguard "—orders like this would emphasize the fact that the battery and military elements are always foremost in the modern ship. "Captain of No. 3 gun," instead of "Captain of the maintop," should designate a petty officer.

This rank and pay would raise the petty officers of the line to the position they deserve, for are not their duties as important as those of a ship's writer, or a yeoman, or a boilermaker? Every petty officer should be made to assume responsibility and be held accountable for the discipline and good conduct of the men in his part of the ship. He should be the man upon whom to call for assistance in military discipline, and there can be no doubt that with such a staff the officer would find his orders obeyed with greater willingness and alacrity than at present.

The blue-jacket, instead of looking to a non-combatant billet— to the position of a clerk or a yeoman—for promotion and good pay, would soon learn to regard the gun-captain as the most important petty officer in the ship, as in all reason he should be. He would recognize the fact that the man who fires an VIII-inch gun in battle, and who helps to discipline the crew, is quite as respectable as one who serves out soap and tobacco, and, very naturally, he would strive for excellence in seamanship and gunnery as the surest means of obtaining promotion to an honorable office. In short, the calling of a man-of-wars-man would be made sufficiently respectable to attract Americans, and they would not seek to get rid of the blue shirt as the only means of securing a good position in the navy, the only way to be trusted and treated like men. That many who leave the navy now—and the very men we need, too— would remain under a system like this cannot be doubted, and the navy would soon find itself raised in tone and esprit to a patriotic and efficient service. Among the yeomen and appointed officers we now find many excellent men. The rate of gun-captain would be quite as acceptable to men of that class, and other rates in connection with the battery and torpedoes, with good rank and pay, would raise the man-of-war element to its proper place afloat. After twenty years' continuous service, the man-of-wars-man should have the option of retiring on a pension.

The Board appointed to consider the question of new ratings doubtless made all the necessary provisions in this line, but their report has not been published. Since this paper was begun the writer learns that the rate of Gun-captain was proposed by the Board.

The beneficial effect upon the apprentice of such a change in the service cannot be overestimated. He would be brought under a better and more elevating influence. He would regard the petty office as one worth working for, and he would learn how to be a good petty officer—a thing that he does not learn and is not taught to-day. His energy and ambition, his better impulses, would come to the front, and decency and good conduct would receive proper encouragement. The vital importance of this part of the subject must not be forgotten. Instead of losing 90 per cent of the apprentices, a large proportion of them would remain, because the service is not without its attractions to many of them. With all its drawbacks there are undoubted charms and advantages in a naval life, and much that makes it objectionable may be removed with great improvement in discipline and efficiency. The navy can be made as attractive as any trade or employment on shore, and men will gladly choose it for life, if their manliness is recognized and cultivated, if faithfulness is rewarded.

The recruiting stations should play a very important part in the preparatory drill and discipline of the enlisted men. Barracks and strict military routine, and thorough instruction with modern arms, provided for that purpose, should be the rule. Gun-captains should be charged with much of the squad drill and instruction, to accustom the men from the start to being controlled by petty officers. Duty at recruiting stations should be regarded as being of the first importance, otherwise valuable opportunities will be lost. An excellent system is now in operation on board the U. S. R. S. Vermont. Petty officers and seamen are taught to drill the recruits, and they do this duty to the perfect satisfaction of the officers in charge. This is an example of modern naval training.

There are some who fear that to make a man military in his habits is to ruin him as a sailor. But there is no ground for such fear. Naval officers should not be thrown into a panic by such a ghost. Why should military instincts and training make a man less efficient as a sailor? Is it necessarily the case with an officer? Why should a slouchy bearing, bad habits, and general untrustworthiness in military matters be conducive to perfection in a sailor? It is not so. On the contrary, military training will make any man more efficient as a sailor—certainly as a modern man-of-wars-man, and that is what is wanted first of all. It will cause him to obey an order the first time, instead of waiting until it is three times repeated; he will obey a petty officer instead of entering into a long argument; he will be more attentive and aspiring, and for this reason it will be far easier to teach him than it is now.

The nautical training of apprentices and seamen need not be neglected. It is not proposed that the instruction in seamanship shall be less thorough. Training ships should be full-rigged as Admirals Porter and Luce declare, because the drill aloft certainly breeds qualities that are invaluable—alertness, fearlessness, and nerve—qualities that cannot be so well secured by any other training or exercise. It may be well to have a training ship on the home coast ·for seamen and landsmen, where men may be sent from the recruiting stations for instruction in things nautical, in matters·pertaining to anchors, chains, hawsers, and wire rope. The English have a ship for this purpose in the Mediterranean.

Until sails disappear entirely from the merchant marine, officers and men in the navy must certainly be taught to handle sails and sailing ships. Every cruising ship should be barque or brig rigged, with light spars and a single suit of sails. In time of peace the necessary instruction may be given to prevent complete ignorance of such matters, and the important advantage of the exercises, physically and professionally, may be secured without undue sacrifice of time. When war comes, the spars and sails may be sent ashore, leaving only the lower masts standing with their military tops. This position regarding sails is a happy mean which will secure, in time of war, all the advantages claimed by the advocates of mastless cruisers, and, in time of peace, all the essential advantages claimed for full-rigged ships. It simply surrenders the idea of using sails as a motive power in time of war.

In all reason, the modern man-of-wars-man should possess, alike, many of the qualities of the sailor and of the soldier. He must know all that the marine knows and much more. To be exact and trustworthy like a soldier or marine, need not make him less skillful with sails, boats, anchors, hawsers and chains; while with guns, torpedoes and explosives he will be more expert, and more to be relied upon. Such a system will attract better men—Americans—who will take more interest in the service if they are made to feel that they form an important part of it, and who will learn seamanship quite as well as the foreigner, or the devil-may-care "Jacky" of the old school.

The seafaring element need not be regarded as the only class from which to recruit modern men-of-wars-men. With proper regard for comfort, pay, and promotion, recruits may be obtained from the interior who will be quite as efficient afloat, after a little training, as officers are, who come from the same districts. As soon as it is noised abroad that the navy has become an American institution, that it seeks a self-respecting and trustworthy class of men, assuring them an honorable career, there will be no trouble in obtaining desirable recruits.

There need be no difficulty in providing for necessary guard duty in case marines are withdrawn from service on board ship. Ships in the navy have done without marines many a time, and there has been no trouble whatever. Men in the Revenue Marine, the Coast Survey, and the Lighthouse Service prove themselves worthy of trust and responsibility. But the man-of-wars-man is a very dangerous character! He must be suspected, to have good discipline! He is necessarily a lower type of man and cannot be trusted—this seems to be the theory. And yet sailors have served on shore with credit, and have performed military duties faithfully when required to do so by their officers.

Select from each division the men who are to form the guard for the day, with a certain number of petty officers to act as corporals. A line officer should be detailed to take charge of the torpedo and electric plants of each ship, and to him should fall the duties at present performed by the marine officer. The guard should receive his special attention, and he should require at all times the strictest performance of military duties. At the same time he would be available for watch duty and general service in case of emergency. May not a line officer be as efficient as a marine officer in matters of ship discipline? Does he lack the education and training required of a disciplinarian? Do his duties unfit him to decide how men shall be trained on board ship, and is he incompetent to train them?

Boatswain's-mates on both decks should exercise military authority, and perform the functions of corporals of the guard at all times, in addition to the men specially detailed for that purpose each day. They should be taught to bring men to the mast for every violation of good discipline. Why should a boatswain's-mate be taught simply to whistle and shout and stand a helpless spectator while the officer of the deck waits in vain for the corporal of the guard to come to his assistance?

Every man in charge of a compartment below decks should be taught to be a sentry in that compartment, and to report all infrac-

tions of regulations within such limits. Why should his responsibility be limited to the skillful application of scrubbing brushes, soap,
and swabs, to decks and paint-work?

It is absurd to say that "Jacky" cannot be made to do these
duties. If the men who now compose the crews of our ships are
never to be trusted in anything, they are not the men we want, and
there could be no better reason for condemning the system that has
brought such men into the service and made them so worthless.
Every man who will not do military duties as faithfully as a marine,
should be dishonorably discharged from the service at once, and
means should be taken to prevent his re-enlistment. As soon as it
is understood that the navy will accept none but good men, such
men will enlist. As soon as men understand that discharge from the
service will inevitably follow military inefficiency, they will do their
duty. There may be some friction at first in introducing a system
that requires men to be self-respecting and trustworthy. It could
hardly be otherwise after insisting for years that the sailor is essentially unreliable. But the service would recover from the shock
much sooner than timid people imagine. The navy will be considered desirable by the better elements when it is made too hot for
the worthless.

A brother officer has suggested that a partial change in the uniform
may influence the bearing of the man-of-wars-man. This is an
excellent idea. Perhaps it would be well to do away with the
present "mustering suit" and "flat cap," and substitute a suit and
cap like those now worn by first-class petty officers. The coat
should have a standing collar, and be fitted to the form so that the
belt could be worn outside. This uniform should be worn while on
guard duty, and upon all full-dress occasions that involve no going
aloft. The change would emphasize the fact that modern men-of-
wars-men must be soldierly, and there would be a natural tendency
for men to brace up and be less slouchy in their bearing. With the
exception of this change the present sailor dress may be retained for
general service.

The effect upon the sailor of requiring him to do military duties
will be to greatly improve his morals. If his commanding officer
will trust him as an orderly, he will prove himself worthy of the trust,
and his respect for his superiors will be increased. If petty officers
are consulted, as a marine officer consults with sergeants and
corporals in matters of discipline, the commissioned officer will find

himself supported at all times by a powerful influence in the midst of the crew. Every petty officer will bring a squad with him to the immediate execution of an order, and there will be a shuffle of feet whenever a command is given.

The standard proposed is not too high. *The higher it is made, the more surely will trustworthy Americans enlist and remain in the navy.* Among 65,000,000 of people the men can be found who will satisfy the requirements of modern men-of-wars-men, and the petty officers will be forthcoming—there are many in the service to-day— who will prove themselves as worthy of the confidence of their superiors as are the non-commissioned officers of an army.

The marine corps is needed for duty at the navy yards and shore stations. If withdrawn from service afloat, the corps would reach even a higher degree of discipline and efficiency than that for which it is justly noted to-day, because battalions would be more permanent, and instruction and drill could be made more thorough and progressive. The corps would be invaluable as a highly trained, homogeneous, and permanently organized body of infantry, ready at all times to embark and co-operate with the navy in service like that at Panama a few years ago. The education of marine officers at Annapolis fits them perfectly for service in connection with the navy. Both the marine and the sailor will be rendered more efficient by such a course.

No reflection is cast upon the marine by advocating his withdrawal from service afloat. *It is simply proposed to stop casting reflections upon the man-of-wars-man,* for it is impossible to dodge the fact that the presence of the marine is equivalent to the charge that the sailor is an inferior man. If this is so, it certainly ought not to be. As claimed in the beginning of this paper, the modern sailor should be the equal, if not the superior, of the soldier in all respects. Then why not try to make him so?

The argument that marines should be retained until we get better men is a very poor one. The best way to get good men is to raise the standard by establishing the fact that marines are not needed. *As long as the marine remains, the officer will not learn to rely upon the sailor, nor to trust and develop the petty officer.* As long as the officer insists that the marine is the only being who stands between him and anarchy on board ship, he condemns the men who should be his best friends, acknowledges that he is afraid of them, and places upon them the brand of inferiority.

The recommendation of the Board of Organization to withdraw marines from ships of war, recognizing that they stand squarely in the way of the military development of the sailor, is the most important reform affecting the personnel that has ever been proposed, the first and most important step toward improving the tone and efficiency of the service; a reform that takes the most intelligent means of attracting Americans to serve their flag afloat by abolishing a pernicious semi-convict system of discipline, and securing to petty officers and blue-jackets their proper duties and their legitimate rewards on board ship; a reform that will tend to create some national feeling in the navy; that will cultivate the manliness, utilize the intelligence, and win the hearty loyalty of 8000 men, by enabling them to feel that they may have a personal pride and interest in the service, instead of being merely its scapegoats.

And the sailor and petty officer having been assigned their proper places in the organization and routine afloat, the commissioned officer, freed from petty duties that degrade his intellect, will bend his mind and devote his energy to the work that properly belongs to an officer—in short, brains and manhood will have the same chance in the navy as in any other profession, which is not the case to-day.

DISCUSSION.

Commodore JAS. A. GREER, U. S. Navy.—I can only say that I fully agree with the views and suggestions so admirably expressed by the writer.

Lieutenant SEATON SCHROEDER, U. S. Navy.—I am glad to have the opportunity of expressing my cordial approval and endorsement of most of the views advanced by Lieutenant Fullam. There is no doubt in my mind that two of the most serious questions we have to deal with in the service are the presence of the marines on board ship, and the performance by commissioned officers of non-commissioned officers' duties.

The idea of individual responsibility is the one most difficult to inculcate among our petty officers, and the serious fact exists that in that feature they do not come up to the mark. The worst of it is that this is often thoughtlessly attributed to the supposed fact that a sailor naturally cannot be trustworthy like a marine. A remarkable conclusion indeed! Of two men taken from the same social class in civil life, the one is assigned to duty on the police force of the ship, accepts that duty and does it, and does it well, whatever may be his natural proclivities as exhibited when on liberty; the other, who may have served in many ships and faced many dangers, who may be remarkably expert with helm, oar, marlinspike, and gun, is invariably brought up to believe that in one essential particular he is lacking, viz., in trustworthiness, responsibility, ability to exert authority; and, that fact being so constantly thrust before him, he naturally accepts it, and loses much of his inherent value in consequence.

Our men can be perfectly relied upon to perform the duties they are trained to do. They are good seamen, skillful gunners, and can march, and handle a musket efficiently; but when it comes to assuming responsibility, commanding attention at drill, enforcing obedience, reporting one of a gun's crew for being drunk or out of uniform at inspection, that is not their business, and it never will be their business until they form their own police force. If one of those very men were rated master-at-arms or ship's corporal, then he would probably do his duty well, just because his shipmates are accustomed to the master-at-arms doing that work. If the marine guards were withdrawn from shipboard, and the guard duty devolved upon the several gun-divisions in rotation, they would quickly get accustomed to the new order of things. At the very first there might be some little difficulty—such as always attends any change—but that would not last long, and the good effect of the change would be felt in every feature of naval life.

In expeditions on shore in a hostile country the blue-jackets are perfectly trustworthy when on guard; why should they not be so at other times? There is certainly no reason why the system in vogue in all armies and in some navies should not be equally successful when applied to our navy. It is not at all necessary that the surveillance over a military body should be exercised by men of a different corps and uniform. I can picture to myself the astonishment that our brothers of the army would feel on being told that it had been decided to have at each post one watch of sailors to preserve discipline, search the non-commissioned officers and privates when returning from leave, etc., etc. Yet that would be no more absurd than it is to have marines performing the same duty on board ship.

The essayist is quite right, I think, in laying stress upon the question of pay. When a barber gets more money than a seaman-gunner, and a bugler more than a gun-captain, it inferentially takes away from the prestige of those positions. If the excellent recommendation be carried out of creating the rating of gun-captain, with the pay of $50 a month, we shall get good men— men from whom we can exact not only good shooting but also power of control over their arms, a personality that will impress itself among their fellows, and command prompt and cheerful obedience to their orders. The rating should not be lightly conferred. A tailor, from the constant sight of a gleaming needle, is apt to be a very good shot with a parlor rifle; but to get good work out of a six, ten, or thirteen-inch gun in action, or even in target practice, requires certain qualities not always possessed by tailors.

In the matter of supervision of routine work there is no doubt that at present the younger officers have to do much that should properly devolve upon the petty officers.. A ship's smartness, both as regards appearance and actual efficiency, is not generally kept up unless the officers give their constant personal attention to the smallest details. In some ways this is harmful to the officers, but less so than it is to the men. The cadets, on leaving the Academy, are an intelligent and highly-trained body, but must profit by contact with the men forward, I think. With regard to the effect on the men of too much supervision from aft, it is at once one of the causes and one of the results of their not being brought up to think more of themselves; and it cannot be changed until we take steps to imbue them with a higher sense of their own worth and of the power that we wish them to exert. In my opinion, one of the first steps to take to secure this, is to withdraw the marines from service afloat, and to let the men, who are naturally their equals in intelligence and their superiors in attainments, do their own guard duty, and incidentally acquire the power and habit of exercising authority when conferred upon them.

'The sailor can do everything that a marine can, and a great many things besides. He now receives just as much instruction with the rifle, has as much small-arm target practice, and is as good a shot. And there the attainments of the marine end. Substitute sailors for marines, man for man, and you would increase by just such a number the number of men on board trained to manœuvre and fight their ships and their boats, and capable of meeting all the emergencies likely to arise afloat.

For fear of being misunderstood, I would like to say that it is not a lack of respect or of admiration for our gallant marines that prompts the emission of the views I have just expressed. I consider them a splendid body of men; the reputation of the corps for steadiness and gallantry under trying circumstances is excellent, and deservedly so. But I think the ship is not the place for them now, except for transportation, like the *Infanterie de Marine* of France, for instance. They not only take up room that could be occupied by men better adapted to service on the sea, but they actually, though innocently, are a stumbling-block in the development of a thoroughly efficient naval personnel.

Ensign A. P. NIBLACK, U. S. Navy.—Lieutenant Fullam has shown up very vigorously a great many of the weak features of our present organization and methods of training and discipline, without in the least overstating the case. With regard to the withdrawal of the marines from service afloat, he has even quite overlooked a most important phase of the argument. The limited berthing space of the. new ships necessitates a resort to a most radical reduction in the number of men aboard each ship, and at the same time requires of those she can carry, largely increased intelligence and the ability of each individual to perform a wide range of more or less important duties. Now, the marine of to-day is just what he was thirty years ago, and probably will be thirty years hence, unless the fundamental idea is modified. Either the marine has got to become more of a sailor, or else he must ultimately go. Whether he goes or not, the sailor has to become more of a military element. Of course, if the marine goes, his duties still remain to be performed by some one else, and, unless that some one else is more versatile than the marine, we have gained nothing either in berthing space or in efficiency. If the every-day enlisted man cannot be trusted to perform the duties now carried on by marines on board ship, then we have the best reason for withdrawing the marines and starting in on first principles to teach and compel the men to perform such duties faithfully. If line officers throw up their hands in fear and horror when such a thing is suggested, it is a good sign that something is fundamentally wrong with their ideas of discipline. A consideration of the laws relating to the Marine Corps will show that no violence is contemplated. Sec. 1616 of the Revised Statutes says, "Marines may be detached for service on board the armed vessels of the United States, and the President may detach and appoint for service in said vessels such of the officers of said corps as he may deem necessary." As a matter of fact, the Marine Corps was organized before the Navy—hence the foregoing. Sec. 1619 says, "The Marine Corps shall be liable to do duty in the forts and garrisons of the United States, on the sea-coast, or any other duty on shore, as the President at his discretion may direct." This really opens up the proper field of usefulness for them, as it would undoubtedly add to the efficiency of our national sea-coast defenses to have them in charge of officers who have had nautical training. The marine is simply no longer necessary on board ship, and it is quite in the line of promotion that this efficient and able corps should be given duties commensurate with their capabilities. We inherited many of our naval traditions from England, and the marine guard on board ship was one of them. In the English service to-day they have two kinds of marines, the "reds" and the "blues"; but the feeling of the necessity for withdrawing them from service afloat is as strong and general as in our own navy. On many, if not all, English men-of-war the ward-room servants belong to the marine guard. The marine is now out of place aboard ship. He takes up too much room, he is not versatile enough, and his presence is the worst kind of a stumbling-block to the military training of the rest of the crew. If these reasons are not conclusive, then the only other thing is to reverse the process and train all our men to be an "improved" type of marine that can paint and coal ship, tar and rattle down, fence, swim, pull an oar, sail a boat, pass a weather-earing, steer a ship, strap a block, or do all of the thousand and one things that only a regular man-of-war's-man knows how to do, adding to it all the military training of the marine.

What we really do want in the way of improved training of our men is to have the term of enlistment made four years, and have the preliminary training of recruits carried out thoroughly on shore before they are drafted off for service afloat. A modern ship, in the first months of her commission, offers few opportunities for the training of men in the rudiments of their profession, and the more that is accomplished at the recruiting station, the better. This applies equally to men in the engineer's force.

The seaman-gunner, the seaman-artillerist, should be the keystone of our present organization, and the scheme now being so efficiently carried out on the Dale for the improvement of the seaman-gunner type is full of promise for the future. With the addition of the Alarm, with her new 6-inch B. L. R., as a gunnery-ship, and the rifle ranges at the Bellevue magazine reservation, it is contemplated that when a man qualifies hereafter he shall require excellence in marksmanship to get his certificate. It would be a wise plan to have a board organized temporarily to draw up an improved plan of instruction for seaman-gunners, with a view to uniformity in future requirements and requalification of those now in the service. The Chiefs of Bureaus of Navigation and Ordnance, the Commandant of the Washington Navy-yard, and commanding officers of Dale and Torpedo Station would certainly be able to outline and carry out a definite and uniform system of instruction and promotion for seaman-gunners. Just now it is not at all uniform.

New watch, quarter, and station bills are needed for our new types of ships. "Coal is king," and with twin screws it is idle to talk of sail-power except for storm purposes. The gun, the ammunition-whip, and the coal-bunker should be the basis of our new organization. Abolish parts of the ship, and have three gun-divisions, embracing 45 per cent of the crew. Have no special duty men whatever taken from the gun-divisions. The engineer's force, with twin screws, requires on an average 25 per cent of the crew. From the remaining 30 per cent, with the assistance of as many of the engineer's force as are needed and can be spared, organize the powder and navigator divisions. This 30 per cent would include petty officers and idlers, messmen, watch petty officers' orderlies, sentries, side cleaners, messenger boys, the mail carrier, steam launch and dingy's crew. (This assumes that the marine guard is withdrawn.) In special ships it might be found desirable to increase or decrease the 30 per cent at the expense or gain of the gun-divisions, but the point here aimed at is to do away with all this excusing men for special duty out of running boats and from the guns' crews. Make a gun's crew a boat's crew for "arm and away" and for running boat. The powder division is the disorganizing factor in our present organization, and its importance is hard to overestimate. With the numerous chains of supplies for all sorts of guns that fire ammunition so rapidly, with the numerous water-tight doors on and below the protective deck to be looked out for, numerous pumps to man in case of fire or accident, with the care of the wounded sent down from the decks above, and with the scattered forces of such a large division to keep in hand, we have some very difficult problems to work out. The only satisfactory solution is to wipe out our present type of watch, quarter, and station bill and begin all over. Lack of space forbids the attempt to outline even what we should have in our new ships.

The next few years will see some changes and improvements in the character of the personnel of our crews, and it behooves every one to awake to a realization that new ships mean enormously increased responsibilities on the part of the officers. It is worth an earnest effort to try and attract and help keep in the service the best class of Americans possible. Lieutenant Fullam has hit so many nails squarely on the head that it is only possible, without going into details, to endorse briefly all that he has said. In his service on the Boston, Vesuvius, Yorktown, and Chicago, he has evidently done some thinking, and time will demonstrate that unless more thought and good-will is shown in the improvement of the status of our enlisted men, we can never hope to have a navy that will merit the respect and confidence of the American people.

Lieutenant C. E. COLAHAN, U. S. Navy.—I have read with great interest the paper written by Lieut. Fullam, on the system of naval training and discipline required to promote efficiency in the navy, and do earnestly endorse and approve the general argument contained therein.

There is, in my opinion, a lamentable neglect in the present system, in not making the position of the petty officer on board ship one of responsibility and importance. The pay of the petty officers of the higher grades, especially those of the line, is clearly inadequate.

The chief petty officers of a vessel of war, and those receiving the highest pay, should be the leading men in the gun-divisions, those who are proficient in all drills and capable of instructing recruits; they should be held responsible for the improvement of those instructed. These petty officers might be styled Division Mates, which rate should replace that of Boatswain's Mate; then would follow the Gun Captains, the grades thus being given the military titles instead of the old titles, which in the modern ship have no significance. These men should have the care of the deck, bright-work, etc., in the vicinity of their guns, as well as of the guns themselves; they should be held to a strict responsibility for the proper keeping of the same. The officer of the watch is now held personally responsible for innumerable petty routine matters that should properly belong to the petty officer; they should not be upon the mind of the officer of the deck at all.

In regard to the presence of the marine guard on board ship, I have merely to say that there is no room for them. The Marine Corps is a remarkably well disciplined and a very efficient corps; so also is the 1st regiment of U. S. cavalry, but neither of them has a place on board a man-of-war, which should have a homogeneous crew of the seaman-military class, each member of which could be called upon for all duties : for that of the service of the guns, that of the sentry on post, or those duties in connection with the sails, anchors, boats, etc. We cannot give room on board ship to a number of men having circumscribed duties only.

It will be admitted that the seaman has quite attained to the proficiency of the marine as a sharpshooter, that his drill with the rifle is quite as good; his intelligence is also about on an equality; then why, in the name of common sense, should he not be quite as efficient in police and sentry duty? By distributing this duty of trust and responsibility throughout the ship's company, the discipline must be bettered.

The whole question contained in the title of the paper under discussion may, I think, be answered in a few words. The pay of the chief petty officers must be increased, or our best men will be lost to us; their position must be made one of responsibility and trust, and their every-day service be of a more military character.

Lieutenant R. C. SMITH, U. S. Navy.—With the general ideas expressed in this paper I am wholly in accord. It is well-timed and forcible. The subjects discussed are demanding the serious thought of every officer who finds himself serving on the new cruisers. Lieutenant Fullam's presentation of the idea that the ship is designed for fighting and for no other purpose is logical and to the point. The *personnel* is a tool to this end that can be no more neglected than can the *matériel*. It is undoubtedly true that the lack of efficiency of men now found in the service is due to the survival of methods which in their day were admirable but which cannot now be utilized. All energies lately have been turned to building up a navy. It is only when results begin to appear that it is discovered that the organization and methods of the steam sailing ships are no longer applicable. The present condition of affairs is clearly shown in the paper. It is totally unsatisfactory. To devise improvements is to reorganize the navy. Mr. Fullam has certainly advanced many practicable and desirable changes. I do not know that I agree with him in putting the disembarking of the marines in the first place. That their influence on shipboard is harmful there can be no doubt, and that it would still be harmful if the seamen were trained to the highest point of military excellence is also probably true. At the same time I believe that there are methods of raising the tone, *morale*, and military efficiency of the blue-jacket which would produce greater results than simply sending the marines ashore. Therefore I put these methods, many of which Mr. Fullam discusses, in the first place, though I agree with him that the ship would be better off at the same time without the marines.

I think I understand the word "military" in a little different sense than that used by Mr. Fullam. He starts by saying that "all other than *military* elements have been reduced to a minimum in naval warfare and in naval architecture," and later that "the strictly *military* duties on board ship are performed by the marines. Here we come to the real cause, direct and indirect, for the absence of *military* instincts among men-of-wars-men—the fact that a company of soldiers is placed on board ship to do the *military* duties, to watch and search and discipline the sailor." The italics are mine. Now, if all the elements of naval warfare are to consist in such military duties as the marines perform on board ship, we shall not progress very far. The inconsistency would be removed by substituting "guard and sentry" for "military" where it relates to marines, though I am afraid it would destroy the point. Guard and sentry duty is not the best means for teaching men to be military, and by *military* I mean attentive to duty, disciplined, and exact in handling all the weapons with which the ship is provided. It is in frequent and progressive drills with these weapons that military efficiency is best promoted. Guard and sentry duty is good, to be sure; but I should like to see the marines done away with afloat, more on account of the injurious effect on the *morale* of the sailor element that their presence as a police force exerts. If they are not needed as guards and sentries, there is evidently no reason, as Mr. Fullam very clearly shows, for retaining them on board ship.

In regard to improving the quarters of the men, Mr. Fullam's suggestion of flush spar decks is a good one. The plan is being followed extensively abroad, and will be followed perhaps in our Bennington. I do not believe the open decks are the sole cause of discomfort in the ships named. In the Yorktown, which is the only one of them with which I am familiar, one great trouble seemed to be that the officers as a whole occupied a very large share of the available space. They were more numerous by a third or a half than in foreign ships of a similar type. When by the necessities of the service the complement of officers has been materially reduced, it would seem that a more economical arrangement could be made by a combination of the messes, thus rendering more space available for the men, and at the same time reducing the non-combatant servant element. It is probable that all or nearly all the officers required for the Yorktown class could find comfortable quarters under the long poop. The space surrendered below would be given to the crew. In the French cruiser Forbin, of the same size and total complement, there were ten officers to fifteen in the Yorktown; they were all quartered under the poop, and the whole of the lower deck was given up to the men. An arrangement of this sort would remedy much of the discomfort that Mr. Fullam mentions. The changes now in progress in the Yorktown are in this direction.

I concur in all the other suggestions for promoting discipline and efficiency, and for attracting a suitable class to the service. It is to be regretted that the writer did not go farther in a plan of drills and interior organization that would promote the same end. The main ideas to inculcate are, that the ship is a fighting machine, and that duties tending to increase her efficiency as such must take precedence of everything else : the buoyancy, *morale*, spirit of the whole *personnel* must be raised; duties that exert a repressive influence should cease; responsibility should be more extended; and lastly, the health, comfort, and happiness of officers and men should be zealously sought.

Lieutenant-Commander E. H. C. Leutzé, U. S. Navy.—Having only a limited amount of spare time, but feeling very strongly on this subject, I would like, if possible, to aid this "good cause" by a few words. I thoroughly agree with the author in everything he says in regard to the blue-jackets and marines. The latter are not only not needed on board of a modern vessel of war, but are doing a positive harm, their presence tending to drive self-respecting men from the service, or, in other words, the class of Americans that we need and must have. The blue-jackets are the men who should be fostered, as they will be the mainstay of the vessel during action. I agree with the author that there should be only one mess of officers, excepting flag and commanding officers, on board a vessel of war. I think it would be beneficial to both the older and younger officers, and would lessen the number of non-combatants. I agree with the author that young officers should be put in more responsible positions on board ship immediately after they have learnt the duties and methods of carrying them on. In regard to the organization of the crew of a war vessel, I would state that the one suggested by the author, namely, by gun-divisions, is in practice on board the U. S. S. Baltimore and U. S. S. Philadelphia, of which latter vessel I have the honor to be the executive officer. I can therefore say advisedly, that even with the present drawbacks, the system gives excellent results; but before it can be perfected we must have the rate of gun-captain as the principal working petty officer of a vessel of war; and other corresponding changes must be made.

Lieutenant H. S. Knapp, U. S. Navy.—To attract Americans of the proper sort, and to hold them in the naval service, a career must be provided for the men, and the conditions of life be made comfortable and self-respecting for them. The career, to my mind, must relate to the gun, and Lieutenant Fullam strikes a key-note when he says, "the men to man such ships should be, pre-eminently, naval gunners—not sailors of the old school." The gun and torpedo, in other words, the military instruments of offense and defense, must become before long the basis of our naval organization; and with that change will necessarily come a change in ratings that will carry with it the abolition of many of them, time-honored but now obsolete. Warships are built primarily to carry guns, and the absurdity of organizing their crews on any other basis than the gun is so evident that I cannot believe our present system will survive much longer. Very few petty officers should have better pay than the captain of a main battery gun or a torpedo-tube. We must have the *best* men we can find for and in these billets; but it is idle to expect the best men in ratings inferior in pay and position to a large number of other ratings in the ship's organization. Right here is the possibility of creating an attractive career for our enlisted men. Let the ratings be given and named, as far as possible, in relation to the weapons of the ship, and let the gun-captain hold a rating not to be secured without effort, but permanent under proper conditions when once attained, and carrying with it responsibility, position, and sufficient pay to retain good men. Fifty dollars per month, or even more, if found necessary, would not be exorbitant. It will be wise economy in peace, and wise insurance against war, to have a trained body of gunners who will make every shot tell. We can't afford to be without them; the fire of modern guns is too expensive, and the supply of ammunition on shipboard too limited. The adoption of such a scheme would, I firmly believe, by opening up a considerable number of attractive billets, serve to keep a far larger proportion of apprentices in the service than at present. As Lieutenant Fullam observes, they form now by far the most intelligent part of our crews, and every means should be adopted to hold them, especially if such measures conduce to efficiency. But, in addition, if we are to get and keep the class of men we want, better provision for their comfort must be made. Their quarters on many of the new ships are painfully inadequate, and the present system of messing abominably bad. Every man should have a comfortable billet, without the necessity of having somebody slinging above or below him. Petty officers should berth and mess apart, both for their better comfort and as a matter of discipline, to make their position more marked. Certain ones, whose duties are especially arduous, should have state-rooms or quarters somewhat similar to those occupied by the junior officers, where they can sleep in peace after "all hands" in the morning, or during the day when off watch. Wet weather clothing is prescribed, not permitted, and yet no place is ordinarily provided. To provide for these matters in the ships already built would doubtless be difficult, but they should be looked to in all new constructions.

The measures mentioned above would, I am convinced, promote discipline and efficiency, apart from the fact that they would operate to bring in and keep a better class of men in the service. With the separation of the petty officers from the remainder of the crew, and with a greater difference than now between their position and pay and those of the rank and file, the former's self-respect and manliness will be fostered, their control of those under their supervision and authority be increased, and they will be in touch with the commissioned officers in matters of discipline, which, unfortunately, is not often the case at present. On the other hand, the commissioned officer will hold the petty officer in higher estimation, and look to him for more assistance in proportion as his position is magnified. More responsibility should be given the petty officer, and he should be held vastly more accountable than now for the proper performance of duty by his subordinates. In this matter the commissioned officers are at fault to a very great extent, and the remedy is as largely in their own hands. The petty officer cannot hold his office in great esteem when his seniors treat him merely as one of the crew, give him little or no responsibility, hold him to little or no accountability, and give him only the slightest backing in his proper exercise of authority.

When the Board of Organization recommended the abolition of the Marine Corps, as far as sea-going duties are concerned, I confess that I was opposed to the step. But a more careful consideration has led me to see the wisdom of their decision. Now that every combatant on board ship is taught to use his rifle, marines are not needed as sharpshooters. I have served in one ship where there were no marines, and orderly duty was done by the sailors, to the entire satisfaction of everybody, as far as I know. I have also seen a quarter-deck guard of blue-jackets render honors quite as acceptably as the marines, barring the absence of the drum. Really, then, there is no apparent reason for this retention except to do guard duty. The army has no distinctive corps for that purpose, and I am unwilling to admit that the men of the navy are worse disposed than those of the sister service. Thus there seems no *necessity* for marines on board ship; and if the bad effect of their presence on the rest of the crew, so earnestly and fully described by Lieutenant Fullam, be conceded, there are slight reasons for, and very urgent reasons against, their retention. I believe this bad effect does exist, though not perhaps quite to the extent that' the lecturer thinks. Therefore, being convinced on this point, and feeling assured that no necessity exists for retaining marines on shipboard, I endorse the opinion of the lecturer that they should be replaced by an equal number of general-service men, no less combatants than they, and at the same time available for all-round duty.

Throughout these remarks, from force of habit, the appellation "petty officer" has been used. It is a misnomer, and I suggest changing it to some such title as "sub-officer," in the interest of the self-respect of the men who come under its designation.

The opinions advanced by Lieutenant Fullam are radical, and will doubtless encounter earnest opposition. But his conclusions represent, in the main, the views of a considerable number of officers, who will be glad to see them so ably and forcibly presented.

Ensign A. A. ACKERMAN, U. S. Navy.—One feels grateful to Lieutenant Fullam for his courageous expression of conviction, supposed to be distasteful to many officers connected with the Marine Corps.

It is an excellent indication of the tone of the service when a junior officer (one of those, who, in personally supervising every petty detail of the work aboard ship, are closest to and best acquainted with the character of the blue-jacket of to-day) announces that he is as capable of policing himself as any soldier could be.

Undoubtedly, if the Marine Corps is dispensed with, there must be a wiser, firmer, and more watchful control exercised by the officers than is now necessary. But, on the other hand, what a commentary on the wisdom, firmness, and energy of the officers of the navy, the present duty of the Marine Corps affords !

The presence of the marines on shipboard is almost as un-American as would be the control, by troops, of citizens already provided with their legal and efficient authorities ; for our ships are efficiently officered. No better amendment to the act prohibiting the enlistment of aliens could be made than one providing that American sailors should constitute their own police, and formulate their own pattern of military excellence on the sea.

When navies were manned by force, and governed by bloodshed and brutality, in the days of the press-gang and cat, then there may have been need of a special corps to carry harsh rules into effect, and protect the authorities against men wronged to desperation. Does the blue-jacket of to-day obey simply through fear of the marine guard ? Does he respect his officers the more because they trust their own authority and governing power the less? The blue-jacket is to-day robbed of quarters, military training, pride and honor, by the Marine Corps. Is it a wonder that he so rarely exhibits spontaneous patriotism ?

I am sorry that the essayist has expressed himself in favor of retaining sails and spars "for exercise," especially as it seems inconsistent with his previous remarks concerning routine exercises. The handling of sails and spars, from the creation of the navy, has formed the very foundation of routine ; it is improbable, should they be retained, that they will not continue to occupy a more prominent position than their merits for exercise warrant ; neither is it probable that a reorganization scheme, having improved service and fighting efficiency for its objects, will be promoted by the continuance of obsolete elements upon which the present organization was founded. With regard to our present routine drills, some are provisional and new ; the rest retain almost all the features that have been added during the past one hundred years, however inappropriate they may be at present ; nevertheless they are excused because "they exercise the crew." No wonder that the sailor-man becomes restless on drill, when day after day he must play a part in so dreary a farce. The worst feature of it is that there is nothing in the drill to indicate what is real and what is farcical ; sometimes the two qualities are hopelessly confused. No wonder these drills lack snap when the incentives of purpose and competition are alike absent.

If it is the sincere intention to devote a certain time to the physical culture of the crew, let the matter be handled systematically, keeping that purpose to the front. Let boxing, wrestling, and fencing masters take the men in hand. Without the accompaniments of storm and urgent need (and these would hardly affect "light spars for exercise "), work aloft will no more develop nerve and courage than traveling on the elevated railway or over the Brooklyn Bridge, though doubtless many have felt trepidation while undergoing this experience for the first time.

The new navy is not a vague possibility. It is here. But unfortunately it lacks a *personnel* benefited by the experience that would have led to its natural and progressive development. Turn in many directions and we are confronted by misfits. Both organization and drill need a radical overhauling, though the confusion of change and condemnation of old standards apparently leads to such demoralization as to bring about a sturdy resistance from some of the most conservative officers in the service. In this connection it may be said that the service cannot help but be conservative, as three-fourths of the officers are ten years longer engaged in subordinate duties than they would be with a steady and regular promotion. The ships are new; the officers are ten years older than they should be to perform their duties to the best advantage.

Lieutenant W. L. RODGERS, U. S. Navy.—I agree heartily with almost all that Mr. Fullam says.

The navy is largely filled with foreigners, who ship only to send their wages to Europe; and when our apprentice-boys leave the training-ships they find themselves in a foreign atmosphere to which they are unaccustomed.

Further, the present system of messing (which satisfies our present navy, composed of what a politician would call the "pauper labor of Europe") is simply disgusting to American-bred boys, and it is no wonder they cannot be retained in the service. The remedy for this state of affairs appears to be in the adoption of the Alien Enlistment Act and an entire change in the mess.

I think it would be practicable to put the men's mess on substantially the same basis as the ward-room mess. The ship should have a large space divided off as a pantry, and this would require much less room than at present, the whole berth-deck being occupied as a kitchen all day long. All the rations could be commuted, and supplies drawn from the paymaster's stores or purchased on shore, as convenient. An officer should have the general supervision of the matter, and a steward, ship's cook, an assistant, and four or five scullions would be ample for a crew of two hundred men or more. Dishes would be washed and mess tables put away after each meal by one or two "hands from each part," or other daily or weekly detail.

It is apparent that further inducement must be offered to get a high class of men in the service. I think, as does Mr. Fullam, that the *pay* at present offered is in most cases sufficient. The increased expenditure necessary should be applied to "flush-decking" all our ships built hereafter, for the greater comfort of the crew. The fighting efficiency of the Baltimore, Yorktown, and the other new ships may be well enough, and the ships altogether suitable for navies depending on conscription. For such a navy as ours, depending on voluntary enlistments, these ships are not commendable, as desirable men refuse to remain in a service where their material comfort is entirely disregarded. Constructors tell me that even with the present poop and forecastle vessels they could do much to increase the comfort of the men, were it not for the condition imposed upon them of granting absurdly large quarters to commanding officers. The Boston and Atlanta are the only ships that I know of where the cabins are of moderate size in proportion to the ship. Consider, for instance, the advantages that the changes just announced in the Bennington will give her over the Yorktown. With a flush deck, the ward-room will doubtless be put on the main deck and the cabin be much reduced in size. Ships of the Bennington's size need no junior officers. The present ward-room and junior officers' quarters should be assigned to the petty officers as their quarters and mess-room. The principal petty officers should be given state-rooms or alcoves, two men in each room. Storekeepers now make it a general practice to sleep in their store-rooms ; this custom should be maintained, as the ventilation is very good. If all cruisers were to have such accommodation for the crew as the Bennington, ships would be comfortable homes, and much more desirable men would enlist. Another great amelioration would be to have a table *always* spread with cold dishes for the engineer's force. This is the rule in the merchant service, and is much missed in the navy by all concerned.

Marines have no longer a function on board ship that can be filled by them alone. No doubt, were the marine guard removed from a ship, there would be some trouble for a little while, but only such as a man recovering from lameness might experience when he first throws away his crutch. On board the coast-survey and small ships in the navy, marine guards are found unnecessary ; so there seems no reason for retaining them on the large vessels. This is tacitly acknowledged by marine officers themselves when they express a desire to have the secondary battery, etc., turned over to the marines ; for, if they now have their well-defined place in the ship's organization, why should they wish for more ? If the marines are not a necessity on board ship, the disadvantages of a double organization and system of discipline, where men cannot be transferred as convenience demands, are perfectly obvious.

Finally, I come to the only point in which I disagree with Mr. Fullam, namely, in regard to sail-power on our new ships. In a single-screw ship, sail-power is necessary as a reserve motor in case of accident to the primary motive power. But in twin-screw ships, one engine is the reserve of the other, and I see no more reason for embarrassing such ships with sail-power as a second reserve motor than I do for putting sweeps on board a single-screw ship as her second reserve. As for sail-power enabling ships to keep their coal on hand, we must recollect that in making a passage under sail at low speed the consumption of coal for lighting, distilling, etc., will be a large proportion of what would be required to make the same passage under steam at a higher speed, and also that the coal endurance of a large ship is quite equal to her provision endurance, and that provisions as well as coal have been declared contraband of war by the French.

No doubt, as Mr. Fullam says, officers and men should know how to sail ships in order to take charge of their prizes ; also, the handling of sailing ships cultivates a certain nerve and resource, but this could be acquired in special sailing-ships such as the English "Cruiser," attached to the Mediterranean squadron, which takes drafts of men from the mastless ships for three months' instruction under sail and then returns them to their old ships.

In conclusion, I would like to repeat that what seems to me most needed to raise the quality of our enlisted men are improvements in their condition and increase of comfort on board ship, and to point out that an amount of money annually expended on increasing the pay of the navy would produce no result whatever; if spent on increasing somewhat the displacement of ships building, so as to give better quarters to the crew, it would probably produce very marked benefits to the service.

Lieutenant J. F. MEIGS, U. S. Navy.—I think that naval officers are, as a rule, in agreement with Lieutenant Fullam as to the necessity for a change in the method of recruiting and training our enlisted men. To retain Americans in the navy we must hold out to them advantages which, at least, approximate to those found in other occupations. The matter of training them will, perhaps, follow easily when we have devised a method by which to get and keep them. When we come to examine the exact steps necessary for this, a considerable diversity of opinion will be found to exist, though as to the nature of these steps there is substantial agreement that an advance of pay and a general betterment of position are necessary. It is my belief that Congress will give us the additional amount of money required to pay certain ratings more, as soon as we have decided what those ratings and their duties shall be, and that the betterment of their condition lies almost wholly within our present power. I may mention that it is my opinion that the creation of a rate of *gun captain*, in the manner suggested by Lieutenant Fullam, would be well received by officers generally. Other, and in some cases similar, changes would follow naturally. It is not strange that we should be unable to agree as to exactly what we want in these matters; that will come as the issue of the publication by the Naval Institute of such excellent papers as that of Lieutenant Fullam.

Lieutenant-Commander C. S. SPERRY, U. S. Navy.—In the days of sails, the training of the crew in handling them was of vital importance. In a modern vessel, the captain, beside the man at the steam-wheel, manœuvres the ship, and upon the organization of the battery the issue of the combat depends. It has long been my settled conviction that the battery should be the basis of the organization, and of the titles and ratings of the petty officers.

As for the petty officers, they should be as few in numbers as consistent with efficiency. The fewer they are in number, and the more rigidly they are held to the performance of their legitimate duty, the greater will be the dignity and comfort of their positions.

Every officer knows the disadvantage of a swarm of rated men on board ship with petty titles, which they consider a sufficient excuse for never seeing the light of the sun except when the executive officer has the deck. These men should have no rates. A sufficient number of people should be detailed to perform their duties, but they should be excused from only as much of the general work of the ship as is absolutely necessary. Certain of them, while performing special, and perhaps extra, duty, should be allowed a specified extra pay. As few men as possible should be unavailable for general work at any time.

Rank and pay commensurate with their importance, and that is second to none, should be given to petty officers exercising military control over men.

As for the presence of the marine guard on board a vessel of war, I fail to see that it degrades the position of the sailor.

Indiscriminate overhauling and searching of the person would be none the less humiliating, as customs go, if done by a blue-jacket police. The question is : shall a vessel with a complement of two hundred men have thirty men on board whose *general* usefulness is very limited, and is usually extended by making them do the work at the battery of the sailor-men whom they crowd out of the complement? Or shall we have thirty able-bodied general service men who can be taught to shoot as straight as a marine, and at the same time consider it their natural mission to work a gun, handle a sail, or go in a boat?

It is doubtful if we can undertake to give in the service the wages that a good, capable American citizen can earn on shore ; but there are possible compensations. Under a liberal system of continuous-service enlistments and retiring pensions, his income, though small, would be secure. As a well-paid, trusted, and responsible petty officer, to be disrated only by sentence of a court-martial, his position would probably be quite equal in dignity to any he is likely to attain on shore, and in many ways would be very attractive.

Quarters should be good, even if a light upper deck must go on, pay and clothes good, and the tenure of rates secure. But all of this is of no use if we wish to keep in the service the Americans whom we shall long for in the day of battle and danger, unless the positions to which they are advanced carry with them real, active *responsibility.*

We all know what kind of shipmates some of the best of our countrymen make when they have little to do and no responsibility—and how they stand behind one when things look black.

We have many fine men now whom we are always glad to shake by the hand as old shipmates. Let us hope we may get more and keep them.

Lieutenant J. C. COLWELL, U. S. Navy.—It is not washing our dirty linen in public to try to improve the personnel of the navy by telling the truth. The American public is neither blind nor collectively fools ; but it is spending many millions of dollars each year with the object of having a navy, and it wants results. It cares nothing about our hobbies. It wants the full value for money expended ; and when some day it takes it into its head to investigate for itself, the service optimist who proclaims himself satisfied with the present efficiency of officers and crews will find that he has not lightened the weight of the hand that will surely fall.

I believe Lieutenant Fullam has done a great service by speaking out in meeting and saying what he believes about the inefficiency of our present system of organization and training. At least it will cause discussion.

Radical changes are what the navy needs, and radical changes will be made for us before many years, unless we indicate those needs and so have a voice in the making. The personnel surely should receive as much attention to develop it as does the matériel. The enlisted man, as we find him in the drafts that come to us, is the person we have to work with, not some theoretical sailor-man who does not make his appearance. His origin is what we must first consider. With a proper organization and a consistent system of discipline, we can form his habits, control his character, and make his training what we choose. If he be a seaman, the chances are that he has been brought up in a merchant ship hailing from some port in the north of Europe where work is hard and pay poor, and he sees that if he does fairly well the work to which he has been accustomed, about the decks or aloft, his pay will be raised and he will be rated a petty officer. The distinction is not valued, but the additional American dollars are, and he devotes himself to being a more or less efficient care-taker of the decks and rigging. He cares nothing for the drills, because he sees other men, like himself, given a rating when they do not understand the difference between a 6-inch and a 6-pdr. gun, and cannot read the figures on the sights of their rifles. He expects to receive his reward in due course. He has no pride in the service because he is there for dollars only, and values the service because he is paid more of them for fewer and less valuable qualifications than he would receive in any other seafaring occupation in the world. If he be an ordinary seaman, apprentice, or landsman, he is list-less at drills because it is not made an object to him to be anything else. He sees the honors bestowed for reasons that ignore the existence of guns on the ship. He takes no pride in the service because he finds placed over him men who do not value the honor beyond the extra pay it brings, and for whom, if he be intelligent, he can have no possible official respect. It results that the officers have to devote much of their attention to doing the duties of petty officers, who think that they are giving a full return for pay received if they man the brooms when sweepers are piped, or answer in person a call for a hand from their part of the ship.

The native element that comes to us is good enough for all the petty offices we have to bestow, if it is taught to think so, and this element will rise to the requirements in a surprisingly short time. I have known a young American, on his first enlistment drawing a landsman's pay, after six months' drill and instruction (not coddling), to be quite competent to drill a squad of recruits at a great-gun; and he was trusted to take charge of a cutter under oars for the instruction of others. In the same ship were petty officers who could not be trusted to do either. After an experience of about a year of our most un-American system, the young man very naturally deserted. The appren-tices, properly handled, are an excellent element; but they are not our sole hope for salvation from the north of Europe. Young men up to the age of 24 or 25 years should be offered the inducements of a career of good pay and preferment according to zeal and ability. One year's training of the right sort would make excellent men-of-war's men, and the training should be with the type of ship and gun they would be required to man should war come upon us.

I heard a distinguished officer of our service once say that the best ship's company he ever commanded was composed mainly of a regiment of moun-taineers from one of the Southern States, prisoners of war, who had been induced to enlist to escape confinement, and that after a few months' hard training he had a crack frigate's company.

To attract the class of men we need, we must show that we are prepared to recognize and reward by advancement, brains, zeal, and good conduct, just as those qualities are recognized in the labor market of our country. Under our present system, a young American with the sea instinct, and possessing all those qualities, but who is unfortunate enough to enlist as a landsman, finds himself in a veritable slough of despond. His brains are unrecognized, his zeal is mistaken, and his good conduct serves but to get him on the liberty list. I have known several such who ended an enlistment in the rate they commenced—still landsmen after three years' good service, with no recognition of any qualities beyond those exhibited as permanent sweeper or berth-deck cook. These men, who had the making of first-class men-of-war's men in them, had they been properly encouraged and advanced, saw one foreigner after another, inferior to themselves in every desirable quality, advanced to petty-officer's rank simply because the talismanic word "seaman" appeared after their names on the muster-rolls, a title that means nothing as to the man-of-war efficiency of the possessor. Young men of that type should be given every encouragement to improve their technical knowledge and to stay with us; but rarely do they serve out an enlistment, and when they do, it is to leave the service with a well-grounded disgust, and to disseminate a dislike for it among the very people we should attract. The apprentices feel the same discouragements. Vastly inferior men may be, and are, placed over them as petty officers, because the seaman-apprentice is too small, or too young, or hasn't been at sea long enough; and then we solemnly stultify ourselves by making the same young seaman a gun captain, and stationing the captain of his part of the ship at the same gun—a military anomaly and a direct blow at discipline. But the apprentice was intelligent and could handle his gun admirably, while it was beyond the understanding of his petty officer. It is safe to say that that young man will not be a good recruiting agent when he drops a service that opens no career to his seamanlike qualities of the best kind. Or, again, when he finds all the prizes of the enlisted force in the way of pay, position, uniform, and consideration bestowed upon the non-combatant element, it is not likely that he will feel kindly towards a navy that ignores his value as a skillful fighting factor. He finds his contemporaries of sedentary tendencies advanced to positions of trust, and put in the way of ratings to writers, yeomen, and clerks, though they may know little about ship or gun and care nothing. *He* had the sea instinct or he would not have come to us; he developed it, and we ought to repay him with the best prizes of pay and position; but we do nothing, and he goes, while the clerk remains with us as the result of that effort to Americanize the navy.

The crews that we have fall very far short of meeting modern requirements, and it is owing to their lack of systematic, consistent, military training. They are not *required* to be attentive, careful, and thoughtful in military drills; they are required to be all this in caring for ship, boat and top. They are told military drills are important, but they see the rewards of service, the petty offices, going, not to the man skillful with the great gun and rifle, who attentively listens to, and intelligently grasps, the instructions of his officers on drill, but to the man who has a reputation for being a good seaman, which means with us not what it ought to mean in a navy fast becoming sailless ; or to the man who toils painfully over paint and ladders, or spends hours polishing up a cutter, which, likely as not, is then forbidden to be used except on show occasions, for fear of its being soiled.

How many of our present petty officers could be safely drafted to a volunteer organization, or ship, as instructors, to form a trained nucleus around which the men fresh from civil pursuits could rally and look up to as models to imitate ?

The gap between the men and the petty officer, not as he is, but as he ought to be, must be made greater, and between the latter and the commissioned officer, less. Better quarters and better pay for the fighting petty officers, and permanency of rank beyond the chances of a change of ship or of commanding officer, will do much to attract and hold the men needed.

A man-of-war that is "kept like a yacht," as we occasionally hear, is a very attractive spectacle, but that is not the end for which we are maintained by the nation. Nobody will care when we go into action whether our ships look like yachts or like coal-barges. The only solicitude will be for military results ; and we are not training our crews to the end required of us. The good "seaman" (and good at nothing else) has an entirely fictitious and mainly sentimental value in these days. He was of prime importance when the captain who could most quickly manœuvre his ship, handle his sails, and repair damages aloft, was thus able to place and keep his battery in its best tactical position. He is no longer needed for that purpose. A couple of men in the conning-tower place the battery where desired, and it is the battery that wins or loses the fight, according as the crew is skilled or unskilled in its manipulation. Individual skill, here and there, we have, in spite of discouragements ; organized general skill, with the best men in the most important places, holding rank and drawing increased pay by virtue of that skill, is what we need. A petty officer made gun captain by virtue of his rank, who can neither shoot straight, handle his gun properly, nor drill a squad, is a very serious misuse of the weapons entrusted to us ; yet this state exists. It neutralizes the advantage of possessing accurate guns, and discourages and demoralizes intelligent subordinates, who know that they possess skill and knowledge superior to the men placed over them. The effect is felt all through the ship.

A plain remedy is to make the battery the basis of organization, with petty officers selected for their military efficiency, bearing titles that indicate their responsibility, drawing pay commensurate with their value, and requiring them to exercise, at all times, authority over their own particular squads, whether at the battery, in the company, or in the necessary work in the part of the ship where their gun is mounted. Impress upon the crew that the battery with its attendant rifle and pistol practice is *first;* back it up by rewarding the attentive and skillful, and the best element will quickly respond. Particularly will the native element come to the front, for there is nothing the American appreciates so much as recognition of his abilities in the race of life, and the possession of responsible authority over those whom he himself has demonstrated to be his professional inferiors.

I quite agree with Lieutenant Fullam as to the desirability of dispensing with the marines on board ship, though I do not believe that it is the first nor the most important move to be made in an effort to improve the *morale* of our crews. The day of the marines passes with that of the old-style seaman, and we no longer need them. As a part of the fighting force of a ship they are gradually becoming an embarrassment. On one ship we find them scattered at the guns, filling subordinate stations; on another, assigned to man the secondary battery, and on another, drawn up in line to be used as sharpshooters when opportunity offers, meanwhile to be shot at and killed by rapid-fire and machine guns at distances beyond the effective range of their rifles. Each of these is a waste of good material, and a poor use to make of the training of a crack corps.

With a crew organized on the battery basis, it would be a simple matter to mount each morning a guard of one or as many gun-crews as might be needed to furnish the sentries and orderlies for the succeeding twenty-four hours. The moral effect on the crew would undoubtedly be great. There would be visible to them independent, responsible command exercised by members of their own body, the petty officers knowing their duties and trusted to perform them, respected by their superiors and obeyed by their inferiors—an object lesson in discipline that could not fail to make its effects visible in every variety of ship's duties.

The officers of subordinate rank feel discouragements of a somewhat similar nature to those indicated for the men. It does not require much brains to see that the rewards go to those who have "sought the beach." Naturally other men go there. If they have brains, they find them recognized, and duty assigned commensurately. On shipboard, the only measure of ability is the inches of gold braid on the cuff, and the only qualification for increasing the measurement is the elapsed time since the uniform was first put on, an interval during which many brains of original promise rust out, or, discouraged, fall into the rut of routine. Routine is not the end, nor is what we call discipline, nor yet is a ship polished into a thing of beauty. They are but means to the end—FIGHTING EFFICIENCY—and unless we attain that end, all the rest is wasted time and labor. If we attain it now, it is in spite of our organization and system of training, not by reason of them.

Our crews should be trained with the tools that they are to use in war. Yards and sails are no longer those tools. The guns, torpedoes, electric plant, signals, boats, wheel, lead, engines and boilers, *are*. There is plenty of scope. Oars, single-sticks, and a gymnasium outfit will provide additional physical training. The latter is seen in perfection and daily use in the French navy, and is being developed in the Italian navy. I am also informed that it is successfully used in the German navy.

Sailing ships no longer carry the bulk of the world's commerce. In time of war they would disappear from the trade routes. Why, then, should our men be organized on the basis of their ability to handle what the navy will never require them to use, and why should time be wasted in teaching them an obsolete naval art? There will be no more sailing prizes brought into port. The cruiser costing from $700,000 to $3,000,000 cannot afford to detach officers and men for prize-crews. Burn, sink, and destroy (excepting, of course, neutral vessels) will be the orders of cruisers in the next naval war, and the only exceptions will be in the case of prizes of value sufficient to warrant a convoy; such prizes can be found only in the great steam-liners which are practically sailless.

Sails for war use have gone out of date; the old-style seaman, the unit of organization in days past, must also go. But a higher type of seaman must take his place, and how to produce him and retain him is the problem that is set us to solve. According to our solution shall we be judged by the people, in the day of trial, and sentiment will not enter into the judgment.

Commander HENRY GLASS, U. S. Navy.—The paper which has just been read calls attention to the existing need of a more thoroughly military organization for our vessels of war than we now possess, and I think this a need which should receive the careful consideration of all officers of the navy. I am in entire accord with most of the ideas advanced by the writer, and with the suggestions for changes in the rating and pay of enlisted men, and for increasing the inducements offered to induce valuable men to remain in the service, all of which I advocated some years since in a paper written for the Institute.

The present organization of the crews of our vessels grew out of the needs of the sailing ship; and while it was perfectly suited to a frigate or sloop-of-war, even in the days of auxiliary steam-power, it has, unfortunately, been retained in the service long after the vessels for which it was intended have become obsolete. What is needed at present is an organization appropriate to the modern battle-ship and fast cruiser and the duties to be performed on board them, and such an organization should be devised and put into use as soon as possible.

I think that all the suggestions made by Lieut. Fullam were more or less fully discussed and recommended for adoption in reports made to the Hon. Secretary of the Navy by the Board on Organization in 1889; indeed, that Board went much further than he does in recommendations to increase the efficiency of the enlisted men, and to induce competent, trustworthy men to remain permanently in the service. As allusion has been made to some of the recommendations of the Board, I am permitted to state in general terms to the members of the Institute what were some of the results of its deliberations.

The Board, taking the ground that an immediate necessity exists for improving the character of the enlisted men of the navy in order to obtain the best results in the vessels of war of the present day, commenced with the recruit on his entry for service, or rather with the naval apprentice.

A large increase in the number of apprentices was recommended; the apprentices when enlisted to be sent to a central station, where they would be retained on shore long enough to be uniformed and given preliminary drills, and some knowledge of naval discipline, when they would be transferred to cruising training ships, for voyages of some months at a time, until sufficiently. advanced for entry into the regular service. It was recommended that apprentices should be required to serve, at least, one enlistment of three years, after attaining their majority, before being entitled to final discharge, and that a portion of the time should be spent at the training station under special instruction.

Two vessels of moderate size, one with full sail-power, both armed with modern service guns, and supplied with torpedo outfit and electric-light plant, were recommended for the training station, to be used in making short cruises with small detachments of apprentices and young recruits, the detachments being shifted from one vessel to the other for different drills and instruction. 'It was recommended that one of these vessels be fitted for special instruction of firemen and coal-passers in handling fires under forced draft.

The enlistment of men without previous naval service was to be discontinued, as far as possible, and minors were not to be enlisted except through the training station, unless in the case of those too old to become regular apprentices, and in such cases a longer term of enlistment than three years, as now authorized by statute, was recommended.

As the crew of a vessel of war should be composed primarily of seamen, it was recommended that the terms "Landsman," and "Ordinary Seaman," should no longer be used; that men for special duties should have rates designating, as nearly as possible, those duties, and that the seamen should be divided into three classes, according to their experience and ability, with corresponding pay. The pay of the first class seaman was fixed at a higher rate than at present, and it was provided that seamen of all classes should receive extra pay for special acquirements; in ordnance work, torpedo work, and marksmanship, at the rate of $2 per month for each specialty. The increased pay for these specialties was to be open to all other men of the crew, with the restriction that men rated and paid for special work should not receive the additional pay for their specialties; thus a gun captain might qualify for torpedo work, or ordnance work, or both, but not for marksmanship. The present increased pay for successive reënlistments was to be continued.

It was recommended that, with certain necessary restrictions, the appointments of all petty officers should be permanent from one enlistment to another. A retiring pension was also recommended, sufficiently large, it was thought, to offer a high premium for continuous faithful service. A recommendation was made that the time between discharge and reënlistment, under honorable discharge or C. S. certificate, should be increased to six months, with pay for three months, as now allowed by law.

The guns' crews were to become the units of the organization, and the special rate of Gun Captain was recommended, with high pay, to be held by leading men in the crew who excel in marksmanship. The gun captain was to be responsible for the men under his command when at the gun or performing duty in any part of the ship; he was to replace the present Quarter Gunner in everything relating to the care and preservation of his gun, and he was to perform the duties of a captain of a top where necessary on account of the rig of the ship.

The gun division was to take the place, for duty on deck, of the watch or quarter watch, and instead of Boatswain's Mate the rate of Division Mate was proposed, to be held by the leading man of the division, who was to command the division in the absence of an officer, when acting as a whole on drill, or in the performance of any other duty. Other rates for special duties were recommended, but enough has been said to show the character of the changes proposed. The efforts of the Board in assigning rates to men in the new organization were directed to making the title. represent, as nearly as possible, the duty to be performed, to reducing the number of titles to as small a number as convenient, and to grading the pay of men for special duties according to their ability and the importance of the duty. The number of rates proposed was, in consequence, smaller than under the present system; but a more equal and regular promotion and increase of pay from grade to grade would, it was believed, be the result.

The rates of pay proposed were intended to emphasize the importance of the military and skilled elements on board ship as being greater than that of the police and clerical force employed.

I think the writer overstates the argument for dispensing with marine guards on board our vessels. Certainly the recommendation of the Organization Board was not due to any feeling that the presence of marines on board vessels of war tended to degrade the seamen, or to render them less efficient in the performance of military duties. It was proposed to withdraw the marines from ships in commission and to increase the strength of the corps considerably, with the idea that larger bodies of men being stationed together at the different navy yards, a higher degree of efficiency would result, and the Navy Department would have always at hand a compact, thoroughly drilled, and organized force to be used where landing parties were needed. At the same time it was supposed that the better class of men whom it is desired to attract to the service could perform efficiently all the purely military duties now done by the marines, and also carry on all the routine duties of the man-of-war's-man.

The concluding paragraph of Lieut. Fullam's paper seems to contain a proposition with which, under no circumstances, could I agree. While the enlisted man should be trained to the highest point of efficiency, and be entrusted with the responsibilties of much important duty, I cannot think that any duty, however trivial apparently, is unimportant in the daily life on board a vessel of war; still less that it can be degrading to either the position or intellect of a commissioned officer. It is only by careful, constant attention to all details of duty on the part of executive and watch officers that the present high standard of discipline in our service can be maintained, and our ships made to present, as they have always done, examples of neatness, cleanliness, and readiness for service.

The responsible commissioned officer may, and should, direct petty officers to carry on duty ; but close supervision should never be relaxed for a moment by the officer of the watch ; and no study of scientific principles, if pursued at the expense of the habit of close attention to all duty, can be to the real interests of the service.

My experience is not in accord with one of the writer's statements that the younger officers are not allowed opportunities of performing responsible duties on board ship. In every ship in which I have served since graduating from the Naval Academy, with one exception, ensigns have always been assigned to duty as regular watch and division officers, and in many of them midshipmen, or naval cadets, have been assigned to such duties, if not under orders from the Navy Department, then by the commanding officers.

, Lieutenant WILLIAM G. CUTLER, U. S. Navy.—While I agree with the essayist that marines are no longer needed afloat, I would rather do so upon the ground that their room is needed for the general service artilleryman, than upon the less tenable ground of the latter's degradation.

What I regard as the most important point of the paper, viz., " Permanency of Rating," the essayist subordinates to the necessity of getting rid of the marines.

When we recognize that the sailor of the future is an artilleryman afloat, and when we rate and pay him in accordance with his value as a fighting factor, we will have made a great step in advance.

The question, " How can a better class of men be induced to man our rapidly building ships?" is of the utmost importance to all patriotic Americans. I respectfully suggest the following inducements :

1st. High pay to the skilled artilleryman.

Assuming the captain of the Boston's 8-inch gun to be what a modern gun-captain should be, there should be no man forward of her mast whose pay exceeded his. A recent article in the daily press complains that the seamen-gunners, after being instructed at Newport and Washington, are accepting positions in steel works and small-arm factories. Can they be blamed for improving upon the $26 of the navy?

2d. Permanency of rating to the extent that no man who, after due trial, has proved himself worthy, shall lose his rating except by sentence of a G. C. M.

3d. Increase the gun-captain's authority, gradually giving him entire charge of detail drills.

4th. Improve the men's quarters and comforts to the utmost extent consistent with the efficiency of the vessel. Procure skilled berth-deck cooks. Berth servants apart from crew, and uniform them distinctively.

5th. Reduce the number of non-combatants to a minimum ; servants by consolidation of officers' messes ; reduction in number of yeomen, writers, etc.

6th. Exercise far greater care in the enlistment of men, especially of men who enlist in ratings.

As to the relations of officers and men, they could not but be improved by the following :

In port relieve the watch and division officer from much of his unnecessary watch duty; not that he may go on shore, but that he may identify himself more closely with his battery and with his men. He and his men will surely be more efficient by knowing and respecting each other. Hold the division officer, and not the gunner, responsible for his battery and all things pertaining thereto.

Require all officers to be on board during working hours. The spectacle of a procession of officers fleeing the ship at 1 P. M. is surely a cause of discontent among the men. For the sake of the men, as well as the officers, naval cadets and ensigns should not be made to do petty officers' work.

Lieutenant PAUL ST. C. MURPHY, U. S. M. C.—I have read with more or less interest Mr. Fullam's argument in support of what he considers necessary to "promote efficiency" in the navy, and while agreeing with him in certain minor matters, think that in the main his zeal misleads him, so far as existing conditions are concerned; that he errs in regard to the malign influence of "marines afloat," and that his view of the U. S. sailor of to-day is too pessimistic.

Marines and sailors have been associated from the beginning of our navy and have shared in common its toils and its glory. Marines have always contributed their full share to victories afloat, and the depressing effect of their surveillance was not discovered in the days when battles were fought and victories won. During my sea service, extending over a period of more than eleven years, I have failed to see in the blue-jackets any evidence of repressed manliness due, in even a remote degree, to the presence of the marines. The espionage complained of is exercised by the marines over their own men quite as much as over the sailors; the same is exercised by guards and sentinels in the army, yet I doubt if the *esprit* of marines or soldiers is impaired thereby. My experience warrants me in believing that the *morale* of the sailor has never suffered from contact with the marine. If sailors are to be made military, the example of a body of subordinate, well-drilled men should be a help rather than a hindrance to the development of a military spirit. Mr. Fullam omits to state that the marines, besides "watching the scuttle-butt," "attending the commanding officer," and "standing by for ceremonies," also stand regular sea-watches with the sailors, are stationed with them at all general exercises, and are noted for promptness and zeal in this sailor work; that they keep clean their own part of the ship, and that they form an important part of the ship's battalion. Marines are not in any sense "idlers," and this fact cannot be unknown to Mr. Fullam, whose acquaintance with the watch, quarter, and station bill must make him familiar with the "principles of modern organization."

If the withdrawal of marines from service afloat is to work the reform predicted by Mr. Fullam, if it is to make the service more attractive, what need for so large an increase of pay?

Mr. Fullam has few, if any, uncomplimentary things to say of the "marines as such," yet the pay of "13 dollars per month, with 3 dollars clothing allowance," does not deter good men from seeking service in the corps, the quota of which is generally full. I am not an advocate of small pay. I would like to see the pay in my own corps increased, especially for non-commissioned officers; but it seems to me that pay has little to do with the matter in question. Of course, pay might be so increased that there would be as much eagerness in seeking billets before the mast as is at present the case in the scramble for place in the civil service of the Government. It is something more than doubtful, however, whether, when the pinch comes, the material so obtained would be superior to the old.

And now a word for the sailor of to-day. I believe Mr. Fullam very much underrates him. The "heterogeneous crowd" to which he alludes as making up the companies of our ships of war is about the same "crowd" that did gallant service for us in 1812, and later, and that has made our flag respected everywhere. The men who followed Farragut, and the men of the Trenton who, with death staring them in the face, cheered the escaping Calliope during the memorable disaster at Samoa, were of the same "heterogeneous crowd." Whatever credit marines have gained (they have gained much and deserve more), has been hand in hand and side by side with these "foreign" tars. The same "heterogeneous crowd" fills and has always filled our army, and has proved itself, afloat and ashore, the material of which good sailors and good soldiers are made.

I fear there will always be an absence of Mr. Fullam's "American" element; that element may be heard from in time of war, but at any other time it cannot be counted on. Perhaps the present element can be made more military; it is possible, and to a certain extent desirable; but there is always the danger of taking the temper out of the steel by the grind of over "sojering." I have failed during my experience to see in the discipline and training of the navy the "unbending" and "un-American requirements" that Mr. Fullam sees; and the element that he considers driven back by them to civil life is, in my opinion, an element pre-eminently fitted for civil life; it is certainly quite unfit for the navy.

I see nothing to blush for in the past and present of the navy, and nothing to despair of in its future.

Captain LOUIS C. FAGAN, U. S. M. C.—The paper written by Lieutenant Fullam, U. S. Navy, is a powerful one, granting his theory that marines are responsible for all evil is correct; but reflection will perhaps show that even if marines were withdrawn from on shipboard, there would not be that rush of young Americans to make up the crews of ships the sanguine theorist supposes. The fact must be ever borne in mind that *life at sea is an unnatural one,* and men cannot be persuaded to exist for any length of time away from the comforts of the shore, no matter how rude these comforts may be.

American lads, as shown by the costly failure of the training system (Mr. Fullam says 90 per cent are lost to the navy), will not stay long in the service; and even if the marines were relegated to the scrap-heaps in our quiet navy-yards, the "boys" could not be induced to remain and be tossed around, even if granted a chance to look through portholes to view the world. The best proof of the bad effects that a· "life on the ocean wave" produces is seen in the discontented moods and wild fancies that enter the minds of intelligent men, when exposed to the trials of mid-watches and a sea diet, and the most polite and elegant officer on shore often becomes a perfect demon to those under him at sea; even rich men owning yachts do not stay on them any longer than is necessary to entertain and display their wealth, notwithstanding the fittings of luxury. No; the object of naval reorganization of the personnel is to supply to a ship of war men who can *do the fighting* in case of hostilities, and uphold the honor of the flag during *peace or war*. Jealousy, or prejudice, or fine-spun theories ought not to interfere with this great object. The days when promotion followed, during a three years' cruise, from midshipman to lieutenant-commander have passed never to return, *unless* we have a successful war; and to make this possible we must foster any system that gives us *warriors*, whether they wear the blue shirt of the sailor or the trim uniform of the "ever faithful." The "bone and sinew of the land" will enter as soldiers under the splendid discipline of the Marine Corps, fostered by naval commanders, when they hesitate to ship as sailors, and it would be a foolish move to displace men who are acknowledged by their bitterest foes to be efficient and faithful, in order that a few officers should be benefited by increased rank and easy billets.

We have in the Marine Corps an organization largely American, and which has been pronounced a strong arm of defense, not only by the veterans who have made the navy all it is to-day, but by the most progressive commanding officers and those highest in broad intelligence, who feel most keenly the necessity for our men-of-war being manned by the best material, regardless of personal wishes of the inexperienced.

One word in conclusion. In every community there must be a visible force to back the law. Take all the bad men out of a town, or of legislative halls, and "police" would still be necessary. How much more important is it, then, that in the unnatural life on board a man-of-war, with many of the passions suppressed among the men (picked out for their animal perfection), the captain should have a force ready to uphold the majesty of the law, and at the same time be able to undertake the management of the most complicated guns and use them against the enemy. The idea of comparing men on board ship with a regiment on shore is preposterous to any observer of human nature, for the evil passions of soldiers on shore, with practical freedom, are given vent, while on board ship, often for weeks, the natural impulses are checked, and unlawful desires fomented and violence precipitated. Then a remedy must be ready to apply.

Taking into consideration a first-class navy, not a make-shift affair, the primary thing to be thought of is, *how can the navy be made most efficient for any emergency?* No doubt departments and officers in plenty could be dispensed with on board ship for the general good. Great wars were conducted, the nation saved, and skillful battles fought without the aid of the Naval Academy, for instance. Yet how foolish and absurd would it be for any man or class of men to advocate the effacement of that noble school, which has given so many level-headed men of brains to the service of the country, simply because the navy, like other navies, could somehow get along without such an institution; and then, perhaps, the men who seek their own advancement through the ruin of others might be spared with great advantage.

Lieutenant E. B. BARRY, U. S. Navy.—It seems to me the main idea under-lying Lieutenant Fullam's essay is a condemnation, not only of the methods employed in our naval service afloat, but, also, of those employing them. I doubt if the service at large will agree with his course of reasoning. Lieutenant Fullam must have been particularly unfortunate in the composition of the divisions he has commanded on board ship; still, I think much of what he complains about is due to his own idea of what constitute *military matters*. What does he mean by this? Does he mean a rigid observance of position and motion throughout all drills, as taught in the German navy, where army discipline holds good? Would he organize our navy on the German army plan? Are these ideas " proper concessions to national traits and institutions "? Men usually follow the lead of their officer; if the one is interested in his drills and exercises, the other will be. The divisional officer is to continue and perfect what is begun in the training squadron and on board receiving ships. The essayist seems to forget that we are in a transition state, and that the entire reorganization of the navy cannot be effected in a few months. He blames the old sailor for not being a soldier. When the efforts of the Navy Department to introduce some sort of system into target practice bear fruit, when the gunnery and torpedo schools have turned out enough qualified men to captain all the guns afloat, and when *modern* instruction can be imparted more generally, then, if no results are produced, it will be time to condemn sweepingly. The vital question concerning petty officers is rating and disrating. When they can be disrated only by sentence of court-martial, many of the evils complained of will disappear. Even this takes time. Let us begin with the rate of Gun Captain; as the essayist says, diffuse these men through the service, and watch the result. There is an old dodge known as " working for a rate "; this consists in " being seen " by the captain and executive officer, but in shirking when they are not on deck. In spite of all opposition, many of these worthless fellows were rated by those in power, who naturally believed their own eyes. This will be done away with if reports of efficiency are made by all watch and divisional officers on board ship, from which to make up a man's record. This will aid the younger officer to " escape the professional wet blanket." I fully agree with that portion of the essay on quarters and comforts for the men; also, on pay and promotion. Give properly qualified petty officers good pay, permanent ratings and a retired pension; recognize that the old-time sailor is giving place not to the *soldier*, but to the gunner, and the problem is solved. I think the question of marines can be settled by a uniform system of punishments prescribed by the Department, and having it strictly enforced. Human nature is ever the same. Uniformity of rewards and punishments will go far to make men contented; stern but even justice is preferable to lax and variable partiality. I think the marine will in time be withdrawn from service afloat as he becomes more and more of a non-combatant aboard ship. Either he is to become one of the gun's crew, the boat's crew, and one of the ship cleaners, or else his place will be taken by men to do these duties. But he will not go because of the bad moral effect exercised by him upon the ship's company, for, in my opinion, this effect does not exist. When that day comes, the divisional officer will find himself just as much responsible for the condition of his battery and guns' crews as he is to-day. He never will be able to throw his own responsibility upon petty officers.

Commander S. W. TERRY, U. S. Navy.—In my judgment, the article we have heard read this evening is not calculated to result in any good to the service. It is not only an attack upon the usefulness of the marine guard, an important adjunct of a man-of-war, but it is an unjust reflection upon the officers charged with the discipline and government of men on board ship, calculated to produce discontent among the men of the navy, and to create doubt and distrust in the people as to the efficiency of the service in general. For my part, I do not think this Institute was ever established for the purpose of disseminating any such doctrine, and for that reason I regret very much that the paper should be published in its Proceedings and distributed through the country and abroad.

At a time when the navy and our people at large are congratulating themselves that a good beginning has been made in building up the navy, the author of this paper comes forward and not only charges that our men are not treated as they should be in the matter of pay, promotion, and employment, but that such of the new vessels as the Charleston, Chicago, Baltimore, and Philadelphia are faulty in design; this is in the face of the fact that the plans of these vessels were prepared by a board of the most experienced experts to be found among officers of the navy.

While, I suppose, there is no one who will claim there is no room for improvement in the condition of the crews of our vessels, few will admit the correctness of their condition as described in the paper read this evening. The type of ships completed and being constructed, and the material composing the crews of the vessels of our navy, will compare favorably, in my opinion, with those of any other country. It is easy to find fault, but not so easy to formulate the remedies. I think we can agree that an efficient naval personnel can only be developed from the raw material by persistent and intelligent training on board ship, and this must be conducted by zealous and capable officers anxious to reach a high degree of efficiency. While other countries were engaged in transforming their navy from an old to a modern type of vessels with modern armaments, our own Government did nothing to keep pace with them. Their experiments and experience were of great value to us, however, and in 1883 we began the building of modern and improved ships. In March, 1883, Congress passed a law providing that no wooden ship should be repaired for a greater expense in hull or machinery than 20 per cent of the cost of a new vessel of the kind. Up to this time much money had been expended in repairing vessels that many regarded as useless, but this law marked the beginning of the end of wooden ships for our navy, and many were soon gotten rid of. Under these circumstances it has not been practicable to afford officers as much sea duty as was customary in former times, a condition, in the opinion of the best judges, unfavorable to an efficient navy. This, in my opinion, was of greater disadvantage to officers of the junior than to those of the higher grades, as many juniors remained on shore duty from three to five years. Some of these younger officers, not long out of the Naval Academy, and with limited experience in control of men, naturally interested in the absorbing question of rebuilding the navy and providing the most improved armament for our ships, have, now and then, occupied themselves in the solution of such problems as this paper deals with, and which, in my judgment, might better be left in the hands of more experienced officers. A navy, like an army, requires time, perseverance, and opportunity, for perfecting itself as a fighting organization.

We have not had for many years a navy worthy the name, and, in consequence, our men may not be up to the standard we hope to attain. It has been impossible to train men for service in the new cruisers up to this time, but in spite of this, and before any fair opportunity is had, we hear the complaint that the men are unsuited and incapable, and that the vessels themselves are faulty in design and not of a kind to attract the American sailor, this latter being represented as exacting as to his compensation and the amount of comfort to be provided in the ship; and, further, it may be inferred that if he should choose to return from liberty, in liquor, a marine should not search him. As reference is made in the essay to the crews of the vessels composing the Squadron of Evolution, and the essayist is also on board one of them, it may be that his experience there affords ground for what he says of the unfitness of the men now enlisted for manning our new ships. I am sorry the brief cruise of the squadron should not have made a more favorable impression on his mind, and I beg to dissent from his conclusions as to the capacity of the crews of the squadron. It appears the squadron sailed in December last and returned in July following, and, judging from the essayist's paper, it might be inferred the men had not reached a very high degree of efficiency. This is not in accordance with the official reports, however.

His propositions seem to be three :

1st. To withdraw the marines from service afloat, their presence on board ship being detrimental to discipline and an implied distrust of the sailor.

2d. That in the matter of " pay, discipline, and promotion," there is not sufficient encouragement to Americans to enlist in the navy.

3d. That the navy is composed of a variety of nationalities, "a heterogeneous crowd"; "all languages are heard on our decks."

Now, let us suppose the marines are withdrawn. Other men, blue-jackets, would take their place, and there would be no gain in room. There will be a necessity for orderlies and sentinels as at present, and a sufficient number would be required to afford reliefs. Then there would be need for a guard on all occasions of ceremony. It is not claimed that the marines do not now perform their duties efficiently, but that sailors will do it as well. If it is proposed to employ sailors promiscuously in the duties now performed by marines, and in ship's work; the result, I maintain, will not be for the good.

We will have neither good sailors nor good marines. Men who pull in boats, coal ship, scrub decks, go aloft, etc., must, as everybody knows, become more or less unmilitary in consequence. Look at the soldier employed on police duty in a garrison or on board ship. Is he not made less a soldier by it? It therefore appears that the principal reason for desiring the withdrawal of marines from service afloat is that their presence is an implied distrust of sailors, etc. I doubt if such an idea ever existed in the mind of the sailor, for everybody knows there is no reasonable foundation for the assumption. The author should know that the duty of searching liberty-men and boats' crews, when it is done, and in well-regulated ships this should rarely be necessary, properly belongs to the master-at-arms and ship's' corporals, and not to the marines; nor are marines themselves exempt from a like search. These petty officers also confine and have charge of men under punishment.

It is a strained inference to suggest that the presence of marines on board ship implies a distrust, or lack of intelligence on the part of the sailor. Such is not the case, and I deem it unworthy of notice. In 1885, when the detachment of marines were sent to the Isthmus of Panama, the guards from the training squadron that I commanded were withdrawn, and for nearly six months the three vessels were without guards. I have no hesitation in saying that the vessels were less efficient without than with the marines, and I hope never to serve in another ship without them. The Marine Corps has formed a part of the navy ever since we had a navy. They have performed excellent service both in peace and in war. Their appearance calls forth praise wherever their duties take them. Their officers are competent and work harmoniously with the navy, and I see nothing to justify the proposition to withdraw them from the service afloat.

Let us consider the proposition that there is not, in the matter of pay, promotion and discipline, sufficient to encourage Americans to enlist in the navy. Any fair-minded person who should compare the rating and pay table now, with what it was twenty-five years ago would be surprised at such a careless statement. I think I can show that our government does more than any other country in the world in the way of inducing men to enlist, and remain in the navy, and the difficulty, though nothing like what is charged, is that the opportunities for employment in this great country are much greater than elsewhere in the world. At the close of the war, in 1865, the seaman received $18 per month. Now he receives $24, with an addition of one dollar per month for each enlistment under a continuous service certificate. This involves an increase of more than one-third, and, in my judgment, is quite sufficient. The seaman in the British navy gets £2, ($9.68). All other ratings have received an increase proportional to that of seamen, with the additional pay for enlistment under continuous service certificate. The number of petty officers and rated men have been largely increased. The following may be mentioned: machinists, boiler-makers, water-tenders, oilers, buglers, schoolmasters, ship's writers, seamen-gunners, seamen-apprentices, tailors, barbers, jacks-of-the-dust, lamp-lighters, caulkers, and additional stewards. All these positions offer increased pay and promotion to meritorious men. The position of boatswain, gunner, carpenter, and sailmaker is open to worthy and capable enlisted men of the service, and these positions are very largely held by men from the service. No country provides a better ration, and few as good, as our men have. The health and comfort of our men are matters that receive constant and careful attention.

The selection of petty officers rests exclusively in the hands of the commander, and they are not deprived of their rates except the commander, in his judgment deems it necessary. What better disposition can be made of this, authority? I confess some surprise at the criticism of little details of duty on board ship, by officers or men. Nowhere else are so many people living in so contracted a space. The cleanliness and good appearance of a ship contribute largely to her efficiency, and, therefore, it is very necessary to look closely, to minor details. Now permit me to enumerate some legislation made to encourage men to enter the navy. A man who has served not less than ten years, and is disabled, may be pensioned upon the recommendation of a board, of three officers, approved by the Secretary of the Navy, (Sec. 4757, R. S.).. A liberal pension is provided for any man injured in line of duty; also provision for his family in case of his death. Interest at 4 per cent upon his savings deposited with the paymaster. A home on board receiving ships, and allowance of full pay during the interval between his discharge and re-enlistment, within three months, for three years. Transportation to his place of enlistment if discharged elsewhere. One-fourth additional pay if detained in the service beyond his term of enlistment. An allowance of an outfit of clothing and bedding to an apprentice on his first enlistment. Upon certain conditions a man may enlist in a petty-officer's rating, and he cannot be deprived of this except by sentence of court-martial. And finally, any man after twenty years' service, and disabled for sea service, is entitled to admission to the Naval Home in Philadelphia, or in lieu thereof, to a pension of half the pay of his rate when last discharged, (Sec. 4756, R. S.). In the matter of punishments, Congress has carefully defined their character, and officers are held to a strict account for any deviation from the law. There are, also, many privileges and, indulgences lodged in the hands of the commanding officer that are seldom, if ever, withheld from the deserving men. Libraries are provided all our ships, and the men are allowed reasonable time for recreation and enjoyment. As to the matter of the Government competing with other employers in the pecuniary compensation to those entering its service, I hold it has gone far enough when our men are paid more than two and three times what other governments pay. No country ever pretended to compensate its men in money for service in its defense. This should not be expected. A man should be ready and willing to serve in the army and navy for a reasonable compensation, and a fair provision for himself or his family in case of injury or death. This much, I claim, this country gives to those in the army and navy; and the security of these benefits is more than equal to the uncertainty offered by the individual employer.

Some officers are inclined to the opinion that there is too great inequality of pay; that if such rates as ship's writers, buglers, yeoman, and other non-combatants are properly paid, the gun captain, boatswain's mate, etc., are underpaid. There may be something in this, but it should be remembered that the price of labor depends upon the demand. There is no market for the seamen, who are called upon for these latter positions, except at sea, while there is a demand for writer, buglers, etc., not only on board ship but on shore. Every one knows how very difficult it is to get a reliable and competent ship's writer, or yeoman. There is a certain qualification required of them which the seaman need not have, hence I think there is good reason for the difference in compensation in these rates.

Now as to the variety of nationality found among the crews of our vessels. This is greatly exaggerated. Native-born Americans are not by nature inclined to a seafaring life, if we except those on our coasts. In so large a territory as we have had undergoing settlement and development, it would be surprising indeed if the men should prefer the unsettled life of a sailor to the regular employment of a farm, of railroading, manufacturing, etc. Nor would any rate of pay that this very prosperous country could afford overcome the natural antipathy of its people to a seafaring life.

The remedy for the ills of the service lies not in the clamor for more pay and more promotions for the men, nor in the agitation of imaginary grievances by naval writers, but in a conscientious and hearty effort on the part of all officers to be content with what we have and strive to raise the service to a higher state of efficiency. It is to be observed that these complaints come from officers and not from the men, that a discontented service will never be an efficient service, and that those who disseminate the seeds of discontent are acting in antagonism to the best interests of the navy and their country.

We have many foreign-born men in the navy, it is true. Some are citizens and some are not; but I see nothing discouraging in this fact. Do we not invite immigration from every country except China? Are not these men good citizens, as a rule? Are they not elected and appointed to office under state and national government, excepting only the offices of President and Vice-President? Was it attempted to restrict enlistment in the army, or navy, to citizens, in the war of 1861? Are there not in the history of that war conspicuous instances of these men performing gallant and meritorious service, and, in many cases, losing their lives for the country? Many such men occupy conspicuous positions in Congress, to whom the navy looks for its very existence. Now, what should constitute citizenship to fit a man to serve in the navy? It is well known that many men come to this country, generally, I suppose, without family, and after acquiring a knowledge of our language enlist in the navy. They re-enlist and serve faithfully and loyally. The character of the service prevents them from becoming citizens in law, for it should be borne in mind that this alone does not carry with it all the benefits of citizenship.

The highest privilege of citizenship in this country is the right of franchise; to have a part in choosing our elective officers. To possess this privilege requires a fixed residence in some state for a specified time, and often other requirements are exacted. These conditions the man serving in the navy is generally prevented from complying with. Faithful and loyal service in the navy for a certain time constitutes, in my judgment, all that should be exacted for service in the navy; and this seems to be the view taken by the government. It is a great mistake to say that "all languages are heard on the decks of our ships," because there may be a few instances where men unfamiliar with our language are enlisted for special duties. It is not the rule, and the matter is entirely under the control of the officers themselves. The inducements the essayist considers necessary to attract Americans to enlist in the navy must strike those responsible for naval administration as extravagant beyond reason. I think I have enumerated what will be regarded by most reasonable men as quite sufficient inducements, but notwithstanding these, we are told the Chicago, Newark, Baltimore, Philadelphia, and San Francisco are 'not provided with sufficient comforts, such as "decks" and "more locker room," to attract Americans. This suggests a degree of fastidiousness undesirable even in the American sailor. It should be remembered that a man-of-war in these days is a combination of many important conditions, involving a harmonious compromise of all, the principal of which are speed, armament, and capacity to keep the sea. It is believed by those most competent to judge that the limit of comfort for officers and men has been reached in these vessels, without sacrificing any of the material requirements of a fighting machine, and when we compare them in the matter of comfort with those of former days there is no room for just complaint.

Now, Mr. Chairman, I confess my surprise and regret that an officer of our navy, who has not yet been charged with the more serious responsibilities of the service, should feel himself justified in drawing such a gloomy picture of the navy as the member has presented in the paper we have listened to this evening. I do not think it will have a depressing influence upon the great majority of officers; it may do the navy harm in the country and abroad. I am glad to say, if I may speak from my own experience, which began in May, 1861, and covers every grade from midshipman to the command of the Training Squadron of the three old-fashioned sailing sloops Portsmouth, Jamestown, and Saratoga, that the time never came for me to separate from the ship I served in that I did not do so with regret. There never was a ship in which my association with the sailors of our navy did not call forth my admiration and praise for their uncomplaining performance of every duty required of them. Whether our men have been called to service in the Arctic, or to the unhealthy climate of the tropics, in whatever sphere or locality duty called them, they have given a good account of themselves.

One word more, Mr. Chairman, and I have done. Every one is familiar with Jack's tendency to growl. We ourselves know that he takes this way of enjoying himself. He means no harm, and generally does not mean what he says. Take away the "shellback's" right to growl, and I fear the navy would lose some of its most valuable men! Can it be that our young friend is developing, a little prematurely, this weakness? Let us hope so. But, seriously, this paper can do the navy no good. It is simply a case of "airing our linen" before the public, to our own injury. The controversy between the line and staff was made "a stench in the nostrils of the public," and when this seems to be subsiding, we have an attack upon the usefulness of the marines on board ship. I think this Institute will consult its right to the respect and good opinion of its readers by not publishing this paper and the discussion thereon. I therefore move, as the sense of the meeting, that its publication be suppressed.

Commander C. D. SIGSBEE, U. S. Navy.—The paper presents a more dismal view of the service than will meet with general acceptance. The fact of the case is that the navy needs reorganization, and has needed it for some time, but no man, nor set of men, has been strong enough to effect it in the face of the opposition of conflicting opinions. When the question of a new navy first began to seriously engage the attention of Congress, that body found it impossible to decide on types and policy, because of the wide divergence of opinion then expressed by naval officers on the subject. Since we are now well advanced in the construction of ships, the question of reorganization is, most naturally, pressing for decision ; for to be well fought, ships must be well organized. Again, we find a divergence of professional opinion that is likely to delay the adoption of active measures. Unfortunately, the opinion of a moment is as emphatically asserted as one that flows from ample consideration.

It is doubtless the case that this paper has been made strong by its able author in order to excite discussion, and thereby mould opinion to a condition approximating unanimity. If discussion fail to convince disputants, it may nevertheless clear away doubt in high places. However that may be, we must not fall into the error of supposing that great measures can always be put into operation on the stroke of a bell. The writer's views relative to the narrow range of duties required of cadets and ensigns apply not only to those grades, but all along through the watch grades of the line, a condition largely due to the backwardness of the navy as a whole. It is to be hoped that the time is not far distant when young officers will be employed in duties commensurate with their scientific education and training. The several points raised by the writer have been subjects of private discussion for some years past, and opinions are already modifying in the general direction of his views.

We have heard much about the kind of seaman needed for future naval duties. It is the common belief that he must be intelligent, active and orderly, besides being skillful to a high degree in special lines. Judging from the talk that we hear, there is not a general comprehension of the fact that these qualities bring good pay on shore, nor that the man who would make a good topman might fall short of our new standard. Most unaccountably, ideas seem to dwell on the old kind of sailor at the old price, with a few rewards thrown in. Formerly, when the great naval powers were in serious need of men-of-war's-men, they took forcibly the first that could be found by the press-gang ; then, with the improvement in social conditions, came enlistment by deception, by conscription, by blarney, and, finally, by allurements or baits, all these being supplementary to voluntary enlistment. Our own most recent naval system, in time of peace, is by means of petty baits—as one dollar per month extra on

re-enlistment, for instance. This system worked well for a while and helped
to fill our complements, but it will not serve to entice the new order of man.
" The laborer is worthy of his hire," and the laborer knows it now if he never
did before. Doubtless many a rollicking ne'er-do-well might be found to go
to sea·to relieve himself of the burden of orderly existence, but we should not
find that kind acceptable. Our training system is good—capable of improve·
ment, undoubtedly, but good nevertheless. The boys are well paid and well
cared for, but there is only a ludicrous future offered them, and the majority
of them find it out. An officer once said to me, " When a boy leaves the training service and goes on board ship for general service, he finds that he has
been decoyed." If any one wish to know why naval apprentices have refused
to stay with us, let him investigate the *Classification and Pay Table* of the Navy
.Register, and bear in mind that the apprentices are prepared for the seaman
and military branch of the service. The table discloses that we have gradually
drifted into a condition of rating and pay in which the military element in. a
military service is heavily discounted. The exhibit shows three grades of
petty officers ranking the common seaman, and even the seaman-gunner. In
the first grade of petty officers are found twelve ratings, only three of which
belong to the seaman branch. These three are of the ship, not of the service ;
they offer no certainty of attainment, are not awarded on board all ships, and
are commonly filled by seamen who have far exceeded the age at which
British men-of-war's-men are retired from service. The pay allowed is $35
per month. In the other classes of that grade the pay is not less than $45
per month, and in six cases is $60 or more. In the second grade of petty
officers are thirteen ratings, of which four belong to the seaman. With one
exception on each side, the same marked difference of pay, unfavorable to the
seaman, is found. In two instances the pay of artificer ratings is twice that
of three of the four seaman ratings. In the third grade of petty officers there
are eleven ratings, of which eight belong to the seaman. With one exception
the artificer ratings get more pay than the seaman ratings. The next grade
has no official status and contains the seaman-gunner, the nearest approach
the service has to the ideal future seaman. This grade also contains the
seaman and the seaman apprentice of the first class. To oppose to them there
are nine special and artificer ratings which carry more pay, with a single
exception, than the seaman rating. For instance, the lamp-lighter gets one
dollar per month more than the seaman, and only one dollar per month less
than the seaman-gunner. Let us fancy two men of equal intelligence starting
out to learn, one the duty of a seaman or a seaman-gunner, and the other that
of a lamp-lighter ; it is no exaggeration to say that any man fit for becoming
a seaman-gunner could learn the duty of a lamp-lighter in a forenoon watch.
Every rating in the table that involves the knowledge of a trade, or the possession of even minor skill, is given more rank or more pay, or both, than that
of seaman. Let the investigator compare the position of captains of parts of
the ship, in the third grade of petty officers, with the ratings that compose the
first and second grades, and let him go back thirty years and compare the
relative amounts of pay at that time ; then let him compute the percentage of

increase of pay for the several ratings, so far as possible, up to the present time. He will find that the seaman has, apparently, had no very active friends. It is shown that the man most difficult to get and to retain is treated as if he were, of all men on board, the least valuable. Let us suppose that two naval apprentices of equal intelligence decide to remain in the service on completing their course in the training ships. No. 1 takes to the service at once and is made a seaman at $24 per month. No. 2, with an eye to windward, delays three years, during which time he learns the trade of machinist. At the end of three years No. 1 has advanced far towards being our ideal seaman; he is thoroughly well drilled, and has a good knowledge of the modern instruments of warfare. If we lose him he is a loss indeed; he must be replaced by an apprentice. In addition to his experience he has the weight of more than one apprentice, because we actually have him and we do not get all of the apprentices. However, he re-enlists and gets $1 per month additional pay as provided for by law. He now gets $25 per month. No. 2 presents himself on board the same ship, with a dozen other machinists, to apply for a place. Being used to the ways of shipboard, and his record being known, he is taken at $70 per month, and becomes a first-class petty officer at once. Suppose the commanding officer, overcome by sympathy, should decide to outrage tradition in favor of justice so far as to make No. 1 a chief boatswain's mate, and that No. 1 could then maintain that rating in his future service. No. 1, with the $1 extra, now gets $36 per month. After this second term of enlistment it will require thirty-four re-enlistments, or one hundred and two years of continuous service, to bring his pay up to that of No. 2 the machinist. If No. 2 had preferred to take a three months' course in a business college instead of learning a trade, he might have enlisted as yeoman at $60 per month, or as ship's writer at $45 per month, and have been a first-class petty officer in either case. In the latter and minor case it would have required nine re-enlistments, or twenty-seven years, for No. 1 to have got the same pay. There may be some peculiar working of the Classification and Pay Table that I fail to consider, but I find it impossible to understand how we can hope to retain intelligent young men in this land of high wages, so long as we require from them seaman's duty at the price we offer. The system that is criticised is the result of attempts to maintain or improve organization by slight modifications from time to time. It is the system of no one man nor any set of men that could be named. It is the system of a service that had been allowed to drift to the rear by a people tired of war.

In any attempt to reorganize our crews it is to be hoped that the leading impulse will not be to adjust old ratings to new conditions, but that the duty to be done, and the policy that will enable it to be done, will receive the first consideration. When conditions are stable, a high regard for tradition serves to maintain a high standard of duty ; but when new conditions of service are forced upon us, we may perhaps meet with greater success by giving more latitude to originality than has been done heretofore. Tradition over-valued induces half-hearted measures. The importance of a rating to the service and the difficulty of filling it, especially in time of war, should, in great degree, help to constitute the measure of class and pay. Compliment will not serve the purpose of pay. As a rule, important rank should go only with the exercise of special authority. Unless a rating require the exercise of military authority, pay should be the inducement rather than rank. The chief object of pay should be to maintain enlistment; the object of rank, to perfect fighting efficiency. Rank too widely diffused levels distinctions, and distinctions of one kind or another are the outward evidences of authority. The distinctions of rank or grade forward of the mast should be better pay, special berthing, messing and uniform, greater responsibility, accountability and privilege, restricted numbers, address by title instead of by surname alone, and permanency of rating. The more important ratings should be of the service rather than of the ship. Special pay might be given for industrial skill and for special technical knowledge, but rank should not be carried so far as to defeat the object of rank. For example, it would only weaken a regimental organization to double or treble the number of its non-commissioned officers.

I doubt that the writer expresses quite what he means when he asserts that all languages are heard on our decks. This would convey an idea of Babel. We have many foreigners, but it would be hard to find those who cannot speak English more or less well. A fair proportion of these foreign-born men are excellent, worthy men, who have learned to appreciate the flag, at least for the benefits that it confers in excess of what they have been used to. It is the American who can do better outside the service.

In respect to dispensing with the marines on board ship, it may be claimed that many thoughtful officers favor that measure, and are prepared to accept the embarrassment that may prevail for a short time in the course of read-justing the situation. If it be true, as some assert, that marines are indispensable in maintaining discipline, then it must be because the standard and habit of the soldier are better for that purpose than those of the sailor. Surely no better argument could be given for recasting the characteristics of the sailor by adopting some of those of the soldier. I will venture to predict that this is precisely what is coming, and that the presence of marines on board ship will become anomalous, and their services be relinquished.

The writer suggests a return to sail as an auxiliary method of propulsion in time of peace. While weight still attaches to the use of square sail for training purposes, many officers believing, contrary to my own views, that it is the surest means to prepare both men and officers to enter upon the new naval duties, it is a waste of verbal ammunition to urge a reaction in favor of sail for propulsion. Sail is as dead as Julius Caesar! If the present efficiency of the coal-pile warrants the abandonment of sail in any considerable degree, the development of the latent efficiency remaining in the coal-pile will banish it altogether for naval cruising purposes. Officers sometimes fall into the error of discussing this point, as if one could rationally assume, at pleasure, either 16 or 20 of the same kind of ounces to the same kind of pound; as if a ship's weight-displacement could readily be made to exceed 100 per cent. A certain percentage of the weight-displacement is absorbed by the hull; the useful displacement remaining may be apportioned among the qualities, according to the service required of the vessel, but the total displacement, in the absence of miracle, is just 100 per cent. Sail and the adjuncts of sail have weight, and if any portion of the displacement be absorbed by sail, just so much must be taken from fighting efficiency for the supreme moment of battle.

Several weeks ago a distinguished naval officer, whose opinions have great weight in the service, published a paper in the Institute, in which he gave the strongest argument that I have heard for reconsidering the sail question. We were warned that a vessel without sail might be despatched to a distant point under steam, to give battle on her arrival, and arrive on her fighting ground unfit for fighting, through lack of fuel. It was argued that this contingency, to a country not possessed of coaling stations, demanded the retention of sail for propulsion. With much hesitation, I permit myself to differ from the conclusion at which he arrived, and for the following reasons : 1. Propulsion under sail is more or less fortuitous, and therefore unsuited to the rapid movements of modern warfare. 2. Fighting efficiency would be sacrificed to only moderate cruising efficiency. 3. It would be better to tow a coal schooner, or be accompanied by a supply steamer a certain part of the distance, as, in general, promising a more expeditious passage. 4. In modern men-of-war it is impossible to give an approximation to the old sail-power. 5. Supply schooners and steamers are our natural recourse for fueling in time of war. They are, in fact, mobile coaling stations. They may be required to rendezvous anywhere. This kind of coaling station must be discovered in order to be destroyed. 6. The *desideratum*, I submit, is not sail power, but a method of fueling in a seaway. In conclusion, it may be said that if the adoption of liquid fuel were possible, the problem of fueling at sea would be solved immediately. It would then only be necessary to tow the supply vessel astern at moderate speed, just enough to keep the tow-line taut, and pump the fuel on board the towing vessel through a slack hose.

Captain HENRY A. BARTLETT, U. S. M. C.—I have had as much experience as any officer in the service with marines afloat, having done duty on over twenty men-of-war, all large ships. Lieut. Fullam in his paper would give the impression that marines were non-combatants. On every ship to which I have been attached they have had a great-gun division. During the war, the marine divisions that I had charge of were, on more than one occasion, held up to the crew at general muster as an example of excellence for their proficiency at the great-guns. On every ship I have served on, the marine guard has been fully instructed in great-gun exercise. On my last ship (the Trenton) the guard was instructed in working all the guns, and we had a battery of 8 in. converted rifles, on carriages of five different designs ; in case of necessity, the marines could have been detailed for any gun division. The marines have been, and in fact always are, instructed in all kinds of military duty, in addition to their regular duties as sentinels. The foundation of the navy was the Marine Corps : they were the original fighting men of the ship. The sailors, in my long experience, have always shirked military duty, and are now, and always will be, lax in everything pertaining to military discipline. They not only try to circumvent the marines on duty, but also the officers and petty officers of the ship. On more than one occasion I have known the chief petty officers of a ship to be the leaders in mutinous conduct, and if there had been no marine guard on board there is no telling what might have become of the ship. I remember well one night aboard the New Ironsides. She had a crew of over 500 men. The ship anchored off Port Wagner. We had been in action that day. About 9 P. M. I was sent for by the 1st Lieut. (as the executive officer was then called). He informed me that there were a number of men drunk; the marine guard were ordered to muster on the spar-deck. I inspected them and found nearly every man perfectly sober. This great crew were nearly all more or less intoxicated, and fighting was going on everywhere. Our chief quartermaster, and nearly every petty officer was in irons that night. Every pair of irons was in use, and fully two hundred sailors were dropped down in the forward hold, and the old frigate under the enemy's guns. What would have happened had there been no marines on board?

Again, I remember well the occasion when the grand old frigate Wabash grounded off Frying Pan Shoal with some 700 souls on board. The marines got out a launch, manned her, carried out an anchor, etc., and Captain De Camp in his official report to the Navy Department said that the ship was saved by the efficiency and good discipline of the marine guard.

A policy of espionage, if there is such, does not tend to bring out vicious qualities, but tends to check them. All enlisted or shipped men must be examined or searched at the gangway; the same course is pursued at all navy-yards and garrisons; if it were not so, liquor would be constantly brought in, and government articles taken out. This inspection applies to marines and sailors alike, and its humiliating effect, that Mr. Fullam speaks of, I have yet to notice. The marine protects the sailor as well as government property. They are the first men called upon in case of trouble either on shipboard or on shore. Many, many times have I been called upon, in my service of nearly thirty years, to assist in quelling riots, to guard public and private property at fires, and even lives on board ship and on shore, both at home and abroad.

The type of men needed, under the conditions of modern warfare afloat, have not as yet been born; the man-of-war's-man of to-day is not competent to do both naval and military duty any more than the soldier is to do both military and naval duties, or even all military duties on shore; this is why the army is divided up into different corps to perform their varied duties. Does the presence of the marine on board ship degrade the whole service? There are usually soldiers of different corps in camps and garrisons—engineers, ordnance, etc., as well as infantry and cavalry. Because the infantry and cavalry do the police and guard duty of the camp or garrison, do they look down on the others; or do the others consider themselves a lower class? In this construction the engineer's force on board ship must be considered; they now receive the largest pay, yet some of them require the most looking after;

they form now a very large part of the crew; does Lieutenant Fullam wish to put sailors over them? In this case, were his argument true, they would be degraded by having another class of blue-jackets over them.

In time of war seamen must be taken from the merchant marine, and we all know that they are the flotsam and jetsam of the sea. The landsman will naturally go into the army. There are few native American sailors to-day. In time of peace it will, no doubt, be well to require the sailor to be naturalized. Our navy in war-time will be recruited by volunteers, and not by drafts, like some of the foreign navies which do not usually have their marines afloat. These sailor volunteers are a roving class, little bound by ordinary ties, and full of spirit (a little more than simple American love for liberty), and therefore they will require a different system of discipline. Until we again become a maritime nation, and even then probably, this will be true.

Congress will not allow men enough to fully man our ships in time of peace; therefore, if we are to have marines on board in time of war, there is need of them in time of peace to man all the guns, as well as to give them the necessary instruction on board that they may be prepared for war.

There are some points (as the upholding of petty officers in orders) in the paper which, if carried out, would increase the discipline and efficiency of the crew; but that the presence of the marine on board is the cause of whatever lack of these there is, I must earnestly protest against; to them rather is due most of the good that at present is obtained.

It is well known that the number of different things that the men, as well as the officers, should have some knowledge of are many more now than formerly, and it is also certain that they cannot have a thorough knowledge of all. The advisability is argued of giving officers different specialties, and the argument applies with as much, if not more, force to the men. Why, then, is it not better to keep the different branches to their own appropriate duties?—the engineers to the steam department, where they will be required in action, with enough knowledge of drills and firing for cases of emergency; the marines to the preparation for action and guard duty; and the sailor to preparation for action and cleaning duty, sail and spar drills, and boat duty, which with our small crews often takes most of the available men from the deck. Why do away with an already trained and tried corps of men (justly admitted to be such), for what seems the almost impossible elevation of the sailor to the military plane of the marine?

Lieutenant H. O. RITTENHOUSE, U. S. Navy.—In a discussion such as that in which the Institute is now engaged, an unguarded expression, or even the use of a simple inoffensive word, seems to raise such an intellectual dust that friends and foes can hardly be distinguished. Thus it often happens that people who are earnestly working in a common cause fall upon each other with great fury, each thinking his adversary to be the very enemy. To some nautical minds, for instance, the terms "soldier," "military," etc., are like the traditional red rag. There are those, for example, who can't tolerate the idea of having any "soldiers" on shipboard, but at the same time regard the

" marine" as indispensable. Others again don't want "soldiers," but are working hard for " gunners."

It is certainly a waste of energy to continue controversies based only upon some narrow signification of a word. Speaking broadly, a soldier is a man with a gun; so also is a marine or a gunner. If we look squarely at the subject, I think we can all advance a long way together without disagreement. A modern ship is an instrument designed to fight under steam without canvas, carrying small guns, great guns, and a variety of medium-sized guns and other more or less powerful weapons, with men to handle them. The ultimate object is to so organize and train these men that the highest efficiency in action may be secured. It matters not (for the discussion) what the men may be called. "Fighters" is good enough. But it does matter how they are organized and trained. Up to this point I am sure we can all go together; and, considering the certainty that the new weapons and ships will impose at least some modifications of drill, if not of organization, and the further fact that no system (however good it may be, or may have been) is good enough if we can get a better; considering this, I say, we can well afford to welcome for discussion any proposed plan for improvement, however much we may differ from the conclusions of its author.

For myself, I think the weakest point in the efficiency and discipline of our ships is with the petty officers. The conditions of their service should be made such that their interests are with the officers rather than with the men under them. They should constitute the strongest links in the chain of discipline. As to the best method of securing good petty officers I do not concur fully with the writer of the essay, who appears to think the removal of the marine guard from the ships is a first essential; while I believe further efforts should be made to obtain them without interfering with the guard.

There is food for thought, however, in his argument. A disorderly ship is among the worst of evils. No one knows this better nor feels it more than the intelligent men of the crew. The question then arises, would they be more reliable and efficient if charged with the full duty of upholding the common welfare, than if bearing little or no responsibility for it? Again, let us suppose that in a military organization on shore, some one company of non-commissioned officers and privates should be selected to do all the post and guard duty, while the others were employed in other ways, and never took their turn at the peculiarly disciplinary duties. How would such a measure affect the efficiency of the body as a whole; and, particularly, how would it affect the non-commissioned officers? Would the result be affected in any way if the special company wore a different uniform? Our military brethren can best answer these questions. A man-of-war at sea is not the same thing as a military camp on shore, and therefore the answers would not necessarily decide us as to the advisability of removing the marine guard; but they might help us to a decision.

Lieutenant-Commander HARRY KNOX, U. S. Navy.—Lieut. Fullam's paper seems to me most excellent and opportune; some may deem it radical, but none can deny that it is earnest and thoughtful.

Our types of ship and armament are undergoing radical change, and it is not unnatural to anticipate the need of change in other directions, even admitting that in the past our methods have been most wise.

I believe with the writer that it would be good policy to require the general service men of the navy to perform all military duties on board ship, and I do not think there would be any trouble in carrying that plan out at once, if law and regulation permitted it.

Our aim should be to bring the general service man up to the plane of modern requirement, and it is the educating effect of the military training that I would ask in his behalf.

The question of the selection of the leading petty officers of the ship by the criterion of marksmanship alone has always presented grave difficulties. My idea is that a leading man under one system of selection would probably be a leading man under almost any system; but he might not be a good shot, at least, not the very best. The best marksman is apt to have an inborn talent, and he may be a person of no force of character, not a leader of men, not the man that one would want for a boatswain's mate or a sergeant. This fact has led me to think that we might have a gun captain who would manage the gun and crew, but who would not necessarily point and fire; though it would, of course, be most desirable to combine the two duties in the one person.

A STUDY OF THE MOVEMENTS OF THE ATMOSPHERE.

. BY LIEUT. E. FOURNIER, OF THE FRENCH NAVY.

(Translated by Prof. J. LEROUX, U. S. Naval Academy.)

The author of the present study prefaces his essay with a few general considerations upon the advances made in meteorology, especially in the study of the mechanics of the atmosphere, which is now coming to the front. Sailors will be the first to benefit by this movement of opinion; meteorology has invariably awakened in them a powerful interest, and this is only natural. " When one spends half of a lifetime upon the deck of a ship continually moving about, one meets the same phenomenon under so many different aspects that he cannot help asking himself what the determinative principle is." (Admiral Mottez.) The writer further states that his efforts have been directed to establishing upon a firm basis the conclusions that terminate his work and form its quintessence, so to speak.

NATURE OF THE AIR—THE ATMOSPHERE.

The terrestrial globe is surrounded by an envelope of air nearly inalterable in its composition at all points and altitudes: 79.20 per cent of azote and 20.80 of oxygen in volume. To those two gases, which are the principal constituent elements of our atmosphere, must be added aerial dust, carbonic acid, and vapor of water, which enter into the mixture in proportions varying according to place, season, and even the hour of the day.

The density of the air also varies with the spot under consideration, the temperature, the fraction of saturation, and, above all, the altitude. As we rise in the atmosphere, this density decreases in a rapid proportion that would assign to the aerial envelope a very near boundary, if close observations did not invalidate its accuracy beyond a certain limit. The measurements made by Bravais upon the aurora borealis permit us to reckon at 200,000 meters the height of the atmospheric column; some writers calculate it as high as 300,000 m.

Pressures.—This immense air-column produces upon every part of the earth a pressure whose mean annual value varies with the point under consideration.

The following table, arranged by Maury and reproduced in the nautical meteorology of Ploix and Caspari, gives the mean barometric height at sea in function of the latitude. We have selected in preference the southern hemisphere, owing to its vast preponderance of sea surface.

Latitudes.	Barometer.	No. of Observations.
From 0° to 5°	760.46	3692
" 5° " 10°	761.48	3924
" 10° " 15°	762.70	4156
" 15° " 20°	763.51	4248
" 20° " 25°	764.59	4536
" 25° " 30°	764.40	4780
" 30° " 36°	763.30	6970
" 40° " 43°	761.69	1703
" 43° " 45°	756.40	1130
" 45° " 48°	752.59	1174
" 48° " 50°	752.33	672
" 50° " 53°	748.78	665
" 53° " 55°	745.73	475
" 56° " 60°	743.95	1126

The annexed curve is a graphic representation of Maury's table; it shows us the existence of a barometric maximum at the tropic and the steep gradients of pressure towards the 42d parallel.

Farther on we shall examine this curve in detail, but we can already perceive that there exists a palpable relation between the latitude and the corresponding barometric height.

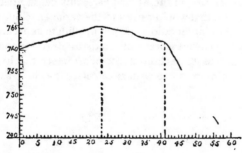

FIG. I.—Curve of the barometric heights in the southern hemisphere.

What forces are held in equilibrium by a column of mercury of Torricelli? Are they simply the weight of the air and the tension of the vapor of water it contains, or rather the sum of molecular and electric attractions of the atmosphere?

It does not seem admissible, given the great thickness attributed to the aerial envelope, that beyond the limits of this same envelope the super-elevated layers are connected with the earth by means of their meré weight. At that height, the extremely rarefied and imponderable air has not ceased to be a gas and to possess all the properties of gas; it must therefore be endowed with an indefinite expansive energy. Thus we cannot conceive of the sudden boundary between a gaseous state, of whatever kind, and an absolute vacuum, except through the adjunction of a force independent of the density of the gas under consideration. This force, which acts as a barrier to our atmosphere and prevents it from disappearing little by little into the interplanetary spaces, is found in the electrical tension of the air, which by actual experiment has been proven to increase progressively in inverse ratio of its density. In that case, electrical attraction constitutes an element in the total barometric pressure, whose principal factors are the molecular weight of the air and the tension of the vapor of water it contains.

Should a distinction be made between the total pressure of air in apparent state of repose and the same pressure when the atmosphere is in motion? We do not think so. Air pressure, with the exception of well characterized cyclonic disturbances, hardly varies a few millimeters above or below the static pressure, according to whether the winds predominating on the surface are cold and dry, or warm and moist. We cannot see, on the other hand, why air moving in a horizontal direction should prevent the atmosphere from exercising all its weight upon the earth's surface, or be capable of developing internal dynamic action sufficient to interfere with gravitation. Many cases are cited where in strong northwest gales in the Mediterranean the barometer stood above the mean. The force of the wind seems, therefore, to have little influence upon the barometric height. It is not so, however, when the wind acts in a direction deviating from the horizontal; it is evident, à priori, that every downward vertical component causes a rise in the barometer, and every upward ascending motion of the air decreases the pressure.

RELATIONS OF BAROMETRIC PRESSURE TO ATMOSPHERIC MOTION.

1st Principle.—It is generally admitted that the differences of pressures are the primary causes of the movements of the air. Researches into those differences, combined with the examination of the rotation of the earth, constitute the study of atmospheric disturbances: as water is set in motion by the differences of levels, and no electric current can exist without potential differences, in the same way, also, air is not displaced nor can there be any atmospheric current except by virtue of a difference of pressure.

Sometimes the above principle is thus interpreted: Every wind has its motive in front of it. That interpretation is not a correct one, for very frequently winds move athwart their determinative causes. It would be more correct to say: To every atmospheric displacement corresponds a difference of pressure that has caused the movement. We shall see in the course of this study what share the active forces that the globe in motion distributes unequally over the different parts of its gaseous envelope, contribute to the aerial circulation.

2d Principle.—The second principle upon which we shall rely is the following: When a depression takes place at a given point, the air is attracted on every side with an energy proportionate to its expansive force; that is, to its density, at least in the generality of cases. From this proposition, which to us appears evident, we can evolve a very important law: Density is greater at the surface; it is therefore by changes of air surface, in the same manner as liquid in two vases communicating at the bottom is leveled, that the equilibrium of pressure will be sought, the points of lesser resistance at the surface acting as adducing channels. The equilibrium would be thus re-established, but the rotation of the earth modifies the direction of the air currents, and for this reason the equalization of pressures is never reached directly.

INFLUENCE OF ELECTRICITY ON THE MOTIONS OF THE ATMOSPHERE.

There is another natural force that may, we think, cause atmospheric disturbances, but only in a limited zone: this is electricity. It has been seen what presumed part it plays in the barometric pressure, and we may infer:

1. That a perceptible loss in the electric tension of the air at any

given point is apt to determine a partial depression, hence a disturbance.

2. That an inrush from the upper strata into our immediate atmosphere (whether the inrush be direct or caused by the clouds that those elevated strata contained) must produce well characterized electrical phenomena, besides considerably lowering the temperature and thus causing snow and hail.

In reality the rôle of electricity in the atmosphere and its influence upon the displacements of the air are almost unknown.

It appears probable, à priori, that two neighboring clouds charged at different potentials are attracted one towards the other in order to equalize their difference, and that this attraction must bring on a displacement in the surrounding atmosphere. But can this effect be felt at great distances, knowing that electric attraction varies in inverse ratio to the square of the distance? We do not think so. The primary condition for a force to make a perceptible impression upon atmospheric inertia is the duration of this same force. Now, clouds cannot long keep electricity at a high potential; if the discharge is not immediate, the expansion goes on incessantly by means of the moist air, which is relatively a good conductor.

The source of electrical production is at the earth's surface, and the reservoir, in the upper regions. Everybody is familiar with the experiment with the Armstrong boiler. The boiler remains charged with negative electricity, while the wet steam that escapes from it is charged at a positive potential.

When water is evaporated on the surface of the globe by the sun's rays, the earth acts as the boiler, and the vapor of water, rising in the atmosphere and condensing, imparts by degrees to the surrounding air the electricity with which it is charged.

To this primary cause of the production of atmospheric electricity must be added another hypothetical one, the friction of the air against the earth's surface; a part of the work thus expended could be found in the upper regions under the form of electrical potential. It is certain that the upper strata of the atmosphere cannot go on receiving the electricity rising from the earth without expansion either by a sudden, tumultuous disruption, or slowly and by degrees; in other words, this accumulated force must find its use. Storms are one of the means of bringing on its return to zero. Is electricity the determinative principle, or else does it simply make use of the cirrus that rise to the upper regions to come down again in contact with the earth?

That is a hard problem to solve, considering the actual state of meteorologic knowledge.

Whatever it may be, and until proof of the contrary is brought forth, we shall continue to think that electrical attractions are the consequences, and not the causes of the motions of the air.

VAPOR OF WATER IN THE ATMOSPHERE.

Vapor of water plays a considerable part in the equilibrium of pressures. In order to justify its importance, it will suffice to mention the formula which expresses the weight P of a volume of air V, at the temperature t, when the exterior pressure is H and the tension of the vapor of water, f.

$$P = V^{\text{liter}} 1.293\, g \times \frac{1}{1 + at}\left(\frac{H - 0.378 f}{760}\right).$$

The formula may be translated as follows:

With even pressures and temperatures, a liter of dry air is heavier than a liter of humid air.

If, therefore, the barometric pressure be the same in two neighboring countries, the one dry and the other humid, everything being otherwise equal, it is safe to say that the air column is less in the first than in the second. Inversely, if the heights be the same, it is because the level of the mercury is higher in the dry region. The tension of the vapor of water forms an integral part of the atmospheric pressure; being absolutely independent of the particular pressure of the air, it varies in a general manner with the temperature, and for a given center, with the surface of evaporation, the emitting power of the soil, and the activity of the strata that contain it.

For a fixed temperature, the expansive force of water vapor attains a maximum that it cannot exceed. It is then clear that as long as this force remains below its limit value, the changes it may undergo react integrally upon the barometric pressure.

When that maximum is reached, if the temperature remains unaltered, the tension of the vapor of water, and consequently the total pressure of the air, remains unaltered.

We have seen that the vapor of water is the means of electricity accumulating in the upper regions of the atmosphere; it is also the vehicle of solar heat, which without it could not rise to any great height. It is a well known fact that luminous rays traverse dry air without perceptibly heating it; obscure rays reflected by the earth raise the temperature of the strata that envelop them. This air, made lighter by surface evaporation and its own dilation, has a tendency to rise to the upper regions. When the phenomena remain the same upon vast spaces, the surfaces of equal expansive force remain parallel with the earth; but the least disturbance, the least difference in the temperature of two neighboring spots may upset the equilibrium and determine a center of activity at the point where the upward pressure is greatest. The fact that heated air ascends is questioned by none; in order to demonstrate it, it is only necessary to place the hand above a lighted candle and to compare the sensation felt with the heat received when placed at the same distance from the light but in a lateral position. This difference results principally from the shock to the skin, of air molecules and carbonic acid endued with great upward velocity. And yet the products of combustion, in even temperatures, are heavier than air.

By the mere fact of its ascension the air becomes cooled; it is then that the vapor of water interferes to give by condensation renewed energy to the flagging movement. This vapor, first condensed in the form of aqueous vesicles or bladders, constitutes the clouds with their ever-varying forms; higher up, those clouds are composed of fine icy needles; it is the region of the cirrus; finally there exists, well within the confines of the atmosphere but at heights yet undetermined, a limit zone of the vapor of water, at the point where the temperature is such that the elastic energy of this vapor is null, and a limit zone of the clouds at the point where the density of the air is insufficient to maintain in suspension these meteors. If the clouds ascend still higher, this is due to the momentum of their ascending motion; and when this momentum is exhausted, they immediately descend towards the earth until they find a proper surrounding density.

Doubtless, at these limit zones stops the fluctuation of the vertical motions of the atmosphere.

INFLUENCE OF THE VAPOR OF WATER UPON THE CHANGES OF TEMPERATURE.

The water disseminated throughout the atmosphere under the forms of clouds or vapor acts like a heat moderator; it lessens considerably the variations of temperature, which would be excessive without its intervening agency. During the day the vapor of water absorbs in its formation a notable portion of the solar heat; at night the clouds hinder the dark radiations from the earth towards the planetary spaces and thus prevent the cooling off; when the sky is clear, the vapor of water yields to the surrounding air, by condensing into dew the heat that it held. But this precious characteristic is not without its drawbacks at the point of view of variations in pressure.

This condensed vapor, in fact, maintains in space a pressure which becomes lessened, and the level of the mercury is lowered by a quantity equal to the decrease in the tension of the vapor of water. Equilibrium can be re-established only by surface atmospheric displacements whose energy is in direct proportion to the depression.

In reality the barometric fall is inappreciable as long as the changes in the physical state of the water take place slowly, by a sort of relaxing process, so to speak, as in the diurnal phenomena we have already pointed out. It is different, however, when these changes are caused by a sudden cooling off similar to that produced by the inflow of the cirrus in our immediate atmosphere. In the latter case the fall is very marked, and the consequences are a surface atmospheric disturbance more or less energetic according to the copiousness of the condensations.

The above exposé may explain the segmentation of cyclonic disturbances into the secondary whirlwinds that form in the proximity of the main depression and move parallel with its track. We shall recur to this subject farther on.

INFLUENCE OF CONTINENTS UPON THE MOVEMENTS OF THE AIR.

When one seeks to infer a general law from pure theory combined with observations of facts, he should as much as possible lay aside all the causes of exceptions. Those causes are numerous in the study that engages our attention; they are the continents.

The ideal globe in which the wind system would possess perfect regularity should be a sphere destitute of land, and whose equatorial plane would constantly blend with the ecliptic plane; the whole would be symmetrical with the equator and any meridian whatever. Those conditions are far from existing, and the problem is much more complex.

Upon continents, causes variable in the extreme, arising from the nature of the soil, its form, its elevation, its emissive power, the vegetation with which it is covered, etc., have a constant tendency to modify the state of the temperature, and as a consequence the system of pressures.

The resistance that continents present to the motions of the air must necessarily exert an influence upon the velocity as well as the direction of the winds; and if, in order to deduce a general law, we were to consider only the limited spaces occupied by these lands, the results obtained would be justly considered doubtful.

When sailing over the oceans, especially the largest ones, one is struck with the harmony that reigns in that apparent chaos, and feels naturally inclined to seek the primary causes.

This is the task we shall undertake, now that we know how to assign the atmospheric disturbances to the differences of pressures, and have specialized the forces that act upon the level of the barometer.

RÉSUMÉ OF THE ATMOSPHERIC MOTIONS OBSERVED ON THE SURFACE OF THE GLOBE.

The wind system is nearly symmetrical in regard to the thermal equator, which oscillates a few degrees with the declination of the sun, north of the geographical equator.

In the regions between the tropics and the equator blow nearly constant winds, which have received the name of "trades," and which blow from the northeast in the northern hemisphere, and from the southeast in the southern hemisphere. The trades of the two hemispheres are separated by a calm belt; their polar limits oscillate with the declination of the sun.

The zone occupied by the trades is sometimes crossed, notably during the shifting period of the isothermal lines, by whirlwinds forming complete circuits, the winds shifting in an inverse direction of the hands of the watch in the northern hemisphere, and with the hands of the watch in the southern hemisphere.

These whirlwinds or cyclones are incited with a motion of parabolic translation whose concavity is turned to the east.

Higher up the trades, begins the belt of variable winds, in which the predominating winds blow from the southwest in the northern hemisphere; from the northwest in the southern hemisphere. (These winds have received the name of "counter-trades.") In this belt, the winds move generally (at least, in appearance to an observer standing still) in a contrary direction to the cyclonic rotation above referred to.

Finally, higher up, a limit not easily defined, but supposed to be near the 40th parallel, the winds blow nearly constant in a direction comprised between north and south, through west. This zone is called the belt of the general western winds.

We shall now try and explain all these movements and deduce a few conclusions relative to navigation in the belt of the general winds.

INFLUX WINDS (VENTS D'ASPIRATION).

When a communication is opened between two rooms of unequal temperature, an inferior current of cold air flows towards the thermal maximum, while an upper warm one flows in an inverse direction. The difference in temperature creates a difference of pressure. In order to re-establish the equilibrium, a displacement of the air is necessary—the cold air, being heavier, flowing below; the warm air, lighter, above. The whole solar action on the surface of the globe may be summed up in these few words.

The most remarkable cases of winds of "aspiration" or influx air currents are the land-and-sea-breezes, the trades, and the monsoons.

At a great many points upon the shores of the oceans there have been noticed two inverse atmospheric motions in the course of the same twenty-four hours. During the day the lands get more heated than the seas; the layers of air that envelop them become dilated, and, rising to the upper regions, create an inflow whose center is at the same time the thermal maximum and the barometric minimum; the sea breeze that rushes in to fill the vacuum begins to blow in the early hours of the afternoon. During the night, on the other hand, the land loses through radiation more heat than the sea, the barometric minimum shifts conjointly with the thermal maximum, and the flow of air seawards begins in the early hours of the morning. In either case the return of the heated air takes place through the upper regions of the atmosphere.

TRADE-WINDS AND MONSOONS.

The trade-winds come from the same causes: the annual mean temperature is highest under the thermal equator; the air, extremely dilated and saturated with vapor of water, produces a barometric minimum—a relative minimum, of course; it is therefore another center of incitement for the strata of air situated to the north and south.

The monsoons are the exclusive results of the disturbing influences due to continents; they consist in land-and-sea-breezes lasting several months.

OBLIQUITY TO THE MERIDIAN OF THE TRADE-WINDS AND MONSOONS.

The trades being occasioned by the difference of pressures, which themselves result from the inequality of temperatures, should always move parallel with the meridian, since this inequality is most marked on the meridian.   But all the parts of the globe revolve to the east with a velocity proportional to the cosines of their latitudes; hence it results that the molecule which possesses at its point of departure a given velocity is behindhand on the successive parallels that it crosses to reach the equator; its relative motion is, therefore, oblique to the meridian.   This is why the north trade blows from a quarter nearing northeast, whilst the south trade blows from near southeast.

MONSOON OF THE INDIAN OCEAN ; RETROGRADE MOTION.

During the northern summer the rarefaction of air produced by the overheating of the vast plains of Central Asia transforms the northeast trade of the Gulf of Bengal into a southwest monsoon.   It must be noted that the extending motion of the monsoons is a retrograde one ; that is, it takes place inversely with the direction of the wind.   Thus, when the southwest monsoon of the Indian Ocean sets in, it is first met with in the northern parts; it progresses backwards; it is observed at Calcutta earlier than in Ceylon, and in Ceylon earlier than on the equator ; every day it is 15 or 16 miles farther to the south.

This motion, so characteristic, defines quite accurately the breezes of "aspiration," and justifies the name.   Let us also observe that the monsoon of the Indian Ocean partially destroys the symmetry of the wind system relatively to the equator.   In the Mozambique Channel the winds in October predominate from south-southwest, south, and southwest, and continue to blow from those quarters down to the equator.   They thus join off the coasts of Africa the southwest monsoon, which reigns north of the equator in the sea of Oman and Gulf of Bengal.   This is not an exceptional case, but is the largest body of air that establishes a communication between the two hemispheres.

IMPETUS DUE TO THE ROTATION OF THE EARTH ; IMPELLED WINDS.

When we sail up the tropics the unequal distribution of heat over the surface of the earth is insufficient to explain the atmospheric disturbances. The temperature of a spot, in fact, is solely a function of the latitude, setting aside the continents; and if it be true that the force of the flow is less as we get farther from the equator, it is no less true that the "aspiration" works from place to place gradually, so that the air column of a given parallel is drawn towards the warm regions through the vacuum created by the moving columns of air of the intermediate parallels.

Why then do we see this motion stop on either side of the equator near the 25th parallel? It is because the mean barometric pressure keeps constantly increasing from the 25th parallel to the equator, and constantly decreasing from the 25th to the polar regions; yet the air of the higher latitudes is colder, less saturated with moisture, and weighs more to the same volume than that of the lower regions. It is then necessary that the thickness of the atmospheric stratum go on decreasing from the limits of the trades to the poles—in other words, that the pole be rarefied. The cause of this decrease is entirely in the centrifugal force occasioned by the rotation of the earth upon its axis.

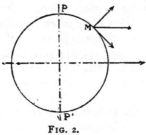

FIG. 2.

Let M be a point whose latitude is l. If F be the centrifugal force developed at the equator, that at the point M will be $F \cos l$, whose tangential component is $F \cos l \sin l$ or $\frac{1}{2} F \sin 2l$.

. This component is not an unimportant quantity, since the centrifugal force at the equator decreases the gravity by $\frac{1}{289}$; nothing at the pole and at the equator, it attains its maximum value under the 25th parallel.

In reality this component has no action on the bodies placed on the surface of the earth ; owing to the elliptical form of our globe, the resultant of the centrifugal forces and gravitation is normal on the surface of still water. Are the conditions the same in the atmosphere ? In other words, is the resultant of all the forces that pass

through a given molecule normal to the envelope that contains it—
an envelope that can be conceived as perfectly moulded on the form
of the globe? This indeed would be the case, and the gaseous
envelope would have no relative motion to the earth, the angular
velocity remaining the same to the extremities of the aerial radius,
if the sun did not determine in the mass vertical indraughts (aspirations) along this radius, and as a consequence, displacements in the
direction of the meridian.

The effect of the sun being to change constantly the level of surfaces of equal expansive forces, the atmosphere cannot remain indifferent to the tangential component of rotation; but at what height
does this force find the most favorable field to exercise its action?

It appears rational that close to the earth, friction, the pressure of
the upper strata, and the cohesiveness of the molecules to each
other, raise an impassable barrier against the tangential component;
but the situation is quite different in the upper regions of the atmosphere, where the air, being lighter and freer, is endued with extreme
mobility, and yields unrestrained to every impulsion. Moreover, its
lineal velocity of rotation, and consequently its centrifugal force, is
greater than on the surface. The air of these regions is therefore
drawn towards lower latitudes, and this impulsion is marked, as
shown by the barometric curve, by an increase in the density of the
atmospheric envelope from the pole to the 25th parallel. The difference in pressure between these two points represents the work
accomplished by the tangential component of rotation.

It is evident that air cannot keep thus accumulating towards the
lower latitudes without seeking an issue. On the other hand, the
rarefied regions of the pole become in turn indrawing centers, and
it is for this reason that we see predominating, from the limits
of the trades up to the higher latitudes, surface breezes that blow
from southwest in the northern hemisphere, from northwest in the
southern hemisphere. These breezes start at the north and at the
south, according to the hemisphere; the rotation of the globe then
modifies their initial direction, because they are in advance, upon
the successive parallels that they cross in reaching the poles. To
this primary cause of the prevalence of the western winds in the
extratropical belt must be added the tendency to the algebraic
equilibrium of the impulsion of the air.

Here a few words of explanation are necessary. If the earth were
isolated from the sun, in order that the period of gyration around the

axis should be constant, it would be necessary that every aerial displacement, in whatever direction, caused by an interior force, should have for a corollary a motion equal in quantity and in an opposite direction. Now, it is admitted that the earth throws off from its bosom towards the interplanetary spaces a quantity of heat at least equal to that it receives from the sun. The forces developed by this luminary may, therefore, be considered as interior forces in the interval when the quantities given out and received are equal : we have then the right to apply to the whole of the forces that govern the atmosphere the theorem of the interior forces, and to establish as a principle that the algebraic sum of .the impulsion given to the aerial envelope tends to zero, and becomes strictly null from one year to another.

The trades of the two hemispheres cover a belt representing nearly two-fifths of the whole surface of the globe. They have a very marked west component, due, as already explained, to the rectilinear velocity eastward, which attains its maximum at the equator. Having reached the center of indraught, these great gaseous masses annul their polar component and, rising in the atmosphere, continue to move westward by virtue of their acquired velocity. The movement is all the more marked, as the height they reach is greater. It is only by slow degrees that they swerve towards the poles to become decidedly southwestward and northwestward, according to the hemisphere. It will be readily seen that in this immense area up or down, the motions of the air in the majority of cases are westward.

In order that the equilibrium of the impulsions be a fact, it is necessary that in the extratropical belt the motions of the air be in a great measure directed eastward.

[Let us note incidentally, that the seasons when the western winds are stronger and most prevalent correspond with the winter period, when the trades are more intense and less intermittent.]

Such are the origins of the counter-trades and western winds in general. It seems difficult to explain otherwise than we have done the distinguishing character of the counter-trades, which is to move in an inverse direction to the breezes of "aspiration" from the warm latitudes to the cold ones. It has been accepted up to the present that these winds are caused by the landing of the return upper trades. Such being the case, how would the surface trades be fed, and how explain the barometric maximum occurring at the very point from which the trades and counter-trades start on their opposite journey? The theory of the return of the two upper currents, on the contrary, furnishes us with a rational explanation of that maximum; thanks to this very meeting, the trade and counter-trade are naturally fed. And now let us study the origin and formation of the normal whirlwind or anticyclone.

But first, we must mention a secondary effect of the tangential component of rotation, an effect that has recently been pointed out by the eminent American meteorologist, W. Ferrel, and which may be expressed by the following formula: All aerial currents flowing eastward tend to the equator, and all currents flowing westward tend to the nearest pole.

When the atmosphere is at rest relative to the earth, the centrifugal force with which it is endued is the same as that of its contact parallel; but if its velocity eastward or westward be greater or less, its centrifugal force, and, as a consequence, the horizontal component of this force, will be superior or inferior to that of the same parallel. The aerial current will therefore assume in regard to the earth a relative motion whose direction will be governed by the difference in the centrifugal forces, towards the equator, if moving eastward, towards the pole, if moving westward. With these conditions, the law of the wind-gyrations appears to be in plain conformity with the phenomena that come under our daily notice. In the northern hemisphere the winds revolve like the hands of a watch; in the southern hemisphere in an inverse direction.

NORMAL WHIRLWIND OR ANTICYCLONE.

The barometric curve points to the existence of a maximum upon the parallels near the tropic; on the other hand, we know that this line is the starting point of the trade going towards the equator and the counter-trade going up towards the pole. The apparent anomaly of this maximum and flow of air in two opposite directions can find an explanation only in the meeting of the superior return currents: the return of the trade in the upper regions of atmosphere, and the impulsion of the polar air towards the equator, also in the upper regions.

How will the meeting of these two upper currents take place, and what gyratory movements will they originate? Both are evidently oblique to the meridian, and both possess a tendency to gyrating due to the tangential component of rotation. Considering only the southern hemisphere, we see that the direction of the polar current is from the southeast, with a tendency to blow from a point of the compass that removes it farther from the equator, east-north-east ; the direction of the equatorial current is from the northwest, with a tendency to blow from a direction nearer to the pole, west-south-west.   If we suppose that the impulsions are destroyed by the meeting of these two adverse currents, there remains present only their tendency to gyrate in the same direction, and the normal whirlwind is formed.   This influx of air from two opposite directions will seek an issue in a vertical direction above as well as below ; hence a descending current that feeds the trade and counter-trade, and an atmospheric inflation·clearly demonstrated by the barometric maximum noted on the curve.

Such are the conditions under which the two upper currents will meet : around the center the breeze revolves in an inverse direction to the hands of a watch, the polar current to the right, the equatorial current to the left.

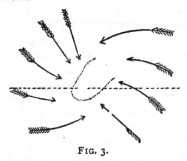

FIG. 3.

Fig. 3 represents the sum-total of the determining forces of the movement in the upper regions of the atmosphere, but the breezes that come down to us on the surface from either side partake at the same time of the enveloping motion of the current to which they belong and the indraught energy in the direction of the radius, occasioned by the difference of pressure existing between the barometric maximum and the surrounding spaces.   It is, in fact, known from experience that the winds that contribute to an area of high pressure have a marked centrifugal tendency.

The wind charts of Commander Brault, published by the " Dépot " and bearing Nos. 3383 and 3384, point out the existence of similar gyratory motions—the one around the Azores during the northern summer, the other around Tristan d'Acunha during the southern one. Around the Azores the breeze revolves like the hands of a watch; around Tristan, in a contrary direction. These two symmetrical gyrations agree perfectly with the barometric maxima described at those two points during the above-mentioned seasons.

SUMMER INFLUENCE.

These movements are principally noticeable during the summer. In this season, in fact, the difference of pressure between the higher and lower latitudes becomes less, the atmosphere expands under the beneficent influence of the solar rays, and the tangential component, whose action as we have said affects the upper regions only, carries towards the equator a light atmosphere, whose energy is feebler than in winter; everything takes place, therefore, as if the density of the atmospheric envelope had increased in the high latitudes. Moreover, the flow of the trades is diminished in consequence of a greater uniformity in the distribution of heat; the upper return trade is therefore less energetic, all the movements are more quiet, and the two upper currents, instead of rushing to fill a vacuum and create formidable whirlwinds, oppose one another from the effect of their acquired velocity, and form a calm area and a barometric maximum upon their meeting parallel.

STABILITY OF THE NORMAL WHIRLWIND.

The volume of those great masses of whirling air has no translatory motion, and there is no reason why it should. If the supply of the equatorial current is more abundant than that of the polar current, it seems that the latter, traveling a greater distance and distributed over a larger area, must be endowed with a greater velocity, and the energy of the two is thus equalized.

ROUTE OF A SHIP IN A NORMAL WHIRLWIND.

A ship running east-southeast in the South Indian Ocean, and meeting the cyclone on the west side, will notice the breeze working from north to south through the west, the barometer remaining high, but lowering somewhat with the northern winds, to rise again with the southwest winds. If the route of the ship should cause it to sail too close to the center, it will meet with calms characterized by a barometric maximum.

All these results perfectly agree with the data furnished by experience.

Conclusions.—These normal whirling motions are not fleeting meteors like the cyclones; their diameter is considerable, and the conditions they determine in a given region may endure for months. These normal whirlwinds or anticyclones cause great changes in the temperature—extreme heat in summer, intense cold in winter; this is caused, apparently, by the dryness of the air columns descending from the upper regions.

ABNORMAL WHIRLWINDS OR CYCLONES.

The cyclone is a whirling motion of the air in which the breezes revolve in the same direction in the same hemisphere. In the center of the vortex, characterized by a great barometric depression, there exists a relative calm; this central portion acts like a flue for the air streams in motion. All these whirlwinds describe a parabolic trajectory whose concavity is turned eastward.

Where do they originate, and to what agency do they owe their formation? Cyclonic movements are met with in all latitudes; they reach the coasts of France and England only after crossing the Atlantic from southwest to northeast, or from west-southwest to east-northeast; but the breezes revolving around their center are far less energetic than in the intertropical regions. Cyclones appear to originate in the neighborhood of the thermal equator, particularly at the time of the change of the monsoons in those belts where there exists a double annual system, and at the time of the displacement of the isothermal lines in the zones where the trades blow free. For instance, in the South Indian Ocean these storms occur most frequently during the months of February and March, at the epoch when, the sun ascending north, the trade winds, weakened and disturbed during the summer period, resume again the energy and regularity they will preserve during the whole winter. In the Gulf of Bengal, cyclones manifest themselves only at the time of the change of the monsoons, and those that make their appearance at the beginning of the northeast monsoon are more powerful and destructive than those that spring up with the southwest monsoon in May.

It seems from the above instances that cyclones are due to a disturbance of the dynamical equilibrium of the atmosphere, a disturbance caused by a change in the conditions of the distribution of solar heat at the surface.

Theories of the Cyclone.—In later years the theory of cyclones has excited a lively interest among the learned, and has given rise to two contrary opinions: the theory of "aspiration" or ascending motion; the theory of descending motion, due to M. Faye. Both theories have their fervent adherents; it is not our province to solve the controversy; we shall simply sketch the main lines of the dispute and point out that neither theory is exempt from scientific criticism.

Theory of Aspiration.—The older in point of date, and also the one that rallies the most adherents, is the theory of aspiration. We defined it in a few words at the beginning of this study.

When a center of depression is created at a given point, owing to a more active evaporation, to the superheating of the air masses that envelop the earth and find their expansion in the upper regions, and owing also to the vacuum that the incipient monsoon creates behind it at the moment when the dying monsoon still keeps on its course—the air is solicited on all sides with an energy proportional to its expansive capacity, *i. e.*, in most cases to its density. As the density is maximum at the surface, the equilibrium of pressure tries to reform by renewals of surface air. But the directions of the inflowing breezes are modified by the rotation of the globe, so that these breezes, instead of directly reaching the center and filling the vacuum, can do it only after a succession of spiral movements. The centrifugal force developed by the curving of air currents contributes to perpetuate the movements by moderating the influx to the center. The gaseous masses that finally penetrate into this flue-chimney must acquire, at least at a certain altitude, a considerable ascending energy, if we are to judge from the great heavy clouds that pour down upon the whole periphery of the vortex after visiting the upper regions. During this journey the liquid drops have turned into fine icy needles. When their energy in centrifugal flight, as well as in height, is well expanded, they incline earthward by the law of gravity, and cause in the spaces surrounding the trajectory of the cyclone atmospheric disturbances, such as thunderstorms, waterspouts, and secondary depressions.

Cyclonic Translation.—Cyclones change place; they move in a parabolic curve whose concavity is turned to the east. But here rests the theory of aspiration, powerless to define the reason why in the intertropical zone the whirl appears to be drawn by the upper air masses instead of following the motion of the lower stratum.

THEORY OF THE DESCENDING MOTION.

M. Faye compares atmospheric disturbances, such as waterspouts, hurricanes, cyclones, etc. (the magnitude of effects depending solely on the energy of the forces at work), to the whirling motions forming in streams, from the inequality of velocity in the surface, liquid threads or currents. "When in a stream there occur differences of velocity between currents flowing side by side, there is a tendency to form at the expense of those inequalities a regular gyratory motion around a vertical axis. The spirals described by the molecules are perceptibly circular and central with the axis; they are, to speak more correctly, like the threads of a screw slightly conical and inclined downwards, so that by following a molecule in its movements one sees it revolve with rapidity around the axis, which it nears little by little with a descending motion much less rapid than the lineal velocity of rotation."

According to the above theory, therefore, the origin of waterspouts, tornadoes, hurricanes, cyclones, etc., would have nothing in common with the physical phenomena we notice on the surface of the globe; it would exclusively proceed from a difference in the lineal velocity of air currents moving above our heads, and *ipso facto* be a pure caprice of the elements. Everything would take place as if the confines of a mass of air in motion and the vacuum were as decided as the surface of a stream of water; as if the density of the air were uniform from the top of a whirlwind down to the bottom, the same as in the hydraulic mass.

We have seen that the origin of cyclones is intimately connected with the régime of surface temperatures.

It is also well known that storms are more frequent in spring, the season when the isothermic lines are suddenly disturbed, than at any other time; that hail falls far more frequently in the daytime than at night, and that the maximum of falls corresponds with the diurnal maximum of temperature, towards 2 o'clock in the afternoon.

Finally, the season of the pamperos in La Plata, and tornadoes on the African coasts corresponds to the period of excessive heats.

It seems to us, moreover, that a vast funnel, as described by M. Faye, can be formed only concurrently with the existence of a vacuum above our heads, and the vacuum, if it exists, is found at an altitude of 300 kilometers. In the contrary case, every attempt at a partial vacuum will be opposed by the upper strata flowing down vertically; it would therefore be necessary to admit that the whole column of air is sucked up by the whirlwind to that prodigious height where the atmosphere is, so to speak, imponderable. On the other hand, the descending column, as represented to us, is composed of air currents inferior in density to the strata through which they must pass. Here is something abnormal that the mind cannot easily conceive. Does it not require a steel point to pierce bronze or iron? And would it not be consistent to suppose that under the pressure of the gaseous masses flowing from top to bottom, the level of the mercury should rise? Now, it is just the contrary that happens.

Those are not the only objections that can be raised against M. Faye's learned theory. In order to explain away gyration, which is always the same in the same hemisphere, we are simply told: "the direction of the rotation of cyclones may be attributed to the fact that in these extremely curved currents, velocity diminishes transversely from the concave edge to the convex." Contradiction is very difficult owing to the impossibility of verifying the assertion, but the explanation is not one that will fully satisfy the mind; it is a statement *ipso facto*.

Finally, one last objection: in all the cyclonic disturbances that visit our shores (Europe), the winds felt at the surface have a decided centripetal tendency. It suffices, in order to make sure of the fact, to cast a glance upon a synoptical chart of isobares when a depression traverses Europe from west to east; the breezes converge towards the center as if a regular suction took place there. In the hypothesis of the descending currents it would seem, on the contrary, that the moving gaseous masses should expand, in diverging, from the base to the periphery.

It should be stated that lately* M. Faye has made an important concession to the theory of "aspiration": he admits, *à priori*, "two cyclones with very different depressions," the ordinary intertropical cyclone and the fixed depressions, "where the succession of phenomena works quietly; it is a question of static meteorology. There are formed towards the periphery more or less convergent breezes (regularly deviated by the rotation of the globe), but no violent gyrations. The air *ascends slowly* in them."

In the following sitting,† M. Mascart, the learned superintendent of the "Bureau Central Météorologique," whose opinion in such

---

* Compte rendu of the Academy of Sciences, July 2, 1888.
† Ibid. July 9, 1888.

matters is too weighty to be passed over in silence, took notice of that declaration in the following terms: "With the exception of the meaning attributed to *fixed depressions*, and the relative importance of the phenomena, this is a new concession that I am happy in recording. I should be pleased to hope that our colleague will go a step further, and, yielding to the evidence of facts established all the world over, recognize that the partial convergency of the wind in depressions is the general rule, as well in cyclones of all kinds as in the mean annual or season effects."

We must add, before finishing this cursory glance at the theory of descending motions, that if the various criticisms we have formulated be swept aside, this theory explains quite satisfactorily the movement of translation of cyclones and their segmentation into secondary whirlwinds. This is one of the most important points where the defenders of the "aspiration" theory fail obviously; none of the explanations upon this point published up to the present, at least as far as we are aware, fully satisfies the mind.

M. Faye has, after all, stirred up a world of new ideas and fixed the minds of the learned upon a question worthy in the highest degree of their attention.

After all, what is of importance to the sailor is not the "why" but the "how" of cyclones. It is to his highest interest that he should be filled with the idea that cyclonic movements of closed circuit are met with in all parts, that he learn to recognize their approaches under the higher latitudes, and manœuvre in view of his objective point instead of resigning himself to kick about on the same spot from sheer ignorance in regard to the phenomenon.

What is also essential to remember touching intertropical cyclones, by far the most dangerous, is that the almost perfect circularity of the wind around the center is admitted for these storms by many eminent meteorologists. M. Faye in this respect is very positive. M. Mohn, a partisan of the adverse school, expresses himself thus: "In the revolving storms of the tropics the greater portion of the movements of the air is performed *circularly* around the center; but the great velocity causes an extraordinarily considerable mass of air to pass constantly from the exterior into the interior of the hurricane, whilst the latter surrounds the vortex. Thus, therefore, there must exist in intertropical cyclones, the same as in our own cyclones, an ascending air current accompanied with all the phenomena that are inherent to it and sustain it."

Such being the case, the rules we are acquainted with are all that can be desired, and it will always be possible, provided the structural strength of the ship will allow it, to sheer off from the center in order to avoid the greater force of the hurricane.

We must, however, make some reservations in regard to one of those rules set down by Commander Bridet in his work on hurricanes in the Indian Ocean. It is known that the approach of the cyclone off the island of Réunion is indicated by a violent freshening of the breeze, besides the fall of the barometer and the leaden gray clouds that sweep over the sky. Admitting as a principle that the center of the disturbance is always at right angles to the wind, Commander Bridet advises captains who notice the trades increasing in freshness without changing direction, to bear away so as to cross the path of the cyclone forward of the center, and to keep on this course until the barometer shows a decided rise, and the wind begins to veer to south and southwest. At this moment, bring to on the starboard tack.

The above manœuvre would be perfect if the air currents described an exact circumference; but it has been pretty well ascertained that, at least within a certain distance, the winds converge partially towards the center of the inflow. The ship that would strictly follow the above rule would be in danger of running into the cyclone, and perhaps passing through the very center that she was trying to avoid. To provide against this grave danger it would be only necessary to bear up two or three points, *i. e.* to steer west-northwest or west instead of northwest; but every ship will not be able to steer that course, and the advice applies only to strongly built vessels; the others will simply come to on the port tacks and maintain their position if the wind veers to the east and north, and put about, on the contrary, if the wind veers to the south and west.

GENERAL WESTERN WINDS.

When we sail higher up than the 40th parallel, the atmospheric circulation on the surface hardly varies except from north to south through west, or the reverse according to the hemisphere.

In these parts (Europe), as has been admirably shown by M. Lephay, the west winds are an indication and the consequence of a depression crossing the Atlantic from southwest to northeast. Around this depression the winds revolve in an inverse direction to the hands of a watch, so that an observer standing still may see the breeze veer from southeast to northwest through west, if he is to the right of the path; from east to northeast, and northwest if he is to the left. It happens in most cases that on the left edge of the cyclone, which is the manageable half-circle, the winds remain feeble, even blow from the west, the velocity of translation being greater than the gyratory motion. The center of the disturbance generally passing south of Ireland going towards Central Russia, it follows that the French coasts are swept by the right edge of the cyclone, on which the breezes blow from southeast to northwest passing through west.

Must we adopt for the whole of the zones of the so-called general western winds the conclusions reached by M. Lephay in regard to the Atlantic, to wit, "that the existence and vagaries of a well established aerial current are proofs of the passage of a cyclone in the neighborhood"? We do not think so. Certain well observed western gales, for instance the strong winds on the Needle Banks during the southern summer, present but a slight resemblance to cyclonic disturbances. On the other hand, what mostly strikes the navigator in the Indian Ocean or the Pacific is the persistence or rather the permanency of the western winds. Some ships have sailed over the immense extent of seas that reach from the first meridian to Tasmania without encountering an eastern breeze even for an hour. There must then exist in our part of the globe conditions that can be explained only through the influence of continents. We are too near the Gulf of Mexico, where cyclones have their origin, and the American continent is too close to allow the winds fair play in the interval that separates us.

In our opinion, the forces we have considered above are the only ones that contribute to the great aerial circulation, and cyclones in general are but accidents that may modify the foregoing laws without impairing them in the least; perhaps even some depressions are formed on the spot by the alternate action of the western and eastern winds. Let us consider only the southern hemisphere. We have seen that the meeting of the upper return trade and the breezes brought down from the pole through the horizontal component of the centrifugal force determines an atmospheric inflation upon a pretty extensive belt; the difference of level between that belt and the cold regions determines a flow of surface air towards the pole.

These winds begin at north, then shift to northwest in consequence of their lineal velocity eastward, and inflect more and more to the west and southwest, as much owing to the tangential component as to the general law of algebraic equilibrium of the motion quantities. If these winds renew themselves constantly with the same gyratory motion, it is because the forces that rule them are constant, independent of the temperature, and because they derive from the rotation of the globe around its axis a movement that is uniform and constant. We even believe that the anti-trades (southwest in the northern and northwest in the southern hemisphere) would be as regular and as permanent as the trades, if their air channel did not go on narrowing, whilst that of the upper wind of impulsion, on the contrary, goes on widening ; in other words, if the relation of surfaces to be attained was not zero or infinity. In the relative conditions in which they are placed, the performance of these two superposed winds can only be a succession of fitful inrushes seeking to realize an impossible dynamic equilibrium.

Let us suppose that we stand still on any spot whatever in a high latitude and observe what takes place on the surface. The winds having originated in the north to die out in the south, after passing through the west, have determined a rarefication in our west and a relative barometric maximum in the east. In virtue of the continuity of the pendular movements, the end sought has been overreached ; therefore it becomes necessary to fill up anew this rarefaction that solicits the breeze on all sides. But while from the north or northwest fresh supplies are constantly arriving, on the southeast the air keeps on rarefying, owing to the flowing towards the pole of the whole lower mass; there is therefore no equality in the struggle about to begin.

Three cases may then present themselves: either the depression will be filled by the northwest winds alone, and the movement will continue as before, or else the southeast winds will be feeble and, after passing around east and north, will die out after a short while before the flow of the northwest winds; or again, the depression may be such that the energy of the eastern breezes may not be unimportant, and the combined evolution of the two opposing winds around the center of depression will create a cyclonic movement that will sweep like the lower stratum from northwest to southeast. As may be presumed, the latter will occur only in the case of a marked depression, for instance, when the western winds have been very violent ; thus one storm begets another of a different nature.

We repeat, that in our part of the world, where the solar action operates through the continents in quite a different manner, the sur-

face meteorologic phenomena must present a different character; the
anti-trade rises no doubt to a certain altitude, where it reigns supreme.
It is, moreover, well known that in mid-Atlantic the west winds are
more prevalent than near land, and that in winter they are uninterrupted. But in the south seas, where these disturbing causes are
absent, where resistance to the movements of the air is almost null,
atmospheric phenomena present a great regularity; this is confirmed
by reports of captains and books of navigation, and we cannot do
better than to compare the alternate action of west and east winds with
the pendular motion of a ship well heeled over: when the reactions
on the immersed side are very energetic, the ship, passing the vertical
line, rolls to windward, but only to roll all the more heavily to
leeward; the roll to leeward corresponds to the west winds, and the
roll to windward corresponds to the east winds.

GALES OF WIND ENCOUNTERED BY THE TRANSPORT LA DORDOGNE.

It must not, however, be supposed, on the strength of what precedes, that sailing over the zones of the western general winds presents no difficulty and that the breezes are invariably favorable. The
winds will sometimes sweep violently over them from the east. In
such case one can easily judge whether he has to deal with a cyclone,
and manœuvre accordingly in order to save time and distance.
This we propose to demonstrate from our own experience.

In a voyage around the world, from the latter part of 1883 to
the middle of 1884, the sailing transport La Dordogne encountered
within a month three successive gales in the southern hemisphere.
Let us analyze them.

On the 11th of September La Dordogne was in 40′ south latitude and
4° 34′ west longitude; the course was east-southeast, with a moderate
breeze from northwest; the barometer, which had been rising for two
days, showed 750.5 mm. During the afternoon watch the weather
became squally, the horizon in the north thickened, the barometer
falling 2 mm. About 6 o'clock, during a heavy squall, the wind
shifted suddenly to east-northeast, blowing a gale. Night passed on,
the barometer still falling, and the wind veering around little by
little to the north, the ship was hove to on the port tack. The
following day, the 12th, at 8 in the morning, the barometer stood at
734 mm., its maximum level, and the wind blew from north-northwest. During the entire day the barometer kept rising and the wind

became more favorable; finally, on the 13th, at 8 in the morning we continued our route east-southeast, with a fair breeze from the west. Later on, the wind backed to northwest, blowing a fine, steady breeze.

Fig. 4 represents the relative movements of the cyclone and the ship. Struck by the cyclone with winds from east-northeast, we sailed over a chord of the dangerous semi-circle. We assumed as the direction of the movement of translation of the whirlwind the position of the wind at the moment of the barometric minimum north-northwest to south-southeast.

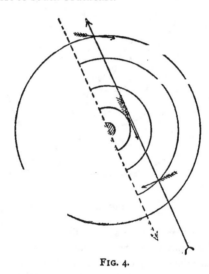

FIG. 4.

It may be seen by the figure that the convergency of the wind is almost null as well at the isobares of the front part of the whirlwind as in the second part. We may conclude from what precedes that we crossed in its dangerous semi-circle a revolving storm going south-southeast.

Gale of December 25th.—On the 24th of December the Dordogne stood at noon in 43° 40′ south latitude and 42° 32′ east longitude, with a light breeze from east-northeast, the barometer at 754. At midnight the wind blew moderate from east-southeast, squally overhead, the barometer indicating 749 and falling rapidly during the night, wind freshening. At 6 in the morning we lay to on the port tack, wind very fresh from southeast, barometer 738. At noon the barometer stood 734, the wind blowing a gale from southeast. From that moment the wind shifted to west and the barometer rose. At 8 in the evening the wind was from the west-southwest, blowing a strong breeze; barometer 745. Finally, at 10 in the morning of the 26th we hauled east-southeast with winds varying from southwest to west.

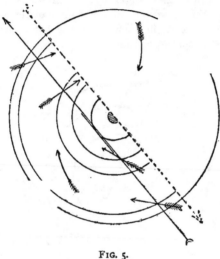

FIG. 5.

As in Fig. 4, we have chosen as direction of the path the position of the wind at the moment of the barometric minimum, but in Fig. 5 the air streams are decidedly convergent. We may again conclude that, sailing over a chord of the manageable semi-circle, we crossed a cyclonic disturbance moving to the southeast.

Gale of January 6, 1884.—For the third time we were warned of the approach of a cyclone through the same indications: rapid falling of the barometer, threatening aspect of the weather, specially in the north; winds from the east, fitful in strength and direction. The captain of the Dordogne, convinced that we were in the presence of a cyclonic disturbance similar to the two preceding ones, decided to pass around the storm on the west and seek north of it winds favoring our course to the east. Here is an abstract from the ship's log :

January 5th, position of the ship 44° 46′ south and 70° 40′ east, running east with a light breeze from west-northwest. During the day the wind shifts about; from west-northwest it veers to southwest, and suddenly jumps to south-southeast, then northeast, and, finally, towards 7 o'clock in the evening settles at southeast, freshening gradually; the barometer has fallen 4 mm. The course is set at north under shortened sail, with orders to keep the wind four points on the quarter, hauling to starboard as the wind veers aft.

This was done. During the entire night, the ship going under close-reefed foresail and fore-staysail, keeps the wind four points on her quarter; at 2 o'clock the barometer stands 475 mm. minimum point, the breeze blowing from south; at 4 o'clock it blows from south-southeast; finally, at 11 o'clock in the morning we resumed our course east, one quarter south, with wind from the southwest; the sea was very choppy, an unmistakable proof of the passage of the cyclone. We lost 70 miles northing, but the winds were favorable, while on the western edge the fluctuation of the barometer was insignificant. Owing to this skilful manœuvring, we saved, perhaps, 36 hours of lying to, and took advantage of favorable winds, which, in cyclonic disturbances in the southern hemisphere, are north of the adverse winds.

We will remark, in passing, that such manœuvre is practicable and unattended with danger only when the ship is assailed in the manageable semi-circle, where the breeze is far less violent than on the other side. Besides, if from the beginning the ship is kept before the wind, there is very little need to fear that the convergency of the winds will drive the vessel into the calm center, owing to the east component of the movement of translation of the whirlwind, which takes the cyclone farther away while the ship heads a course near to north.

We cannot cite many examples of a similar manœuvre, but there is as much value in one single experience as there is in a whole series, and there is no possible reason why a successful manœuvre of the Dordogne would not prove equally so with any other ship.

What is most important to remember in regard to the preceding examples is that:

1. The whirlwinds of the high latitudes may form a complete circuit, *i. e.*, have their share of east winds, which winds are adverse when sailing eastward.

2. The extra-tropical whirlwinds of the southern hemisphere follow a track lying between east and south, generally close to southeast.

The manœuvre of the able commander of the Dordogne is worthy of imitation, but how is one to know of the approach of the storm? By winds from the eastern quarters fitful in force and direction; by the abnormal gyration of the breeze; by a marked fall of the barometer; by the cirrus overspreading of the sky. We do not hesitate to say that the latter indication is perhaps the best.

Whether the cirrus are the causes or effects of cyclones; whether they are incited by the inflow of humid air into the central section, which on rising in the upper regions becomes cooled and then pours down over the periphery; or whether they are the source of an active agency and electricity on which these meteors feed—is a matter of little importance to us. What is far more interesting is to recognize them in time; for their intrusion in our immediate atmosphere, to which they tend by law of gravitation, will cause abundant precipitations, and it may happen that when we wish to examine the sky it is covered by a veil of low clouds.

For two years past we have studied attentively on the coasts of France the cirrus formation, and we can affirm that there is no better forecast, including the barometer, of storms, gales with rain, hurricanes, and bad weather in general. The abundance of coming rain and the violence of the wind may be reckoned by the apparent thickness of the layer of the cirrus; even in summer, when the fierceness of the sun's rays dispels the clouds, we have never observed a non-concurrence between the appearance of small, white, dense, and very high clouds and the subsequent coming of atmospheric disturbances.

RÉSUMÉ AND CONCLUSIONS.

There exist in the belts of variable winds and general winds two inverse gyratory motions: the first, a normal whirlwind of vast extent caused by the meeting of two adverse upper currents. Its characteristics are slight barometrical fluctuations with a central maximum. The west winds are to the north of the east winds in the northern hemisphere; they are to the south of the east winds in the southern hemisphere.

The second, whose range of energy is much less, is the abnormal whirlwind. We think that it is a cyclone either formed on the spot, as we have explained, or in the intertropical regions, whence it follows the polar branch of its path. It is characterized by considerable barometric oscillations, with a central minimum. The west winds are to the south of the east winds in the northern hemisphere; they are to the north of the east winds in the southern hemisphere.

Independently of those two rotatory motions, the winds in the extra-tropical belts generally vary from north to south passing through the west, or inversely according to the hemisphere. These motions are due as much to the rotation of the globe as to the tendency towards the algebraical equilibrium of the impulsions of the air.

SELECTION OF A SAILING PARALLEL.

The first problem to solve for a captain on a long eastward voyage is the selection of the parallel he intends to follow. Without declaring ourselves in favor of this or that latitude, we shall simply remark that there exists a direct relation between the energy, the frequency of the anti-trade upon a given parallel, and the corresponding accelera-tion of the barometric curve; the greater this acceleration, the more numerous the chances of fresh breezes varying from the distant pole to the west. In crossing the Indian Ocean or the South Pacific we should then advise the selection of the belt in which the fall of the barometric pressures is most steep. With this idea in view, it will be found necessary to ascend at least as high as the 42d parallel, and avoid the zone comprised between 46° 30' and 50°.

CONCLUSIONS.

Once the route parallel determined upon, the captain need pay no attention to the gyration of the winds so long as they are favorable; he must take as much advantage of them as possible in reaching his destination. But if fortune should turn and he meet with adverse winds, several contingencies will present themselves. If the east breezes are feeble, the thermometer remaining high (see the mean on the curve), there is no cause for fear; those breezes will not last, and will soon become western after passing through north. If the winds are persistent, irregular in violence and direction, and accompanied by a marked fall of barometer, and if at the same time the sky becomes overspread with very high, small clouds, make ready for a revolving storm with winds from the eastern quarters at the outset.

When the cyclone reaches you, either the winds at the outset will be very near northeast, in which case you are in the so-called dangerous semicircle, and you must bring to on the port tack to await patiently until the breeze, which is bound to veer to north and then to west, has become sufficiently favorable to allow you to steer your course, or else the winds at the outset will be nearly southeast. In the latter case you are in the manageable semicircle, and you may sail to the north of the disturbance, where you will find favorable winds. You will have to round the cyclone on the west by standing to the north (perhaps one or two points west of north, in case the winds should delay in veering south); keep the breeze broad on the quarter, hauling course to starboard as the wind veers aft, until you are steering your course with winds varying from southwest to west. Finally, if with the winds southeast at the outset you find yourself unable to sail north, bring to on the starboard tack in order to keep the ship's head to the sea, and ride as easily as possible.

ADDRESS OF HON. JAMES R. SOLEY,

Assistant Secretary of the Navy,

On the Occasion of the Unveiling of the Jeannette Monument at the United States Naval Academy,
Annapolis, Maryland, October 30, 1890.

We have come here to-day to pay the final tribute to a little band of officers and seamen of the Navy, who, nine years ago, gave up their lives, in a toilsome and difficult enterprise, an arduous labor of exploration and scientific research. It is a date well chosen, for it is the anniversary of the day upon which closed the record of their undertaking. They entered on their task with no certain prospects of success, sure only of the perils and difficulties they were to encounter, and looking for no reward save that which comes from the consciousness of duty done. Their expedition made its contribution to our knowledge of the lands within the Arctic circle, but, like so many others that had gone before it, failed of its great and ultimate purpose. But this same expedition, failure though it was in its attempt to penetrate to the Pole, or even to gain a latitude beyond its predecessors, yet, in its bright example of sustained endeavor, of lofty steadfastness of purpose, of dangers met unflinch-ingly and hardships borne without complaining, was crowned with a success far surpassing the triumphs of scientific discovery, and worthy of all that we can do to commemorate and celebrate it here.

The expedition of the Jeannette was a new experiment in Arctic navigation. Of the three main gateways to the Pole, only that whose opening was at Behring Strait had been as yet untried. Many were the attempts which had left their record of suffering and disaster on the shores to the northward of Baffin's Bay. The passage by the North Atlantic had its ardent advocates, and exploration here had reached far north, but had stopped at Franz Josef Land. The pathway leading north from Behring Strait, if pathway there might be, was unexplored. Herald Island was known, and beyond it glimpses had been obtained of Wrangel Land, but whether this half-seen country was a mere dot upon the frozen ocean, or another Greenland stretching in a wide expanse which reached the proportions of a continent, was hardly guessed at. Towards Behring Strait flowed the warm current of Japan, and its influence might perhaps reach beyond the parting of the continents, making a navigable highway through the ice-pack that filled the great spaces around the Pole.

To solve this problem, to penetrate if possible by this conjectured entrance to the farthest north, was the object of the voyage of the Jeannette. The enterprise was set on foot by the public spirit of James Gordon Bennett, and sustained and carried out solely by his generous liberality. It was Bennett who gave the vessel in which the new venture was to be made, and who selected the man that was to make it. The leader, Lieutenant George De Long, was a volunteer for the service; indeed, had it not been for his burning eagerness to make the trial, it might, perhaps, never have been made. Already he had served his apprenticeship in Arctic navigation, when, in an open steam-launch, the Little Juniata, he made his way across Melville Bay to Cape York in search of the fugitive crew of the Polaris. It needed but one interview to convince Bennett that he had found the leader for his enterprise—a man in whom were united the skill of a practiced seaman, the trained experience of an ice-navigator, the self-reliance, the ready resource, the capacity for organization and discipline of an accomplished naval officer, and the impetuous faith and zeal of an explorer.

The project of an Arctic expedition, now become a settled purpose, was followed up by its promoter with his untiring energy, seconded by the professional judgment of its future commander. After long search, the Pandora, a ship already tested in Arctic voyages, was selected for the work, purchased, and renamed the Jeannette. She came to San Francisco and was taken to the navy-yard, the Secretary having been empowered by Congress to assume charge of the ship for the projected expedition. Under De Long's direction she was fitted out with stores supplied by Bennett's liberal hand, and strengthened to enable her the better to withstand the rude shocks of navigation in the Arctic seas. Her officers and crew, chosen with care and judgment, were gradually assembled; and certainly no ship was ever sent on the perilous work of Arctic exploration with a company braver, more earnest, or more ready in bearing and enduring all things, than that which the Jeannette bore away from San Francisco, buoyant with hope, with a confidence born of a high courage and the strength of disinterested purpose, and eager to win success, however great the dangers of the struggle.

It was on the 9th of July, 1879, that the ship sailed out through the Golden Gate of the Pacific, and two months later that she entered the frozen gate of the Arctic, from which she was destined never to escape. Striking, on the 6th of September, a lead in the ice-pack that offered what seemed a fair promise, the Jeannette was soon beset, fast in the ice, imprisoned beyond all possibility of deliverance by any effort that she or her people could make. Here she was to remain, frozen in, helplessly drifting with the pack, for twenty-two wearisome months. Of these the first fifteen were hardly months of drift; the movements of a year covered the space of a scant hundred miles. The zigzag course of the ice-floe in which the ship lay cradled was hither and thither by slow, short drifts—here a little and there a little, a step to-day in advance, a step to-morrow on the one hand or the other, a step the third day back to the point of starting. In November of 1880 observations showed the ship to be in the same spot where she had been recorded in the previous April. " Thus it goes," said De Long in his journal, " east one day, west the next, north one week, south the next. When will this come to an end?" One day's fair steaming in open water would have covered all the explorations of fifteen months of aimless drift.

Do you realize, my friends, what must have been the strain upon De Long and his companions, thrown in upon themselves during that long imprisonment, filling nearly the space of two whole years? Day by day as he rose and looked about him at the dull prospect, outlined only by those frozen shapes which grew monotonous even in their incessant change, struggling against the overwhelming sense of failure, powerless against forces which no human will could master, what cruel discouragements must he have suffered! Month after month rolled by, autumn became winter, winter gave place to spring, spring to summer, and so on through a second autumn and a second winter—a summer and winter, a spring and autumn of the calendar,

marked only by the long gloom of the Arctic night and the
unbroken, ice-reflected glare of the Arctic sunlight.

> " Thus with the year
> Seasons return ; but not to me returns
> Day, or the sweet approach of even or morn,
> Or sight of vernal bloom, or summer's rose,
> Or flocks, or herds, or human face divine ;"—

nought to be seen hour after hour but the wide expanse of the
eternal pack, stretching away in dreary sameness to the fog-bound
horizon ; no change or prospect of a change ; no compensations
even of half-attained success, or satisfying gleams of hope that the
ice-drift was bearing the ship one step onward in her course.

How do they bear themselves, these men, who sailed from home
full of high hopes and eager aspirations, now that their hopes are
crushed and confidence has yielded to the expectation of defeat?
Read the record in De Long's journal, where every day's events are
faithfully transcribed,—events so trifling that their magnified importance shows only by heightened contrast the dullness of this dull round
of daily existence,—and see how courage and self-devotion can uplift the
mind, even in the darkest hours. It is a narrative touching in its manly
simplicity and unconscious pathos. " There can be no greater wear and
tear," says De Long, in his frank, artless way, " on a man's mind and
patience than this life in the pack. The absolute monotony ; the un-
changing round of hours ; the awakening to the same things and the
same conditions that one saw just before losing one's self in sleep ; the
same faces ; the same dogs ; the same ice ; the same conviction that
to-morrow will be exactly the same as to-day, if not more disagreeable ;
the absolute impotence to do anything, to go anywhere, or to change
one's situation an iota ; the realization that food is being consumed
and fuel burned with no valuable result, beyond sustaining life ; the
knowledge that nothing has been accomplished thus far to save this
expedition from being denominated an utter failure ; all these things
crowd in with irresistible force upon my reasoning powers each night,
as I sit down to reflect upon the events of the day ; and but for some
still, small voice within me that tells me this can hardly be the ending
of all my labor and zeal, I should be tempted to despair."

All this while the crew are kept at work, busily following the day's
routine, and by exercise and occupation keeping away disease and
discontent. Their leader never wearies in his task. He guides their
occupations, watches their health, sustains their manful spirit. Obser-

vations are taken and records kept; investigation and experiment, narrow as is the field of research, are followed up; the daily work of the ship, conformed to naval regulation and tradition, pursues its course. Each Sunday the Captain reads the service, and once a month, the Articles of War. To-day it is a bear hunt, to-morrow the killing of a seal or gull—trivial incidents, but all that breaks the monotony of life, except when the grinding and tearing of the floes pressing upon the ship serves as the muttered menace of catastrophe that any day may bring. Once at least, during the first winter, the ship narrowly escapes destruction, and all is put in readiness for a retreat upon the ice; but these alarms pass by, and for a twelve-month all is quiet.

The officers and crew are worthy of their commander. In that long imprisonment in the ice-pack, and throughout the five hundred miles of the retreat, each and every one stands out, shining with the light of his own serene courage, his own unwavering endurance. To their virtues the Captain's journal bears sincere and willing testimony. First of all come his two right-hand men, the executive officer and the engineer, of whom he writes: "Aided by Chipp and Melville, whose superiors the navy cannot show, with their untiring energy, splendid judgment, and fertility of device, I am confident of doing all that man can do to carry the expedition to a successful termination." Chipp is the model executive officer; calm and earnest, reserved, always with work to do, and always doing it quietly, steadily, and surely. "He says but little," writes the Captain, "but I know where he is, and how reliable and true." Not less reliable and true are the other officers: Danenhower, the navigator, struck down by an affliction that robbed him even of the poor consolations that might be drawn from that barren life, but bearing his confinement with heroic fortitude and patience; Melville, whom his commander finds "more of a treasure every day, always self-helpful and reliant, a tower of strength in himself"; and last of all, Ambler, the good physician, cheerful and kindly, caring wisely and watchfully for the health of his companions, and brightening their spirits by the gentle influence of his sunny temper. Collins, the meteorologist, and Newcomb, the naturalist, pursue faithfully their vocations, observing, collecting, recording, under the unaccustomed burden of naval routine and discipline. In the next rank come Dunbar, the ice pilot, grave and serious, whose experience proves a help at all emergencies; Sweetman, the carpenter, untiring, ingenious, a sure and constant reliance; and Jack Cole, the

rugged boatswain, whose reason finally succumbed to hardships which
his stalwart frame repelled. Equally devoted and capable are the
crew, at their head Nindemann, easily foremost, whose energy and
self-devotion win from the Captain the highest praise, and for whom,
as well as for Sweetman, he asks a medal of honor. For each and
all the others, down through the list, to the two Alaskan Indians and
the Chinese cook and steward, there is naught to be said but praise;
their courage, their patience, their helpfulness and willing obedience,
the cheerful front with which they face misfortune, the steadfast resolution that bears them past disaster. "Another crew, perhaps," said
one of their officers, ' may be found to do as well, but *better*—never."

With the approach of the second winter the drift of the pack took
on a uniform direction, and, gradually gathering force, bore the ship
slowly to the northwest, traversing during the next six months ten
times the space covered in the year gone by. Here was relief at last
in progress, no matter whither; and in May the eyes of the ship's
company were gladdened by the first welcome sight of land. They
called it Jeannette Island. It was a desolate rock in the Arctic,
nothing more—grim and unapproachable; but it was at least one
point that raised itself aloft, breaking the outlines of the sea of ice,
and marking the first result, small though it was, of all the efforts of
the explorers. Presently another land opened out beyond it, Hen-
rietta Island, a second barren rock surmounted by its broad ice-cap.
This last Melville was sent out to explore, his party moving by boat
and sledges over the loose, broken pack. They reached the land
and took possession in the name of the United States; for no human
eye had ever before this day looked upon the island, or, if it had, no'
record had been made of its discovery.

It was now early in June, and De Long's hopes of release began to
rise; but they were destined to be rudely shattered. Release was
coming, but not as he would wish. The ice-field cradling the Jeannette, now drifting swiftly past the islands, still heading to the northwest, came in collision with other fields, and then began a second
battle of the ice-floes, such as had nearly sunk the ship the year before.
Driven by the winds and currents this way and that, the huge masses
crashed together, and in a moment were rent asunder, piling up vast
broken ridges here, and there sinking and disappearing. Hour after
hour the crushing and tearing, the upheavals of shapeless floebergs,
continued with a deafening uproar, and the ice around the ship was
hurled about in tumultuous confusion. The pack splitting apart,

suddenly opens lanes in all directions, and these as suddenly close when the floebergs, propelled by resistless, unseen forces, are dashed upon each other. Each moment the ship seems on the verge of destruction. For a few hours she escapes, but presently a lane opens where she lies, and for an instant she is again afloat. It is but for an instant. The floes that have been torn apart have only recoiled, as it would seem, for one final spring; and returning, they crush the vessel like grain between the upper and the nether millstone. They hold her fast just long enough for the crew to take refuge on the ice with boats and sleds and stores; and then separating, they leave her, a crushed and shattered craft, to plunge into the depths below.

And now, with the sinking of the Jeannette, comes the most marvellous chapter in this strange history. Thirty-three men, with a pack of dogs, three boats, five sleds, and sixty days' provisions, cast upon the drifting pack, started forth to reach the land, five hundred miles away. Never was there such a retreat as this. The nearest point on the Siberian coast was the Lena Delta. Between it and them lay three or four islands, the New Siberian group; but except for these, only a moving pack, and a stretch, how great or little none could say, of open water.

No time was lost. De Long, still strong in his steadfast faith, supported loyally by officers and men, organized his band, assigned to each his duties and his post, and after a few days' careful preparation, setting his face southwards, began the weary march. Each day the pioneers laid out the route, and sleds and boats were drawn over the ice to the determined point. The work was work with sleds and ice, but not the sleds and ice of a boy's winter sport. There was no child's play in clambering over these distorted hummocks, dragging up and down the great whale-boat and the cutters, and sledge-loads of provisions, or charging across the moving floes, leaping from one to another as the intervening leads were opening or closing, often with barely time to escape from one uncertain foothold to a second hardly less uncertain. Their first week was a week of cruel discouragement—of sickness and distress, of mishaps without number, of toilsome struggling with heavy loads over rough masses of broken ice, which opened in wide cracks and lanes of water almost beneath their feet. The progress made was less than a mile a day, and even this must be passed and repassed, with load after load, boats, sleds, provisions, sick, until the whole had been dragged, ferried, tumbled into place. And when, at the week's end, the Captain obtained an observation, he found that the ice-drift had brought all to naught, and far outrun even their painful efforts; so that, instead of five miles to the south, the seven days had sent them thirty miles to the northwest.

Only to Melville and Ambler did De Long make known his secret. The others were allowed to remain in happy ignorance that all their labor had only made their case more hopeless. Still they plodded on towards the southwest, now over great blocks and ridges, now upon sliding, shifting floes. One day the ice would be in motion all about, swinging and swirling this way and that, opening in their path broad lanes of water only to be traversed by ferrying in the boats, or by flying bridges made from the floating cakes. Another day they would cross a stretch of confused floebergs, which was the hardest work of all. Another would find them floundering through deep snow, and ponds and water-holes, tired and wet, with footgear worn to shreds, but ever struggling to reach the open water which they knew must be somewhere upon the ice-pack's outer edge.

So for five weeks they continued their slow advance, making good little more than a mile a day. But the ice-drift was now helping them, and pushed them onward two miles while they were marching one. So that at least there was something for hope to feed on during this dreary time; and still more courage came to them from the sight of land ahead. It cost them eighteen days of toil to reach it, and at the last they were nearly carried past it by the swift currents; but, charging with a rush across the broken floes, they leaped the last space and landed safely on the ice-foot.

The new-found land, for it was their discovery, was named Bennett Island, and upon its shores they rested for a week. Then after gathering fresh strength, they turned their steps once more to the southward. They had not yet cleared the pack, but the water was free enough to work the boats. In the first cutter was De Long with Ambler. The second cutter was assigned to Chipp, while Melville, with Danenhower still in part disabled, commanded the whale-boat. Young ice now checked their progress, and it was only by breaking their way, or tedious tracking along the pack, sometimes even, when the lanes closed up, by dragging boats over the floes, that they could make their progress good. But each day the water became more open, and another month brought them to the islands of New Siberia. Here the three boats, drawn into the mill-race between the islands, were whirled along in the swift current, surrounded by the whirling floes, and found themselves at last upon the open ocean.

The dangers of the past were now forgotten. Resting a short space at the islands, De Long with his three boats headed for the Lena Delta, whence he might make his way southward to the Russian settlements. One more stop was made at the Island of Semenovski, on Sunday the 11th of September. It was the last time that all would be together, although they knew it not, and De Long, as was his wont, held divine service and read the Articles of War. On Monday morning all embarked and sailed with prosperous breezes until the afternoon, when the wind freshened to a tempestuous gale. As night came on—a fearful night in which it seemed as if no boat could live—the three crews, which had for two long years faced together the perils of the Arctic, were driven apart, never to meet again.

Of Chipp, the generous and devoted first lieutenant of the Jeannette, and his people in the second cutter, the last that was seen was on this night, when the others parted from them, struggling in their little boat against the fury of the tempest. What happened to them no man knows, for months of tireless search in every nook and corner of the coast failed to reveal a trace of what might have been their fate. For them we must believe that no morn ever broke upon that night of horror.

We do know that the whale-boat, a stauncher craft, was only saved by the skill and vigilance that guided her movements through the perils of the storm. At daybreak she was still uninjured, and a wise foresight directed her to the eastward along the coast into the great bay that receives the southernmost of the Lena's outlets. Here three days of chance sailing in passages among the Delta islands brought Melville and his party to the native settlements, where, as their force was spent, they landed and found rest and shelter that slowly brought them back to health and strength.

The first cutter, in which were De Long and Ambler with Collins and eleven men, survived the night, though losing mast and sail. Moving with difficulty in their disabled boat, they only reached the coast on Saturday, when they grounded at the Delta's most northern point, and near the mouth of the main northern outlet. Wading ashore through the thin ice that for a mile seawards covered the shallow beach, they landed with their instruments and records, and with four days' scant provisions, set out on their southward march.

It was a cruel fate that marked out De Long's landing place. The Lena Delta in October is a desolate waste, snow-covered, a labyrinth of intersecting streams, impassable to boats by ice, but ice as yet too thin to bridge them. From the landing place on the north coast to the Delta's head, where the river branches out to its divergent mouths, a space of eighty miles in a line due south, there is no human habitation. The point that Melville reached lies far off to the southeast. Along the coast a little way, not more than thirty miles from where the cutter stranded, is another village, the northern-most of the winter settlements. But of this village De Long knew nothing. His wanderings, and the untiring searches that others made to find him, have mapped out for us the lands and waters of the Delta, but to him, at landing, they were all unknown. His map, if map it could be called, had names that seemed the names of places where men dwelt, but they were blind guides. They pointed out only one hope, that by following the main river and going south, he would in time reach a settlement. Of habitations to the west, along the coast and within easy reach, they gave, alas, no clue.

So with the same unfaltering resolution which had carried him, and those whose lives rested upon his word, through so many perils, he started with them on the southward journey. The long confine-ment in the boat, the hardships and exposures of the retreat, had sapped their energies, and they were in no condition for forced marching. One of the men, Ericksen, crippled and weak beyond his fellows, kept all the party back, when every moment was vital to their safe deliverance. Tenderly they cared for him, waited for him, carried him, weak as they were, when he could no longer walk him-self, until at length his death relieved them of their burden. But by this time their own strength was nearly spent, and toil as they might, they could not make up for the days already lost.

We need not dwell upon the events of that fatal journey. Its history is written day by day in the Captain's journal, where, in letters firm and sharp to the very end, the whole record is preserved. It tells how a man of unconquerable will, firm in his faith, kept up for six weeks of agony of mind and body the struggle with a remorse-less fate that stared him in the face. It tells of weary marches and counter-marches; of days of painful toil, fighting against the pitiless gusts of wind that swept those dreary plains ; of bitter nights, with-out shelter and without rest; of weakness and illness creeping over the men by slow degrees; of hunger gradually turning to starvation.

It tells how hopes were raised by tracks of deer, by footprints of men even, only to be blasted, for, as mischance would have it, natives had gone that way but a few days before, and others found the records of the band after De Long had passed. It tells how, after three weeks of unsuccessful effort, the Captain sent Nindemann and Noros off for succor; how for ten days thereafter the main body struggled on, hoping each day that the expected help would reach them; and at last, how for ten days more, unable to move, having reached the parting of the streams at the Delta's head, they waited, giving up one by one, until only the Captain and the Surgeon remained alive. And then the record ceases.

Read the piteous story of these last ten days, the final ending of that glorious retreat, which had lasted one hundred and thirty days since the ship went down—a story of little else than death:

"Friday, October 21st—131st day. Kaack was found dead about midnight between the doctor and myself. Lee died about noon. Read prayers for sick when we found he was going.

"Saturday, October 22d—132d day. Too weak to carry the bodies of Lee and Kaack out on the ice. The doctor, Collins, and I carried them around the corner out of sight. Then my eye closed up.

"Sunday, October 23d—133d day. Everybody pretty weak. Slept or rested all day and then managed to get enough wood in before dark. Read part of divine service. Suffering in our feet—no footgear.

"Monday, October 24th—134th day. A hard night.

"Tuesday, October 25th—135th day.

"Wednesday, October 26th—136th day.

"Thursday, October 27th—137th day. Iverson broken down.

"Friday, October 28th—138th day. Iverson died during early morning.

"Saturday, October 29th—139th day. Dressler died during night.

"Sunday, October 30th—140th day. Boyd and Goertz died during night. Mr. Collins dying" * * *

The pencil drops—the eye is glazed, the arm is palsied, the fingers stiffen and grow cold. The fight has been fought, the struggle is ended. The others, one by one, have passed away. There is no more that human effort can accomplish.

> "O, let him pass! He hates him,
> That would upon the rack of this tough world
> Stretch him out longer."

De Long and Ambler lie side by side. We know not to which of them first came the whispered summons of the death angel, but we know that they were the last to answer. What was it, think you, that made these two hold out to the last? Was it that they were made of tougher fiber, endowed with greater health and vigor—greater than their seasoned shipmates, the hardy seaman, the Alaskan hunter? I think not. It was that in them lay that unconquerable purpose to keep their charge, to fulfil, even to the bitter end, their sacred trust. And thus it happened that De Long, his brother officer still faithful at his side, only surrendered his commission when the Almighty had disbanded his command.

There is more yet to be told of this expedition, but we may not dwell upon it here: how Danenhower first attempted a desperate search for his lost companions; how Melville heard from Nindemann and Noros of his Captain's cruel plight, and how, before he had regained his strength, he braved alone the Arctic storms and cold and hunger, traversing the Delta down to the very coast, freely putting his life again at hazard, in the faint hope that he might bring relief; how a second time he scoured the plains in March, and found the bodies, and, laying them in the earth to rest, built over them the cairn and cross, in whose likeness this stone has been erected here; how the Russian Government, faithful to its old tradition of friendship, lent its aid freely to the explorers; how the later relief parties came out, and how the generous patron of the expedition gave lavishly of his time and fortune to carry on the search and to lighten the sufferings of the survivors. But it is not of these things that we are speaking here and now. It is not to the living, but the dead, that this day is consecrated.

Dead they are indeed, but dying as they did they left behind them a renown that to the Service they loved and died for remains, and will remain forever, a priceless heritage. That long retreat, over five hundred miles of drifting ice and open ocean, a retreat matchless in the records of Arctic achievement, shines out, even through the dark tragedy at its close, with the triumphant splendor of a victory won. On the long roll of the world's explorers are no brighter names than those of De Long and his gallant company of the Jeannette. They fell not, warriors though they were, in war, nor was the fate of nations trembling upon the issue of their struggle. But it is not in war alone that martyrs win their crowns; nor is it only in the clash of arms and the din of battle that is revealed the beauty of heroic death.

It matters not whether their bones lie here, or at their homes, or on the bleak Siberian coast, where they gave up their lives; "the whole earth," said Pericles, "is the sepulchre of illustrious men." But it is fitting that here should be their monument. It stands here for us Americans, who hold our Navy and our Country dear, as a memorial of what her sons have done, and as an earnest of what they will do hereafter. It stands here for you, the comrades of those young officers who fell, to give you added strength and courage, when you too find yourselves the victims of relentless fate, and driven to the edge of the black chasm of despair. It stands here, last of all, for you, Cadets of the Navy, that daily you may have before your eyes this bright example of heroic virtue—virtue which in the past has been the pride and glory of your Service, and which it will rest with you to transmit in undimmed luster to the generations yet to come.

Heart-rending as is the burden of that song, borne to us upon the wings of the Siberian wind, it is not the mournful music of defeat. As with their great Norse ancestors, the sea-heroes, the death-chant of our own sea-heroes rings only with a joyous strain of triumph; and in their triumph we triumph also.

> "Pæans, sing high! What would we with a dirge?
> Proudly we weep our brave."

There they stand, the martyrs of the Lena Delta, the men, who, through high courage, overcame disaster; surrounded by that goodly company of brave explorers, whose memory, like theirs, remains enshrined within these walls—Collins, a victim of the poisonous miasma of the Isthmus; Strain, perishing from hunger and fatigue under the burning sun of the tropics; Talbot, sailing in his open boat over fourteen hundred miles of the Pacific to bring succor to his shipwrecked comrades; and Herndon, issuing unharmed from the perils of the Amazon, but yielding his life, a cheerful sacrifice, to save the passengers upon his sinking ship.

There they stand,

> "with the rays
> Of morn on their white shields of expectation,"

illumined by the brightness of their own imperishable fame, serenely waiting for the final muster; while from the chorus of uplifted voices come the ringing notes of an Io Triumphe, re-echoing throughout all the ages, proclaiming in eternal harmonies the glory of those who fought their fight out to the end, and who, through death, achieved undying victory.

PROFESSIONAL NOTES.

TARGET PRACTICE AT THE NAVAL ACADEMY.

By LIEUTENANT-COMMANDER C. S. SPERRY,

Head of Department
of
Ordnance and Gunnery.

U. S. NAVAL ACADEMY,
ANNAPOLIS, MD., *March* 24, 1890.

The competition for the marksman's badge for small-arm firing shall be governed by the following rules :

1. It shall not be awarded to any one who has not fulfilled all of the following conditions contained in the circular of the Bureau of Ordnance, May 25th, 1889, regulating target practice in the service.

2. " While on shore, firing the service rifle, with service sights and ammunition, at an A army regulation target, 170 yards distant, make 30 out of a possible 50, with 10 consecutive shots, all fired within a period of 7 minutes. The position of the firer during this practice must be standing and without artificial rest.

3. " The firer being in a boat which is afloat in smooth water, do the same thing, except that the position may be any desired ; an artificial rest may be used, and the score to be made shall be 15 out of a possible 50.

4. " While on shore, put 4 out of 6 shots fired consecutively from a service revolver into a surface the size of the bull's eye of the army B target (or into a rectangle 24 inches high by 18 inches wide) at 30 paces range, within $1\frac{1}{2}$ minutes."

5. The firing of the Third class, in the Spring, will be at a range of 170 yards at the army A target.

6. The firing of the same class during the Summer, with the rifle and revolver, will be at the ranges, and under the conditions, of paragraphs 2, 3, and 4.

7. The badge will be awarded as soon as possible after the commencement of the term to the cadet of the Second class who has fulfilled the above conditions, and whose total score for the Spring and Summer practice is the highest percentage of his possible score.

8. In scoring the revolver practice, all hits within the bull's eye shall count 5, and all other shots shall count 0.

9. The percentage under paragraphs 2, 3, 4, and 5 shall be calculated separately and added.

10. In case there is a tie, it shall be decided by firing one or more groups of 10 shots, as may be necessary for a decision, under the conditions of paragraph 2.

W. T. SAMPSON,
Captain, U. S. N., Superintendent.

In the Spring of 1890 medals were offered by the Superintendent to the cadet of the First class making the best score with the Hotchkiss rapid-fire guns, and to the cadet of the Second class making the best score with the Hotchkiss magazine rifle.

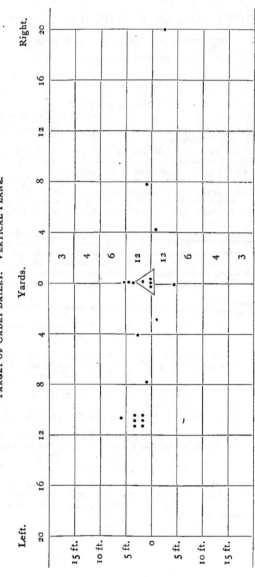

TARGET OF CADET BAILEY. VERTICAL PLANE.

The first medal was awarded in June to Cadet Bailey.

The firing was from the Standish, at the service great-gun target, the range being determined by trial shots. Cadets Bailey, Coleman, and Schofield tied in a score of 138 out of a possible 144. Upon shooting off the tie, Cadet Bailey won, having fired 22 shots—12 at 1400 yards, 5 at 1050, and 5 at 950, scoring 258 out of a possible 264, with 4 direct hits. The firing was upon several different days. At the target practice from the Wyoming, before the Board of Visitors, Cadet Bailey showed that his success was not accidental, by hitting the target several times at 1000 yards with a 32-pounder smoothbore.

The average of the entire First class was 75 per cent of the maximum score on the service vertical target, and at the examination drill alluded to, the target was struck eight times by the heavy guns, Cadet Latimer finally sinking it with a shell from the 8-inch M. L. R. when only a corner of one wing showed above water. The range was determined by Cadet Coleman with the 1-pounder R. F. gun with great accuracy, his shots frequently striking the target.

Each cadet fires during the Second class year five rounds from a heavy gun, and during the First class year about twenty rounds. This amount of practice does not seem sufficient to account for such accurate firing, and it is believed to be due to the steadiness engendered by the constant use of ammunition at all company and battalion drills, as well as to small-arm range practice. Volley firing by company and battalion at a target certainly cultivates nerve. There is no more effective safeguard against accident with arms than the constant use of blank or other ammunition, and it is an open question whether the dummy cartridge is not useless and dangerous. A blank cartridge, if the arm is not properly handled, will betray carelessness by its explosion with comparatively little danger, but the harmless dummy tells no tale, though the ball cartridge, from which a careless person will hardly distinguish it, most certainly will.

Thousands of the cartridge shells can be reloaded as blanks with a small charge with very little trouble and expense, and are invaluable in teaching care in the use of ammunition and steadiness in firing.

The competition for the small-arm medal is in progress.

NAVAL ATTACK ON A PROTECTED HARBOUR.

By Commander Egerton, R. N., H. M. S. Orlando.

(Reprint from the Journal of the United Service Institution of New South Wales.)

"What is a protected harbour, why do we protect a harbour, and what is it protected against?"

The first of these questions—"What is a protected harbour?"—can be easily answered, for it is merely a name given to a harbour which is defended by guns, torpedoes, or vessels of some kind—perhaps one, perhaps the other, perhaps a combination of each—but which cannot be called a fortress, and cannot in any way compare with such places as Portsmouth, Portland, Cherbourg, Kiel, or Cronstadt and many others of similar importance, where nature and the skill of the engineer have done so much to render them strong that nothing short of a powerful fleet of ironclads would ever dream of attacking them; and even after weeks, or perhaps months, of most persistent and severe fighting, at enormous cost, the result would be doubtful.

There are numbers of harbours which are protected by guns and so on, some more, some less, according to the value of the property behind them, either from a strategical or monetary point of view. Such harbours abound around the coast of the British Isles and the Colonies, such for instance as Falmouth,

Holyhead, Belfast, Dundee, Peterhead, Aden, Singapore, Sydney, Melbourne, Hobart, Wellington, and so on.   These are protected harbours.

The next question is, " Why do we protect a harbour?"

This seems to be a plain question, but I am not sure that it has been very clearly answered.   All harbours are not protected, and many are only protected by one or two guns, and yet the value of the property, of which the harbours are the doorways, is very considerable ; and why has it only of late years occurred to people's minds that their harbours should be protected ?   As a proof that this is really the case, it is only necessary to contemplate the enormous sums of money which are *now* being expended for the defense of imperial ports and harbours.   They have been needed just as much for years past as they are now. The only answer that seems to me to meet the case is that in years past it has been assumed that the navy would protect the harbours ; and I think that the naval manœuvres which have taken place for the last few years at home have done much good in bringing it home to the people's minds that it is very easy for a squadron of five or six ships to slip past another on the lookout for it, even in such narrow waters as the English Channel.   It is true they may not be left unmolested for long ; but it does not take very long, with modern ordnance, to do incalculable damage.

Many people think or have thought that in the event of war a ship or ships would be told off to protect their harbours for them, and the more the value of the property (private property, mind you), the larger would be the force told off to protect them.   I do not think this would or could be the case ; surely an admiral would not weaken his squadron by dividing it up into small bits and placing a bit here and a bit there.   No: the defense of the harbour is not the function of the navy ; we are meant for attack of ships more particularly, but still sometimes it may be forts.   We are not trained to defense in its passive form.   We never practice *with* the military, to whom the defense is entrusted, and, with the exception of some half-dozen coast-defense ships, we are not prepared for it ; we are of little use boxed up in a harbour, and our place is at sea.   It would be more economical, more permanent, and more satisfactory in every way to mount guns on shore than to place them in a ship which has the disadvantage that it can be sunk and cannot be hidden, without having any counteracting advantages, since there is not room for the ship to move about.

To defend the approaches to our channels and harbours is quite another thing, and is most decidedly one of the functions of the navy ; here we have scope to move about, and can use the powers with which we are furnished to their best advantage, and if we are in sufficient numbers and do our work thoroughly, the ports or channels we are protecting may not have to fire a shot.

It would seem, therefore, that the real object of protecting a harbour is to delay the enemy until succor arrives.   That is, the enemy, having slipped past the protecting squadron, arrives off a port ; he has double or treble the number of guns that are mounted on shore ; but what the defense has to do is to delay him as much as possible to prevent his bombarding the city, town, depot, or whatever the property may consist of ; to prevent his levying enormous sums of money, to prevent his finding shelter for his ships or replenishing them with coal, ammunition, or provisions, until succor arrives.   The aim, then, of protecting a harbour is to delay the enemy.   Hence our business in attacking a protected harbour is to do our utmost to take it as speedily as possible, knowing that delay to us means most probably defeat.

The third question is, " What do we protect a harbour against?"

Opinions on this question will vary much, no doubt, and therefore there is no harm in raising it, as it will lead to discussion, which is always useful. If a harbour aspires to beating whatever enemy may be brought against it, without any regard to a protecting navy, then I ask where is the line to be drawn ? for, if a harbour is protected against a possible enemy of (say) four cruisers, and six are sent against it, what is to be done then?   You cannot call out, " It isn't fair, I won't play."   You have disregarded or thrown over the protecting navy, and so you cannot hope for succor in that direction, for they may possibly throw you over; so where do you draw the line between a light battery and a first-class fortress? The first may defeat a fillibuster or armed cruiser; the second says, "Come, if you dare." But if it be accepted that the object of protecting a harbour is to delay the enemy, then the amount of that protection will depend upon the naval force protecting the approaches to your coast, the area of seaboard they have to cover, and the distance you are from the base of a possible enemy. On this station the first is not as large as it ought to be, but provision has been made for this, and more ships are to be added; the area of seaboard is enormous, which means that you should be better prepared to delay an enemy for a longer time than ports in the English Channel, for instance; and, on the other hand, you are a long way from the base of a possible enemy, New Caledonia being, I think, the nearest possible base, which is at a distance of 1100 miles.

Before attacking a harbour we should carefully consider what resistance we might expect to find. The natural outline of a harbour or haven has so much to do with the defense which may be expected to be met with, that any harbour to be attacked would have to be considered on its own merits; but, as a general rule, wherever a harbour or haven exists we may expect to find submarine defenses, and we shall endeavor to find out beforehand whether these are kept constantly in position, whether they are real or only on paper, and whether, if our attack is sudden, we may expect to find them laid down at all. All this information we should look to our Intelligence Department to supply. We shall expect to find an inner series of batteries so arranged as to bring a heavy cross-fire on the mine-field, and which will hamper our movements over every inch of ground we gain towards our final goal. There will be outer batteries to harass us and prevent our making arrangements for the attack on the inner defenses without a struggle, and also to prevent our lying off to bombard the town, arsenal, depot, or whatever it may be which is being protected. These outer defenses will have to be silenced before the more serious work of attacking the inner defenses can be attempted. We may safely reckon on meeting all this, because it is generally admitted that, with sea room, forts alone will not stop a fleet. The inner defenses will give us the most trouble, and the deeper they are, the more trouble they will give us, because we shall be confined in our area of movement, and every defended yard will mean delay, which, for reasons I have given, we particularly wish to avoid.

There is yet another obstacle which will give us much anxiety and which we may reasonably expect to encounter. I mean torpedo-boats.

Experience has shown that a purely passive defense is rarely successful, whether on land or at sea. No entrenched camp would be left without outposts to watch the approach of an enemy, or without sorties to harass him when he has made his appearance, and this can best be done by an active naval force and a good supply of torpedo-boats. They are as useful as cavalry for scouting. They are dreaded at nights. The terrible weapon they carry and their power of continuing in flotillas render them in every way a most powerful and necessary item of the defense, and so we must be prepared in our attack to guard against them. Therefore, having given the defending fleet the slip, we proceed to do our work of destruction on a harbour which we know to be defended by an outer line of batteries, a deep submarine defense, an inner line of batteries guarding the submarine defense and inner harbour, and, lastly, torpedo-boats.

The nature and number of vessels sent to attack a protected harbour will, of course, depend upon the individual harbour to be attacked, its natural formation, its armament, its probable submarine defenses, the depth of those defenses, the depth of the water, its distance from succor; and, perhaps most important of all, the general opinions of the fighting qualities of its defenders, their numbers, their training and their amount of practice and organization will all influence this discussion; but I think that this work will generally be relegated to cruisers, as ironclads will not be available. These cruisers will be fast ships, with a large coal capacity, a good armament, and a certain amount of protection. And I propose to say a few words on a modern cruiser, and will take for my type a belted cruiser such as the Orlando, with which most of you are familiar—a ship which would be admirably suited for work of this description.

The Orlando is 300 feet long, 56 feet beam, and 5600 tons displacement, with a maximum draught of 23 feet, and is capable of steaming at a speed of nearly 18 knots for a few hours, and has a radius of action of 4500 miles at 10 knots. She is built of steel throughout, and by referring to the diagram you will see that all the vitals of the ship are protected. The protection consists of a belt of 10-inch armour, steel-faced one-third of its thickness, 200 feet in length and 5 feet 6 inches in depth, about 18 inches being above the water-line at a load draught. On the top of this belt, and resting upon it, is a steel deck of two-inch plates, that is, two inches thick, going the whole length of the ship, being increased to 3 inches at the extremities where the ship is unprotected by armor, and taking a shape thus ⌐— — —⌐ running down towards the ends and assisting very materially to strengthen the ram and take the thrust. Everything that is vital to the ship is kept below this armoured deck, and every hatchway through it is protected by a water-tight door of the same thickness. Below this deck are the bunkers and double bottoms, all of which assist to protect the engines and other important items. Above the armoured deck are the upper bunkers, about 17 feet broad, and running the whole length of the engines and boilers. These, when filled with coal, offer a good protection. The ship above the armoured deck may be riddled and the water wash freely over it, but it cannot get below on account of the water-tight doors before mentioned; and also around every hatchway are high coamings, 3 feet high and 9 inches broad, called coffer-dams. Another noticeable feature in the ship is that instead of having stanchions to support the guns, conning tower, and heavy weights above (and if one is shot away renders excessive strains on the remainder, which may fail to support the weight), they are supported by steel bulkheads, which have the advantage that they may be riddled with holes but will still support much weight. All the guns are protected by 1-inch steel shields. The conning tower is 12 inches thick, one-third of which is steel. From this tower the ship can be steered; guns, engines, magazines, and every part of the ship communicated with by means of voice-pipes and telegraphs; the guns and torpedoes can be fired by electricity from the directors, and the whole of these telegraphs and voice-pipes are carried down through an 8-inch armoured tube, being a solid tube of 24 inches bored out 8 inches in the center. The wires leading to the guns are kept below the water-line behind the armour, and when they rise above it, are duplicated and separated at least 3 feet apart, so that the same projectile will not be likely to damage both wires. There are three steering stations below the armour deck, that is, below the water-line, any of which can be communicated with; and the rudder is placed very low down in the ship, being entirely below the water-line. The whole ship is divided into an immense number of water-tight compartments, every shell-room, bunker, store-room, and magazine being itself a water-tight compartment. There are two engine-rooms divided by a water-tight bulkhead, and two sets of boilers divided in a similar manner. The armament consists of two 9.2-inch B. L. guns, ten 6-inch B. L. guns, six 6-pounder and ten 3-pounder Q. F. guns, and seven machine guns, twelve Whitehead torpedoes with two discharges, six E. C. mines ready for use, which number would be increased in war-time, and 500-pound mines are in store, ready for issue as required, 1200 pounds of wet gun-cotton, 360 pounds of dry, one mile of unarmoured cable, a mile and a half of single core, armoured, 1000 yards of seven-core cable, with batteries, and all other necessaries for mining work, with a crew of 500 men. A squadron of four or six of such ships, with perhaps an auxiliary or two, would constitute a most formidable enemy, and would tax the resources of a good many protected harbours to their utmost.

... ... These cruisers will
... ... and a certain
... ... dern cruiser,
... ... with which
... ... ted for work of

... ... displacement,
... ... at a speed of
... ... 4,000 miles at 10
... ... the diagram you
... ... protection consists
... ... plates, 200 feet in
... ... above the water-
... ... about it, is a steel
... ... the length of the
... ... ship is unpro-
... ... down towards
... ... sea and take the
... ... armoured deck,
... ... door of the same
... ... all of which
... Above the armoured
... the whole length
... offer a good pro-
... and the water wash
... water-tight doors
... coamings, 3 feet
... ... feature in
... the guns, conning
... renders excessive
... weight they are sup-
... may be riddled
... ... are protected by
... one-third of which
... magazines,
... voice-pipes and
... from the direc-
... are carried down
... bored out 8
... below the water-
... and sepa-
... be likely to
... the armour deck,
... with; and the
... below the water-
... water-tight com-
... being itself a
... by a water-
... manner. The
... 3 L. guns, six 6-
... twelve White-
... use, which
... are in store,
... pounds of dry,
... ... 1000
... for mining
... ... ships, with
... enemy, and
... utmost.

relegated to cruisers, as ironclads will not be available. These cruisers will be fast ships, with a large coal capacity, a good armament, and a certain amount of protection. And I propose to say a few words on a modern cruiser, and will take for my type a belted cruiser such as the Orlando, with which most of you are familiar—a ship which would be admirably suited for work of this description.

The Orlando is 300 feet long, 56 feet beam, and 5600 tons displacement, with a maximum draught of 23 feet, and is capable of steaming at a speed of nearly 18 knots for a few hours, and has a radius of action of 4500 miles at 10 knots. She is built of steel throughout, and by referring to the diagram you will see that all the vitals of the ship are protected. The protection consists of a belt of 10-inch armour, steel-faced one-third of its thickness, 200 feet in length and 5 feet 6 inches in depth, about 18 inches being above the water-line at a load draught. On the top of this belt, and resting upon it, is a steel deck of two-inch plates, that is, two inches thick, going the whole length of the ship, being increased to 3 inches at the extremities where the ship is unpro-tected by armor, and taking a shape thus ⌐⸺ ⸺ ⸺⌐ running down towards the ends and assisting very materially to strengthen the ram and take the thrust. Everything that is vital to the ship is kept below this armoured deck, and every hatchway through it is protected by a water-tight door of the same thickness. Below this deck are the bunkers and double bottoms, all of which assist to protect the engines and other important items. Above the armoured deck are the upper bunkers, about 17 feet broad, and running the whole length of the engines and boilers. These, when filled with coal, offer a good pro-tection. The ship above the armoured deck may be riddled and the water wash freely over it, but it cannot get below on account of the water-tight doors before mentioned; and also around every hatchway are high coamings, 3 feet high and 9 inches broad, called coffer-dams. Another noticeable feature in the ship is that instead of having stanchions to support the guns, conning tower, and heavy weights above (and if one is shot away renders excessive strains on the remainder, which may fail to support the weight), they are sup-ported by steel bulkheads, which have the advantage that they may be riddled with holes but will still support much weight. All the guns are protected by 1-inch steel shields. The conning tower is 12 inches thick, one-third of which is steel. From this tower the ship can be steered; guns, engines, magazines, and every part of the ship communicated with by means of voice-pipes and telegraphs; the guns and torpedoes can be fired by electricity from the direc-tors, and the whole of these telegraphs and voice-pipes are carried down through an 8-inch armoured tube, being a solid tube of 24 inches bored out 8 inches in the center. The wires leading to the guns are kept below the water-line behind the armour, and when they rise above it, are duplicated and sepa-rated at least 3 feet apart, so that the same projectile will not be likely to damage both wires. There are three steering stations below the armour deck, that is, below the water-line, any of which can be communicated with; and the rudder is placed very low down in the ship, being entirely below the water-line. The whole ship is divided into an immense number of water-tight com-partments, every shell-room, bunker, store-room, and magazine being itself a water-tight compartment. There are two engine-rooms divided by a water-tight bulkhead, and two sets of boilers divided in a similar manner. The armament consists of two 9.2-inch B. L. guns, ten 6-inch B. L. guns, six 6-pounder and ten 3-pounder Q. F. guns, and seven machine guns, twelve White-head torpedoes with two discharges, six E. C. mines ready for use, which number would be increased in war-time, and 500-pound mines are in store, ready for issue as required, 1200 pounds of wet gun-cotton, 360 pounds of dry, one mile of unarmoured cable, a mile and a half of single core, armoured, 1000 yards of seven-core cable, with batteries, and all other necessaries for mining work, with a crew of 500 men. A squadron of four or six of such ships, with perhaps an auxiliary or two, would constitute a most formidable enemy, and would tax the resources of a good many protected harbours to their utmost.

TIME OF FLIGHT.	DISTANCE WITHIN WHICH AN OBJECT 20 FT. HIGH MUST BE KNOWN, TO HIT.	TIME TAKEN TO COVER THIS DISTANCE AT 15 KTS. SPEED. TARGET 300 FT. LONG.
12.0 Sec.	19 Yds.	14 Sec.
3.5 "	92 "	23 "
8.1 "	31 "	16 "
3.5 "	92 "	23 "

Now, I think no question need arise as to the preparedness for war of a man-of-war "in commission ; a ship in commission is always prepared for war at the very shortest notice ; there is nothing more to be done when going into action than is done regularly once a week, when the whole ship's company are exercised at what is called General Quarters." Every man has his appointed station, knows what he has got to do, and is exercised at doing it 53 times in the year ; besides this, the individual guns' crews and ammunition parties are drilled separately very much oftener.

General quarters is also exercised at night at least four times a year, the bugle sounding to quarters when least expected. This is drill only. The amount of actual practice is considerable. Every quarter is divided into at least two parts ; eight rounds per gun are fired at a target, and a large number of rounds of machine and quick-firing guns. Also once a year prize firing takes place, which is conducted thus : The target is 40 feet long and 15 feet high, and is moored ; three buoys are laid down parallel to the target, 800 yards apart, the center buoy being 1400 yards from the target at right angles to it. The ship steams along the base-line marked by these buoys at 8 knots speed. Firing commences as the ship passes the first buoy, which is 1600 yards from the target, and ceases as the ship passes the last buoy, which is also 1600 yards : nothing but hits count. As the ship passes the center buoy, No. 1 falls out, and No. 2 takes his place, the gun being worked with the diminished crews. The time occupied is about six minutes. Guns larger than 8 inches are allowed two runs, the time being twelve minutes. In 1885 and up to 1886, in different ships in different parts of the world, sixty-four guns of ten-inch caliber and upwards fired 417 rounds at the prize-firing target under the conditions already stated, and made 136 hits in thirteen minutes.

Last year the Cordelia with 6-inch guns fired in this time (six minutes) nine rounds, and hit the target nine times. The Impérieuse fired her nine-inch guns eight times, and hit six times in the twelve minutes. So you will see that, though there is much talk about the impossibility of making good firing from a moving platform, in practice extremely good firing can be made. On this point I have something to say further on. Lastly, but perhaps as important as any, is the half-yearly night practice, firing with machine and quick-firing guns and rifles at a triangular target 12 feet long and 5 feet high, with and without the electric light. The value of this practice cannot be over-estimated, and the difference between the apparent and actual number of hits is most astonishing, which only shows how much practice in this direction is needed.

I have entered rather freely into the description of a modern cruiser, because, speaking as I am to defenders, it may be of interest to them to know what kind of an enemy they may be called upon to tackle, and what his capabilities of doing damage may be ; and I have spoken of the practice to show that we are accustomed to working our guns, that they are not mere ornaments, and that we do not act upon the principle that if we fire them now we shall not have them to fire when required. Unless a large amount of practice is carried out, it is hopeless to expect that any thing like good shooting can be made ; it is remarkable what a difference it makes even with men who are well drilled, when the gun is loaded, especially if the gun is of a new type.

THE ATTACK. (See diagram.)

I have endeavoured to plot out here a supposed attack on the outer defenses of an imaginary harbour in order to facilitate my explanations.

The attack is made by four belted cruisers of the Orlando class and two second-class cruisers of the Marathon class.

The number of guns they mount being :

$$\left.\begin{array}{ll} 4 \text{ Orlando's} & 8 \text{ 9.2-in.} \\ 2 \text{ Marathon's} & 52 \text{ 6-in.} \end{array}\right\} \text{ B. L. R.}$$

74 Q. F.
36 machine.

The guns mounted variously for protecting the harbour are :

$$\left.\begin{array}{ll} 4 & \text{9.2-in.} \\ 20 & \text{6-in.} \end{array}\right\} \text{B. L. R.}$$

20 Q. F.

20 machine

$$\left.\begin{array}{ll} 8 & \text{7-in.} \\ 2 & \text{9-in.} \end{array}\right\} \text{M. L. R.}$$

74, and 4 torpedo-boats.

A general outline of the attack may be something of this sort :

The Admiral-in-command will arrange his movements so as to be off the port at daylight, so that he has a fair prospect of being able to get almost within striking distance without being seen or signalled. He will have communicated to his captains the exact nature of the tactics he intends to adopt, and everything will be ready for immediate action. He forms his four large ships, not into single line ahead, because a shot missing one ship might strike the next astern, and also because, as he intends to attack end on, the leading ships would mask the fire of others ; not into single line abreast, because the broadside of the inner ships will be masked by the outer, and he might be enfiladed by another battery ; but he forms into quarter line, i. e., each ship four points abaft the beam of the ship next ahead, in which formation he has his right-ahead and broadside fire clear. He will not form in quarter line to port because a ricochet might then strike his sternmost ships, but he will form them in quarter line to starboard, so that a ricochet, which always goes to the right, would be clear of all his ships ; each ship in this formation is on a different bearing and at a different distance to the others. In this formation he steams at a speed at which he will not have to force his engines, and which all his squadron can maintain for a time without fear of straining anything ; this speed will be about 15 knots, at which rate he will cover 500 yards a minute, and, allowing ample time between each round, he can make certain of being able to fire one round per gun every 1000 yards, i. e., every two minutes. He will detach the two smaller vessels, which carry six 6-in. guns and ten quick-firing guns each, to harass the two outlying batteries by keeping up a rapid fire on them with his quick-firing and machine guns, and prevent any man daring to expose himself above the parapet or in any exposed positions for sighting his guns or using range-finder. He will select some well-defined bearing on which to steer, and will have noted down certain positions on this line, which he will find by a cross-bearing and from which he has measured off accurately the distance of the batteries on the chart, so that he can elevate his guns with very fair precision. He will probably open fire with his 9.2-inch guns at 6000 yards or so, and with his 6-inch at 5000 or 4000, reserving his fire as each gun is loaded until his bearing comes on, when he knows that his distance is correct. He will approach in this manner to about 2000 or 3000 yards, depending upon circumstances. When the water is deep, and rough weather frequent, a systematic mine defense is difficult ; the area to be defended is so large, and the ships may approach in so many uncertain paths, that unless previous definite information has been obtained of the existence of mines so far at sea, he would not be afraid of them, and would manœuvre his squadron to the best advantage to develop their gun fire, knowing well the difficulty of maintaining even a small group effective so much in the open ; but correct information with regard to the mining matters is always the most difficult to obtain, especially when it can be confined to some half-dozen persons to say whether real or dummy mines have been laid. In our attack here the Admiral thinks it more prudent to alter his course when between 2000 and 3000 yards off, which he will do immediately after firing a broadside, under cover of the smoke. He then alters course together to starboard, and drops back into a 2-point bearing formation, because, had he retained the 4-point bearing forma-

... off the port
... get almost within
... have communi-
... decide to adopt, and
... his four large ships,
... might strike the
... the leading ships
... because the broad-
... might be enfiladed
... each ship four points
... he has his right-
... quarter line to port
... but he will form
... always goes to the
... formation is on a dif-
... this formation he
... and which all
... anything; this
... yards a minute,
... made certain of being able
... minutes. He will
... and ten quick-firing
... up a rapid fire on
... any man daring
... for sighting
... bearing on
... this line, which
... covered off accurately
... move his guns with
... guns at 6000
... fire as each gun
... distance is cor-
... yards, depending
... weather frequent, a
... is so large, and
... previous defi-
... so far at sea,
... to the best
... of maintaining
... information
... to obtain,
... say whether
... Admiral thinks
... yards off,
... cover of the
... dgoing back into a
... bearing forma-

The guns mounted variously for protecting the harbour are :

$$\left.\begin{array}{r} 4 \quad 9.2\text{-in.} \\ 20 \quad 6\text{-in.} \end{array}\right\} \text{B. L. R.}$$

20   Q. F.
20   machine

$$\left.\begin{array}{r} 8 \quad 7\text{-in.} \\ 2 \quad 9\text{-in.} \end{array}\right\} \text{M. L. R.}$$

74, and 4 torpedo-boats.

A general outline of the attack may be something of this sort :

The Admiral-in-command will arrange his movements so as to be off the port at daylight, so that he has a fair prospect of being able to get almost within striking distance without being seen or signalled. He will have communicated to his captains the exact nature of the tactics he intends to adopt, and everything will be ready for immediate action. He forms his four large ships, not into single line ahead, because a shot missing one ship might strike the next astern, and also because, as he intends to attack end on, the leading ships would mask the fire of others ; not into single line abreast, because the broadside of the inner ships will be masked by the outer, and he might be enfiladed by another battery ; but he forms into quarter line, *i. e.*, each ship four points abaft the beam of the ship next ahead, in which formation he has his right-ahead and broadside fire clear. He will not form in quarter line to port because a ricochet might then strike his sternmost ships, but he will form them in quarter line to starboard, so that a ricochet, which always goes to the right, would be clear of all his ships ; each ship in this formation is on a different bearing and at a different distance to the others. In this formation he steams at a speed at which he will not have to force his engines, and which all his squadron can maintain for a time without fear of straining anything ; this speed will be about 15 knots, at which rate he will cover 500 yards a minute, and, allowing ample time between each round, he can make certain of being able to fire one round per gun every 1000 yards, *i. e.*, every two minutes. He will detach the two smaller vessels, which carry six 6-in. guns and ten quick-firing guns each, to harass the two outlying batteries by keeping up a rapid fire on them with his quick-firing and machine guns, and prevent any man daring to expose himself above the parapet or in any exposed positions for sighting his guns or using range-finder. He will select some well-defined bearing on which to steer, and will have noted down certain positions on this line, which he will find by a cross-bearing and from which he has measured off accurately the distance of the batteries on the chart, so that he can elevate his guns with very fair precision. He will probably open fire with his 9.2-inch guns at 6000 yards or so, and with his 6-inch at 5000 or 4000, reserving his fire as each gun is loaded until his bearing comes on, when he knows that his distance is correct. He will approach in this manner to about 2000 or 3000 yards, depending upon circumstances. When the water is deep, and rough weather frequent, a systematic mine defense is difficult ; the area to be defended is so large, and the ships may approach in so many uncertain paths, that unless previous definite information has been obtained of the existence of mines so far at sea, he would not be afraid of them, and would manœuvre his squadron to the best advantage to develop their gun fire, knowing well the difficulty of maintaining even a small group effective so much in the open ; but correct information with regard to the mining matters is always the most difficult to obtain, especially when it can be confined to some half-dozen persons to say whether real or dummy mines have been laid. In our attack here the Admiral thinks it more prudent to alter his course when between 2000 and 3000 yards off, which he will do immediately after firing a broadside, under cover of the smoke. He then alters course together to starboard, and drops back into a 2-point bearing formation, because, had he retained the 4-point bearing forma-

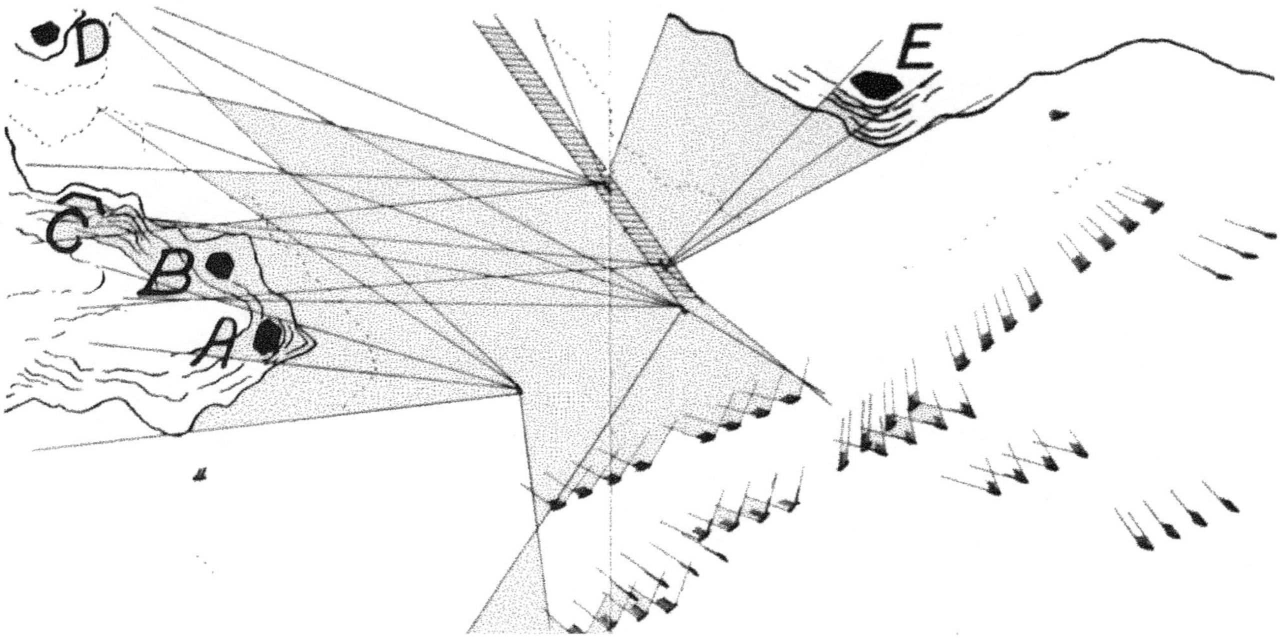
A
B
C
D
E

tion, his ships would show an unbroken line to the fire of battery A. As he approaches a 2000 yards range, he gets into single line ahead, so that in case of accidents, if one ship should be blown up by a mine, the next will probably escape, and when one has passed, the next may pass, probably. By this time he has warmed up to his work, and so have the defense; they are within a good shooting range of one another, and the whole of the guns will bear; perhaps the men are getting a little excited as the fire improves, so now he orders Director Firing, a system used extensively in the navy, and which I will endeavour to explain.

The distance at which it is intended to deliver the fire being accurately known, the elevation due to that distance is previously given to the guns. The bearing of the battery from the ship when at that distance being also known, the guns are trained on to that bearing ;—all this is capable of accurate arrangement beforehand. The guns are joined up for firing by electricity, and the firing-key is in the hands of the officer at the *director*. This instrument is simply a telescope converged and securely clamped on the same point and at the same distance as the guns. Everything is perfectly quiet except the ship, which is bowling along and gradually bringing the cross-wires of the telescope on to the selected batteries. As the point passes the cross-wires, the whole of the converged fire is delivered electrically simultaneously, and the result is according to the care with which the guns and instruments have been placed in the ship, and according to the accuracy with which the bearing, distance, and speed, which have been estimated beforehand, have been adhered to. This firing has been expressly designed to meet the case of ships passing moderately close at high speeds. If, instead of a fort, it was another ship we were engaging, each travelling (say) 12 knots, the space of time required to pass over each other's arc of broadside fire at a distance of 1000 yards would be about 2 minutes. Letting each gun fire independently would be ineffective, as half the guns would never fire at all, owing to the smoke of those which were firing, and owing to the difficulty experienced by the captain of a gun in training on to such a rapidly moving object. The confusion of each gun firing separately is avoided, and perfect silence reigns in the ship, with each gun's crew lying down round its gun. The effect of a broadside thus delivered, if well directed, is immensely in excess of the same number of guns hitting in succession. If the ship is being fired at at the time, this firing possesses the advantage that only one man is necessary to fire the broadside; this would be an officer selected for his coolness under fire and his knowledge of the director, and even he would be in the strongest part of the ship and quite hidden from view and protected; all others would be lying down under cover. Another and most useful method of using this firing is to give the guns the necessary elevation and let the guns' crews keep the training on. The captain of the gun has only one thing to look out for, *i. e.*, his training; the officer at the director will look out for the elevation, and the motion of the ship will be made use of for elevating the guns; with central pivot mountings and cogged gear for training, the motion is so smooth and easy that there is no difficulty in keeping the training accurately on; but of course this manner of using the director is only possible when the object fired at is distinctively visible.

To return to our attack.

As we approach the 5-fathom line we must alter course, which we do under cover of smoke again, and bring our stern guns to bear. All the time we have been engaging these batteries, our Q. F. and machine guns have been keeping up a continuous fire on the embrasures, parapets, or any loop-hole which may have presented itself, the Q. F. guns throwing shell at the rate of five or six a minute, which would be comparatively slow; and with the accurate fire made by these guns it is considered that they would do very considerable execution. The squadron now retires, the two smaller vessels are recalled, and stock is taken of damage done. It is sometimes said that one good shot from a heavy gun may disable a ship. Well, I am doubtful about that; but even supposing it to be the case, the chances are very much against this one particular good shot ever coming off; and taking the case of the Huascar and Almirante Cochrane, the latter hit the former some eighteen or twenty times at close range (200 yards sometimes), and yet after the action the Huascar was capable of steaming 200 miles into port.

Well, we find that the two smaller vessels have suffered most and will take a little time to repair damages.. Several guns have been disabled in the larger vessels, and one has a hole through her bunker. She is rather heeling over on one side, but otherwise still good for fighting. On the side of the defense, battery H is completely silenced; battery E has suffered severely, apparently, but still shows fight; her 6-inch guns on hydro-pneumatic mountings are still game to fight, but as they fire so slowly and make such bad shooting while we are on the move, we do not dread them very much; still we want to silence them and to give B battery a turn, so we return to the attack on much the same lines as before, only making our direct attack on B, and leaving the second-class cruisers to get ready for the work which must take place at night.

In this second attack we succeed in silencing E battery, and have two or three more of our guns disabled. B has been knocked about a good deal, but sufficient has been done to enable us to haul off and prepare for our direct attack on the harbour. The Admiral does not think it wise to land any men, since he is not sure when the defending squadron may appear on the scene.

I am aware that it has been stated that ships cannot make good practice when under weigh, and will prefer to run the risk of anchoring, but this argument has been based on the method adopted at the bombardment of Alexandria, which, unfortunately, is a bad precedent to draw conclusions from in some instances. The glare of the sun behind the forts, the half-hearted enthusiasm of the defenders, and the guns of the ships engaged not being of the most modern type, more particularly their mountings, when training and elevating was not the simple, smooth and accurate motion which it is now—all these reasons were taken into account, and the ships were anchored. So far the action has been an artillery duel between a certain number of ships and a certain number of guns mounted on batteries of various sorts. The number of guns of attack will always be in excess of the defense at the commencement of an action, for sufficient ships will be sent to ensure at least twice as many guns being brought to bear on the defenses as they can bring to bear upon the attack.

Let us, therefore, consider the *pros* and *cons* of such a duel.

To begin with : The attack, whatever its strength may be, will always have the following advantages :

First. They come into action fresh, and not worn out by false alarms and days, perhaps weeks, of suspense and uncertainty, which it is possible may have been the lot of the defenders.

Secondly. They know what they intend to do, what tactics they intend to use, what guns they will have brought against them, and where they are situated.

Thirdly. They will choose the time and conditions of the weather which will be most favourable to themselves and worst for the defense.

And apart from these unquestionable advantages to the attack, there are other considerations which may be open to argument from an artillery point of view. The question of range-finding is one of them :—the enemy approaches from out of range to fairly close quarters, at a speed of 15 knots, at which rate he covers 500 yards a minute; he steers upon a well-defined line, so that the helmsman, who is securely protected and is under the immediate eye of the officer in command, can see what to steer for accurately ; the navigating officer stands by, his compass also protected, and tells the officer in command when certain prearranged bearings come on, which give the position of the ship accurately, and from which the range of the forts or batteries is known by measurement on the chart: thus the ships know the range accurately, the guns are laid for elevation due to this range beforehand, and fired either electrically or by word of command. On the side of the defense the men using the range-finders are more or less exposed. The methods of firing at a rapidly moving target are seldom if ever practiced on account of interfering with shipping, etc., etc., and I would ask you to notice this small table:

Nature of gun. Inch.	Range. Yards.	Time of flight. Seconds.	Distance within which an object 20 feet high must be known to hit. Yards.	Time taken to come this distance at 15-knot speed, allowing 100 yards for length of ship. Seconds.
9.2 B. L. R.	6000	12.00	19	14
"	2000	3.5	92	23
6 B. L. R.	4000	8.08	31	16
"	2000	3.5	92	23

From which it will be seen that range-finders will hardly be of much use, for there is not sufficient time to lay the gun and fire it after the range is ascertained, before it will be again necessary to alter the elevation. It will, therefore, appear that the only way to overcome this difficulty will be to lay the guns for some point over which the ship will pass, and fire as they come on, which, you will observe, is the same principle as that employed by the attack, with this difference, that the attacking ships are sure to pass over the point they steer for; but are not so sure to pass over the point for which the defense guns are laid. On this point, therefore, I think the argument is in favour of the attack.

Now take the question of training and sighting. All new naval gun-mountings admit of being able to fire the gun while in the act of elevating or training, or even adjusting the sights. I am not sure on this point, but I don't think land mountings admit of this. Take the case of guns mounted on hydro-pneumatic disappearing carriages—what sights are used? Scott's sights have to be shipped, the gun laid, and the sights unshipped again before the gun is fired. Surely these will be no use against a rapidly moving target. Reflecting sights have never, I believe, been tried against a rapidly moving target, and ordinary target sights generally smash up by jar of the recoil; but if used, the men sighting the guns have to be exposed, and with the rapid and accurate firing of the Q. F. and machine guns, I don't think that a very steady aim would be taken. On board ship the men are not exposed, and can see along their sights, a much more simple and, I believe, more suitable method for fighting purposes than trusting to reflection or delicate instruments which won't stand the shock of the recoil.

The next point is: Can the guns of the defense be concentrated on any individual ship at the will of the officer in command? On board each ship this power of concentration can be and is done, as I have already explained, by director firing, electrically, and otherwise; but in many cases this ability to concentrate the fire would be far more applicable in firing at ships than in ships at batteries, because in a ship, so many guns and vitals are centralized in such a small place, whilst in a well-defended harbour the batteries are much dispersed, and, in fact, single guns dotted about here and there in good positions would be less liable to destruction than guns mounted in more imposing batteries; but whether the guns are in batteries or single, they should be capable of concentration. To do this, means that the organization must be exceptionally good, and communication perfect. I am assuming that every gun is in free communication, either by telephone or telegraph, with the position from which the defense is being conducted, for semaphore or flag signalling, or lights at night, will be quite out of the question, as the smoke would obscure it, and machine-gun fire would be too hot for the signallers. You will remember that I explained that all communications on board ship are protected by an 8-inch armoured cylinder above water, and are behind the armour below water. Are the communications on shore protected, or are wires suspended from posts used? If so, these will be almost certain to come to grief, and the communication will therefore break down, and control will cease. If they do break down, the attack will gain another advantage. The last point to consider in this duel is the visibility of the object aimed at, and here I will admit that the defenders have decided advantage, at any rate at the commencement of the action, for they can conceal their batteries to a great extent, while a ship cannot be concealed in the daytime. As the action progresses, the batteries will gradually become more visible as the foliage or earthworks get knocked away. Taking all these matters into consideration, I do not think that the attacking fleet will suffer much so long as they have room to move about, and not at all during their direct advance; and when they come to close quarters, their superiority in number of guns, especially the Q. F. and light guns, will tell enormously in their favor, and leave no doubt in my mind as to the ultimate success of the attack in silencing the batteries facing the open sea.

Having sufficiently silenced the outer defenses, the fleet will now anchor in some sheltered spot, if possible, to make their preparations for the attack on the mine-field, which, it is supposed, exists; at any rate it would be rash to go nearer the shore without having carefully examined the ground. The first thing to be done when the ships anchor will be to put the slip on their cables ready for getting away at a moment's notice in case of necessity; then hoist their boats out and prepare their torpedo-net defense. If they have had the opportunity before arriving off the port, they will have seized some steamer and put a crew on board and have anchored her well outside the sphere of action; the boats belonging to this steamer will now come in useful to make up for the boats which have been damaged in action. Gun-cotton charges for blowing up a boom, explosives, sweeps, creeps, and countermines, sheers over the bows of the steamboats for clearing obstructions, fittings to enable the boats to jump a boom, and all the various necessaries for a torpedo attack, have got to be put in the boats, the two longest operations being preparing the countermine, launching and rigging the net defense. All the time these preparations are going on will be a most anxious time, as it will be the opportunity for torpedo or gunboats, if they exist, to worry, if not actually to attack the enemy, and force him to keep a portion of his men at their quarters and prevent him, as much as possible, from getting his torpedo nets into position before darkness sets in. It may be necessary, if the Admiral is not certain of his ground, to anchor his ships before dark in the positions they are to occupy at the commencement of the attack; but it is better, if it can be done, to bring the ships into those positions after dark, as it will prevent the defense knowing the accurate range, and it will be more difficult for the torpedo and gun boats of the defense to make up their minds as to the best method of attack. But in a strange harbour, with no leading lights or with the leading lights purposely placed in wrong positions, it is a difficult matter to place ships in a definite position after dark. The Admiral will in such a case probably determine to be under weigh and get his ships near their positions just as darkness sets in, move up five or six hundred yards after dark, anchoring with a slip on his cables as before, and at once start his search-light, keeping a good lookout for torpedo-boats.

The positions in which he will anchor his ships after dark will depend upon the number of search-lights they carry. His aim will be to throw an effective beam of light completely round his squadron, upon which, with the assistance of his guard-boats, he relies. For preventing a torpedo-boat attack being successful, his nets will be in position, but it is possible that some of the booms supporting the nets may be shot away, so that he takes every precaution he can to guard against the Whitehead torpedo. He must also have sufficient search-light to blind the eyes of the gunners in the batteries. These lights he will not keep fixed, but will direct their rays first on, then off, the batteries, which is most confusing. The light for guarding against torpedo-boats will be worked round slowly, and then back again. It would be better to keep fixed rays, provided there were sufficient lights, but as there are not, each light must be worked through an arc of 90 degrees. The ship which is told off to make the first advance through the cleared channel will remain under weigh, and will do one of two things—either put out all lights in the ship and keep up a continuous fire of shell or blank cartridge (the object being to make much smoke and so to prevent the defense from seeing what is going on), or he will throw his search light in the eyes of the defense, taking care not to allow the rays to fall on his own boats which are advancing : the direction of the wind will decide which is the wisest course to take. The boats will be sent away immediately after dark.

Guard-boats, armed with Q. F. guns, whose duties it will be to keep about, 'above 1500 yards or 2000 from the ship, and not to come closer; to look out for and attack guard-boats of the defense; to give warning by signal if a torpedo-boat is seen passing them; and to protect the countermining boats when they commence operations. They must on no account follow up a torpedo-boat which is attacking, because the ships must have full liberty to fire on all boats approaching them.

Creeping-boats, armed with a machine gun and explosive creeps, fitted with what is called jumping gear, an inclined plane over the bows, to enable the boats to jump a boom. It is astonishing what a boat will jump when so fitted; a boom consisting of four baulks of timber, each one foot square and lashed together, has been jumped in this way, and even more astonishing results than this have been obtained ; but I regret I have not the details by me. The boats will also be provided with small sheers over the bows, with a creep suspended from them, and hanging a foot or two below the surface of the water, to catch nets or entanglements which may be laid down with a view to fouling the propellers of steamboats. Their duties will be to creep for and blow up electric cables to which the mines are connected ; some will be told off to work in the channel to be cleared, and others to go close in shore in the most likely direction for grappling the 4-core or other multiple cables. One such cable being cut may disable about 26 mines.

Boats for sweeping. These boats work in pairs, with a sweep or weighted rope between them. Their object will be to sweep for mines floating below the surface, and when grappled, a gun-cotton is hauled underneath it and exploded, thus destroying the mine. Each pair of boats, following the leading boats, will work in behind one another, so that no portion of the channel to be cleared is left out; they will drop small buoys on their outer flanks, to show what portion has been cleared.

I should have explained that in selecting a channel to be cleared the Admiral will carefully consider in what direction it is best to proceed, taking into account the depth of water, the tides, the distance from the possible position of a test-room, and so on. He will mark off a straight channel as being the shortest, and will not clear more than about 120 yards in width. He will arrange position lights of some kind as guides to the boats advancing, to see that they are keeping the channel laid down for clearance. The steamer, which, you remember, had been taken prisoner, will be brought up, and as few hands as possible, probably volunteers for the work, put into her, with directions to steam through the channel, with the object of enticing the enemy to explode his observation mines, over which she will pass, or to bump any mines to be exploded by contact ;—she will have had a certain amount of water admitted into her to bring her down to the draught of the largest ship to go through the channel. The officer in command will have directions to put his helm hard over directly a mine explodes beneath him, so that should he sink, he will not block the channel. He will be provided with boats to enable the crew to escape. Boats with special charges will be sent on to make a breach in the boom, should one exist; and finally the countermine boats will advance. Time will not admit of my entering into all the numerous details of the fitting of these boats : suffice it to say that their object is to lay large mines (500 pounds), and explode them in the channel to be cleared, at such distances apart that no mines within range of 90 feet from each countermine will be effective after the explosion. The countermines may therefore be laid 180 feet apart, and if two parallel lines are laid with that distance between them, the channel cleared will be 260 feet wide.

Countermining is the most rapid and certain way of clearing a channel; but it requires a large amount of material in proportion to the space cleared. A line, or two parallel lines, will clear a channel 760 yards long in less than five minutes. A service launch carries one line of twelve mines, with sinkers and buoys for marking the channel. The mines are arranged to drop automatically; and the boats towing countermine boat carry the firing batteries for one end, and the advance ship will fire from the other end, all the buoys marking the channel. The next pair of boats then run their mines, and ship advances as before. The ship will keep up a heavy fire during the running of the countermines, in order to cover them as much as possible by smoke. As one ship advances, so another will follow in her wake, and thus the attack will go on until the channel is cleared or daylight appears, when the boats will be recalled, and the ships must trust to their guns until night falls again, or they must retire, as it would be particularly awkward to be caught by the defending fleet in such a predicament. This is where the advantage of having a deep defense comes in. I do not think it will often be possible to continue the attack on a mine-field in daylight unless there are no guns or no gunners left to man them, for the fire of light guns would be too overpowering for boats to work, even though they were covered by a brisk fire from the ships. To retire must mean that most of the work must be done over again, for the defenders would certainly take the opportunity of removing all the buoys placed by the attack and laying down more mines instead, so that perhaps the wisest course would be to retire to the entrance of the mine-field when he can get away quickly, if need be, and where he can keep guard over that portion of the field which he has already cleared. And here we will leave him, as I have trespassed far too much on your time already, and have said quite enough, I hope, to give a general outline of an attack on a protected harbour, and to raise a few points for discussion. But before I sit down, I have a few words to say which will not detain you many minutes. I do not think that in any line of fighting is practice so essential as in that of attack and defense of harbours, for without practice, organization is impossible. The grievous errors that have been made both by the attack and defense in all the experiments which have been made at Milford Haven, Hayling Island, and so on, must be seen to be realized; they are only brought out by practice, and can only be remedied by organization : all this can be done in peace time. It is no use having the material unless we know how to use it to its best advantage. Submarine mines are no use unless they can be laid out quickly and maintained efficiently. It is easy to say we will put a boom across here and boat mines out there; but has it ever been tried out there? Where it has been tried it has been found that the difficulties are very great.

Are there definite orders as to how and where guard-boats and torpedo-boats of the defense are to withdraw, and due precautions to prevent firing upon friends?

Do the search-lights playing on the mine-fields interfere with the gunners working the guns in the battery, and can the light be directed from some position away from the light, as it is well known that those working the lights cannot see where to direct them?

Has the officer in charge of the firing of observation mines ever practiced his duties at night with the atmosphere charged with smoke?

Has the officer in charge of firing batteries definite orders ?

Is it clearly laid down where military duties end and naval begin ?

Can guard-boats be provided immediately, and are the guns ready to go in them, and do the crews know their duties ?

These are a few of the practical questions which have to be settled by trial and observation, and I am sure that Col. De Wolski will agree with me, after our Milford Haven experiences, that unless these attacks are practiced in time of peace at night, and under cover of smoke, the mistakes which will certainly be made when the real test comes cannot be overestimated.

Sydney is defended by guns and submarine mines; there is the naval element here for its defense, and there is a squadron on the station for attack; there are all the necessaries at hand for attack and defense, and I am sure that if the two meet in friendly contest, more practical experience would be gained by all concerned in one night than by years of theoretical study.

[From the *Electrical Engineer*, October 1, 1890.]

The introduction of the electric light on board ships, and especially on war vessels, with a full equipment of generators, has naturally afforded an opportunity for the introduction of other electrical devices, notable among which are electric motors for ventilation, for the training of guns, for the hoisting of shot, etc. But a new departure has recently been taken in the application of electricity to warlike purposes, in the introduction of range-finders depending upon a few simple principles, among which that of the Wheatstone bridge is prominent. The public is already familiar with some forms of Lieutenant Fiske's range-finders, which he has recently, however, simplified to a considerable extent, so much so that the readings can now be taken direct from a graduated dial. The description in this issue of Lieutenant Fiske's new range-finder shows it to be a neat application of the modified Wheatstone bridge principle ; and not the least notable point in connection with it is, that the errors due to the variations of the conditions from actual theory are compensated for by the very construction of the apparatus. With all these refinements and means placed at the service of the commander of a modern war vessel, it is indeed problematical whether the carrying on of warfare would be a pleasant undertaking in the future, considering the enhanced probability of a shot taking effect. The fact may also be noted here that the U. S. S. Baltimore is probably the first naval vessel in which a telephone service has been established, so as to enable the commander to communicate from the conning tower with the various parts of the ship.

REVIEWS.

SUBMARINE MINES IN RELATION TO WAR. By Major G. S. Clarke, C. M. G., R. E. Woolwich: Printed at the Royal Artillery Institution, 1890.

Major Clarke in his concluding paragraph says: ". . . It had been sought to raise discussion upon the various points connected with the war uses of submarine mines, merely because such discussion is at present entirely wanting. All the other elements of defense, all questions of tactics by sea and land, have been subjected to searching and critical enquiry in endless books, pamphlets, and lectures. Nothing but good results from free discussion, by which alone can vagueness of thought and hallucination be swept away. The same method—the ordinary method of all science—must be applied to submarine mines, in order that they may be enabled to take their proper place in the Imperial Armoury. If an authoritative basis can be arrived at, they will fall at once into their legitimate sphere." He adds that his opinions are those of a "non-expert." Lawyers are held to be unfit jurymen, probably because their training is that of partisans, and in the consideration of legal technicalities they are apt to lose sight of the equity of the case. For the same reason it is well to have the opinion of a "non-expert" on the use of submarine mines, free from the bias begotten in the evolution of the mine-field.

Quoting from the Chatham Text-Book of 1873, Major Clarke says submarine mining defense is stated to be suitable for: (a) First-class fortresses, such as Portsmouth; (b) Mercantile harbors, such as Liverpool; (c) Undefended harbors, such as Belfast; (d) Small harbors, such as Whitby; (e) Open towns, such as Brighton; (f) Open beaches, such as Sandown.

The defense of open towns and open beaches against a fleet armed with long-range guns hardly needs serious consideration. To sow the sea with mines for the very ordinary radius, for bombardment, of 6000 yards, would be a serious task. As for an undefended harbor, that is, undefended by ships or guns, whether or not the mine-field is destroyed is simply a question of the disposition of the enemy.

In considering the case of defended harbors, the first question is as to the form of the port. If it is in the nature of a *cul-de-sac*, such as Plymouth, the inner harbor being commanded by guns, would any naval commander enter? Could even a battle-ship occupy herself in bombarding the town from such a position with her limited supply of ammunition, while batteries with a practically unlimited allowance were pounding away at her? It certainly is not likely that a naval commander would run into such a trap, even if cordially invited, and therefore mines would serve no purpose.

When the harbor is of such a shape that a fleet, once past the defenses of a narrow channel, finds itself in wide and safe waters, uncommanded by guns, and free to continue its operations, the use of mines is obvious, since every effort must be made to stop the passage of the enemy into the inner waters.

The time required for the construction and emplacement of modern heavy guns is so considerable that little can be done for the protection of a port, by guns, on the outbreak of war, unless they are already in position; and in such a case it may be protected from the entrance of an enemy's fleet by mines for perhaps a few days, giving time for the arrival of a protecting fleet. The guns, once in position, are always ready for instant use. The laying of a mine-field takes considerable time and can never be commenced until war is at hand. Major Clarke considers it a valuable accessory in certain cases, but he protests against the absurdity of the pretension of the Chatham text-book and against the overgrown mine-field, with special batteries to cover it, a special flotilla to patrol it and special search-lights to illuminate it, the latter probably interfering seriously with the work of the main batteries in case of a night attack.

It is to be considered, moreover, that a really efficient mine protection involves the destruction of the commerce of the port. Merchant vessels will risk running a blockade, but will hesitate to attempt the narrow channel of a mine-field, if one is left, at night.

It is well for us in this country to bear the lesson in mind. We cannot allow the commerce of our ports to be destroyed; and for defense we must have heavy guns, commanding not only the entrance, but the harbor itself, where possible, a fleet of battle-ships to prevent a blockade, and clear channels for the entrance of friendly vessels.   C. S. SPERRY,

                                        Lieut.-Comdr. U. S. Navy.

ARCHITECTURE NAVALE.—THÉORIE DU NAVIRE. Par J. Pollard et A. Dudebout, Ingénieurs de la Marine, Professeurs á l'École du Génie Màritime. Vol. I.

This work on Theoretical Naval Architecture, as contrasted with Shipbuilding, is by two of the lecturers at "L'École d'Application ·du Génie Maritime," where the government students selected for the French Construction Corps take a course of two years.

Only the first volume has been received as yet; the remaining three are, however, in press.

The subjects to be treated in the complete work are : Calculations of the geometrical elements and properties of ships' forms, statical and dynamical stability, the rolling of ships in smooth and rough water, resistance and propulsion, the steering of ships, and the vibration of screw steamers.

Volume I. deals only with calculations of geometrical elements and properties, including methods and forms used in determining statical stability at given angles of inclination. Pp. 1–71 treat of methods used in calculation. The Trapezoidal and Simpson's Rules are explained, and their absolute and relative accuracy discussed. Woolley's Rule, the Method of Differences, graphic differentiation and integration, and mechanical integration are also treated.

Pp. 72–114 are devoted chiefly to the calculations found on an ordinary displacement sheet, dealing with the displacement in upright condition, and the corresponding positions of the center of buoyancy and the metacenters. In this portion of the work there is given, with full explanations, a complete copy of the regulation French tables and forms used in such calculations.

Pp. 115–252 discuss methods which deal with the ship when inclined, either longitudinally or transversely. Under transverse inclinations, the authors explain five English and seven French methods of calculating transverse righting moments, and in addition, several methods of determining righting moments from models, etc.

The remainder of Vol. I, pp. 253–348, is devoted to the mechanics of floating bodies—including ships. All the elementary theorems of the subject appear to be here, and many which are more interesting to the mathematician than to the average naval architect.

As may have been inferred from the preceding brief summary, this work collects and compares many methods which have been published separately in transactions of societies and in various technical journals. Hence, though containing matter of interest to all who have to do with ships, it will be of peculiar value to the naval architect as a work of reference.

The chief original matter in the present volume consists of comparisons between the various methods explained. These are well drawn, and the distinctive advantages and disadvantages of each method are clearly and impartially pointed out.

A valuable feature of the present volume is a " Notice Historique et Bibliographique," being a list of 468 books and papers—chiefly English and French —upon the various subjects covered by the present work. The books and papers are classified according to dates and subjects, with short historical and critical notes upon the books. D. W. TAYLOR,
Assistant Naval Constructor, U. S. N.

HAND-BOOK OF PROBLEMS IN EXTERIOR BALLISTICS. PART I.—DIRECT FIRE. By Captain James M. Ingalls, 1st Artillery. Artillery School Press, 1890.

Captain Ingalls says in his introduction : " It is proposed to give practical solutions of all those problems of Exterior Ballistics which are likely to be useful to the artillerist, with examples fully worked out in the manner which a considerable experience has shown to be the most simple and concise."

The intention has been well and fully carried out, and the book is an excellent guide for practical work. The formulas used are those of Siacci. Certain auxiliary formulas for the convenient approximate solution of certain cases have been added, and the subject of Probability is treated. The book includes Ballistic and convenient Auxiliary tables. C. S. S.

BIBLIOGRAPHIC NOTES.

AMERICAN CHEMICAL JOURNAL.

VOLUME XII, No. 4, APRIL, 1890. W. M. Burton and L. D. Vorce have determined the atomic weight of magnesium, finding the number 24.287 (O $=$ 16). H. W. Hillyer describes a convenient form of gas generator for laboratory purposes. Reviews and Reports: Recent progress in industrial chemistry. Notes: Chemical examination of baking powders. Density, color, and spectrum of fluorine.

VOLUME XII, No. 6, JUNE. Gibbs and Hare continue the study of the action of definitely related compounds on animals.

VOLUME XII, No. 7, JULY. W. A. Noyes concludes his determination of the atomic weight of oxygen, finding the number 15.896. Reviews and Reports: Electrolytic dissociation. C. R. S.

JOURNAL OF THE AMERICAN SOCIETY OF NAVAL ENGINEERS.

MAY, 1890. Notes on modern boiler-shop practice. Notes on the effect of temperature on certain properties of various metals and alloys. Tubulous boilers.

In this article, contributed by Assistant-Engineer Leonard, U. S. N., the principal requisites of a successful tubulous boiler are first given, followed by a general description with illustrations of each of the following types: Herreshoff, Belleville, Ward, Towne, and Thorneycroft. From the data of evaporative tests made at different times by boards of U. S. naval engineers, a table is added to compare the relative weight and space occupied by the three types of boilers—tubulous, locomotive, and Scotch. The following approximation is made from the table, on the assumption that the evaporation varies directly as the combustion, 25 pounds of coal per square foot of grate being used as a unit. The locomotive boiler should receive a favorable correction of about 20 per cent in columns 2, 3, and 5, on account of high rate of combustion. On the other hand, the Belleville boiler should receive an adverse correction of about 10 per cent, on account of low combustion. All the other tubulous boilers given, greatly exceed the Scotch in these advantages of weight and space.

Type.	1 Combustion.	2 Evaporation per cub. foot of space.	3 Weight per H. P.	4 Wgt. per sq. foot heating surface.	5 Wgt. per lb. of water evaporated.
Belleville,	.5	.5	2.02	2.1	2.5
Herreshoff,	1.	.95	.72	.6	.9
Towne,	1.	1.2	1.12	.87	1.3
Scotch,	1.	.44	2.4	1.64	2.3
Locomotive,	3.9	.31	3.7	1.25	3.5
Ward,	2.2	.58	1.27	.5	1.53

The machinery of the torpedo-boat Cushing. The White steam " turn-about " life-launch.

AUGUST. An account of the experiments made on the double-screw steel ferryboat Bergen. The manufacture and inspection of iron and steel boiler tubes.

In this paper is given a general description of the process of manufacturing lap-welded boiler tubes of the kind used in the service.

On the designing of continuous-current dynamo machines.

A reprint of an article by Mr. Andrew Jamieson, read before the Institute of Engineers and Shipbuilders in Scotland.

Speed trials of fast ships. Report upon trials of three steamers—Fusiyama, Colchester, and Tartar. Tubulous boilers, (a continuation of the discussion of this paper in the preceding volume, by Mr. Miers Coryell). Instructions for determining the coal consumption on vessels of the French navy.   J. K. B.

JOURNAL OF THE ASSOCIATION OF ENGINEERING SOCIETIES.

MAY, 1890. The requirements of specifications for steel and iron.

A paper read before the Civil Engineers' Club of Cleveland, by Mr. Ritchie, in which are considered some of the most important points to be looked after in the inspection of steel.

JUNE. Recent progress of the metric system.

AUGUST. Ferroid, a new artificial stone. Compound locomotives.
                                                        J. K. B.

THE ENGINEER.

JUNE 7, 1890. The Justin dynamite shell. Economy of high expansion.

JUNE 21. Robertson's separator for steam boilers. The adaptation of steam machinery. Petroleum in steam boilers.

A paper on the results of using oil in the boilers of the steam-heating plant, at the Lansing Agricultural College. Previous to its use the boilers were badly incrusted with a hard scale. The kerosene seemed to work between the scale and the boiler-plate in such a manner as to loosen large flakes. Quantities of it are also found at the bottom of the boiler, in the shape of fine mud, which is readily removed through the blow-off pipe. At the present writing the boilers contain less scale than at any previous time during the past four years, and the small amount present seems soft and gradually disappearing. No injurious effect on the iron was perceptible.

JULY 5. The Forbin (a description of one of the recent additions to the French navy). The new navy.

JULY 19. The engines of the Mariposa. Coloring of metals by oxidation. Directions for working aluminum. Compound locomotives.

AUGUST 2. Notes on the use of hydraulic cements. Machine design for steel works. American locomotive boilers.

AUGUST 16. Rate of combustion in locomotives. Electric motors compared to locomotives. The progress of electric traction. Heat-energy.

· SEPTEMBER 13. Electric launches. Case-hardening. The trial of the San Francisco.

SEPTEMBER 27. A broken crank-shaft. The education of engineers. Dangers of electric lights on English steamers. Ocean racing. Link motion vs. automatic cut-off efficiency. J. K. B.

INSTITUTION OF MECHANICAL ENGINEERS.

OCTOBER, 1889. On the results of blast furnace practice with lime, instead of limestone, as a flux. First report of the Research Committee, on the value of the steam jacket. Further experiments on condensation and re-evaporation of steam in a jacketed cylinder.
J. K. B.

MECHANICS.

MAY, 1890. The indicator, and its errors. The strength of fly-wheels. Notes on the action of lubricants.

JUNE. A special report of the Twenty-first Semi-Annual Convention of the American Society of Mechanical Engineers.

In addition to a report on the business transacted before the Society, reprints are made of the following papers read before the meeting : Standard method of conducting duty trials of pumping engines. An automatic absorption dynamometer. Steam engine governors. A use for inertia in shaft governors. A universal steam calorimeter. Engine tests. The measurement of the durability of lubricants.

AUGUST. A scientific boiler test.

An elaborate report of a test made by Mr. Michael Longridge, of a Babcock & Wilcox boiler, in connection with a Lowcock economizer. J. K. B.

NORTH OF ENGLAND INSTITUTE OF MINING AND MECHANICAL ENGINEERS.

AUGUST, 1890. Report of the French commission, on the use of explosives in the presence of fire-damp in mines. J. K. B.

PROCEEDINGS OF THE INSTITUTION OF CIVIL ENGINEERS.

VOLUME CI. On the action of quicksands. The Calliope graving dock, Auckland, N. Z. On the probable errors of surveying by vertical angles. Wire rope. The deflection of spiral springs. Notes on a new method of distributing triangulation errors.
J. K. B.

PROCEEDINGS OF THE ROYAL ARTILLERY INSTITUTION.

APRIL, 1890. Siacci's method of solving trajectories and problems in ballistics, Part III, by A. G. Greenhill, M. A., F. R. S., and Mr. A. G. Hadcock, late R. A. Sound velocity applied to range finding, (communicated by Captain G. G. Aston, R. M. A.).

MAY.   Fire control in fortresses, by Captain F. G. Stone, R. A. Submarine mines in relation to war, by Major G. S. Clarke, C. M. G., R. E.

AUGUST.   A range and training indicator, by Captain L. C. M. Blacker, R. A.

SEPTEMBER.   Notes on the handling of artillery in the field, with especial regard to the supply of ammunition, by Lieutenant-Colonel N. L. Walford, R. A.   Control of artillery fire in action, by Captain W. J. Honner, R. A.

JOURNAL OF THE MILITARY SERVICE INSTITUTION.

JANUARY, 1890.   Danger from lack of preparation for war, by General Gibbon.

MARCH.   Development of submarine mines and torpedoes, by Lieutenant James C. Bush, U. S. A.

MAY.   Development of submarine mines and torpedoes (continued).

THE RAILROAD AND ENGINEERING JOURNAL.

JUNE, 1890.   The Brooklyn dry dock.   The latest Italian cruiser.

A general description and illustration of the latest addition to the Italian navy, named the Fieramosca, which is classed as a torpedo ram.   On the trial trip the speed obtained was 18.6 knots, the engine developing 7500 horse-power.   The main battery includes two 25 cm. guns, weighing 25 tons each, mounted in pivot, and six 15 cm. guns, mounted in broadside.   There is a secondary battery of machine guns and six torpedo-tubes.   Owing to the excellence of her armament, her speed, and the ease with which she was handled on trial, it is expected that this ship will be very formidable in action, more so, perhaps, than some of the very heavy battle-ships for which the Italians have shown a preference heretofore.

Preserving wooden piles.   United States naval progress.

JULY.   The new cruiser Latona for the British navy.   The engines of the U. S. S. Maine.   An automatic back-pressure relief valve. United States naval progress.

AUGUST.   Experiments with locomotive boilers.   The use of aluminum in the construction of instruments of precision.   The accident to the City of Paris.   Experiments showing the rate of combustion in locomotives on the B. & O. Railroad.

SEPTEMBER.   Isthmus canals.

An abstract of the report of the commission of French engineers, appointed to examine the Panama Canal.   The estimated expense of completing the canal, upon the plan they propose, is about $120,000,000, the principal items being $60,000,000 for excavation, and $22,000,000 for locks.   To this must be added the general expenses, which will bring the total amount required up to about $180,000,000, representing the amount that will have to be provided, in addition to that already invested in the company.

The new battle-ships.   Collisions at sea.   J. K. B.

THE SCHOOL OF MINES QUARTERLY.

APRIL, 1890. Examination of mineral properties. Indian grass oils. Electrical engineering at Columbia College.

JULY. The wind problem in gunnery.　　　　　J. K. B.

REVUE DU CERCLE MILITAIRE.

APRIL 20, 1890. Notes on the equipment and subsistence of the soldier in the field.

APRIL 27. One year in Tunisia. The military institutions of China.

MAY 4. The French armament, as viewed from a Russian stand-point. Military institutions of China (ended). The war of Senegal.

MAY 11. The new Belgian rifle. Messenger pigeons and dove-cotes (giving interesting points in regard to training, feeding, and liberating, or starting off messenger pigeons, and the important part —every day more marked—that they may play in time of war).

MAY 18. The Russian officer in the army and in society.

JUNE 1. The Danish rifle, model of 1889. The envelope-tube of the barrel of the German rifle.

JUNE 8. Works of the geographical service for 1889.

JUNE 15. The field stadiometer.

JUNE 29. Note on a recording device of regulated firing; Demer-liac system.

JULY 6. The discussion on the two years service in Germany. Mountain artillery. The Chinese navy.

JULY 13. Mobilization and manœuvres of the fleet in England.

JULY 20. The curvigraph, its civil and military use.

JULY 27. New firing regulations in Germany. The curvigraph, etc.

AUGUST 3–10. The Trans-saharian railway.

AUGUST 17. A glance at our naval manœuvres (French). The Trans-saharian railway (ended).

From present indications, it will not be many years before Lake Tchad and the Upper Niger will be in direct rapid communication with the shores of the Mediterranean. The completion of this great undertaking will, no doubt, add greatly to the influence of the French in Africa.

The new firing regulations in the German army.

AUGUST 24. The Victoria torpedo. The new firing regulations in the German army (continued). A glance at our naval manœuvres (ended).

AUGUST 31. The latest improvements in the European navies. A new type of revolver for officers.

SEPTEMBER 7. A curious mode of fabrication of metallic tubes. The new firing regulations in the German army (ended).

SEPTEMBER 14. The fortifications of the Saint-Gothard. The latest improvements in the European navies.   J. L.

REVUE MARITIME ÈT COLONIALE.

APRIL, 1890. A study of the use of the sextant for observations of precision. Oceanography (statics), (continued). The history of a fleet of former times.

MAY. Trajectory of a projectile in case the resistance of air is proportional to the cube of velocity. The history of a fleet of former times (ended). Oceanography.

JUNE. Trajectory of a projectile in case the resistance of air is proportional to the cube of velocity. The English naval man-œuvres of 1889 (an analysis of an article in the Engineering of March 28, 1890). A statistics of wrecks and other mishaps at sea for the year 1888. A Note on Commander Fleuriais' "top" (a professional article). Exploration of the western Soudan.

JULY. Budget of the German navy for the year 1890–91. The navies of antiquity and the mediæval age. Studies of comparative naval architecture. Historical studies of the military marine of France (continued). Biographical notice on Rear-Admiral Le Blanc. Prizes awarded the authors of the best memoirs published in the "Revue."

AUGUST. The navies of antiquity and the mediæval age. Studies of comparative naval architecture. A method for the immediate adjustment of the standard compass in landfalls with, or without, interruption of the ship's course, by Capt. Fournier. Approxima-tion with which a longitude is determined from an observation of the occultation of a star by the moon. Historical studies of the military marine of France (continued).   J. L.

RIVISTA MARITTIMA.

APRIL, 1890. A study on the naval needs of Italy, by Captain G. Bettòlo. The English naval manœuvres of 1889, by C. A. Views of Sir Frederick Abel on smokeless powders (trans.). Col-lisions at sea (trans. from the French).

MAY. New method of determining the velocity of projectiles within the gun, by Prof. Frölich (trans. and notes).

JUNE. Historical study on submarine warfare, by Lieut. Ettore Bravetta. Economical study of electrical conductors, by Giorgio Santarelli, engineer. Spontaneous combustion of coal, by Chief Engineer N. Soliani.

JULY AND AUGUST. A study of the character of Christopher Columbus, by Salvatore Raineri. Historical study on submarine warfare (continued). Graphic determination of the stability of ships, by Giuseppe Rota, naval engineer.

SEPTEMBER. Study on modern naval tactics, by Lieut. G. Ronca.

Study on the compass deviations on Italian ships-of-war, by Capt. A. Aubry. Fire ships and infernal machines in naval warfare (historical), by Lieut. Ettore Bravetta.

RIVISTA DI ARTIGLIERIA E GENIO.

APRIL, 1890. Smokeless powder. Study on field redoubts, by Captain Spaccamela Pio. Miscellaneous: The Fiske range-finder; Rapid-firing guns, system Skoda; Experiments with the 10 cm. rapid-firing, the 75 mm. field gun, and the mountain guns, Canet system. J. B. B.

REVISTA MILITAR DE CHILE.

JUNE, 1890. Trial of guns of the Krupp and the De Bange systems at Batuco, by Lt.-Col. J. C. Salvo. Tactics necessitated by the introduction of the small caliber long-range rifle (trans.), by Lt.-Col. Sofanor Parra. On the military position of Chile, by Col. José A. Varas. Instructions for target firing, by Col. E. de Canto.

AUGUST. A visit to the Krupp gun factories, by Col. Diego Dublé, A. Instructions for target firing (continued). On the military position of Chile (continued).

REVISTA TECNOLÓGICO INDUSTRIAL.

APRIL, 1890. Sanitary details of the city of Barcelona. The artificial harbor of Barcelona. Royal decree on electric installations.

JUNE. The dangers of the electric light. Theory of the steam engine.

BOLETIM DE CLUB NAVAL.

JANUARY, 1890. The question of armor-plated turrets. 7-inch guns, De Bange system. Historical points on the history of the Brazilian navy.

FEBRUARY. Hints on the study of naval tactics. The question of armor-plated turrets.

MARCH. The Brazilian Lloyds. Historical points on the history of the Brazilian navy.

APRIL. The armament of the Aquidaban. The cruiser Almirante Tamandaré. Hints on the study of naval tactics (continued).

TRANSACTIONS OF THE TECHNICAL SOCIETY OF THE PACIFIC COAST.

JANUARY–MARCH, 1890. Notes on the dry-dock and coffer-dam at the navy-yard, Mare Island, by Otto V. Geldern (illustrated).

APRIL. Coffer-dams and floating caissons, by Randall Hunt, C. E. (illustrated).

TRANSACTIONS OF THE CANADIAN SOCIETY OF CIVIL ENGINEERS.

OCTOBER–DECEMBER, 1890. Cantilever bridges, by C. F. Findlay. The Colonial Government dry-dock, St. John's, Newfoundland, by H. C. Burchel. The Esquimault graving dock works, British Columbia, by W. Bennett.

JOURNAL OF THE FRANKLIN INSTITUTE.

DECEMBER, 1889. American isthmian canal routes, by Daniel Ammen, Rear-Admiral United States navy. Philosophy of the multi-cylinder, or compound, engine : its theory and its limitations ; by Robert H. Thurston. On the dilatation and compressibility of water, and the displacement of its maximum density by pressure (translated from the French by B. F. Isherwood).

JANUARY, 1890. The electrical exhibits from the United States at the Paris Exposition. Philosophy of the multi-cylinder, or compound, engine (continued). A rapid method for phosphorus in iron and steel, modified from a method proposed by Dr. Thomas M. Drown ; by G. L. Norris.

FEBRUARY. A test of an Otto gas engine. Philosophy of the multi-cylinder, or compound, engine (continued).

APRIL. Electric railways. The Hollerith electric tabulating system.

MAY. "Water ram" in pipes. Colt's new navy revolver, combined with a new cartridge pack for reloading.

JUNE. The purification of water by means of metallic iron. What does a steam horse-power cost ? by Thos. Pray, Jr.

JULY. Report of the U. S. Naval Board on the Thomson system of electric welding. On the electro-deposition of platinum.

SEPTEMBER. Electricity in warfare, by Bradley A. Fiske.

J. B. B.

THE STEAMSHIP.

MAY, 1890. A new mode of mounting guns.

A successful trial of Sir W. G. Armstrong, Mitchell & Co.'s new mode of mounting guns to be fired *en barbette,* took place off the Isle of Wight, on board H. M. S. Handy, a vessel especially appropriated for gun trials. Particular importance attached to the tests on this occasion, the invention to be tested being designed to meet a defect which has been much felt in regard to the existing method of mounting heavy guns in barbette ships. In the design tried, the gun not only returns automatically into the firing position after each discharge, according to the Vavasseur recoil system, but is capable of being elevated so as to fire at angles up to forty degrees, the caliber of the piece, in this instance, being 9.2 inches, and the weight 22 tons. The carriage on which the gun is mounted is also fitted with a steel shield six inches thick, which is attached to the mounting and trains with it. The construction is such that the port through which the gun fires is completely filled with the gun at all angles of elevation, thus preventing the entrance of projectiles or splinters. The mounting is intended for use in barbette batteries on the upper deck, and no similar carriage has hitherto been provided with any screen capable of resisting the fire of anything more than machine guns, whereas the shield now devised effectually protects the gun and gunners from all rapid-fire guns at present in use in the service. At the trial which took place March :9, fifteen rounds were fired at angles ranging up to the maximum of 40 degrees, with perfect success in every respect.

The strength of ships.

A paper on the above subject, with special reference to distribution of shearing stress over transverse section, was read at the annual meeting of the Institution of Naval Architects, by Professor Jenkins.

The spontaneous ignition of coal cargoes. The effects of stress on steel.

JUNE. The screw propeller.

A paper read before the Institution of Civil Engineers by Mr. Sydney Barnaky, and dealing with the determination of the best dimensions to be given. The author submitted the results of recent experimental research, so far as they bore upon the solution of this problem. The experiments made by Mr. Thorneycroft with model screws were described, and the author's method of tabulating the results was explained.

On steamship propulsion. Marine engine governors. Various theories of the screw propeller.

In this article Mr. James Howden takes exception to almost all the leading ideas advanced on the subject by Professors Rankine, Cotterill, Greenhill, and Dr. Froude, the paper being directed expressly to the manner in which the screw acts on the water in propulsion, and the motions thereby imparted to the water in obtaining its propulsive effect.

JULY. On the dynamics involved in the lines and speeds of ships. Internal corrosion of vessels. The corrosion of iron and steel.

AUGUST. The Légé torpedo. Launch of the cruisers Pallas and Phoebe. Formulas for model experiments. J. K. B.

THE STEVENS INDICATOR.

VOLUME VII, No. 2. Notes on the action of lubricants. Testing machines. Instruments for illustrating the action of the polar planimeter. Ericsson and his Monitor. Cost of lubricating car journals.

No. 3. The fabrication of twelve-inch mortars. Notes on the friction of engines. The indentification of dry steam. The examination of lubricating oils. A new recording pressure gauge.
J. K. B.

TRANSACTIONS OF THE AMERICAN SOCIETY OF CIVIL ENGINEERS.

JANUARY, 1890. The results of investigations relative to formulas for the flow of water in pipes.

FEBRUARY. Cast iron, strength tests, and specifications.

An article by Prof. J. B. Johnson, in which are given the results of some experiments at the Washington University Testing Laboratory, on the testing of cast iron for strength and resilience in tension, compression and cross breaking; the methods and appliances there used; a new machine for cross breaking tests; an explanation of the high cross breaking modulus of cast iron, and, finally, a proposed set of specifications for engineers, to use for cast iron products.

APRIL. The railroad ferry steamer Solano.

JUNE. Calculations of the mean horse-power of a variable stream, and the cost of replacing the power lost, by a partial diversion of the flow. Observations on the Forth Bridge. J. K. B.

TRANSACTIONS OF THE AMERICAN INSTITUTE OF MINING ENGINEERS.

Aluminum and other metals compared. Notes on the energy and utilization of fuel, solid, liquid, and gaseous. The Heroult process of smelting aluminum alloys. Aluminum in wrought iron and steel castings. Preliminary note on the thermal properties of slags.

J. K. B.

UNITED SERVICE GAZETTE.

MAY 31, 1890. The Icarus court-martial. The Dover manoeuvres. A year of lifeboat work. The Victoria torpedo. On a system of signalling between men-of-war and merchant vessels. America's naval policy.

JUNE 7. The summer manoeuvres. The Dover manoeuvres. The defense of Australia. Naval supremacy.

JUNE 14. The Royal United Service Institution. American warship design. The exigencies of modern warfare.

JUNE 21. Naval uniform. The Channel tunnel.

JUNE 28. Mobilization of the Royal Naval Reserve. Royal naval engineer officers.

JULY 5. Launch of war vessels. The Coast Guard. Spontaneous ignition and explosions in coal bunkers.

JULY 19. Appointments for the naval manoeuvres. The royal naval engineer officers—II.

JULY 26. The naval manoeuvres, 1890. Royal naval engineer room artificers.

AUGUST 2. The naval manoeuvres. The navy as a profession.

AUGUST 9. The naval manoeuvres. The strength of the navy. The royal naval engineer officers—III.

AUGUST 16. The naval manoeuvres. Submarine torpedo-boats.

AUGUST 30. The naval manoeuvres. Lessons from the naval manoeuvres.

SEPTEMBER 6. Coaling at sea. Next year's naval manoeuvres.

SEPTEMBER 13.

" A disagreeable surprise has been caused by the report, that at the recent naval manoeuvres three-quarters of the whole number of the vessels engaged were, at different periods, reported as suffering from defects of machinery, most of which were trivial, but some of a more serious nature. Considering the consequences likely to arise from a breakdown in case of actual hostilities, the Admiralty order is warmly welcomed, directing that high-speed trials shall be made once a quarter, each trial to be of not less than twenty-four hours' duration, whenever circumstances will permit, with the engines working at from one-half to two-thirds of their specified natural draft. During each trial the engines are to be run for a period of five hours at their highest speed with natural draft, but the horse-power developed is not to exceed the specified natural draft power."

The German naval and military manœuvres. The French naval manœuvres.

One effect of the French naval manœuvres this year has been to show up the smallness of the supply of fuel kept in the coal dépôts at Brest. These contained only enough for one coaling of the squadron, after which the supply was exhausted. A French naval critic points out, in the *Nouvelle Revue*, that the harbor of Brest, being the only solid base of operations for the Northern Squadron, ought evidently to be provided with coal enough to last six months.

The proposed naval exhibition.

SEPTEMBER 20. Attack on the Thames and Medway defenses. The defense of Australia. The navy. Letter from Admiral of the Fleet Thomas Symonds.

The writer opposes the present system of heavy armaments, advocating lighter guns and consequent higher freeboard. Reference is made to the recent manœuvres, where "half the heavy armament of two battle-ships, Inflexible and Hero, is disabled . . . . and continued silent for many hours, and when soon afterwards a cracked steam-pipe, upon which the working of the hydraulic engines depends, obliged the armor-clad Rodney to return to port." "The serious error which is being made in continuing to overburthen our ships with guns that are, not only, too heavy to be efficiently worked by preventer gear (in case of the failure of the steam or hydraulic engines), but are likewise unreliable in several other respects for naval warfare, is only too manifest."

SEPTEMBER 27. The defense of Australia. The Royal Naval Exhibition. The Sardegna.

By the launch of the Sardegna, which took place at Spezia on Saturday last, another magnificent battle-ship has been added to the Italian navy. She is a sister ship to the Rè Umberto, built of steel, with 18 inches of compound armor on her barbettes ; a length of 400 feet ; breadth, 76 feet 9 inches ; tonnage, 13,251 ; and she is expected to attain a speed of 19 knots. When fully equipped she will carry four 67-ton breech-loading Armstrong guns, eight 15-centimeter, and six 12-centimeter, and ten quick-firing and machine guns. Her cost will be one million sterling.

LE YACHT.

APRIL 12, 1890. Editorial: Remarks on Mr. White's address before the Association of Naval Architects. Notes from foreign ship-yards. Review of the merchant marine. Description of the French armored cruiser Le Charner, with plans.

APRIL 19. Editorial on reform in the French navy. Argument

in favor of squadrons of evolution. Review of the merchant marine. Torpedo-boats built in '89, with plans of torpedo-boat of the first class, built for Denmark by Thorneycroft.

APRIL 26. Description of the Magenta. Review of the merchant marine.

MAY 3. Editorial on the manœuvres of the French squadron escorting President Carnot; criticism on the ships and their personnel. Refrigerating rooms on board ship. Review of the merchant marine. Notes from foreign ship-yards.

MAY 10. Editorial on the work done by French arsenals. Review of the merchant marine.

MAY 17. Editorial criticising centralization in the French naval government. The torpedo-boats built in '89, with plans. Notes from foreign ship-yards.

MAY 24. Editorial comparing the Piemonte with the Forbin. Review of the merchant marine.

MAY 31. Editorial on the competition between guns and armor. Review of the merchant marine. Notes from foreign ship-yards.

JUNE 7. Notes from foreign ship-yards. Review of the merchant marine.

JUNE 14. Editorial on shore duty in the French navy. Notes from foreign ship-yards. Armor trials of the plates of the Chilian man-of-war Capitan Prat, Description and plans of the armored ship Tréhouart.

JUNE 21. Editorial on the separation of the Board of Admiralty from the Minister of Marine and Supreme Council. Review of the merchant marine. Plans of Yarrow torpedo-boats of first and second class.

JUNE 28. Notes from foreign ship-yards. Review of the merchant marine. Review of Captain Mahan's book, " Influence of Sea Power upon History."

JULY 5. Editorial on L'Association Technique Maritime. International Marine Conference. Cut of stern-wheel gunboat for African river service. Notes from foreign ship-yards.

JULY 12. Editorial on the naval appropriation of 1890. Review of the merchant marine.

JULY 19. Editorial on the July manœuvres of the French squadron. Review of the merchant marine.

JULY 26. Editorial on the report of M. Gerville-Reache, on the appropriation for 1891. Review of the merchant marine. Cut of the French cruiser Forbin.

AUGUST 2. Editorial comparing the French and English methods of mobilization. Article on battle-ships. Review of the merchant marine. Engine-room signals.

AUGUST 9. Editorial on the German navy. Review of the merchant marine. The English second-class cruiser Latona, with cut. Armor tests of Schneider plates for the Danish government; four cuts showing impact of shot.

AUGUST 16. Accidents to the French fleet for the year. Resistance of steel.

AUGUST 23. Editorial on English summer manœuvres.

SEPTEMBER 6. Editorial on naval manœuvres. Review of the merchant marine. A. C. B.

REVIEWERS AND TRANSLATORS.

Lieut.-Commander C. S. SPERRY,
Lieutenant J. B. BRIGGS,
Lieutenant A. C. BAKER,
P. A. Engineer J. K. BARTON,

Asst. Naval Constructor D. W. TAYLOR,
Prof. C. R. SANGER,
Prof. J. LEROUX.

OFFICERS OF THE INSTITUTE.

1891.

Elected at the regular annual meeting, held at Annapolis, Md.,
October 31, 1890.

PRESIDENT.

REAR-ADMIRAL S. B. LUCE, U. S. N.

VICE-PRESIDENT.

COMMANDER HENRY GLASS, U. S. N.

SECRETARY AND TREASURER.

ENSIGN HERMAN G. DRESEL, U. S. N.

BOARD OF CONTROL.

LIEUTENANT-COMMANDER C. S. SPERRY, U. S. N.
*LIEUTENANT-COMMANDER H. KNOX, U. S. N.
LIEUTENANT H. O. RITTENHOUSE, U. S. N.
LIEUTENANT R. G. PECK, U. S. N.
PASSED-ASSISTANT ENGINEER J. K. BARTON, U. S. N.
PROFESSOR N. M. TERRY, A. M., PH. D.
ENSIGN C. N. KNEPPER, U. S. N.
ENSIGN H. G. DRESEL, U. S. N. (ex officio.)

*Declined. ENSIGN C. N. KNEPPER was elected to fill the vacancy.

THE PROCEEDINGS

OF THE

UNITED STATES NAVAL INSTITUTE.

Vol. XVI., No. 5.	1890.	Whole No. 56.

U. S. NAVAL INSTITUTE, ANNAPOLIS, M. D.

INTRODUCTION.

THE ANNAPOLIS ARMOR TEST.

THE competitive test of armor plates made at Annapolis on September 18th and 22d last, was in many respects specially noteworthy, Not that anything positively new was made manifest, nor as our ultra-enthusiastic journals will have it, that the armored fleet of Great Britain has been proved disastrously vulnerable. The test has added a great deal to the amount of data previously existing, whose comparison and coördination are necessary for direction in the true line of armor development.

To the artillerist there are several points connected with this test that make it of the greatest interest. 1st. Because a new material for armor has sustained its first public and crucial trial. This point is of the greatest interest in view of the result obtained. 2d. Because in the test, guns, projectiles, and plates were most admirably matched, so that the test furnishes a very maximum of reliable information. It can readily be understood that the choice of too high a caliber and projectile energy would have so completely pierced every plate at each shot that no margin would have been left for comparing effects, and, on the other hand, were the striking energy too light, so that all the plates had thoroughly resisted, no clear distinction could have been made. A wise measurement of the true power needed resulted in just taxing the plates sufficiently beyond their resisting powers to make a judgment simple and plain concerning them. 3d. Because not only was the test a public one in the amplest sense, but that the official report is clear and concise, and has been made public at the earliest moment. In this case it is not necessary to depend upon journalistic accounts, which, however well intentioned, almost invariably err in the omission of some really important detail, or in giving wrong data, or unintentionally giving a false coloring to results obtained. Nothing can be more distracting than to see a carefully written statement of a test where initial velocity is given and the distance of the plate from the gun is omitted, or where a gun caliber is given, and the weight and description of projectile are omitted; or where the minutest detail is given of the effects of a shot on the plate, whilst nothing is found regarding the effect of the plate on the projectile. The official report of this test is as remarkable for the completeness of its record as for its brevity. It contains exactly what is needed by the artillerist for study—nothing more and nothing less.

To those who have not made an intimate study of armor contests the comparative importance assigned to this test doubtless seems exaggerated. A thickness of plate of 10½ inches is not much in comparison with the 16- and 19-inch plates that have been the prominent subjects of contest abroad; and the 6-inch gun is a weak weapon compared with the 15- and 17-inch monsters that have been swung against plates in Europe.

In truth, though, this test cannot be belittled on the score of light plates and projectiles. The striking velocity of the 6-inch shells was higher than that of any test heretofore made, so that the power of the plates to resist punching was thoroughly tested. In addition to this, by bringing the 8-inch gun to bear against the already heavily punished plates, its shells having a striking velocity far below that of the 6-inch, whilst its total energy was enormously greater, the power of the plates to resist cracking, or disintegrating, or showing any of the defects attributed to racking was thoroughly tested. It is true that the plates were light as compared with the heavy belt plates of European battle ships, but the importance of the test cannot be belittled on this account without directly leading to an awkward conclusion that no artillerist would feel prepared to accept. Having given certain qualities of metal that, by proper treatment, will make a more or less perfect armor plate, the manufacturer encounters difficulties in developing these qualities, which increase very rapidly as the plate becomes thicker. These plates were furnished by the best manufacturers in the world, who not only have long been rivals, but whose business interests would be affected by the result. If, then, these plates were not thick enough to be of serious account, it is inevitable that compound armor must be condemned out of hand, for the compound plate failed where the others succeeded, and just so far as difficulties in perfect manufacture may be overcome, exactly to that degree does the inherent excellence of the material become manifest in the result.

It will not do to accept the results of this test as being true generally. Although the compound plate failed completely; it must be remembered that ever since the commencement of the rivalry between steel and compound plates, the comparative superiority has always been in doubt. It is but a few years since, in a competitive test in Russia, a steel plate fared as badly in comparison with a compound one as was the reverse in Annapolis. It is only by comparison of tests and of development that any true idea of the ultimate value of the materials of which the plates are composed can be determined.

Whilst it would be quite out of place here to attempt a detailed comparison of tests made against different steel and compound plates, it may be possible to point out a way by which anyone so inclined may, when he secures the data concerning any test, get a fairly correct idea of the relative resisting powers of different plates.

It is necessary, first, to take into consideration the scale of measurements, as it were; for it is readily seen how endless arguments will occur unless there be some common standard of reference. For example, in England compound plates beat English steel, where a projectile having 1975 feet striking velocity was used. At Annapolis, with the same thickness of plates, same caliber of gun, but 2075 feet striking velocity, French steel plates beat compound. Now, it would be interesting to know whether the steel at Annapolis was better than the steel in England, and if so, whether it was better than the compound plate in England, and whether the English Cammell plate was better than the American Cammell plate. Then again, how is one to compare the results of 6-inch projectiles on 10-inch plates with those of 11-inch projectiles on 15-inch plates? There is a method, not at all abstruse, which will give a fairly true basis of comparison. It is only necessary to prelude the description with the warning that a knowledge of this method and of the results obtained does not make one an expert judge of armor, and that it requires an expert to properly weigh all the circumstances of any test. By the use of this method one could not well prescribe what should be the conditions of a test to be made, but since tests are prepared and carried out by experts, due regard has been had to the proper composition of the elements, so that, results translated or compared by this true method will be nearly correct.

Before wrought iron passed out of service as failing in resisting power, plate manufacture had been so perfected throughout Europe that not only was it remarkably constant in its resisting power, but this power was practically the same wherever the plate was made. That is, no manufacturer could claim any particular superiority in resisting power for his product. Projectiles had so far developed that not only steel but chilled iron ones would readily pierce these plates without being either broken or deformed. Thus it became possible to compute by rigid formulas the exact amount of energy required to send a projectile of a certain caliber through a plate of a given thickness.

As compound and steel plates came in to replace iron ones, two great difficulties were met with in attempting to compare results obtained. The development in manufacture was slower and much more difficult, so that very great variations were met with in the resisting power of similar plates. So great, that it was found impossible to formulate the results in any way that would be even remotely satisfactory. The best that could be done in the way of conveying an intelligent idea was to adopt wrought iron as a standard of reference, which its regularity permitted, and to speak of a steel or compound plate as being equivalent to a certain greater thickness of wrought iron. It must be remembered that a mere expression of a certain number of foot-tons of energy conveys no intelligible idea, even to the expert, for it requires different energies to force shells of different calibers through a plate of a given thickness. It was more definite to say, for instance, of a 10-inch compound plate that it was equivalent to 12½ inches of wrought iron; for, at least, by this means one could get at some idea regardless of caliber of projectile.

The other difficulty was that the power of the plates was sufficient to smash or distort the best projectiles that could be brought against them. That a plate should be capable of smashing a projectile is creditable to the plate within certain limits; but we evidently cannot determine the comparative resistance of two plates if both break their projectiles, for we have no standard of measurement. There is an immense difference between making a projectile test the strength of a plate and making a plate test the strength of a projectile.

Projectiles gradually improved in strength until in the Holtzer shell they seem to be developed to a point where the most powerful armor cannot break or deform them, as a rule. This acquisition brings us at once within a fair and safe measuring distance of our standard, and we may in a single expression convey to the mind a definite idea, where not only can we omit the caliber of projectile, but also the thickness of plate itself, and thus we arrive at a direct comparison of armor systems. The method of arriving at this expression is as follows:

Assume a steel plate of 9 inches thickness, against which a Holtzer shell is fired with a certain energy, and assume that this shell barely pierces the plate. Take one of the many well-known formulas for penetration of wrought-iron plates (the Gavre formula recommends itself by its simplicity and flexibility), and compute the energy necessary barely to send *that same* projectile through a wrought-iron plate of the same thickness. Suppose it to be found that this energy is 1000 foot-tons.

Now compute the actual energy with which the projectile struck the plate and assume that this energy is 1070 foot-tons. Then 1070 divided by 1000 equals 1.07, and we translate the expression by saying that the steel plate is a seven per cent plate. That is, that it requires seven per cent more energy to pierce it than would be required for a wrought-iron plate of the same thickness. The beauty of this expression is that, no matter what be the caliber of projectile brought against the plate, it is always a seven per cent, as compared with the same projectile against wrought iron, and no matter what may be the thickness of plate, the per cent always holds against a corresponding thickness of wrought iron. So then, by this expression we may grade plates in resisting power, whether they be steel, compound, or nickel-steel; and we may directly classify the nineteen-inch plate attacked by the fifteen-inch shell with the nine-inch plate attacked by the six-inch shell.

It is well to point out two features that interfere with the absolute truth of this expression: 1. The projectile must not be smashed or distorted by the plate, as the resulting per cent value will be much too high. If one looks at any authority of five years ago, it will be found that both steel and compound plates are said to be from twenty-five to thirty per cent higher in resistance than iron plates. The advent of perfected projectiles upset this estimate completely, and forced the manufacturers to renewed and strenuous exertions. 2. The plate must not split or smash under the blow. This frequently happens and gives constant renewal to the old argument of racking versus punching, but to those who have carefully followed and studied the course of armor development it seems certain that as any system of armor improves, racking effects, which at first are invariably predominant, are gradually overcome until in the end they disappear entirely, whilst the resistance to punching is maintained or increased. The Annapolis test brings these features out very prominently, for the nickel steel showed both punching and racking resistance; the steel plate showed punching resistance, but was racked; the compound plate failed in both features.

This method of classifying armor is here put forward for the first time. The method is by no means novel, but no utilization of the resulting expression seems to have heretofore been made, and it appears to be particularly well adapted to common use.

A few citations will bring out the character of the expression, and doubtless give a clearer idea of the strength of the plate exhibited at Annapolis.

In 1888 a lot of 6-inch Holtzer shells were delivered to the English Government under the stipulation that when fired against a 9-inch compound plate they should not break or be notably deformed. Under test they readily pierced the plate, and it is found by computation that the test was what might be called a twelve per cent one; or, as the plate was pierced with considerable to spare, it could not be classed as high as twelve per cent.

In the early part of 1888 an important test of armor plates took place at Portsmouth, England, at which the plates and 6-inch projectiles were exactly assimilated to those of the Annapolis test, the striking velocity being 100 feet less. This, by the method of classification, was a seven per cent test, and was well resisted by a Cammell compound plate, and a Vickers steel one. The test at Annapolis was an eighteen per cent one, and the Cammell plate broke down under it. Here, by this single expression a deal of light is thrown upon the action of the Cammell plates in England and here.

Not long ago, at St. Petersburg, an 11-inch projectile was fired against a 15½-inch Cammell compound plate, the projectile barely getting its point through. The test was a five per cent one, and compares excellently with the one quoted as at Portsmouth, although plates and projectiles were widely different. Let it be remarked that neither of these two plates was noticeably racked, nor were the projectiles hurt. Both plates apparently would take the same classification, which would be more than seven per cent.

At Gavre a 19.7 inch Schneider plate was tested with a 15½-inch projectile, both standing well up to the work. The test was a fourteen per cent one, and the projectile pierced with considerable to spare.

At Le Bouchet a 5½-inch Schneider plate received a thirteen per cent test, which exactly limited its classification, the shell barely going through.

The examination of a great number of tests made under proper conditions leads to the conclusion that this method of classification is very fairly correct. Under it, the steel and nickel-steel plates tested may be classed fairly as eighteen-per-cent plates, while the Cammell plate must fall below them, although taking the Portsmouth and Annapolis tests together, it is fair to place it well above seven per cent, and presumably quite close to twelve.

In closing, it may be well to add a few remarks concerning the English sentiment in favor of compound armor, since much of an uncomplimentary nature has been said of late. The Annapolis test, beyond all doubt, gave a decided victory to steel over compound plates; but it will not do to jump to the conclusion that compound plates are worthless. For nearly fifteen years the struggle has been going on without any decision as to supremacy. Compound armor has, at times, been far in the lead—so far, that even France, the home of steel plates, opened its doors very wide indeed to compound ones. Leave the ships of Great Britain entirely out of consideration, and there will remain far more ships built with compound than with steel armor. Nor should we be carried away by the idea that the British artillerists are averse to steel. From the very inception of armor, steel has been constantly tried, and with such excellent results and constancy, that it could scarcely be called revolutionary were England to throw compound armor overboard to-morrow.

The United States is especially fortunate in coming to the study of the best system to adopt, after projectiles had reached the stage where they could outmatch armor in withstanding punishment. Before that the contest between steel and compound armor was indifferent and unsatisfactory. Now all is different, and one system, perhaps both, must soon go to the wall. The United States has entered to aid in solving the grand problem; and, whilst we must bear in mind that "one swallow does not make a summer," we may properly boast that our first test was not only thoroughly excellent, but that one of the plates clearly showed a greater power of resistance than has ever been developed heretofore.

EDWARD W. VERY.

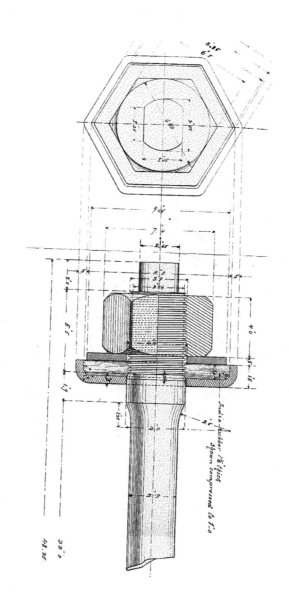

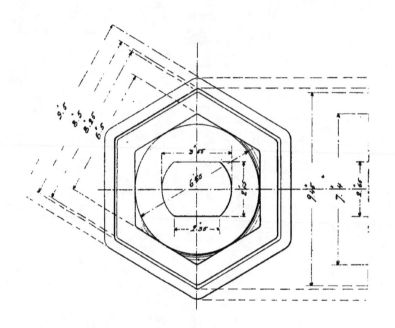

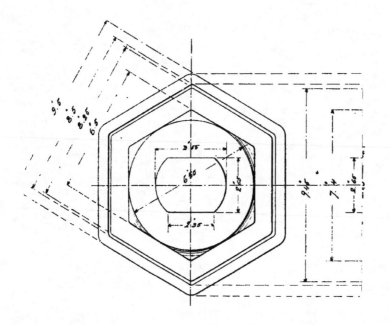

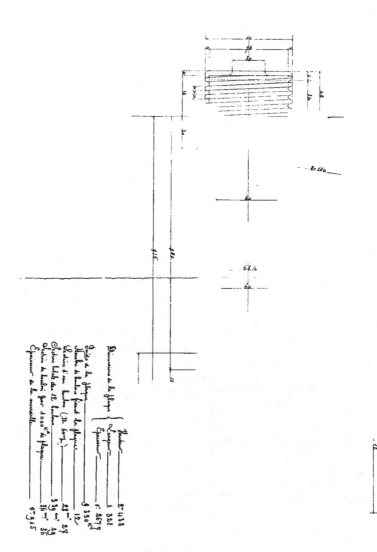

Dimensions de la plaque

Hauteur 2ᵐ418
Largeur 1ᵐ822
Épaisseur 0ᵐ267

{
Poids de la plaque 9 330ᵏᵍ
Poids de la fonte par plaque 12ᵗ
Pertes à un 28ᵐᶜ 27
Pertes L.L. du 12 l'autre (D. 6ᵐʲ) 39ᵐᶜ 29
Pertes L.L. par 1000ᵏᵍ de plaque 36 36
Épaisseur de la muraille 0ᵐ915
}

Dessin № 730/90

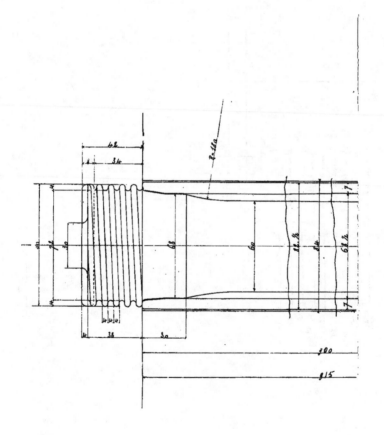

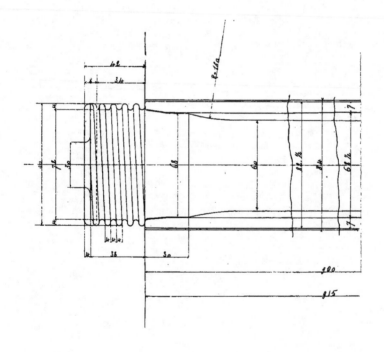

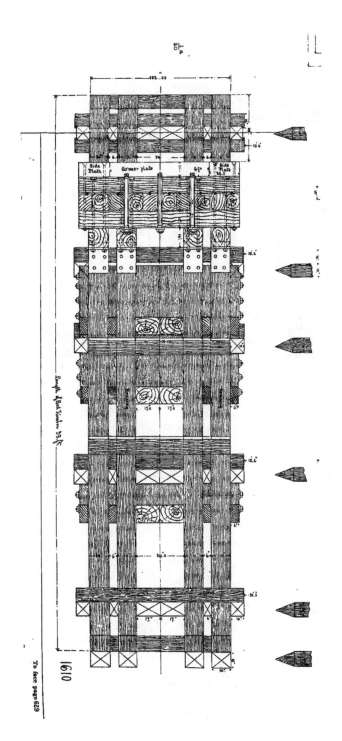

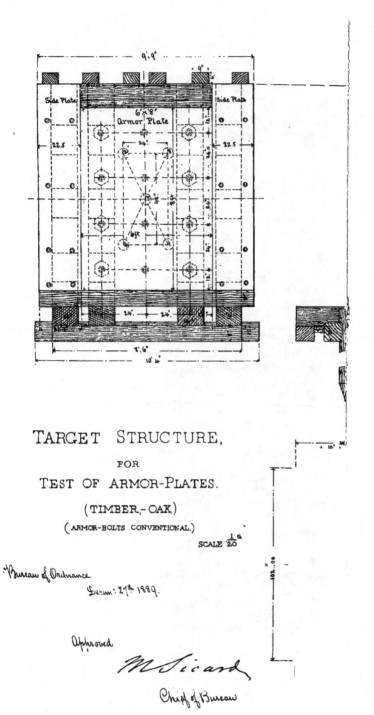

TARGET STRUCTURE,

FOR

TEST OF ARMOR-PLATES.

(TIMBER,- OAK)

(ARMOR-BOLTS CONVENTIONAL.)

SCALE $\frac{1}{20}$th

Bureau of Ordnance
Decem: 27th 1889.

Approved

M Sicard

Chief of Bureau

REPORT OF THE BOARD ON THE COMPETITIVE TRIAL OF ARMOR PLATES.

NAVAL ORDNANCE PROVING GROUND,
ANNAPOLIS, MD., OCTOBER 11, 1890.

IN obedience to the Department's order of July 11, 1890, the Board met at the Naval Ordnance Proving Ground, September 18, 1890, for the purpose of conducting a competitive test of three armor plates, one of steel and one of nickel steel manufactured by Schneider & Co., Le Creusot, and one compound plate manufactured by Cammell & Co., Sheffield.

The Board made a careful preliminary inspection of the plates and their bolts and backing. They were all erected in accordance with a drawing furnished by the Bureau of Ordnance, to the Inspector of Ordnance-in-Charge at the Proving Ground. The steel plate and nickel steel plate were secured to the backing with twelve bolts 2.36 inches in diameter. The compound plate was secured to its backing by eight bolts 3.19 inches in diameter.

Plate.	No. of bolts.	Diameter of bolts.	Total bolt cross-section.	Weight of plate.	Bolt section per ton of plate.
			Sq. inches.	Lbs.	Sq. inches
Steel	12	2.36″	52.29	20776 (actual).	5.66
Nickel steel	12	2.36	52.59	20679 (actual).	5.69
Compound	8	3.19	64.32	20992 (calculated).	6.86

The plans of the backing, the arrangement, shapes, and method of securing the armor bolts are shown in the photographs appended.

Caliper measurements were taken by Lieutenant-Commander Dayton along the edges of the plates before they were secured. A table of these is appended.

CALIPER MEASUREMENTS OF ARMOR PLATES.

Commencing at lower right-hand corner, up right edge, across top, and down left side, at intervals of one foot.

Point of measurement		Schneider all steel.	Schneider nickel stee l.	Compound.
		Inches.	Inches.	Inches.
Lower right corner . . .		10.56	10.22	10.65
Right side	1	.61	.45	.63
	2	.61	.65	.76
	3	.58	.65	.81
	4	.53	.64	.70
	5	.52	.49	.76
	6	.51	.45	.80
	7	.53	.38	.70
Upper right corner . . .		.55	.24	.70
Top	1	.59	.38	.69
	2	.60	.41	.71
	3	.61	.41	.77
	4	.63	.41	.75
	5	.63	.38	.74
Upper left corner. . . .		.55	.27	.67
	1	.56	.34	.70
	2	.51	.38	.59
	3	.55	.42	.67
	4	.53	.43	.71
	5	.58	.43	.67
	6	.63	.41	.69
	7	.63	.57	.76
Lower left corner. . . .		.50	.37	.76
	1	...	...	.83
	2	...	...	.82
	3	...	.30	.75
	4	...	...	.76
	5	...	...	.73
Mean		10.57	10.42	10.72

The surfaces of the plates were scoured clean.

The plates were arranged on chords of a circle with the gun pivot as the centre, and the muzzle of the gun 28 feet distant from the centre of the plate toward which it was pointed, the axis of the gun being then normal to the surface of each plate.

The gun was a 6-inch B. L. R., Mark III., 35 calibers long, mounted on a service centre-pivot carriage. The charge used was 44½ lbs. for each round; brown prismatic powder manufactured by Messrs. Dupont. The striking velocity was 2075 feet per second.

The projectiles were all Holtzer 6-inch Armor-piercing Shell, with ogival of 2 calibers radius, and were brought up to the standard weight of 100 lbs. by filling them with sand and fragments of iron. These fragments of iron weighed about 2 oz. each.

The gun was mounted on a heavy wooden platform protected by a structure of timber roofed with iron plates and a pile of sand-bags in front, through which the muzzle projected.

The pointing was done by means of central cross-hair sights in the axis of the bore.

Bomb-proofs had been erected at convenient places on the grounds. The one nearest the gun, at a distance of 50 feet, protected the firing party. The gun was fired by means of friction primer and a long lanyard.

The Board caused photographs to be taken of each plate before and after each shot; of all three plates in a group after four shots each, and, also, after five shots each; of the recovered projectiles, and of such fragments of plate as could be identified.

At 11 A. M. the Board began the firing trial.

ROUND 1.

1ST SHOT AT THE STEEL PLATE.

Gun,	6″ B. L. R.	Elevation, — 2° 50′.	
Projectile,	Holtzer A.-P. Shell, 100 lbs.	Striking velocity, 2075 f. s.	
Charge,	44½ lbs. brown prism.	Striking energy, 2988 ft.-tons.	

The point of impact was 24 ins. from the bottom and 24 ins. from the right-hand edge of the plate. The projectile entered and remained in the plate, with its base projecting 6.50 ins. and its body apparently intact. A bulge, 1 inch high and 17.5 ins. diameter, was raised on the face of the plate, with a well-developed fringe projecting one inch. The visible portion of the hole in the plate was funnel-shaped, extending to a depth of 3 or 4 ins. Six short radial surface cracks were produced in the bulge. One small piece of fringe scaled off.

The side plates were uninjured; one bolt slightly started. The backing was undisturbed and no bolts were started. The rotating band was stripped off the projectile. Its base plug could still be easily turned.

(Photograph No. 4.)

ROUND 2.

1ST SHOT AT THE COMPOUND PLATE.

Gun,	6″ B. L. R.	Elevation, — 2° 50′.	
Projectile,	Holtzer A.-P. Shell, 100 lbs.	Striking velocity, 2075 f. s.	
Charge,	44½ lbs. brown prism.	Striking energy, 2988 ft.-tons.	

The point of impact was 24 ins. from the right-hand edge and 24 ins. from the bottom of the plate. The projectile perforated the plate and lodged in the backing, with its base at a distance of 21.5 ins. from the face of the plate. Eight fine radial cracks were produced, varying in length from 24.5 ins. to 55 ins. from the centre of the hole. One of these extended to the right edge of plate, at which point it seemed to extend through the steel face. The right-hand corner of the plate, along the through crack mentioned above, was set out to a distance of 0.10 in. The steel face around the hole was scaled off to a diameter of from 14.5 to 19 ins., and to a depth of from 1.3 to 2.25 ins. The interior of the shot-hole showed a rough surface.

The side plates were uninjured. Two bolts of right side plate near point of impact were slightly started. One tie-bolt of the backing, about 8 ins. from the point of impact, was started 5 ins. to the rear. The rear side of bolt heads in first row of vertical bolts started $\frac{1}{10}$ inch from the braces.

The two inner lower braces of the backing were started upward and backward 0.25 in. The projectile remained buried in the backing with its base visible through the hole.

(Photograph No. 5.)

ROUND 3.

1ST SHOT AT THE NICKEL STEEL PLATE.

ᐟ Gun,	6″ B. L. R.	Elevation, — 2° 50′.
Projectile,	Holtzer A.-P. Shell, 100 lbs.	Striking velocity, 2075 f. s.
Charge,	44½ lbs. brown prism.	Striking energy, 2988 ft.-tons.

The point of impact was 24 ins. from the right-hand edge and 24 ins. from the bottom of the plate. The body of the projectile remained in the plate. The rear end, 7.5 ins. long, broke off at a distance of 4 ins. in from the face of the plate and rebounded 30 feet to the front. A front bulge was raised, 17 ins. diameter and 1.25 ins. high. There was no fringe.

Two short radial surface cracks were produced in the bulge. A small piece of the bulge, one inch deep, was scaled off from the side of the shot-hole.

The side plates were uninjured. The head of one bolt was split in two.

The bolt in backing nearest the point of impact was driven 30 ins. to the rear. The backing was undisturbed.

(Photographs Nos. 6 and 17.)

ROUND 4.

2D SHOT AT THE STEEL PLATE.

Gun,	6″ B. L. R.	Elevation, — 2° 50′.
Projectile,	Holtzer A.-P. Shell, 100 lbs.	Striking velocity, 2075 f. s.
Charge,	44½ lbs. brown prism.	Striking energy, 2988 ft.-tons.

The point of impact was 24 ins. from the left-hand edge and 24 ins. from the bottom of the plate. The projectile penetrated to a depth of 12 inches and rebounded entire to a distance of 26 feet to the front and 3 feet to the right. It was shortened 0.10 in. A front bulge was raised, 18.5 ins. diameter and 1 in. high, with a well-developed fringe projecting 0.75 in. Seven short radial cracks were developed in the bulge, two of them being connected by two short surface hair cracks. One small piece of fringe scaled off. The shot-hole was regular with a smooth surface, and showing star-shaped cracks at inner extremity.

The side plates were uninjured. Two bolts were started. One nut of a side plate bolt was broken off. The backing was undisturbed.

(Photographs Nos. 7 and 17.)

ROUND 5.

2D SHOT AT THE COMPOUND PLATE.

Gun,	6″ B. L. R.	Elevation, — 2° 50′.	
Projectile,	Holtzer A.-P. Shell, 100 lbs.	Striking velocity, 2075 f. s.	
Charge,	44½ lbs. brown prism.	Striking energy, 2988 ft.-tons.	

The point of impact was 24 ins. from the left-hand edge and 24 ins. from the bottom of the plate. The projectile perforated the plate and lodged in the backing, with its base visible at a distance of 15 ins. from the face of the plate. The steel face was scaled off the plate around the shot-hole, irregularly, to an average diameter of 14 ins. and depth of 1.25 ins.

Six radial cracks were produced, all extending to the edges of the plate and apparently quite deep. A surface crack was produced about 3.5 feet above lower edge of plate, extending nearly across the plate.

Two connecting radial cracks, produced by the first shot, deepened and were extended entirely across the plate through both shot-holes, throwing out the lower part of the plate from

0.6 in. to 1.8 ins. Where this crack reached the right edge of the plate, the steel face showed a vertical curved fracture of a minimum thickness of 0.3 in. and separated 1.9 ins. at the top.

The interior surface of the shot-hole was rough and broken, showing a disintegration of the metal, many small loose fragments of which were lying in the hole. The cracks developed by the preceding shot were all increased in size, the lower part of the plate being much racked.

The side plates were uninjured. Two bolts were slightly started.

The backing showed but slight set-back. One tie-bolt in backing nearest impact was driven 4 ins. to the rear. The rear side of bolt heads of first vertical row were started 0.10 in. from the braces.

(Photograph No. 8.)

ROUND 6.

2D SHOT AT THE NICKEL STEEL PLATE.

Gun,	6″ B. L. R.	Elevation, — 2° 50′.
Projectile,	Holtzer A.-P. Shell, 100 lbs.	Striking velocity, 2075 f. s.
Charge,	44½ lbs. brown prism.	Striking energy, 2988 ft.-tons.

The point of impact was 24 ins. from the left-hand edge and 24 ins. from the bottom of the plate. The projectile penetrated and remained in the plate apparently intact, its base projecting 1.6 ins. The rotating band was still in place, but sheared flush with the body of the shell.

A front bulge was thrown up, 16 ins. in diameter and 1.10 ins. high. A piece was scaled off for about two-thirds the circumference of the bulge, of an average diameter of 12.55 ins. and average depth of 1.10 ins.

The side plates were uninjured, two bolts slightly started. The backing was uninjured. The tie-bolt nearest the point of impact was driven 12 ins. to the rear.

(Photograph No. 9.)

ROUND 7.

3D SHOT AT THE STEEL PLATE.

Gun,	6″ B. L. R.	Elevation, — 2° 50′.
Projectile,	Holtzer A.-P. Shell,	Striking velocity,
	100 lbs.	2075 f. s.
Charge,	44½ lbs. brown prism.	Striking energy,
		2988 ft.-tons.

The point of impact was 24 ins. from the right-hand edge and 24 ins. from the top of the plate. The projectile penetrated to a depth of 12.5 ins. and rebounded entire to a distance of 30 ft. It was shortened 0.14 in., and the diameter of the bourrelet was increased 0.01 in.

A front bulge was found, 18.5 ins. in diameter and 1 in. high, with a well-developed fringe projecting 1 in. Five short radial surface cracks were developed in the bulge, and two small pieces of the fringe were scaled off. The shot-hole was smooth and regular, showing star-shaped cracks at its inner extremity.

The side plates were uninjured, one piece of bolt 7 ins. long was broken off and thrown out to the front. The backing and its bolts were undisturbed..

(Photographs Nos. 10 and 17.)

ROUND 8.

3D SHOT AT THE COMPOUND PLATE.

Gun,	6″ B. L. R.	Elevation, — 2° 50′.
Projectile,	Holtzer A.-P. Shell,	Striking velocity,
	100 lbs.	2075 f. s.
Charge,	44½ lbs. brown prism.	Striking energy,
		2988 ft.-tons.

The point of impact was 24 ins. from the right-hand edge and 24 ins. from the top of the plate. The projectile perforated the plate and lodged in the backing, with its base 23.75 ins. from the face of the plate.

Ten deep radial cracks were produced, some of them being connected by cross-cracks. A large piece of steel face in upper right-hand corner was slightly separated from the wrought iron back, and a smaller piece was in the same condition below the shot-hole.

Old cracks in lower part of plate were increased in size, and the face of the plate was quite shattered.

The interior surface of the shot-hole was rough and broken, showing disintegration of the surrounding metal, many small fragments of which had been shaken off and were lying in the hole.

The side plates were not injured. The nuts of five bolts in the left side plate were unscrewed from 0.25 in. to 0.50 in.

The backing was uninjured.

(Photograph No. 11.)

ROUND 9.

3D SHOT AT THE NICKEL STEEL PLATE.

Gun,	6″ B. L. R	Elevation,	2° 50′.
Projectile,	Holtzer A.-P. Shell,	Striking velocity,	
	100 lbs.	2075 f. s.	
Charge,	44½ lbs. brown prism.	Striking energy,	
		2988 ft.-tons.	

The point of impact was 24 ins. from the right-hand edge and 24 ins. from the top of the plate. The projectile penetrated and remained in the plate apparently intact, its base projecting 4.5 ins., and its rotating band half torn off. A front bulge was raised, 18 ins. in diameter and 1.5 ins. high.

Two very short radial surface cracks were produced in the bulge. Two small pieces scaled out of bulge. The side plates were uninjured; the bolt heads were slightly bent.

The backing was undisturbed, and no bolts were started.

(Photograph No. 12.)

ROUND 10.

4TH SHOT AT THE STEEL PLATE.

Gun, 6″ B. L. R. Elevation, 2° 50′.

Projectile, Holtzer A.-P. Shell, Striking velocity,

   100 lbs.   2075 f. s.

Charge, 44½ lbs. brown prism. Striking energy,

          2988 ft.-tons.

The point of impact was 24 ins. from the left-hand edge and 24 ins. from the top of the plate. The projectile penetrated to a depth of 12.5 inches and rebounded, breaking into three large and several small pieces. The head was entire, the body in halves, the transverse break at 7 ins. from the base of shell.

A front bulge was raised, 17 ins. in diameter and 1 in. high, with a well-developed fringe projecting 1 in.

Six short radial surface cracks were produced in the bulge, and one small piece of the fringe was scaled off. The surface of shot-hole was smooth and regular, showing star-shaped cracks at its inner extremity.

The side plates were uninjured.

The backing was undisturbed. One bolt, broken by previous shot, was driven to the rear 2 ins.

(Photographs Nos. 13 and 17.)

ROUND 11.

4TH SHOT AT THE COMPOUND PLATE.

Gun, 6″ B. L. R. Elevation, 2° 50′.

Projectile, Holtzer A.-P. Shell, Striking velocity,

   100 lbs.   2075 f. s.

Charge, 44½ lbs. brown prism. Striking energy,

          2988 ft.-tons.

The point of impact was 24 ins. from the left-hand edge and 24 ins. from the top of plate. The projectile perforated plate and backing, breaking into three large and several small pieces. The largest piece passed over the earth backing, and lodged in the butt 60 feet away.

Three deep cracks were developed below shot-hole. All the steel face above this and the preceding shot-hole separated from the wrought-iron back, and the greater part of it was thrown off in seven large, and many small fragments, the larger fragments weighing, respectively, 29, 58, 92, 125, 157, 178, and 380 lbs.

' A part of the steel face below the preceding shot-hole sep-arated from the wrought-iron back and was thrown off.

All previous cracks widened.

That part of the wrought-iron back which was exposed on the upper part of the plate showed a rough, broken appearance and contained three narrow cracks. The shot-hole was very rough on the interior surface, showing much disintegration of the metal, with small fragments of which it was nearly filled.

The remainder of the plate was badly racked, and showed 17 large cracks through steel face. The side plates were uninjured.

There was an irregular hole torn in the backing by the shot. No bolts were started.

(Photographs Nos. 14, 16, and 17.)

ROUND 12.

4TH SHOT AT THE NICKEL STEEL PLATE.

Gun,	6″ B. L. R.	Elevation, 2° 50′.
Projectile,	Holtzer A.-P. Shell, 100 lbs.	Striking velocity, 2075 f. s.
Charge,	44½ lbs. brown prism.	Striking energy, 2988 ft.-tons.

The point of impact was 24 ins. from the left-hand edge and 24 ins. from the top of the plate. The projectile broke 3.5 ins. from face of plate, and 6.5 ins. from its base, its head remaining in the plate.　The rear half rebounded 30 feet, burying itself in the sand-bags around the gun.

A front bulge was raisèd, 16 ins. in diameter and 1 in. high. One fine crack was developéd in the upper part of the bulge.

The visible interior surface of the hole was smooth and regular.

The side plates were uninjured, bolt heads slightly bent. The backing was undisturbed, and no bolts were started.

(Photographs Nos. 15 and 17.)

After the preceding twelve rounds had been fired, the Board adjourned for two days, until an 8-inch gun, with which it was decided to continue the firing, could be mounted.　The firing with this gun took place on Monday, September 22, 1890.

The gun was mounted on the same platform as the 6″, with its pivot 6½ feet to the rear of the 6″ pivot.　The muzzle of the gun was 30 feet distant from the plates.

The charge was 85 lbs. of brown prismatic powder, made by Dupont.　The striking velocity was 1850 feet per second.

The projectiles were Firth Armor-piercing Shell, weighted to 210 lbs. by filling them with sand.

ROUND 13.

5TH SHOT AT THE STEEL PLATE.

Gun,	8″ B. L. R.	Elevation,　0° 00′,
Projectile,	Firth A.-P. Shell, 210 lbs.	Striking velocity, 1850 f. s.
Charge,	85 lbs. brown prism.	Striking energy, 4988 ft.-tons.

The point of impact was at the centre of the plate.　The projectile penetrated to a depth of 15 ins. and rebounded, broken in three large pieces.　A front bulge was raised, 21 ins. in diameter and 1 in. high.　There was no fringe.　A piece scaled off from the lower right-hand edge of hole radially for 5 ins.

Four through cracks radiating from the shot-hole were developed, two above and two below, each one extending diagonally through one of the 6″ shot-holes to top and bottom of face respectively, and forming an irregularly-shaped X. These cracks above the centre hole varied in width from 0.25 to 0.50 in., and below the centre hole, from 0.16 to 0.50 in.

The upper right-hand crack reached the top at a point 13 ins. from the side. The upper left-hand crack reached the top 5.5 ins. from the side. The lower right-hand crack reached the bottom 13.25 ins. from the side, and the lower left-hand crack reached the bottom 8 ins. from the side of the plate.

The side plates were uninjured and the bolts intact.

The backing was driven back bodily 0.10 in.

(Photographs Nos. 19, 22, and 23.)

ROUND 14.

5TH SHOT AT THE NICKEL STEEL PLATE.

Gun,	8″ B. L. R.	Elevation,	0° 00′.
Projectile,	Firth A.-P. Shell, 210 lbs.	Striking velocity, 1850 f. s.	
Charge,	85 lbs. brown prism.	Striking energy, 4988 ft.-tons.	

The point of impact was the centre of the plate. The projectile entered and broke 5.25 ins. from the face of plate, part of the head remaining in the hole.

A front bulge was raised, 16.5 ins. in diameter and 0.25 in. high. The circumference of the hole was scaled out to the following radial distances from the centre:

At top,   7   ins. out, maximum depth, 4 ins.
At right side, 6   ins.   "   "   "   2 ins.
At bottom,   6.5 ins.· "   "   ::   2 ins.
At left side, 7.25 ins. "   "   4 ins.

A hole of considerable length was visible on the side of the interior of the hole, between the projectile and interior surface of the plate.

There were no cracks visible on the plate.

The side plates were uninjured, and no bolts were started.

The front timber of the backing, in the line of the shot-hole, was apparently broken in two.

The fragments of projectile that were recovered are shown in the photograph.

(Photographs Nos. 20, 22, and 23).

ROUND 15.

5TH SHOT AT THE COMPOUND PLATE.

Gun,	8″ B. L. R.	· Elevation,	0° 00′.
Projectile,	Firth A.-P. Shell, 210 lbs.	Striking velocity,	1850 f. s.
Charge,	85 lbs. brown prism.	Striking energy,	4988 ft.-tons.

The point of impact was the centre of the plate. The projectile perforated the plate and backing, and disintegrated much of the metal about the hole. Most of the front plate was scaled off in 24 large and a great number of small pieces, the larger pieces varying in thickness from 3.75 to 5.50 ins.

Parts of the plate were left complete along each edge, as shown in the photographs, but much of it separated, some of the fragments of the steel being nearly on the point of falling off. The exposed wrought-iron back showed 18 cracks of varying length.

The metal about the interior of the hole was rough and broken. The wrought-iron surface was rough and irregular. The metal around all shot-holes showed considerable disintegration. Several concentric cracks appeared in the wrought-iron back about the 8″ shot-hole.

The side plates were uninjured.

The projectile broke an irregular hole through the backing, but all damage was local.

The shell was recovered entire, having penetrated the earth

backing -about 15 ft. It was shortened 0.24 in., and the diameter of the bourrelet was increased 0.015 in.

(Photographs Nos. 21, 22, and 23.)

The Board then adjourned until the plates could be removed from the backing so as to expose to view their rear faces.

The Board met at the Naval Ordnance Proving Ground on October 10, 1890, and examined the backs of the plates and the fronts of the backing, with the following results:

BACK OF STEEL PLATE.

(Photographs Nos. 24 and 24a.)

No. 1 Shot. Back bulge 3.5 ins. high and 16 ins. diameter.

No. 2 Shot. Back bulge 3.0 ins. high and 15 ins. diameter.

No. 3 Shot. Back bulge 2.6 ins. high and 16 ins. diameter, with a piece broken out.

No. 4 Shot. Back bulge 2.6 ins. high and 15 ins. diameter.

No. 5 Shot. Back bulge 6.25 ins. high and 22 ins. diameter, with two pieces broken out.

Diagonal cracks from front of plate all through to back.

BACKING OF STEEL PLATE.

(Photograph No. 25.)

No. 1 Shot. Indented 2.75 ins.

No. 2 Shot. Indented 2.40 ins.

No. 3 Shot. Indented 2.00 ins., two pieces embedded in the backing.

No. 4 Shot. Indented 2.40 ins.

No. 5 Shot. Indented 5.25 ins., and one piece embedded in the hole.

BACK OF NICKEL STEEL PLATE.

(Photographs Nos. 26 and 26a.)

No. 1 Shot. Back bulge 5 ins. high and 21 ins. diameter.

No. 2 Shot. Back bulge 6 ins. high and 21 ins. diameter.

No. 3 Shot. Back bulge 4 ins. high and 19 ins. diameter.

No. 4 Shot. Back bulge 4 ins. high and 18 ins. diameter.

No. 5 Shot. A crater in back of plate was formed, of an average depth of 3.5 ins. and 21 ins. diameter.

BACKING OF NICKEL STEEL PLATE.
(Photograph No. 27.)

No. 1 Shot.   Indented 5.25 ins.
No. 2 Shot.   Indented 4.75 ins.
No. 3 Shot.   Indented 3.50 ins.
No. 4 Shot.   Indented 3.50 ins.
No. 5 Shot.   A hole 10½ ins. deep in the wood, and 6 ins. deep
                to a fragment of metal left in the backing from
                the crater in the plate.

BACK OF COMPOUND PLATE.
(Photographs Nos. 28 and 28a.)

No. 1 Shot.   Back bulge 6.4 ins. high and 20.5 ins. diameter.
No. 2 Shot.   Back bulge 6.25 ins. high and 18 ins. diameter.
No. 3 Shot.   Back bulge 6 ins. high and 20 ins. diameter.
No. 4 Shot.   Back bulge 7.25 ins. high and 19 ins. diameter, a
                large piece of plate broken out.
No. 5. Shot.   Back bulge 7.5 ins. high and 21 ins. diameter, a
                piece of plate broken out. ·

BACKING OF COMPOUND PLATE.
(Photograph No. 29.)

No. 1 Shot.   Base of projectile 11 ins. in from face of backing.
No. 2 Shot.   Base of projectile 4.75 ins. in from face of backing.
No. 3 Shot.   Base of projectile 13.1 ins. in from face of backing.
No. 4 Shot.   Through hole: piece of plate embedded in side of
                hole.
No. 5 Shot.   Through hole: piece of plate embedded in side of
                hole.
        No bolts in any of the plates were injured.

NOTE.—The word "front," when applied either to the plate
or fragments of plate or projectile, refers to the side of the plate
toward the gun.

The swellings of the plate caused by the striking of a projectile are denominated, respectively, the "front bulge," and the
"back bulge."

        *   *   *   *   *   *   *   *   *   *

SUMMARY.

The Compound Plate was perforated by all projectiles, and its steel face was destroyed. Two of the shells passed completely through both plate and backing.

Both Steel Plates kept out all projectiles, the All-Steel Plate showing slightly greater resistance than the Nickel Steel Plate, but the former was badly cracked by the 8-inch shell, while the latter remained uncracked. The Board, therefore, places the three plates tested in the following order of "relative merit:"

1. Nickel Steel.   2. All Steel.   3. Compound.

L. A. KIMBERLY,
Rear Admiral, U. S. Navy,
President of the Board.

E. O. MATTHEWS,
Captain, U. S. Navy, Member.

W. R. BRIDGMAN,
Commander, U. S. Navy, Member.

W. MAYNARD,
Lt. Comdr., U. S. Navy, Member.

J. F. MEIGS,
Lieutenant, U. S. Navy, Member.

B. H. BUCKINGHAM,
Lieutenant, U. S. Navy, Member.

W. H. H. SOUTHERLAND,
Lieutenant, U. S. Navy, Member.

F. F. FLETCHER,
Lieutenant, U. S. Navy, Member.

PHILIP R. ALGER,
Ensign, U. S. Navy, Member.

A. A. ACKERMAN,
Ensign, U. S. Navy, Member.

ROBERT B. DASHIELL,
Ensign, U. S. Navy, Recorder.

**Attention Scanner:
Foldout in Book!**

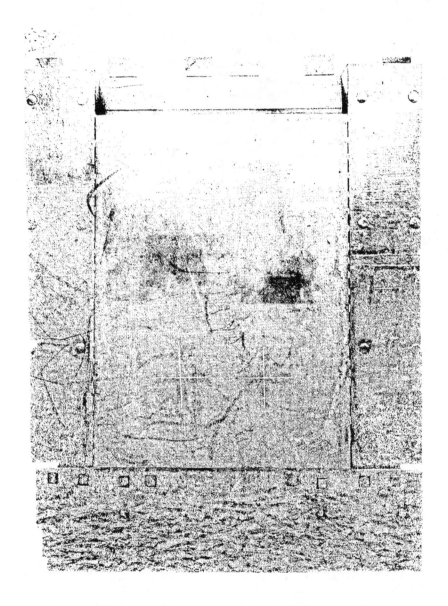

No. 1. STEEL PLATE BEFORE FIRING.

(Schneider et Cie.)

No. 2. COMPOUND PLATE BEFORE FIRING.

(Cammell & Co.)

No. 2. COMPOUND PLATE BEFORE FIRING.

(CAMMELL & CO.)

No. 3. NICKEL-STEEL PLATE BEFORE FIRING.

(Schneider et Cie.)

No. 4. STEEL PLATE AFTER FIRST SHOT.

(SchneIder et CIe.)

No. 5. COMPOUND PLATE AFTER FIRST SHOT.
(Cammell & Co.)

No. 6. NICKEL-STEEL PLATE AFTER FIRST SHOT.
(Schneider et Cie.)

No. 7. STEEL PLATE AFTER SECOND SHOT.

(Schneider et Cie.)

No. 8. COMPOUND PLATE AFTER SECOND SHOT.

(Cammell & Co.)

No. 9. NICKEL-STEEL PLATE AFTER SECOND SHOT.
(Schneider et Cie.)

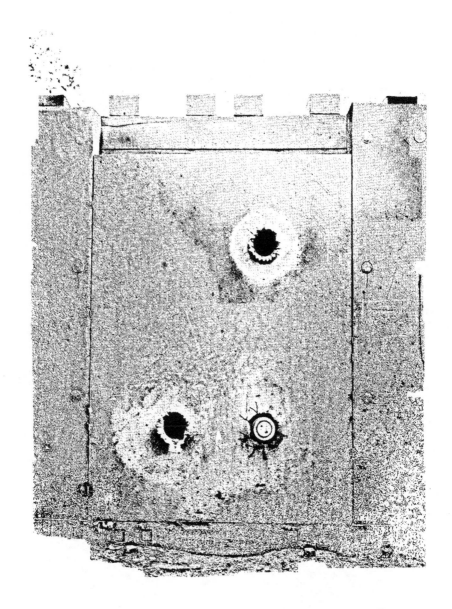

No. 10. STEEL PLATE AFTER THIRD SHOT.
(SCHNEIDER ET CIE.)

No. 11. COMPOUND PLATE AFTER THIRD SHOT.
(CAMMELL & CO.)

No. 12. NICKEL–STEEL PLATE AFTER THIRD SHOT.

(Schneider et Cie.)

No. 13. STEEL PLATE AFTER FOURTH SHOT.

(Schneider et Cie.)

No. 14. COMPOUND PLATE AFTER FOURTH SHOT.

(CAMMELL & CO.)

No. 15. NICKEL-STEEL PLATE AFTER FOURTH SHOT.

(Schneider et Cie.)

No. 16. COMPOUND PLATE AFTER FOURTH SHOT.

No. 1   RECOVERED SIX-INCH PROJECTILES.

**Attention Scanner:
Foldout in Book!**

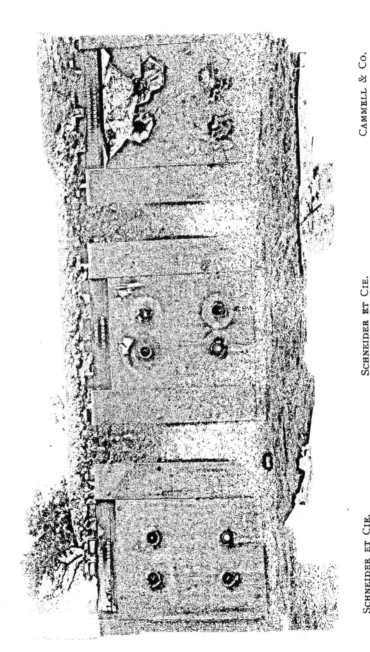

SCHNEIDER ET CIE.   SCHNEIDER ET CIE.   CAMMELL & CO.

No. 18. VIEW OF PLATES AFTER FOUR SHOTS EACH.

No. 19.   STEEL PLATE AFTER FIFTH SHOT.

No. 20. NICKEL–STEEL PLATE AFTER FIFTH SHOT.

**Attention Scanner:
Foldout in Book!**

No. 21. COMPOUND PLATE AFTER FIFTH SHOT.

**Attention Scanner:
Foldout in Book!**

**Attention Scanner:
Foldout in Book!**

FIRED AT STEEL PLATE.   FIRED AT COMPOUND PLATE.   FIRED AT NICKEL-STEEL PLATE.

No. 23.   RECOVERED EIGHT-INCH PROJECTILES.

**Attention Scanner:
Foldout in Book!**

No. 24. BACK OF STEEL PLATE.

(Schneider et Cie.)

**Attention Scanner:
Foldout in Book!**

No. 24a. BACK OF STEEL PLATE.

**Attention Scanner:
Foldout in Book!**

No. 25. BACKING OF STEEL PLATE.

**Attention Scanner:
Foldout in Book!**

No. 26. BACK OF NICKEL-STEEL PLATE.

(Schneider et Cie.)

Attention Scanner:
Foldout in Book!

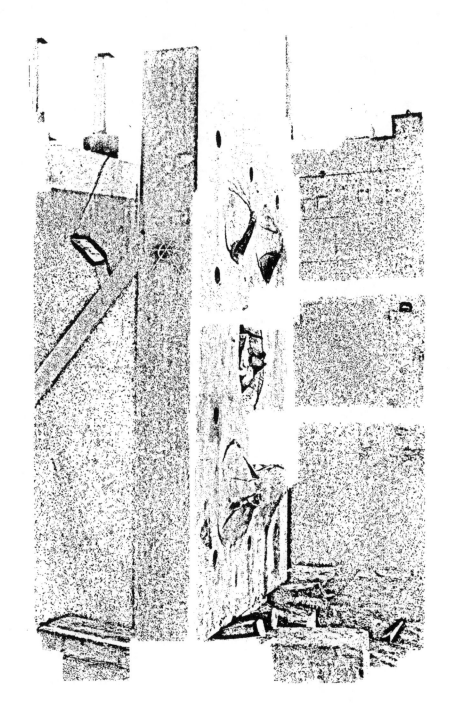

No. 26a. BACK OF·NICKEL-STEEL PLATE.

No. 27. BACKING OF NICKEL-STEEL PLATE.

**Attention Scanner:
Foldout in Book!**

No. 28. BACK OF COMPOUND PLATE.

(CAMMELL & CO.)

**Attention Scanner:
Foldout in Book!**

No. 29. BACKING OF COMPOUND PLATE.

APPENDIX.

Vol. XVI., 1891.

PRESS OF ISAAC FRIEDENWALD CO

BALTIMORE, MD.

UNITED STATES NAVAL INSTITUTE.

1873–1891.

ORIGIN, PROGRESS, AND OBJECT.

HISTORICAL.

The inaugural meeting of the Institute was held October 9, 1873, at which Rear-Admiral John L. Worden, U. S. N., presided, and the late Commodore Foxhall A. Parker, U. S. N., read a paper entitled " The Battle of Lepanto." The object of the association as then stated was for the advancement of professional and scientific knowledge in the Navy by affording a medium for the free interchange of serious thought and the debate of important subjects concerning naval science and practice. In 1874 a Constitution was adopted and the membership increased to seventy-five. The first volume, entitled "The Papers and Proceedings of the United States Naval Institute," appeared in 1875. During the succeeding few years the Institute grew in membership and in the estimation of the Army and Navy, and by the adoption of the Constitution of June 2, 1884, it took a firmer hold and marked out its future success.

In addition to its membership the Institute has a large number of subscribers and exchanges, so that now the published quarterly edition numbers 1350 copies.

Since its organization, the field of usefulness of the association has become enlarged, and besides the first object, already mentioned, its work includes the following:

1st. By means of the large number of associate members, comprising Army officers, those of the Revenue Marine, naval architects of distinction, consulting and mechanical engineers of high standing, and the leading manufacturers of steel and other material for the Navy, it promotes good feeling and mental stimulation, which are broadening and beneficial to all parties.

2d. Its members and subscribers, including libraries and colleges, disseminate important authentic information relating to naval science and to the subject of national defense.

3d. The Professional and Bibliographic Notes furnish a compendium of valuable information, taken from home and foreign sources, to which most members have not access.

4th. The publications of the Institute become a repository for important historical records concerning the naval service, as well as for valuable reports of experiments in a form convenient for reference.

PUBLICATIONS.

' Quarterly, or as much oftener as the Board of Control may decide," the Proceedings are published, and one copy is mailed to every member. Up to the present time fifty-six (56) numbers have been issued. Full sets may be obtained from the Institute by application to the Secretary and Treasurer.

GENERAL INFORMATION.

Membership.—Under the Constitution (Article VII.) the active members of the Institute are divided into three classes, styled respectively: Regular, Life, and Associate Members. The provisions for Honorary Membership are embodied in Section 4, Article VII.

Officers of the Navy, Marine Corps, and all civil officers attached to the naval service are entitled to become regular or life members, without ballot, on payment of dues or fees to the Secretary and Treasurer, or to the Corresponding Secretary of a Branch. The Prize Essayist of each year becomes a life member without payment of fee.

Associate members are elected from officers of the Army, U. S. Revenue Marine, foreign officers of the naval and military professions, and from persons in civil life that are interested in the purposes of the Institute. Associate members not to exceed one hundred (100) may be elected to life membership. Section 7, Article VII., Constitution gives manner of election of associate members.

For information on any subject of business with the Institute, address the Secretary and Treasurer, U. S. Naval Institute, Annapolis, Md.

DUES, FEES, SUBSCRIPTION.

No initiation fee is required. Regular and associate members pay $3 on joining the Institute, and upon the first day of each succeeding January. Life members pay $30 in lieu of all dues, which sum is invested with the Reserve Fund and held in perpetuity, to guarantee the future interests of such members.

When a member becomes in arrears for more than one year, sending the publications to him is discontinued. Every person admitted to membership is considered as belonging to the Institute, and liable for the payment of dues until he shall have informed the Secretary and Treasurer at Annapolis, Md., of his desire to withdraw, when, if his dues have been fully paid, his name shall be erased from the list of members.

Annual subscription to Proceedings for non-members, $3.50. Single copies of Proceedings, $1 each. Copies bound in cloth are furnished to members and subscribers at $1 per annum extra. Volumes bound in half morocco or half calf at $0.90 per cover.

Remittances should be made payable to the order of the Secretary and Treasurer, U. S. Naval Institute.

Additions to the Library.

Contributions of books on professional subjects will be received and added to the Library. Members of the Institute and others interested in its purpose are invited to secure and send to the Institute such books or reports. The Institute Office at Annapolis, Md., is open every business day from 9 A. M. till 1 P. M., and from 3 to 5 P. M. Members visiting Annapolis, Md., are invited to the office of the Institute.

OFFICERS OF THE INSTITUTE.

1891.

PRESIDENT.

REAR-ADMIRAL S. B. LUCE, U. S. N.

VICE-PRESIDENT.

COMMANDER HENRY GLASS, U. S. N.

SECRETARY AND TREASURER.

ENSIGN H. G. DRESEL, U. S. N.

BOARD OF CONTROL.

LIEUT.-COMMANDER C. S. SPERRY, U. S. N.
LIEUTENANT H. O. RITTENHOUSE, U. S. N.
LIEUTENANT R. G. PECK, U. S. N.
PROFESSOR N. M. TERRY, A. M., PH. D.
P. ASST. ENGINEER J. K. BARTON, U. S. N.
ENSIGN CHESTER M. KNEPPER, U. S. N.
ENSIGN H. G. DRESEL, U. S. N., (ex officio).

EDITOR.

ENSIGN H. G. DRESEL, U. S. N.

Presidents of the Naval Institute since its organization.

1873—1874.—REAR-ADMIRAL J. L. WORDEN, U. S. N.

1875—1876—1877.—REAR-ADMIRAL C. R. P. RODGERS, U. S. N.

1878.—COMMODORE F. A. PARKER, U. S. N.

1879—1880—1881.—REAR-ADMIRAL JOHN RODGERS, U. S. N.

1882.—REAR-ADMIRAL C. R. P. RODGERS, U. S. N.

1883—1884—1885.—REAR-ADMIRAL THORNTON A. JENKINS, U. S. N.

1886—1887.—REAR-ADMIRAL E. SIMPSON, U. S. N.

1888—1889—1890.—REAR-ADMIRAL STEPHEN B. LUCE, U. S. N.

BRANCHES.

WASHINGTON, D. C.

CORRESPONDING SECRETARY.

LIEUTENANT JOHN H. MOORE, U. S. N.,
1503 Pennsylvania Avenue.

NEW YORK.

CORRESPONDING SECRETARY.

LIEUT.-COMMANDER L. CHENERY, U. S. N.,
University Club.

NEWPORT, R. I.

VICE-PRESIDENT.

COMMANDER T. F. JEWELL, U. S. N.

CORRESPONDING SECRETARY.

PROFESSOR C. E. MUNROE, CHEMIST U. S. TORPEDO CORPS.

LIST OF MEMBERS.

HONORARY MEMBERS.

Arranged in Order of Election.

Hon. B. F. Tracy, Secretary of the Navy, (ex officio).

Chief-Justice C. P. Daly, President of the American Geographical Society, New York.

President Chas. W. Eliot, LL. D., Harvard University.

Mr. John D. Jones, President Atlantic Insurance Company, New York.

Captain Alfred Collet, Répétiteur de l'École Polytechnique, Paris.

President D. C. Gilman, LL. D., Johns Hopkins University.

LIFE MEMBERS.

Brown, A. D.,	Commander,	U. S. N.	Prize Essayist, 1879.
Belknap, Chas.,	Lieut.-Comd'r,	U. S. N.	Prize Essayist, 1880.
Very, E. W., late	Lieutenant,	U. S. N.	Prize Essayist, 1881.
Kelley, J. D. J.,	Lieutenant,	U. S. N.	Prize Essayist, 1882.
Calkins, C. G.,	Lieutenant,	U. S. N.	Prize Essayist, 1883 and 1886.
Chambers, W. I.,	Lieutenant,	U. S. N.	Prize Essayist, 1884.
Farquhar, N. H.,	Captain,	U. S. N.	Prize Essayist, 1885.
Hutchins, C. T.,	Lieut.-Comd'r,	U. S. N.	Prize Essayist, 1887.
Reisinger, W. W.,	Lieut.-Comd'r,	U. S. N.	Prize Essayist, 1888.

Ackland, W. A. D., Commander, R. N., Broad street, Oxford, England.

Allen, R. W., Paymaster, U. S. N.

Bailey, F. H. P., Passed Assistant Engineer, U. S. N.

Barclay, Chas. J., Commander, U. S. N.

Barker, Albert S., Commander, U. S. N.

Bartlett, Henry A., Major, U. S. M. C.

Bicknell, G. A., Lieutenant-Commander, U. S. N.

Bixby, W. H., Captain, Engineer Corps, U. S. A.

Bolles, T. Dix, Lieutenant, U. S. N.

Brown, Austin, No. 1426 F street, Washington, D. C.

Brownson, W. H., Commander, U. S. N.

Brush, George R., Surgeon, U. S. N.

Burgdorff, T. F., Passed Assistant Engineer, U. S. N.

Canet, Gustave, Ingénieur Chef du Service d'Artillerie dé la Ste. des Forges et Chantiers de la Mediterranée, 1 Rue Vignon, Paris, France.

Capps, Washington L., Assistant Naval Constructor, U. S. N.

Carrington, H. S., Secretary, Sir Joseph Whitworth & Co., Limited, Ashton Road, Openshaw, Manchester, England.

Centre, Robert, No. 18 West 21st street, New York, N. Y.

Cooper, Theodore, Consulting Engineer, 35 Broadway, New York, N. Y.

Coryell, Miers, Consulting Engineer, 21 E. 21st street, New York, N. Y.

Dayton, J. H., Lieutenant-Commander, U. S. N.

Duncan, W. B., Jr., No. 11 Pine street, New York, N. Y.

Elger, Francis, LL. D., Director, H. M. Dockyards, Admiralty, London, S. W., England.

Emmens, Stephen H., Emmensite Explosives, New Stanton, Westmoreland Co., Pa.

Evans, E. T., Manager, Lake Superior Transit Co., Buffalo, N. Y.

Evans, R. D., Commander, U. S. N.

Fletcher, A, North River Iron Works, 266 and 267 West st., New York, N. Y.

Floyd, Richard, care of H. E. Matthews, 120 Sutton street, San Francisco, Cal.

Gardner, H. W., Providence Steam Engine Co., Providence, R. I.

Gates, H. G, Ensign, U. S. N.

Gledhill, M, Manager, Sir Joseph Whitworth & Co., Limited, Ashton Road, Openshaw, Manchester, England.

Griffis, Wm. Elliott, D. D., 638 Fremont street, Boston, Mass.

Habirshaw, W. M, F. C. S, No. 159 Front street, New York, N. Y.

Hagenman, John W., Lieutenant, U. S. N., 350 S. 5th street, Reading, Penna.

Hanford, Franklin, Lieutenant-Commander, U. S. N.

Harris, U. R., Lieutenant, U. S. N.

Hicks, B. D., Old Westbury, Queens Co., N. Y.

Huger, D. E., Colonel, Mobile, Alabama.

Hunsicker, J. L., 230 Washington street, Buffalo, N. Y.

Hunsiker, Millard, Civil Engineer, Carnegie, Phipps & Co., Limited, Pittsburgh, Penna.

Jackson, J. B., care of B. F. Stevens, No. 4 Trafalgar Square, London, England.

Keim, G. DeB., 1122 Spruce street, Philadelphia, Pa.

Kempff, Louis, Commander, U. S. N.

Kirby, Frank E., Supt. Detroit Dry Dock Co., Detroit, Michigan.

Knox, Harry, Lieutenant-Commander, U. S. N.

Leary, John D., Shipbuilder, Eagle and Prevost streets, Brooklyn E. D., N. Y.

Lyeth, C. H., Lieutenant, U. S. N., Martinsburg, West Virginia.

Madden, Thomas P., Occidental Hotel, San Francisco, Cal.

Mason, T. Bailey Myers, Lieutenant, U. S. N.

McCook, J. J, Colonel, No. 120 Broadway, New York, N. Y.

McCrackin, Alex., Lieutenant, U. S. N.

McGiffin, Philo N, Lieutenant, I. C. N., care of Consul Gen'l, Shanghai, China.

Meigs, John F., Lieutenant, U. S. N.

Merrell, John P., Lieutenant-Commander, U. S. N.

Merry, John F., Lieutenant-Commander, U. S. N.

Metcalf, William, Miller, Metcalf & Parkin, Pittsburgh, Penna.

Michailovitch, Alex., H. I. H. Grand Duke of Russia, St. Petersburg, Russia.

Moore, John H., Lieutenant, U. S. N.

Nelson, Thomas, Commander, U. S. N.

Palmer, N. F., Jr., Quintard Iron Works, New York, N. Y.

Paul, Allen G., Lieutenant, U. S. N.

Peck, G., Medical Director, U. S. N., Elizabeth, N. J.

Peck, R. G., Lieutenant, U. S. N.

Perkins, George H., Captain, U. S. N.

Phoenix, Lloyd, 21 E. 23d street, New York, N. Y.

Picking, Henry F., Captain, U. S. N.

Pinn, J. F., Japan Herald, Yokohama, Japan.

Pond, Charles F., Lieutenant, U. S. N.

Quick, George, Fleet Engineer, R. N., Hidcote Manor, Campden, Gloucester-
shire, England.

Quintard, George W., Quintard Iron Works, New York, N. Y.

Rand, Stephen, Jr., Paymaster, U. S. N.

Reamey, L. L., Lieutenant, U. S. N.

Rognetta, F. B., Colonel, 181 Via Nazionale, Rome, Italy.

Rohrer, Carl, Lieutenant, U. S. N.

Rowland, Thomas F., Continental Iron Works, Brooklyn, N. Y.

Ruschenberger, C. W., Lieutenant, U. S. N.

Saldanha da Gama, L. F. de, Captain, Brazilian Navy, Bibliotheca da Marinha,
Rio-de-Janeiro, Brazil.

Schroeder, Seaton, Lieutenant, U. S. N.

Selfridge, Thomas O., Captain, U. S. N.

Slack, Wm. II., No. 1714 18th street, Washington, D. C.

Smith, Joseph A., Pay Inspector, U. S. N.

Stayton, Wm. H., Robison, Bright, Biddle & Ward, No. 150 Broadway, New
York, N. Y.

Steers, Henry, Naval Architect, 10 East 38th street, New York, N. Y.

Tanner, Z. L., Lieutenant-Commander, U. S. N.

Thomas, Chas. M., Commander, U. S. N.

Thurston, R. H., Director of Sibley College, Cornell University, Ithaca, N. Y.

Tyler, Geo. W., Lieutenant, U. S. N.

Ubsdell, J. A., Superintendent Eads Jetties, Port Eads, La.

Ward, Aaron, Lieutenant, U. S. N.

Watrous, C., No. 140 Pearl street, New York, N. Y.

Weed, George E., Morgan Iron Works, New York, N. Y.

Wetmore, M. Boreum, No. 67 Madison Avenue, New York, N. Y.

Wilkinson, Ernest, Rooms 33 and 35 Atlantic Building, 930 F street, Washington, D. C.

Wood, Spencer S., Ensign, U. S. N.

Wright, R. K., care of U. S. Consul, Aspinwall, U. S. of Colombia.

REGULAR MEMBERS.

Ackerman, Albert A., Ensign, U. S. N.

Ackley, S. M., Lieut.-Comr., U. S. N.

Adams, J. D., Lieutenant, U. S. N.

Addicks, W. R., Bay State Gas Co., Room 4 Rialto Bldg., Boston, Mass.

Alger, P. R., Professor, U. S. N.

Allderdice, Winslow, The Warren Tube Co., Warren, Ohio.

Allibone, C. O., Lieutenant, U. S. N.

Almy, J. J., Rear-Admiral, U. S. N., 1019 Vermont Ave., Washington, D. C.

Ames, Howard E., P.-Asst. Surg., U.S.N.

Ammen, D., Rear-Admiral, U. S. N., Ammendale, Md.

Anderson, E. A., Ensign, U. S. N.

Anderson, E. B.

Andrews, Philip, Ensign, U. S. N.

Ashbridge, R., P.-Asst. Surgeon, U.S.N.

Atwater, C. N., Lieutenant, U. S. N.

Babcock, W. C., Lieutenant, U. S. N.

Badger, C. J., Lieutenant, U. S. N.

Badger, O. C., Commodore, U. S. N., 1517 20th street, Washington, D. C.

Baird, Geo. W., P.-Asst. Engin'r, U.S.N.

Baker, Asher C., Lieutenant, U. S. N.

Baker, C. H., Chief Engineer, U. S. N.

Balch, G. B., Rear-Admiral, U. S. N., Dawsonville, Montgomery Co., Md.

Barber, F. M., Commander, U. S. N.

Barnette, W. J., Lieutenant, U. S. N.

Barroll, H. H., Lieutenant, U. S. N.

Barry, E. B., Lieutenant, U. S. N.

Bartlett, C. W., Lieutenant, U. S. N.

Bartlett, J. R., Commander, U. S. N.

Barton, J. K., P.-Asst. Engr., U. S. N.

Bassett, F. B., Ensign, U. S. N.

Bassett, F. S., Lieutenant, U. S. N.

Batcheller, O. A., Commander, U. S. N.

Baxter, W. J., Asst. Naval Constructor, U. S. N.

Beardslee, L. A., Captain, U. S. N.

Beecher, A. M., Ensign, U. S. N.

Beehler, W. H., Lieutenant, U. S. N.

Belden, Samuel, Lieut.-Comr., U. S. N.

Benham, A. E. K., Rear-Admiral, U.S.N.

Benham, H. K., Ensign, U. S. N.

Bennett, F. M., Asst. Engr., U. S. N.

Bernadou, J. B., Ensign, U. S. N.

Berry, R. M., Lieut.-Comr., U. S. N.

Berwind, E. J., Lieutenant, U. S. N., No. 55 Broadway, New York, N. Y.

Bishop, Joshua, Commander, U. S. N.

Bitler, R. O., Ensign, U. S. N.

Blish, J. B., Ensign, U. S. N.

Blocklinger, G., Lieutenant, U. S. N.

Book, G. M., Lieut.-Commander, U.S.N.

Bostick, E. D., Lieutenant, U. S. N.

Boush, C. J., Lieutenant, U. S. N.

Bowles, F. T., Naval Const'r, U. S. N.

Bowman, C. G., Lieutenant, U. S. N.

Bowyer, J. M., Lieutenant, U. S. N.

Boyd, A. A., No. 11 E. Genesee street, Auburn, N. Y.

Bradbury, Chas. A., Lieutenant, U.S.N.

Bradford, R. B., Commander, U. S. N.

Bradshaw, G. B., Naval Cadet, U. S. N.

Braine, D. L., Rear-Admiral, U. S. N.

Brand, C. A., Naval Cadet, U. S. N.

Breed, Geo., Engineering Department, Edison Genl. Electric Co., 68 Broadway, New York, N. Y.

Bridgman, W. R., Commander, U. S. N.

Briggs, John B., Lieutenant, U. S. N.

Brittain, C. B., Ensign, U. S. N.

Brown, George, Rear-Admiral, U. S. N.

Brown, R. M. G., Lieutenant, U. S. N.

Buckingham, B. H., Lieut., U. S. N.

Buehler, W. G., Chief Eng'r, U. S. N.

Buford, M. B., Promontory Box, Elder County, Utah.

Bull, J. H., Lieutenant, U. S. N.

Bullard, W. H. G., Ensign, U. S. N.

Bunce, F. M., Captain, U. S. N.

Burrage, G. H., Ensign, U. S. N.

Caldwell, W. M., 724 Twenty-first st., Washington, D. C.

Calhoun, G. A., Lieutenant, U. S. N.

Capehart, E. E., Ensign, U. S. N.

Carney, R. E., Naval Cadet, U. S. N.
Carpenter, C. C., Captain, U. S. N.
Carter, S. P., Rear-Admiral, U. S. N.,
    1316 Conn. Ave., Washington, D. C.
Casey, Silas, Captain, U. S. N.
Chadwick, F. E., Commander, U. S. N.
Chandler, L. H., Ensign, U. S. N.
Chase, J. V., Naval Cadet, U. S. N.
Chase, V. O., Ensign, U. S. N.
Chenery, Leonard, Lieut.-Comr., U.S.N.,
    University Club, New York, N. Y.
Chester, C. M., Commander, U. S. N.
Clark, C. E., Commander, U. S. N.
Clason, W. P., Lieutenant, U. S. N.
Cline, H. H., Chief Engineer, U. S. N.
Clover, R., Lieut.-Comr., U. S. N.
Codman, John, Captain, P. O. Box 3015,
    New York, N. Y.
Coffin, J. H. C., Lieutenant, U. S. N.
Coghlan, Jos. B., Commander, U. S. N.
Cogswell, J. K., Lieutenant, U. S. N.
Colahan, C. E., Lieutenant, U. S. N.
Colby, H. G. O., Lieutenant, U. S. N.
Cole, W. C., Naval Cadet, U. S. N.
Coleman, N. T., Naval Cadet, U. S. N.
Collins, J. B., Lieutenant, U. S. N.
Colwell, J. C., Lieutenant, U. S. N.
Converse, G. A., Commander, U. S. N.
Conway, W. P., Lieutenant, U. S. N.
Cooke, A. P., Captain, U. S. N.
Cook, Simon, Lieutenant, U. S. N.
Cooley, M. E., Professor, University of
    Michigan, Ann Arbor, Mich.
Cooper, Geo. F., Ensign, U. S. N.
Cooper, Philip H., Commander, U. S. N.
Cotton, Chas. S., Commander, U. S. N.
Couden, A. R., Lieut.-Comr., U. S. N.
Courtis, F., Lieut.-Comr., U. S. N.
Cowles, William, Consulting Engineer,
    foot of 43d street, Brooklyn, N. Y.
Cowles, Wm. S., Lieutenant, U. S. N.
Craig, J. E., Commander, U. S. N.
Cramer, S. W., U. S. Assay Office, Char-
    lotte, N. C.
Cramp, C. H., 1736 Spring Garden st.,
    Philadelphia, Pa.
Cresap, J. C., Lieutenant, U. S. N.

Crocker, F. W., Lieut.-Comr., U. S. N.
Cromwell, B. J., Captain, U. S. N.
Crose, W. M., Ensign, U. S. N.
Culver, A. E., Lieutenant, U. S. N.
Cutler, W. G., Lieutenant, U. S. N.
Dalrymple, E. W., 20 Commercial st.,
    Monroe, Jasper Co., Iowa.
Daniels, D., Lieutenant, U. S. N.
Davis, C. H., Commander, U. S. N.
Davis, C., Naval Cadet, U. S. N.
Delehanty, D., Lieutenant, U. S. N.
Denfeld, G. W., Lieutenant, U. S. N.
Derby, R. C., 22–24 Bellevue Avenue,
    Newport, R. I.
De Steiguer, L. R., Nav. Cadet, U. S. N.
Dewey, George, Captain, U. S. N.
Dickins, F. W., Commander, U. S. N.
Diehl, S. W. B., Lieutenant, U. S. N.
Dillingham, A. C., Lieutenant, U. S. N.
Dombaugh, H. M., Lieutenant, U. S. N.
Downes, John, Lieutenant, U. S. N.
Doyle, R. M., Lieutenant, U. S. N.
Drake, F. J., Lieutenant, U. S. N.
Dresel, H. G., Ensign, U. S. N.
Driggs, W. H., Lieutenant, U. S. N.
Dunlap, A., Lieutenant, U. S. N.
Dunn, H. O., Lieutenant, U. S. N.
Durand, Geo. R., Commander, U. S. N.
Dyer, G. L., Lieutenant, U. S. N.
Dyer, N. M., Commander, U. S. N.
Eaton, J. G., Lieut.-Comr., U. S. N.
Eldredge, H., Ensign, U. S. N.
Eldridge, F. H., P.-Asst. Eng'r, U. S. N.
Ellicott, J. M., Ensign, U. S. N.
Elmer, H., Commander, U. S. N.
Emanuel, J. M., P.-Asst. Eng'r, U. S. N.
Engard, A. C., P.-Asst. Eng'r, U. S. N.
Evans, G. R., Ensign, U. S. N.
Farragut, L., 113 E. 36th street, New
    York, N. Y.
Febiger, J. C., Rear-Admiral, U. S. N.,
    Easton, Md.
Fechteler, Aug. F., Lieutenant, U. S. N.
Fermier, Geo. L., Naval Cadet, U. S. N.
Fernald, F. L., Naval Const'r, U. S. N.
Field, H. A., Ensign, U. S. N.
Field, W. L., Lieutenant, U. S. N.

Field, Wiley R. M., Ensign, U. S. N.

Field, T. Y., Colonel, U. S. M. C.

Fiske, Bradley A., Lieutenant, U. S. N.

Fletcher, F. F., Lieutenant, U. S. N.

Fletcher, W. B., Ensign, U. S. N.

Flynne, Lucien, Lieutenant, U. S. N.

Folger, W. M., Commander, U. S. N., Chief Bureau of Ordnance.

Ford, J. D., P.-Asst. Engineer, U. S. N.

Ford, W. G., Jr., Cropsey and 17th Avenues, Bath Beach, N. Y.

Forney, James, Lieut.-Col., U. S. M. C.

Forsyth, Jas. M., Commander, U. S. N.

Fox, C. E., Lieutenant, U. S. N.

Franklin, W. B., Ensign, U. S. N.

Franklin, S. R., Rear-Admiral, U. S. N., 1338 19th st., Washington, D. C.

Fullam, W. F., Lieutenant, U. S. N.

Galt, R. W., P.-Asst. Engineer, U. S. N.

Garrett, L. M., Ensign, U. S. N.

Gartley, A., Naval Cadet, U. S. N.

Garvin, J., Lieutenant, U. S. N.

Gause, J. T., Harlan & Hollingsworth, Wilmington, Del.

Gearing, H. C., Lieutenant, U. S. N.

Gheen, E. H., Lieutenant, U. S. N.

Gibbons, J. H., Ensign, U. S. N.

Gibson, W. C., Lieut.-Comr., U. S. N.

Gillmore, J. C., Lieutenant, U. S. N.

Gillpatrick, W. W., Lieut.-Commander, U. S. N.

Gilman, A. H., Pay Director, U. S. N.

Gilman, H. K., 1st Lieut., U. S. M. C.

Gilmore, F. P., Lieut.-Comr., U. S. N.

Glass, Henry, Commander, U. S. N.

Gleaves, A., Lieutenant, U. S. N.

Glennon, J. H., Lieutenant, U. S. N.

Goodrich, C. F., Commander, U. S. N.

Gorgas, A. C., Med. Director, U. S. N.

Gorgas, M. C., Ensign, U. S. N.

Graham, J. D., Commander, U. S. N.

Graham, S. L., Lieutenant, U. S. N.

Grant, A. W., Lieutenant, U. S. N.

Greene, B. F., Professor, U. S. N., West Lebanon, N. H.

Greene, S. Dana, Edison General Electric Co., Broad St., New York, N. Y.

Greene, F. E., Lieutenant, U. S. N.

Greer, J. A., Rear-Admiral, U. S. N.

Griffin, T. D., Lieutenant, U. S. N.

Grimes, J. M., Lieutenant, U. S. N., 506 Knoxville Ave., Peoria, Ill.

Gross, C. J., 1225 E. North Ave., Baltimore, Md.

Gunnell, F. M., Med. Director, U. S. N., 600 20th St., Washington, D. C.

Haeseler, F. J., Ensign, U. S. N.

Halford, Wm., Gunner, U. S. N.

Hall, M. E., Lieutenant, U. S. N.

Hall, Reynold T., P.-Asst. Eng'r, U.S.N.

Halsey, W. F., Lieutenant, U. S. N.

Hannum, W. G., Lieutenant, U. S. N.

Hanscom, J. F., Naval Const., U. S. N.

Harber, G. B., Lieutenant, U. S. N.

Harkness, W., Professor, U.S.N., Naval Observatory, Washington, D. C.

Harlow, C. H., Ensign, U. S. N.

Harmony, D. B., Rear-Admiral, U. S. N.

Harrington, P. F., Commander, U. S. N.

Hartrath, A., Asst. Engineer, U. S. N.

Hayden, E. E., Ensign, U. S. N., 1802 16th st., Washington, D. C.

Hayward, G. N., Ensign, U. S. N.

Hazlett, I., Lieut.-Comr., U. S. N.

Heald, E. D. F., Lieut.-Comr, U. S. N.

Hemphill, J. N., Lieut.-Comr., U. S. N.

Henderson, R., Lieutenant, U. S. N.

Heywood, Chas., Colonel, U. S. M. C.

Hichborn, P., Naval Const., U. S. N.

Higgins, R. B., Asst. Engineer, U. S. N.

Higginson, F. J., Commander, U. S. N.

Hill, F. K., Ensign, U. S. N.

Hobson, R. P., Naval Cadet, U. S. N.

Hodgson, A. C., Lieutenant, U. S. N.

Hoff, A. B., Naval Cadet, U. S. N.

Hoff, W. B., Commander, U. S. N.

Hollis, Ira A., P.-Asst. Eng'r, U. S. N.

Holman, Geo. F. W., Lieut., U. S. N.

Hosley, H. H., Lieutenant, U. S. N.

Howard, W. L., Ensign, U. S. N.

Howison, H. L., Captain, U. S. N.

Howland, C. H., Providence Journal, Providence, R. I.

Hubbard, J., Lieutenant, U. S. N.

Hubbard, N. M., 926 New York Life Building, Omaha, Neb.

Hughes, E. M., Lieutenant, U. S. N.

Hughes, W. S., Lieutenant, U. S. N.

Hunicke, F. H., 709-711 Lucas Ave., St. Louis, Mo.

Hunker, Jacob J., Lieutenant, U. S. N.

Huntington, R. W., Major, U. S. M. C.

Huse, H. McL. P., Lieutenant, U. S. N.

Hutchins, Hamilton, Lieut., U. S. N.

Hutchison, B. F., Naval Cadet, U. S. N.

Ingate, C. L. A., 2d Lieut., U. S. M. C.

Ingersoll, R. R., Lieutenant, U.S.N.

Iverson, A. J., Lieut.-Comr., U. S. N.

Irvine, John C., Lieutenant, U. S. N.

Jaques, W. H., South Bethlehem, Pa.

Jayne, J. L., Ensign, U. S. N.

Jenkins, F. W., Jr., Ensign, U. S. N.

Jenkins, T. A., Rear-Admiral, U. S. N., 2115 Penna. Ave., Washington, D. C.

Jewell, T. F., Commander, U. S. N.

Jones, H. W., Asst.-Engineer, U. S. N.

Jouett, J. E., Rear-Admiral, U. S. N.

Kafer, J. C., P.-Asst. Engineer, U. S. N., care Morgan Iron Works, foot of E. 9th st., New York, N. Y.

Kane, T. F, Captain, U. S. N.

Kellogg, Wainwright, Lieut., U. S. N.

Kennedy, Duncan, Lieutenant, U. S. N.

Kiersted, A. J, Chief Eng'r, U. S. N.

Kimberly, L. M., Rear-Admiral, U. S. N.

Kinkaid, T. W, Asst. Eng'r, U. S. N.

King, W. R., P.-Asst. Engineer, U.S.N.

Knapp, H. S., Lieutenant, U. S. N.

Knepper, C. M, Ensign, U. S. N.

Knight, A. M., Lieutenant, U. S. N.

Lasher, O. E., Lieutenant, U. S. N.

Lawrence, F. W., Chaple Station, Longwood, Brookline, Mass.

Lee, S. P., Rear-Admiral, U. S. N.

Lefavor, F. H., Lieutenant, U. S. N.

Leroux, J., Professor, U. S. Nav. Acad.

Leutze, E. H. C., Lieut.-Comr., U.S.N.

Lillie, A. B. H., Lieut.-Comr., U. S. N.

Linnard, J. H., Naval Const'r, U. S. N.

Little, W. McC., Lieutenant, U. S. N., University Club, Madison Square, New York, N. Y.

Lloyd, E., Lieutenant, U. S. N.

Longnecker, E., Lieut.-Comr., U. S. N.

Loring, C. H., Chief Engineer, U. S. N.

Low, W. F., Lieutenant, U. S. N.

Luby, J. F., Ensign, U. S. N.

Lucas, L. C., Naval Cadet, U. S. N.

Luce, S. B., Rear-Admiral, U. S. N., Newport, R. I.

Lyon, H. W, Lieut.-Comr., U. S. N.

MacDougall, W. D, Naval Cadet, U.S.N.

Mackenzie, M. R. S., Lieut.-Com'dr, U. S. N.

Macomb, D. B, Chief Eng'r, U. S. N.

Magee, G. W, Chief Eng'r, U. S. N.

Mahan, A. T., Captain, U. S. N.

Mahan, D. H., Lieutenant, U. S. N.

Mahoney, J. E., 2d Lieut., U. S. M. C.

Manney, H. N., Lieut.-Comr., U. S. N.

Manning, C. H., P.-Asst. Engineer, U. S. N., Amoskeag Manufacturing Company, Manchester, N. H.

Mansfield, H. B., Lieut.-Comr., U.S.N.

Marble, Frank, Ensign, U. S. N.

Marion, Henri, Asst. Prof., U. S. Naval Academy, Annapolis, Md.

Marvell, Geo. R., Naval Cadet, U. S. N.

Marsh, C. C., Ensign, U. S. N.

Mason, N. E., Lieutenant, U. S. N.

Matthews, E. O., Captain, U. S. N.

Maxwell, W. J., Ensign, U. S. N.

Maynard, W., Lieut.-Comr., U. S. N.

Mayo, H. T., Lieutenant, U. S. N.

McAlister, A. A., Chaplain, U. S. N.

McCalla, B. H., Commander, U. S. N.

McCann, W. P., Rear-Admiral, U. S. N.

McCarteney, C. M., Lieutenant, U. S. N.

McCrea, Henry, Lieutenant, U. S. N.

McDonald, J. D., Ensign, U. S. N.

McElroy, G. W., P.-Asst. Eng'r, U.S.N.

McGowan, J., Jr., Commander, U. S. N.

McGregor, C., Commander, U. S. N.

McLane, Allen, 1500 Vermont Avenue, Washington, D. C.

McLean, T. C., Lieutenant, U. S. N.

McLean, W., Lieutenant, U. S. N.

McNary, I. R., Chief Engineer, U. S. N.

Mead, W. W., Lieut.-Comr., U. S. N.

Menocal, A. G., Civil Engineer, 44 Wall street, New York, N. Y.

Mentz, G. W, Lieutenant, U. S. N.

Mercer, Samuel, Captain, U. S. M. C.

Merriam, G. A., Lieutenant, U. S. N.

Merriman, E. C., Commander.

Miller, F. A., Lieut.-Comr., No. 11 Farragut Place, Morristown, N. J.

Miller, J. M., Lieutenant, U. S. N.

Miller, J. N., Captain, U. S. N.

Miller, J. W., General Manager, Providence and Stonington Steamship Co., New York, N. Y.

Miller, M., Commander, U. S. N.

Miner, R. H., Lieutenant, U. S. N.

Mitchell, G. G., Naval Cadet, U. S. N.

Mitchell, Henry, Professor.

Mitchell, Richard, Lieutenant, U. S. N.

Moore, C. B. T., Lieutenant, U. S. N.

Moore, E. K., Lieutenant, U. S. N.

Moore, T. M., 78 Summer street, Buffalo, N. Y.

Morgan, Jos., Jr., Chief Engineer, Cambria Iron Co., Johnstown, Penna.

Morgan, Stokely, Ensign, U. S. N.

Morrell, H., Lieutenant, U. S. N.

Moser, J. F., Lieutenant, U. S. N.

Moses, L. H., Naval Cadet, U. S. N.

Much, G. W., Naval Constructor, 1510 S. Broad street, Philadelphia, Pa.

Muir, W. C. P., Ensign, U. S. N.

Munroe, Chas. E., Professor, Chemist, U. S. Torpedo Station, Newport, R. I.

Murdock, J. B., Lieutenant, U. S. N.

Nazro, A. P., Lieutenant, U. S. N.

Nelson, H. C., Medical Insp'r, U. S. N., Westminster, Md.

Nelson, V. S., Lieutenant, U. S. N.

Nes, D. S., York, Pa.

Neumann, B. S., Naval Cadet, U. S. N.

Neville, N. C., Naval Cadet, U. S. N.

Newcomb, S., Professor, U. S. N., 941 I street, Washington, D. C.

Newell, J. S., Lieut.-Comr., U. S. N.

Newton, J. T., Lieutenant, U. S. N.

Niblack, A. P., Ensign, U. S. N.

Nichols, F. W., Lieutenant, U. S. N.

Nichols, H. E., Lieut.-Comr., U. S. N.

Nichols, S. W., Commander, U. S. N.

Nicholson, R. F., Lieutenant, U. S. N.

Nickels, J. A. H., Lieutenant, U. S. N.

Niles, Kossuth, Lieutenant, U. S. N.

Niles, N. E., Lieutenant, U. S. N.

Nixon, L., Cramp's Shipyard, Philadelphia, Pa.

Norris, G. A., Lieut.-Comr., U. S. N.

Norton, A. L., Ensign, U. S. N.

Norton, C. F., Lieutenant, U. S. N.

Norton, C. S., Captain, U. S. N.

Nostrand, W. H., Dobb's Ferry, Westchester Co., N. Y.

Offley, C. N., Naval Cadet, U. S. N.

Oliver, James H., Lieutenant, U. S. N.

O'Neil, C., Commander, U. S. N.

Osterhaus, Hugo, Lieutenant, U. S. N.

Paine, S. C., Lieutenant, U. S. N.

Parker, J. F., Lieutenant, U. S. N.

Parker, J. P., Lieutenant, U. S. N.

Parks, W. M., P.-Asst. Eng'r, U. S. N.

Parmenter, H. E., Ensign, U. S. N.

Patch, N. J. K., Lieutenant, U. S. N.

Patton, J. B., Naval Cadet, U. S. N.

Peary, R. E., Civil Engineer, U. S. N.

Pegram, J. C., P. O. Box 1101, Providence, R. I.

Pendleton, E. C., Lieut.-Comr., U.S.N.

Perkins, H., 25 Exeter st., W. Boston, Mass.

Perry, Thomas, Lieut.-Comr., U. S. N.

Peters, G. H., Lieutenant, U. S. N.

Phelps, T. S., Rear-Admiral, U. S. N.

Phelps, W. W., Naval Cadet, U. S. N.

Pillsbury, J. E., Lieutenant, U. S. N.

Poe, C. C., 131 E. Congress street, Detroit, Mich.

Porter, Theodoric, Lieutenant, U. S. N.

Potter, E. E., Captain, U. S. N.

Potter, W. P., Lieutenant, U. S. N.

Potts, T. M., Lieutenant, U. S. N.

Poundstone, H. C., Ensign, U. S. N.

Poyer, J. M., Ensign, U. S. N.

Pratt, W. V., Naval Cadet, U. S. N.

Prime, E. S., Lieutenant, U. S. N.
Prindle, F. C., Civil Engineer, U. S. N.
Qualtrough, E. F., Lieut., U. S. N.
Radford, C., Naval Cadet, U. S. N.
Ramsay, F. M., Commodore, U. S. N.,
　Chief Bureau of Navigation.
Read, J. J., Commander, U. S. N.
Reeder, W. H., Lieutenant, U. S. N.
Reiter, Geo. C., Commander, U. S. N.
Remey, G. C., Captain, U. S. N.
Remey, W. B., Colonel, Judge Advo-
　cate General, U. S. N.
Reynolds, E. L., 37 East 50th street,
　New York, N. Y.
Rhoades, W. W., Lieut.-Comr., U.S.N.
Rice, John Minot, S. B., Ph. D., Pro-
　fessor, U. S. N., Northborough, Mass.
Richards, B. S., Lieut.-Comr., U. S. N.
Rittenhouse, H. O., Lieut., U. S. N.
Robertson, A. H., Ensign, U· S. N.
Robeson, H. B., Captain, U· S. N.
Robie, E. D., Chief Engineer, U. S. N.
Robinson, L. W., Chief Eng'r, U· S. N.
Rock, G. H., Naval Cadet, U. S. N.
Rodgers, C.R.P., Rear-Admiral, U.S.N.,
　1721 I street, Washington, D. C.
Rodgers, F., Captain, U. S. N.
Rodgers, J. A., Lieutenant, U. S. N.
Rodgers, Raymond P., Lieut., U. S. N.
Rodgers, T. S., Lieutenant, U. S. N.
Rodgers, W. L., Lieutenant, U. S. N.
Roelker, C. R., P.-Asst. Eng'r, U· S. N.
Rogers, C. C., Lieutenant, U. S. N.
Roller, J. E., Lieutenant, U. S. N.
Rooney, W. R. A., Lieutenant, U. S. N.
Roosevelt, N. L., 44 Pine street, New
　York, N. Y.
Roper, Jesse M., Lieutenant, U· S. N.
Ross, A , Lieut.-Commander, U. S. N.
Ruhm, T. F., Naval Cadet, U. S. N.
Rush, R., Lieutenant, U. S. N.
Safford, W. E., Ensign, U. S. N.
Salter, T. G. C., Lieutenant, U· S. N.
Sampson, W. T., Captain, U. S. N.
Sargent, N., Lieutenant, U. S. N.
Savage, Thomas, Boatswain, U. S. N.,
　734 3d st., N. W., Washington, D. C.

Sawyer, F. E., Lieutenant, U. S. N.
Schley, W. S., Captain, U. S. N.
Schofield, F. H., Naval Cadet, U. S. N.
Schouler, John, Commander, U. S. N.
Sebree, U., Lieut.-Comr., U. S. N.
Seely, H. B., Captain, U. S. N.
Selfridge, J. R., Lieutenant, U. S. N.
Semple, Lorenzo, Ensign, U. S. N.
Sharp, A., Lieutenant, U. S. N.
Shearman, John A., Lieutenant, U.S.N.
Shepard, E. M., Commander, U. S. N.
Shoemaker, W. R., Ensign, U. S. N.
Sicard, M., Captain, U. S. N.
Signor, M. H., Naval Cadet, U. S. N.
Sigsbee, C. D., Commander, U. S. N.
Simpson, E., Ensign, U. S. N.
Skelding, H. T., Paymaster, U. S. N.
Skerrett, J. S., Commodore, U. S. N.
Sloan, R. S., Oswego, Oswego County,
　N. Y.
Smith, J. T., Lieutenant, U. S. N.
Smith, R. C., Lieutenant, U. S. N.
Smith, Sidney F., St. Paul's School,
　Concord, N. H.
Snow, A. S., Commander, U. S. N.
Soley, J. C., Lieutenant, U. S. N., P. O.
　Box 3188, Boston, Mass.
Soley, Hon. Jas. R., Asst. Secretary of
　the Navy, Navy Dept., Wash., D. C.
Southerland, W. H. H., Lieut., U.S.N.
Sperry, C. S., Lieut.-Comr., U. S. N.
Sprague, F. J., 16–18 Broad street, New
　York, N. Y.
Stahl, A. W., Asst. Naval Constructor,
　U. S. N.
Stanton, J. R., Paymaster, U. S. N.
Stanton, O. F., Captain, U. S. N.
Stanworth, C. S., Ensign, U. S. N.
Staunton, S. A., Lieutenant, U. S. N.
Steele, R. W., Naval Const'r, U. S. N.
Stevens, T. H., Rear-Admiral, U. S. N.,
　1604 19th street, Washington, D. C.
Stevens, T. H., Lieutenant, U. S. N.
Stevenson, H. N., P.-Asst. Engineer,
　U. S. N.
Stewart, R., Jr., Asst. Engineer, U.S.N.
Stirling, Yates, Commander, U. S. N.

Stockton, C. H., Lieut.-Comr., U. S. N.
Stoney, G. M., Lieutenant, U. S. N.
Strauss, Joseph, Ensign, U. S. N.
Street, G. W., Asst. Naval Constructor, U. S. N.
Strong, E. T., Lieut.-Comr., U. S. N.
Strong, W. C., Lieutenant, U. S. N.
Stuart, D. D. V., Lieutenant, U. S. N.
Sturdy, E. W., Lieutenant, U. S. N.
Swinburne, Wm. T., Lieut.-Commander, U. S. N.
Symonds, F. M., Lieut.-Comr., U. S. N.
Taussig, E. D., Lieutenant, U. S. N.
Tawresey, John G., Asst. Naval Constructor, U. S. N.
Taylor, D. W., Naval Const'r, U. S. N.
Taylor, H. C., Commander, U. S. N.
Terhune, W. J., Naval Cadet, U. S. N.
Terry, N. M., A. M., Ph. D., Prof. U. S. Naval Academy.
Thackara, A. M., No. 1300 Chestnut st., Philadelphia, Pa.
Thomas, C., Lieutenant, U. S. N.
Tilley, B. F., Lieut.-Comr., U. S. N.
Tilton, McL., Lt.-Col., U. S. M. C.
Tisdale, R. D., Ensign, U. S. N.
Todd, C. C., Lieut.-Comr., U. S. N.
Train, C. J., Commander, U. S. N.
Treadwell, T. C., Naval Cadet, U. S. N.
Turnbull, F., Lieutenant, U. S. N., Morristown, N. J.
Turner, T. J., Med. Director, U. S. N.
Turner, W. H., Lieutenant, U. S. N.
Twining, N. C., Naval Cadet, U. S. N.
Underwood, E. B., Lieutenant, U. S. N.
Vail, Holman, Lieut.-Comr., U. S. N.
Vansant, W. N., Asst. Naval Constructor, U. S. N.
Varney, W. H., Naval Constructor, U. S. N.
Veeder, T. E. D. W., Lieut., U. S. N.
Very, S. W., Lieut.-Comr., U. S. N.
Vreeland, C. E., Lieutenant, U. S. N.
Wadhams, A. V., Lieutenant, U. S. N.
Wadleigh, G. H., Commander, U. S. N.

Wadsworth, H., No. 45 Beacon street, Boston, Mass.
Wainwright, Richard, Lieut., U. S. N.
Walker, Asa, Lieut.-Comr., U. S. N.
Walker, John G., Rear-Adm'l, U. S. N.
Waring, H. S., Lieutenant, U. S. N.
Watson, E. W., Lieut.-Comr., U. S. N.
Weaver, W. D., Asst. Engin'r, U. S. N.
Webb, T. E., Naval Const'r, U. S. N.
West, C. H., Lieut.-Comr., U. S. N.
White, E., Commander, U. S. N.
White, W. P., Ensign, U. S. N.
Whitehead, William, Captain, U. S. N.
Whittelsey, H. H., Ensign, U. S. N.
Williams, C. S., Ensign, U. S. N.
Williams, P., Naval Cadet, U. S. N.
Wilner, F. A., Lieut., U. S. N.
Wilson, Byron, Captain, U. S. N.
Wilson, J. C., Lieutenant, U. S. N.
Wilson, T. D., Chief Constructor, U. S. N., Navy Dept., Wash., D. C.
Winder, William, Lieutenant, U. S. N.
Winn, J. K., Commander, U. S. N.
Winslow, F., Lieutenant, U. S. N, Raleigh, N. C.
Winslow, Herbert, Lieutenant, U. S. N.
Winterhalter, A. G., Lieut., U. S. N.
Wise, F. M., Lieutenant, U. S. N.
Wise, Wm. C., Commander, U. S. N.
Wood, E. P., Lieut.-Comr., U. S. N.
Wood, W. M., Lieutenant, U. S. N.
Woodbridge, W. E.
Woodward, J. J., Asst. Naval Constructor, U. S. N.
Woodworth, S. E., Lieutenant, U. S. N.
Wooster, L. W., P.-Asst. Engineer, U. S. N., Gibbsborough, N. J.
Worden, J. L., Rear Adm'l, U. S. N.
Worthington, W. F., P.-Asst. Engineer, U. S. N.
Wright, Benjamin, Ensign, U. S. N.
Yates, A. R., Captain, U. S. N.
Young, Lucien, Lieutenant, U. S. N.
Ziegemeier, H. J., Naval Cadet, U.S.N.

ASSOCIATE MEMBERS.

Abbot, F. V., Captain, Engineers, U. S. A., 3 Southern Wharf, Charleston, S. C.

Abbot, Henry L., Brigadier-General, Engineers, U. S. A., No. 39 Whitehall street, New York, N. Y.

Angstrom, A., Civil Engineer, Cleveland Shipbuilding Co., Cleveland, Ohio.

Archbold, Geo., M. D., Chemical Laboratories, Navy Yard, Washington, D. C.

Aspinwall, Wm. H., 25 East 10th street, New York, N. Y.

Azoy, Anastasio C. N., care of J. E. Ward, 113 Wall street, New York, N. Y.

Baba, R., Lieutenant, I. J. Navy, Japanese Legation, Washington, D. C.

Babcock, W. I., Manager, Chicago Shipbuilding Co., Chicago, Ill.

Balbach, E., Jr., 233 River street, Newark, N. J.

Balch, G. T., Colonel, Auditor, Board of Education, 143 Grand street, New York, N. Y.

Barr, F., Captain, U. S. R. M.

Barstow, G. F., Major, U. S. A.

Benet, Laurence V., Military Engineer, Hotchkiss Ordnance Co., 21 Rue Royale, Paris, France.

Bennett, T. G., President, Winchester Repeating Arms Co., New Haven, Conn.

Billings, Geo. E., care of Hall Brothers Shipbuilding Co., 28 California street, San Francisco, Cal.

Bogert, J. L., Flushing Iron Works, Flushing, Queens Co., N. Y.

Bole, J. K., Otis Iron and Steel Co., Cleveland, Ohio.

Bostrum, A. O., Navy Department, Washington, D. C.

Brooke, J. M., Professor, Virginia Military Institute, Lexington, Va.

Brough, John, M. E., P. O. Box 214, Vallejo, California.

Burr, Edward, Lieutenant, Engineers, U. S. A., Cascade Locks, Oregon.

Burr, J. H. Ten-Eyke, Cazenovia, Madison, N. Y.

Campbell, J. B., Captain, 4th Artillery, U. S. A.

Canfield, A. Cass, Commodore, C. S. Yacht Club, 60 West 54th street, New York, N. Y.

Carbaugh, H. C., Captain, Asst. Judge Advocate, U. S. A.

Chase, Constantine, Lieutenant, 3d Artillery, U. S. A.

Collins-Stanforth, F. S., Union League Club, New York, N. Y.

Colwell, A. W., Colwell Iron Works, foot of W. 27th street, New York, N. Y.

Comly, Clifton, Major, Ordnance, U. S. A.

Copeland, C. W., C. E. and M. E., 24 Park Place, New York, N. Y.

Cowles, Alfred H., McRea Block, Lockport, N. Y.

Cowles, Eugene H., 47 Windsor Avenue, Cleveland, Ohio.

Craft, Elijah R., 2–4 Stone street, New York, N. Y.

Dagron, J. G., Engineer of Bridges, Baltimore and Ohio R. R., Baltimore, Md.

Davenport, R. W., Supt. Midvale Steel Co., Philadelphia, Pa.

Davin, A., Lieutenant de Vaisseau, de la Marine Française, 92 Boulevard Raspail, Paris, France.

Davis, D. P., N. Y. Safety Steam Power Co., 30 Courtlandt street, New York, N. Y.

D'Oremieulx, Leon F., Secretary, S. C. Yacht Club, 7 E. 32d street, New York, N. Y.

Drayton, Percival L., Union Club, New York, N. Y.

Durfee, Wm. E., M. E., Birdsboro, Berks Co., Pa.

Dutton, S. E., M. E., 212-214 California street, San Francisco, Cal.

Edison, Thomas A., Orange, N. J.

Edson, Jarvis B., M. E., 87 Liberty street, New York, N. Y.

Emery, C. E., Civil Engineer, 22 Courtlandt street, New York, N. Y.

Eunsôn, R. G., Room 53, 171 Broadway, New York, N. Y.

Falsen, C. M., Lieutenant, Norwegian Navy, Horten, Norway.

Forbs-Leith, Alex. J., 40 Park Avenue, New York, N. Y.

France, J. R., Arlington Manufact'g Co., 86 Leonard street, New York, N. Y.

Gaskin, Edward, 173 W. Terry street, Buffalo, New York.

Gatling, R., M. D., Gatling Gun Co., Hartford, Conn.

Gilpin, F. M., 251 South 4th street, Philadelphia, Pa.

Graves, Miles W., Treasurer, Pratt & Whitney Co., Hartford, Conn.

Greenough, G. G., Captain, 4th Artillery, U. S. A.

Hale, Irvin, Lieutenant, Engineers, U. S. A.

Hall, Henry P., The Cloister, New Haven, Conn.

Halsey, James T., care of Cooke L. & M. Co., Paterson, N. J.

Hand, S. Ashton, Hand Manufacturing Co., Toughkenamon, Pa.

Handbury, T. C., Major, Engineers, U. S. A., P. O. Drawer 50, Portland, Oregon.

Hardon, Henry Winthrop, Attorney-at-Law, 52 Wall street, New York, N. Y.

Harmon, O. S., 318 Monroe street, Brooklyn, N. Y.

Hartshorne, Joseph, Supt. Pottstown Iron Works, Pottstown, Pa.

Haug, John, M. E., 206 Walnut Place, Philadelphia, Pa.

Hayes, A. A.

Hedden, E. F., First Asst. Engineer, U. S. R. M.

Herr, Rev. Benj. L., 172 Court street, Binghamton, N. Y.

Hewins, E. H., Colonel, Supt. N. E. Weston Electric Light Co., 18 P. O. Square, Boston, Mass.

Hillman, Gustav, Naval Architect, 470 Greene Avenue, Brooklyn, N. Y.

Hoffman, J. W., 333 Walnut street, Philadelphia, Pa.

Holzapfel, M., Quayside, Newcastle-on-Tyne, England.

Howland, M. Morris, University Club, Madison Square, New York, N. Y.

Hughes, R. P., Colonel, Inspector General, U. S. A., Governor's Island, N. Y.

Humphrey, E. W. C., 412 Center street, Louisville, Ky.

Hunt, W. P., President, South Boston Iron Works, Boston, Mass.

Hyde, Marcus D., 630 Commercial street, San Francisco, Cal.

Ingalls, J. M., Captain, 1st Artillery, U. S. A.

Jamar, M. F., 1st Lieutenant, 13th Infantry, U. S. A.

James, Nathaniel T., 416 California street, San Francisco, Cal.

Johnson, Isaac G., Steel Manufacturer, Spuyten Duyvil, New York.

Kennon, L. W. V., Lieutenant, 6th Infantry, U. S. A., Adjutant General's Office, War Department, Washington, D. C.

Kent, William, Passaic, N. J.

Kittelle, Geo. W., Buford Hotel, Charlotte, N. C.
Knight, Albert, Civil Engineer, P. O. Box 211, Butte City, Montana.
Krupp, Fried., Cast Steel Works, Essen, Germany.
Lang, W. M., Admiral, I. C. N., care Griffin & Co., The Hard, Portsmouth, Eng.
Laureau, L. G., M. E., 60 Washington Square, South, New York, N. Y.
Leavitt, E. D., Jr., 2 Central Square, Cambridgeport, Mass.
Leu, Fred., 68 West street, Greenpoint, L. I.
Livermore, W. R., Major, Engineers, U. S. A.
Loring, B. W., Oswego, Tioga Co., N. Y.
Lottin, Captain, French Army, French Legation, Washington, D. C.
Low, Philip B., 28 South street, New York, N. Y.
Lowe, A. Y., Lieutenant, U. S. R. M.
Lyman, Thomas C., 8 East 65th street, New York, N. Y.
Lyon, Henry, M. D., 34 Monument Square, Charlestown, Mass.
MacMurray, J. W., Major, 1st Artillery, U. S. A.
Macomb, M. M., First Lieutenant, 4th Artillery, U. S. A.
Matlack, J. R., Jr., C. E., 403 Market street, Philadelphia, Pa.
Marx, J. L., Lieutenant, R. N., Arle Bury, Arlesford-Hants, England.
McClellan, Wm. B., Sec'y, Dorchester Yacht Club, 52 Monadnock St., Dorches-
    ter, Mass.
McLaughlin, J., 2041 5th Avenue, New York, N. Y.
McMahan, John Mabry, care Brown, Shipley & Co., London, England.
Mendenhall, T. C., Professor, Supt. U. S. C. & G. Survey, Washington, D. C.
Metcalf, H., Captain, Ordnance, U. S. A.
Miller, H. W., Morristown, N. J.
Miller, P. P., 293 Court street, Buffalo, N. Y.
Millis, John, Lieut., Engineers, U. S. A., 1 Prytania street, New Orleans, La.
Morgan, C. Leslie, 70 South street, New York, N. Y.
Murphy, Peter R., 97 Church street, Albany, N. Y.
Nakamura, S., Lieutenant, I. J. Navy, Japanese Legation, Washington, D. C.
Nordhoff, C., Ensenada, L. C., by San Diego, California.
Norton, Jas. A., M. D., 338 Sandusky street, Tiffin, Ohio.
Oliphant, A. C., Trenton, N. J.
Oliver, Wm. Letts, 328 Montgomery street, San Francisco, Cal.
Otis, Chas. A., Room 13, No. 80 Broadway, New York, N. Y.
Potter, John A., Supt. Homestead Steel Works, Munhall, Pa.
Pratt, N. W., The Babcock & Wilcox Co., 30 Courtlandt st., New York, N. Y.
Quinan, W. R., Manager, California Powder Co., San Francisco, Cal.
Randolph, L. S., M. E., 1214 Bolton street, Baltimore, Md.
Raymond, C. W., Colonel, Engineers, U. S. A., 1428 Arch st., Philadelphia, Pa.
Richards, J. W., E. M., Ph. D., Lehigh University, S. Bethlehem, Pa.
Ricketts, Pierre de P., E. M., Ph. D., 49th street and 4th Ave., New York, N. Y.
Roepper, C. W., The Solid Steel Co., Alliance, Ohio.
Ropes, J. C., 99 Mount Vernon street, Boston, Mass.
Ruckman, J. W., Lieutenant, 5th Artillery, U. S. A.
Russell, A. H., Captain, Ordnance, U. S. A.

Saito, M., Lieut., I. J. N., 16 Jamachi Gochome, Shiba, Tokio, Japan.

Satterlee, C. B., Lieutenant, 3d Artillery, U. S. A.

Schneider, Henri, Au Creusot, Saône-et-Loire, France.

Schuyler, Roosevelt M., University Building, Washington Sq., New York, N. Y.

Scott, Irving M., Union Iron Works, San Francisco, Cal.

Scudder, E. M., Attorney and Counsellor, 54 Wall street, New York, N. Y.

Sears, W. H., Civil Engineer, 150 Ellison street, Paterson, N. J.

See, Horace, M. and C. Engineer, 1 Broadway, New York, N. Y.

Shibayama, Y., Captain, I. J. N., care of Japanese Legation, Washington, D. C.

Simpson, J. M., Captain, Chilian Navy.

Sloan, T. O'Connor, M. E., Ph. D., 361 Broadway, New York, N. Y.

Sloat, Geo. V., Chief Engineer, Old Dominion Steamship Co., New York, N. Y.

Smedburg, W. R., Lieut.-Colonel, 316 California street, San Francisco, Cal.

Snow, J. H., General Supt. National Transit Co., 26 Broadway, New York, N. Y.

Sperry, Chas., M. E., Port Washington, Long Island, N. Y.

Stephens, Wm. P., Yachting Editor, " Forest and Stream," 39 Park Row, New York, N. Y.

Stevenson, Chas. A., Roosevelt & Howland, 55 Beaver street, New York, N. Y.

Stickney, J. S., Office of the New York Herald, New York, N. Y.

Stratton, E. Platt, C. E., 4 Hanover street, New York, N. Y.

Stueler, Rudolph, 159 Front street, New York, N. Y.

Tams, J. Fred, 48 Exchange Place, New York, N. Y.

Taylor, Harry, Lieutenant, Engineers, U. S. A.

Thompson, R. M., Box 1325 and 37 Wall street, New York, N. Y.

Turtle, Thomas, Captain, Engineers, U. S. A.

Uberroth, Preston H., Lieutenant, U. S. R. M.

Vanderbilt, A., 113 Wall street, New York, N. Y.

Waldo, Leonard, M. D., Electrical Engineer, Scoville Mf'g Co., Waterbury, Conn.

Walke, Willoughby, Lieutenant, 2d Artillery, U. S. A.

Washington, H. S., A. B., S. F., care of Brown, Shipley & Co., London, E. C., England.

Webber, W. O., Civil Engineer, 68 Sears Building, Boston, Mass.

Wellman, S. T., 1080 Wilson Avenue, Cleveland, Ohio.

West, Thomas D., Thos. D. West Foundry Co., Cleveland, Ohio.

Wheeler, F. M., Mech. Engr., 93-95 Liberty street, New York, N. Y.

Whistler, G. N., 1st Lieutenant, 5th Artillery, U. S. A.

White, J. F., S. B., Buffalo Chemical Works, Buffalo, N. Y.

Williams, Albert, Jr., U. S. G. and Coast Survey Office, Washington, D. C.

Wilson, A. E., Captain, Chilian Navy, care of Thomas Purves, Valparaiso, Chili.

Wisser, J. P., Lieutenant, 1st Artillery, U. S. A.

Woodall, James, Shipbuilder, foot of Allen St. and Locust Point, Balto., Md.

Woodbury, John McGaw, M. D., 120 Fifth Avenue, New York, N. Y.

Wotherspoon, W. W., Lieutenant, 12th Infantry, U. S. A.

Yamano-ouchi, M., Lieutenant, I. J. Navy, 28 West Parade, Newcastle-upon-Tyne, England.

Zalinski, E. L., Captain, 5th Artillery, U. S. A.

MEMBERS DECEASED SINCE MARCH, 1887.

Archbold, Samuel, Oct. 21, 1890.*
Baker, S. H., Commander, U. S. N., Oct. 30, 1888.
Beaumont, H. N., Surgeon, U. S. N., April 30, 1887.
Bessels, E., M. D., 1888.
Black, Chas. H., Lieut.-Com., Jan. 20, 1891.
Boyd, Robert, Captain, U. S. N., July 30, 1890.
Bridge, E. W., Lieutenant, U. S. N., Aug. 29, 1889.
Case, A. L., Oct. 17, 1890.*
Chandler, R., Rear-Admiral, U. S. N., Feb. 11, 1889.
Coffin, J. H. C., Professor, U. S. N., Jan. 8, 1890.
Danenhower, J. W., April 20, 1887.
Darrah, W. F., Asst. Engineer, U. S. N., Feb. 25, 1889.
Davids, H. S., Chief Engineer, U. S. N., Feb. 8, 1888.
Delmater, C. H., Feb. 7, 1889.*
Denney, Wm., 1887.*
Ericsson, John, New York, March 8, 1889.†
Faron, E., 1888.
Forbes, Hon. R. B., Dec., 1889.*
Gatewood, Richard, Naval Constructor, U. S. N., Dec. 15, 1890.
Gibbons, Charles, Jr., 1888.
Huntington, C. L., Commander, U. S. N., Oct. 14, 1890.
Kelly, J. P., Chief Engineer, Jan. 27, 1890.
Kenyon, A. J., Chief Engineer, July 27, 1888.
Lee, C. S., March, 1888.
Le Roy, W. E, Rear-Admiral, Dec. 10, 1888.
Livingston, G. B., Lieut.-Comr., Sept. 19, 1890.
McCarty, R. H., P.-Asst. Surgeon, U. S. N., April 12, 1890.
McGowan, W. C., P.-Asst. Paymaster, U. S. N., Dec. 25, 1887.
McRitchie, D. G., Lieutenant, U. S. N., Aug. 12, 1888.
Meatyard, E. B., 1889.
Meyers, F. Bailey, New York, 1888.
Michaelis, O. E., Major, U. S. A., May 1, 1890.
Miles, Chas. R., Lieutenant, U. S. N., Jan. 14, 1889.
Mullany, J. R. M., Rear-Admiral, U. S. N., Sept. 17, 1887.
Nicoll, W. L., Chief Engineer, U. S. N., July 2, 1887.
Pearson, F., New York, Dec. 1890.
Peck, Chas. F., Aug. 12, 1890.
Rowan, S. C., Vice-Admiral, U. S. N., March 31, 1890.
Schaefer, H. W., Lieutenant, U. S. N., May 11, 1889.
Sharrer, W. O., Lieutenant, U. S. N., Sept. 8, 1889.
Simpson, E., Rear-Admiral, U. S. N., Dec. 1, 1888.
Smith, W. D., Chief Engineer, U. S. N., Sept. 10, 1887.
Snyder, H. L., Chief Engineer, June 30, 1887.

* Life Member.   † Hon. Member.

Stewart, W. A. W., March, 1888.
Talcott, C. G., Asst. Engineer, U. S. N., July 25, 1889.
Totten, G. M., Lieut.-Comr., U. S. N., May 27, 1888.
Wells, C. H., Rear-Admiral, U. S. N., Jan. 28, 1888.
Willamov, G., 1888.
Wright, M. F., Lieutenant, U. S. N., March 4, 1890.

ANNUAL REPORT OF THE SECRETARY AND TREASURER OF THE UNITED STATES NAVAL INSTITUTE.

To the Officers and Members of the Institute.

Gentlemen :—I have the honor to submit the following report of the affairs of the Institute for the year ending December 31, 1890:

Itemized Cash Statement.

Receipts during Year 1890.

Items.	1st Quarter.	2d Quarter.	3d Quarter.	4th Quarter.	Totals.
Advertisements............	$160 00	$180 00	$20 00	....	$360 00
Dues.	1,181 46	312 98	254 00	$234 04	1,982 48
Sales................	131 29	116 30	13 42	150 14	411 15
Subscriptions..	251 25	319 40	134 10	117 60	822 35
Life-membership fees	120 00	30 00	30 00	30 00	210 00
Binding, extra...............	20 80	4 00	4 60	2 00	31 40
Interest on bonds..	37 41	26 33	45 50	....	109 24
Totals.........	$1,902 21	$989 01	$501 62	$533 78	$3,926 62

Expenditures during Year 1890.

Items.	1st Quarter.	2d Quarter.	3d Quarter.	4th Quarter.	Totals.
Printing publications	$300 00	$1,047 30	$763 81	$11 25	$2122 36
Postage..........	28 17	31 95	34 14	29 81	124 07
Freight........	6 87	8 98	3 95	5 47	25 27
Messenger........	103 00	125 00	135 00	140 00	503 00
Office furniture..........	... .	68 40	....	....	68 40
Office expenses..	2 20	5 20	05	1 59	9 04
Stationery	1 38	41 61	4 50	18 74	66 23
Drawing..........	4 00	....	....	....	4 00
Purchase of bonds.............	127 25	....	....	122 18	249 43
Over credit..	3 00	....	....	....	3 00
Binding, extra.	17 60	....	....	31 10	48 70
Secretary	....	....	....	30 00	30 00
Telegram	....	....	....	27	27
Subscription Army and } Navy List .... }	....	....	....	2 00	2 00
Purchase of back numbers..	....	....	....	7 70	7 70
Notary's fees...........	....	....	....	1 50	1 50
Expenses Branch office	....	....	....	2 00	2 00
Totals...................	$593 47	$1,328 44	$941 45	$403 61	$3,266 97

SUMMARY.

Balance of cash unexpended for the year 1889......................... $1058 27
Total receipts for 1890.. 3926 62

Total available cash for 1890...$4984 89
Total expenditures for 1890.. 3266 97

Cash unexpended January 1, 1891...$1717 92
Cash held to credit of Reserve Fund 159 44

True balance of cash on hand January 1, 1891 ....$1558 48
Bills receivable for sales of No. 55.. 142 10
    "   "   " dues, 1890 .... 360 00
    "   "   " back dues... 378 00
    "   "   " binding... 22 00

        Total assets January 1, 1891...$2460 58
Bills for No. 55, outstanding...

$1000 of the available cash is on deposit in the Seamen's Bank for Savings, New York, drawing interest.

	Year 1889.	Year 1890.	Increase.
Receipts.................	$3447 38	$3926 62	$479 24
Expenditures............	3104 39	3266 97	162 58
Balance...........	$342 99	$659 65	$316 66

RESERVE FUND.

List of bonds deposited for safe-keeping in the Farmer's National Bank of Annapolis, Md.:

United States 4 per cent registered bonds $900 00
District of Columbia 3.65 per cent registered bonds................ .... 2000 00
    "   "   "   "   coupon   " 150 00

                                            $3050 00
Cash in bank uninvested.......... .... 159 44

        Total Reserve Fund........ .... $3209 4
Annual interest on bonds...... .... 114 47
Number of new life members.. 7

During the year four District of Columbia bonds, 3.65 per cent, face value $200, were purchased for $249.43. District of Columbia coupon bonds, 3.65 per cent, to the value of $1000, were sent to the Treasurer of the United States in exchange for $1000 in District of Columbia 3.65 per cent registered bonds. These latter are registered in the name of the United States Naval Institute.

MEMBERSHIP.

During the year 1890 the Institute lost 16 members by resignation and 14 by death, two of the latter having been life members. 27 new members' names were added to the rolls, 3 becoming life members; 4 regular members became life members.

The membership of the Institute to date, January 1, 1891, is as follows:

Honorary members 6
Life   "   101
Regular   "   558
Associate   "   165

Total number of members........ 830

CIRCULATION.

Members...................... 830
Subscriptions 256
Exchanges................... 71
Average sales per quarter..................... 130

· Total circulation per quarter 1287

PUBLICATIONS ON HAND.

The Institute had on band at the end of the year the following copies of back numbers of its Proceedings:

Whole Nos.	Plain Copies.	Bound Copies.	Whole Nos.	Plain Copies.	Bound Copies.
1	198	...	14	6	...
2	201	...	15	2	...
3	59	...	16	220	...
4	149	...	17	3	...
5	122	...	18	91	...
6	6	...	19	110	...
7	12	...	20	127	...
8	35	...	21	230	...
9	42	...	22	273	...
10	4	...	23	175	...
11	215	...	24	198	...
12	56	...	25	1141	...
13	6	...	26	203	27

	Plain. Copies.	Bound. Copies.			Plain. Copies.	Bound Copies.
Whole Nos. 27	290	27	Whole Nos. 42		123	14
28	5	...	43		283	3
29	222	27	44		270	10
30	252	4	45		207	17
31	23	53	46		215	18
32	5	173	47		193	18
33	14	162	48		181	18
34	181	24	49		211	17
35	104	66	50		170	17
36	251	25	51		223	18
37	167	20	52		191	18
38	247	2	53		533	35
39	176	1	54		206	12
40	623	111	55		225	18
41	244	19				

2 Vol. X., Part 1, bound in half-morocco.
1 "   " 2, "   "
1 Vol. XIII., Part 2, bound in half-morocco.
1 No. 34, bound in half-morocco.
4 "   "   " calf.
6 "   "   full sheep.

The archive set complete, Vol. I. to Vol. XIII. inclusive, bound in full turkey.

Very respectfully,

H. G. Dresel, Ensign, U. S. N.,
Secretary and Treasurer.

Annapolis, Md., *January* 1, 1891.

CONSTITUTION AND BY-LAWS.

ARTICLE I.—TITLE.

This organization shall be called the United States Naval Institute.

ARTICLE II.—OBJECT.

Its object is the advancement of professional, literary, and scientific knowledge in the Navy.

ARTICLE III.—HEADQUARTERS.

The Headquarters of the Institute shall be at the United States Naval Academy, Annapolis, Md.

ARTICLE IV.—OFFICERS.

The officers shall be as follows :
A President.
A Vice-President.
A Board of Control.
A Secretary and Treasurer.
A President and Corresponding Secretary for each Branch,

ARTICLE V.—ELECTION OF OFFICERS.

SEC. 1. There shall be a meeting of the Institute at Headquarters on the second Friday in October of each year, of which at least two weeks' notice shall be given, at which meeting all the foregoing officers, except those of Branches, shall be elected by ballot in open session, and a majority of votes given by presence or proxy shall elect ; regular or life members only being eligible for office.

SEC. 2. Absent members who have the constitutional right to vote may vote by proxy at such elections, and in the same manner on all questions involving changes in the Constitution and By-Laws, and upon questions involving the expulsion of members and the election of honorary members. On all other questions voting must be by actual presence. Life members shall have the full right of regular members to vote on every question. Honorary and associate members shall not have the privilege of voting. All proxies must be signed by the member whose vote is to be represented.

SEC. 3. Members elected to the position of officers of the Institute will assume their respective duties at the date from which elected.

SEC. 4. Casual vacancies in the officers of the Institute may be temporarily filled by the Board of Control.

ARTICLE VI.—DUTIES OF OFFICERS.

SEC. 1. The President shall preside at business meetings of the Institute, or its Branches, at which he may be present.

SEC. 2. In the absence of the President at Headquarters, the Vice-President shall preside.

SEC. 3. The Board of Control shall consist of seven members in good standing, regular or life, and its duties shall be the management of all the financial and administrative business of the Institute, including the censorship, printing, and control of its publications. The Secretary and Treasurer, shall be, *ex officio*, a member of the Board, its medium of communication and the recorder of its transactions. The regular meetings of the Board of Control shall be held upon the first and third Saturday of each month. A special meeting shall be called by the Secretary and Treasurer upon the written application of two members of the Board. A quorum shall consist of three members. In the absence of both the President and Vice-President at business meetings of the Institute, a member of the Board of Control shall preside. It shall be the duty of this Board to appoint a committee of three of its own members to audit and certify the books and accounts of the Secretary and Treasurer at least once every quarter.

SEC. 4. The Secretary and Treasurer shall keep a register of the members in which shall be noted all changes; an authenticated copy of the Constitution and By-Laws in force; a journal of the Proceedings of the Institute; a separate journal of the transactions of the Board of Control; a receipt and expenditure book; an account-current with each member. Under the authority of the Board of Control, he shall be the disbursing and purchasing officer of the Institute and the custodian of the funds, securities, and assets, and it shall be his duty to furnish members with receipts for dues paid. He shall attend to all correspondence and keep a record thereof, give due notice of meetings of the Institute and Board of Control, have charge of the stenographer and copyists employed to prepare records of the Proceedings, and he shall distribute all publications. · The books of account of the Institute shall always be open to inspection by any member. Papers accepted by the Board of Control shall be read by the Secretary and Treasurer when the author cannot be present.

ARTICLE VII.—MEMBERSHIP.

SEC. 1. The Institute shall consist of regular, life, honorary, and associate members.

SEC. 2. Officers of the Navy, Marine Corps, and all civil officers attached to the Naval Service, shall be entitled to become regular or life members, without ballot, on payment of dues or fees to the Secretary and Treasurer, or to the Corresponding Secretary of a Branch. Members who resign from the Navy subsequent to joining the Institute will be regarded as belonging to the class described in this Section.

SEC. 3. The Prize Essayist of each year shall be a life member without payment of fee.

SEC. 4. Honorary members shall be selected from distinguished Naval and Military Officers, and from eminent men of learning in civil life. The Secretary of the Navy shall be, *ex officio*, an honorary member. Their number shall not exceed thirty (30). Nominations for honorary members must be favorably reported by the Board of Control, and a vote equal to one-half the number of regular and life members, given by proxy or · presence, shall be cast, a majority electing.

SEC. 5. Associate members shall be elected from officers of the Army, Revenue Marine, foreign officers of the Naval and Military professions, and from persons in civil life who may be interested in the purposes of the Institute.

SEC. 6. Those entitled to become associate members may be elected life members, provided that the number not officially connected with the Navy and Marine Corps shall not at any time exceed one hundred (100).

SEC. 7. Associate members and life members, other than those entitled to regular membership, shall be elected as follows: Nominations shall be made in writing to the Secretary and Treasurer, with the name of the member making them, and such nominations shall be submitted to the Board of Control, and, if their report be favorable, the Secretary and Treasurer shall make known the result at the next meeting of the Institute, and a vote shall then be taken, a majority of votes cast by members present electing.

SEC. 8. The annual dues for regular and associate members shall be three dollars, payable upon joining the Institute, and upon the first day of each succeeding January. The fee for life membership shall be thirty dollars, but if any regular or associate member has paid his dues for the year in which he wishes to be transferred to life membership, or has paid his dues for any future year or years, the amount so paid shall be deducted from the fee for life membership.

SEC. 9. No member of the Institute shall be dismissed except by recommendation of the Board of Control, and by a two-thirds vote of the members · of the Institute voting at any regular or called meeting, of which at least one month's notice shall be given. Without the recommendation of the Board of Control, no member can be dismissed except by a three-fourths vote. In both the above cases there must be a total vote of at least a majority of all those members entitled to a vote, the voting to be either by presence or proxy. Members two years in arrears shall be dropped. Those dropped for non-payment of dues can regain their membership by paying two years' arrearage of dues, but the Board of Control may adjust any special case upon its merits.

ARTICLE VIII.—RESERVE FUND.

The amount now invested ($3050.00) in United States and District of Columbia bonds shall be placed to the credit of a Reserve Fund. All moneys received from life-membership fees shall, as soon as practicable, be invested in United States bonds, or bonds guaranteed by the United States, and shall be added to said fund, which shall be held in perpetuity to guarantee the future interest of said life members. The interest of said fund may, however, be used for the current expenses of the Institute.

ARTICLE IX.—MEETINGS.

SEC. 1. The regular time of holding meetings of the Institute shall be the second Friday of each month, but if there should be no paper accepted by the Board of Control to be read, professional subject to be discussed, or executive business to be transacted, the monthly meeting may be omitted.

SEC. 2. Special meetings of the Institute shall be called by the Secretary and Treasurer when directed by the Board of Control.

SEC. 3. Notice of regular or special meetings shall state the title of papers to be read, with the name of the author, and mention the executive business that will be brought before the meeting.

SEC. 4. A stenographer may be employed when authorized by the Board of Control.

ARTICLE X.—PAPERS AND PROCEEDINGS.

SEC. 1. Quarterly, or as much oftener as the Board of Control may decide, the papers read before the Institute and its Branches, together with the discussions growing out of them, shall be published. Papers on intricate technical subjects of such a character as not to be appreciated on merely casual investigation, and articles too extended to be read at one meeting of the Institute, may be published as a part of the Proceedings when authorized by the Board of Control; and there may also be published, under the heads of Editorial and Professional Notes, such comment and information as may be deemed of value to the service.

SEC. 2. One copy of the Proceedings when published shall be furnished to each regular, life, honorary, and associate member, to each corresponding Society of the Institute, and to such libraries and periodicals as may be determined upon by the Board of Control.

SEC. 3. Copies of the Proceedings and complete sets may be sold at a charge fixed by the Board of Control, and the Board shall also fix the price of annual subscription for others than members.

SEC. 4. A receipt and expenditure account of the Institute's publications, showing the number on hand, shall be included in the report of the Secretary and Treasurer of each year.

SEC. 5. The Board of Control shall decide the size of the edition of each number of the Proceedings to be published, and also the number of reprints.

ARTICLE XI.—ANNUAL PRIZE ESSAY.

SEC. 1. A prize of one hundred dollars, with a gold medal, shall be offered each year, for the best essay on any subject pertaining to the Naval Profession.

SEC. 2. The award for the above-named prize shall be made by the Board of Control, voting by ballot and without knowledge of the names of the competitors; and the time and manner of submitting such essays shall be determined and announced by said Board.

SEC. 3. In the event of the Prize being awarded to the winner of a previous year, a gold clasp, suitably engraved, will be given in lieu of a gold medal.

ARTICLE XII.—BRANCHES.

SEC. 1. The Board of Control is empowered to appoint Corresponding Secretaries for all Naval Stations, both ashore and afloat, where there is no organized Branch; also for Branches where a vacancy exists owing to the resignation of the Corresponding Secretary before a meeting can be called to elect a successor.

SEC. 2. The officers of a Branch shall be a Vice-President and a Corresponding Secretary.

SEC. 3. The Vice-President shall perform the same duty for the Branch as prescribed for the President of the Institute.

SEC. 4. The Corresponding Secretary of a Branch shall keep a register of the members residing within the limits of the Station, and an account-current with each. He shall keep a journal of the proceedings of the Branch and a copy of the Constitution and By-Laws. He shall give due notice of all meetings of the Branch, and shall have control of the stenographer whenever it is deemed necessary to employ one. He shall forward to the Secretary and Treasurer of the Institute all papers read before his Branch, and keep him informed of all new members and their addresses, and of all business relating to the Institute. He shall have charge of the Branch library and of all books and papers, and shall receive and distribute publications. He shall keep a receipt and expenditure book, shall collect dues from members on the Station and give receipts therefor. He shall be authorized to expend such part of the funds in his possession for stationery, postage, printing, and for other incidental expenses as may be deemed necessary. He shall at the end of each month render to the Secretary and Treasurer a detailed statement of moneys received and expended, with vouchers for expenditures, and shall forward to the Secretary and Treasurer the funds remaining on hand, retaining only sufficient to defray the estimated current expenses of the Branch for the ensuing month.

SEC. 5. Monthly meetings of each Branch may be held upon such dates as the Branch shall decide, but if there is no paper to be read or business to be transacted at the appointed date, the Corresponding Secretary may omit the call for the regular meeting. Special meetings may be called when necessary.

ARTICLE XIII.—Copyright.

The Proceedings shall be copyrighted in behalf of the Institûte by the Secretary and Treasurer.

ARTICLE XIV.—Amendments.

No addition or amendment to the Constitution shall be made without the assent of two-thirds of the members voting; the By-Laws, however, may be amended by a majority vote. Notice of proposed changes or additions shall be given by the Secretary and Treasurer at least one month before action is taken upon them. A total vote equal to at least half the number of regular and life members shall be required.

BY-LAWS.

ARTICLE I.

The rules of the United States House of Representatives shall, in so far as applicable, govern the parliamentary proceedings of the Society.

ARTICLE II.

1. At both regular and stated meetings the routine of business shall be as follows:

2. At executive meetings, the President, or, in his absence, the Vice-President, or, in the absence of both, a member of the Board of Control, shall call the meeting to order, and occupy the chair during the session; in the absence of these, the meeting shall appoint a Chairman.

3. At meetings for the presentation of papers and discussion, the Society shall be called to order as above provided, and a Chairman shall be appointed by the presiding officer, reference being had to the subject about to be discussed, and an expert in the specialty to which it relates being selected.

4. At regular meetings, after the presentation of the paper of the evening, or on the termination of the arguments made by members appointed to or voluntarily appearing to enter into formal discussion, the Chairman shall make such review of the paper as he may deem

proper. Informal discussion shall then be in order, each speaker being allowed not exceeding ten minutes in the aggregate, unless by special consent of the Society. The author of the paper shall, in conclusion, be allowed such time in making a résumé of the discussion as he may deem necessary. The discussion ended, the Chairman shall close the proceedings with such remarks as he may be pleased to offer.

5. At the close of the concluding remarks of the Chairman, the Society shall go into executive session, as hereinbefore provided, for the transaction of business as follows:

1. Stated business, if there shall be any to be considered.
2. Unfinished business taken up.
3. Reports of Officers and Committees.
4. Applications for membership reported and voted upon.
5. Correspondence read.
6. Miscellaneous business transacted.
7. New business introduced.
8. Adjournment.

PROCEEDINGS

OF THE

UNITED STATES

NAVAL INSTITUTE.

INDEX VOLUMES I–XV.

PUBLISHED QUARTERLY BY THE INSTITUTE.

ANNAPOLIS, MD.

PRESS OF ISAAC FRIEDENWALD,
BALTIMORE, MD.

THE PROCEEDINGS

OF THE

UNITED STATES NAVAL INSTITUTE.

INDEX, VOLUMES I.—XV.

PREPARED BY PROF. A. N. BROWN, LIBRARIAN NAVAL ACADEMY.

JAQUES, W. H.   The works of the Bethlehem Iron Company.   **15** : 531.

JEANNETTE, The cruise of the Alliance in search of the.   C. P. Perkins.   **11** : 701.

JEFFERS, W. N.   The armament of our ships-of-war.   **1** : 105.

JEFFRIES, B. J.   Color-blindness and its dangers on the sea.   **6** : 37.

JENKINS, F. W.   A proposed method of ranging guns, applicable to flat trajectories.   **12** : 477.

JEWELL, T. F.   Deep sea sounding.   **4** : 37.

JUNIATA, Corrosion of the copper of the.   C. E. Munroe.   **12** : 391.

KAFER, J. C.   Forced draught by revolving screw in chimney. (P. N.)   **11** : 334.

—— Repairing a broken crank with wire rope.   (P. N.)   **7** : 107.

KELLEY, J. D. J.   Our merchant marine.   (P. E.)   **8** : 3.

KENNEDY, D.   The converted eight-inch M. L. rifle.   **3** : 47.

—— *translator.*   Rooms for recreation and places of refuge for the crews of the several squadrons.   A. Pothnau.   **5** : 135.

KIMBALL, W. W.   Ericsson's submarine torpedo system.   **7** : 339.

—— Hovgaard's submarine boats.   (P. N.)   **14** : 249.

—— Machine guns.   **7** : 405.

—— Magazine rifles.   **8** : 695.

—— Magazine small arms.   **7** : 231.

—— Nordenfelt machine guns.   **7** : 189.

—— Report on the Panama canal, 1885.   Reviewed by W. F. Worthington.   (P. N.)   **13** : 679.

KNIGHT, A. M.   Review of Metcalf's Steel: its properties; its uses in structures and heavy guns.   (P. N.)   **14** : 445.

KRUPP, A.   Krupp and DeBange.   A review.   A. Gleaves.   (P. N.) **14** : 776.

—— Sketch of the life of.   A. Gleaves.   (P. N.)   **14** : 775.

KRUPP's trials of a new powder.   A. Gleaves.   (P. N.)   **15** : 389.

KUNSTADTER steering gear, Report on.   (P. N.)   **12** : 248.

LANDING exercise, "Swatara," 1882.   (P. N.)   **8** : 499.

LANDING parties.   Beehler, W. H.   The march to Witu by the landing party of the German corvette Gneisenau.   **12** : 625.

—— Mason, T. B. M.   Boat guns as light artillery for.   **5** : 207.

—— Navy and marines on shore in the Egyptian campaign of 1882. **8** : 566.

McCRACKIN, A. The turning circle: application of coast survey methods to the determination of. **11** : 265.

MACH, E., *and* WENTZEL, J. A contribution to the mechanics of explosions. Translated by K. Rohrer. (P. N.) **14** : 258.

MACHINE GUNS. *See* Guns.

MAGNETISM, The effect of great cold upon. (P. N.) **7** : 84.

MAHAN, A. T. Letter on printing the War College lectures. **15** : 57.

—— Naval education. **5** : 345.

—— The necessities and objects of a naval war college. **14** : 621.

MAHAN, D. H. A new method for carrying and lowering, and for detaching boats; also a suggestion for defending ships against auto-mobile torpedoes. **13** : 397.

—— Review of Graham's Tactics of infantry in battle. (P. N.) **13** : 491.

—— Signalling the position of the helm. (P. N.) **13** : 497.

—— Three considered as a tactical unit. **14** : 343.

—— *translator.* French protected cruisers. (P. N.) **13** : 693.

MANNING, C. H. Types of circulating pumps for surface condensers. (P. N.) **7** : 335.

MANNING of our navy and mercantile marine. S. B. Luce. **1** : 17.

MARION, H. Carrier pigeons. (P. N.) **14** : 780.

MASON, T. B. M. Boat guns as light artillery for landing parties. **5** : 207.

—— Lessons from the future. **2** : 57.

—— Life-saving at sea. **4** : 77.

—— The 100-ton gun. **2** : 101.

—— *translator.* The employment of torpedoes in steam launches against men-of-war. C. C. Arnault. **6** : 79.

MEIGS, J. F. Curves of pressure in guns. **11** : 743.

—— On the combustion of gunpowder in guns. **11** : 321.

—— The tactics of the gun as discoverable from type war-ships. **14** : 655.

—— The velocity of the wind. **2** : 17.

—— The war in South America. **5** : 461.

—— *and* INGERSOLL, R. R., *translators.* Researches on the effects of powder. By E. Sarrau. **10** : 1.

MENOCAL, A. G. Channel improvement, Washington Navy-yard. **8** : 627.

MEN-OF-WAR. *See* Ships; Cruisers.

NECROLOGY.　Ames, Sullivan D.　7 : xviii.
—— Aston, Albert.　8 : xix.
—— Augur, John P. J.　11 : xvi.
—— Bowdon, Frank W.　11 : xvi.
—— Breese, Kidder R.　8 : xix.
—— Browne, Samuel T.　8 : xx.
—— Cabaniss, Charles.　8 : xxi.
—— Clark, Lewis.　12 : xvi.
—— Collins, Frederick.　8 : xxi.
—— Dorr, Ebenezer P.　8 : xxii.
—— Field, Edward A.　11 : xvii.
—— Fillebrown, Thomas S.　11 : xvii.
—— Fox, Gustavus V.　10 : xv.
—— Gibbs, Benjamin F.　9 : xvi.
—— Gorringe, Henry H.　12 : xvi.
—— Grant, Ulysses S.　12 : xviii.
—— Green, Henry L.　10 : xvi.
—— Greene, S. Dana.　11 : xviii.
—— Handy, Henry O.　11 : xix.
—— Harwood, Andrew A.　11 : xix.
—— Hunter, Henry C.　8 : xxii.
—— Jones, James H.　7 : xvii.
—— Karney, Thomas.　12 : xviii.
—— Kennedy, Charles W.　10 : xvii.
—— Lenthall, John.　9 : xvi.
—— Lewis, Callender I.　10 : xvii.
—— Marston, John.　12 : xviii.
—— Middleton, Edward.　10 : xviii.
—— Miller, Peter.　10 : xviii.
—— Newell, Harman.　7 : xvii.
—— Nokes, Noval L.　10 : xix.
—— Noyes, Boutelle.　10 : xix.
—— Nye, Haile C. T.　12 : xviii.
—— Parker, Foxhall A.　5 : 569.
—— Patterson, Carlile P.　8 : xxiii.
—— Peck, Ransome B.　12 : xix.
—— Remey, Edward W.　12 : xix.
—— Rinehart, Benjamin F.　11 : xx.
—— Rodgers, John.　8 : 251; 9 : xvii.
—— Sands, Benjamin F.　10 : xx.

RECREATION rooms for crews. A. Pothnau. **5** : 135.

REISINGER, W. W. Torpedoes. (P. E.) **14** : 483.

REPAIRS on the French transport Shamrock. E. H. C. Leutzé. (P. N.) **15** : 146.

RESISTANCE, Torpedo experiments upon the. (P. N.) **12** : 641.

RESPONSIBILITY. (P. N.) **9** : 335.

RÉVEILLE, J. Determination of currents by a series of four altitudes. (P. N.) **12** : 449.

RICE, J. M. Notes on Sladen's gunnery. (P. N.) **8** : 191.

—— Wave motion and the resistance of ships. **7** : 447.

RICHARDS, W. *and* POTTER, J. A. The Homestead Steel Works. **15** : 431.

RIELÉ hydraulic testing machine. J. H. Glennon. (P. N.) **14** : 453.

RIFLES. *See* Guns.

RIGGING and equipment of vessels of war. **8** : 483.

RITTENHOUSE, H. O. The navigator's position indicator. **13** : 147.

—— Notes on intercepting, chasing, etc. **11** : 723.

RODGERS, J. Uses of astronomy. **5** : 449.

—— In memoriam. **8** : 197.

—— Soley, J. R. Rear Admiral John Rodgers. **8** : 251.

RODGERS, T. S. The navy six-inch B. L. R. **12** : 77.

RODGERS, W. L. Notes on the naval brigade. **14** : 57.

—— What changes in organization and drill are necessary to sail and fight effectively our war-ships of the latest type? **12** : 361.

ROELKER, C. R. The boiler power of naval vessels. **5** : 275.

—— The economy of compound direct acting pumping engines. **13** : 211.

ROGERS, C. C. The bombardment of Alexandria. **8** : 523.

—— Naval intelligence. **9** : 659.

ROGERS, W. A. The coefficient of safety in navigation. **7** : 205, 353.

ROHRER, K. Gun-cotton, its history, manufacture, use. **15** : 463.

—— *translator*. Compressed gun-cotton for military purposes, particularly regarding its use in shell. By M. von Foerster. **12** : 563.

—— A contribution to the mechanics of explosions. By E. Mach and J. Wenzel. (P. N.) **14** : 258.

—— Sprengel's explosives. (P. N.) **15** : 368.

ROSS, A. Aids in the practical work of navigation. **7** : 461.

—— Deck chart-board. **9** : 759.

SHIPS. Bowles, F. T. Towing experiments on models to determine the resistance of full-sized ships. 9 : 81.
—— Calkins, C. G. What changes in organization and drill are
necessary to sail and fight effectively our war-ships of the
latest type? (P. E.) 12 : 269.
—— Fitz Gerald, C. C. P. Possible effect of high explosives on
future designs of war-ships. (P. N.) 15 : 144.
—— Gilmore, F. P. Ship-building and its interests on the Pacific
coast. 15 : 443.
—— Gleaves, A. Proposed non-sinkable battle-ship. (P. N.)
14 : 599.
—— Hichborn, P. Sheathed or unsheathed ships? 15 : 21.
—— Iron and steel for vessels of war. 8 : 267.
—— Leroux, J. A study on fighting ships by K... 14 : 699.
—— Rigging and equipment of vessels of war. 8 : 483.
—— Rodgers, W. L. What changes in organization and drill are
necessary to sail and fight effectively our war-ships of the
latest type? 12 : 361.
—— Schrœder, S. A U-bow section and a long buttock line.
8 : 387.
See also Fleets, Cruisers.
SIGNALS. Miller, J. W. Open letter to Maritime Conference of
1889. (P. N.) 15 : 588.
—— Wainwright, R. Naval coast signals. 15 : 61.
See also Collisions.
SIGSBEE, C. D. A fallacy in composite great-circle sailing. (P. N.)
11 : 332.
—— Graphic method for navigators. 11 : 241.
—— Progressive naval seamanship. 15 : 95.
—— Sigsbee's improved parallel rule. (P. N.) 8 : 321.
SIMPSON, E. The navy and its prospects of rehabilitation. 12 : 1.
—— A proposed armament for the navy. 7 : 165.
SIMPSON's timber dry-docks. C. H. Stockton. 13 : 221.
SINCLAIR, A. Cruise of the Vandalia in the Pacific in 1858. 15 :
347.
SMALL arms. Kimball, W. W. Magazine. 7 : 231.
—— Lee, J. P. The Lee system of. 7 : 325.
SMITH, W. E. Distribution of armor in ships of war. (P. N.) 11 :
777.
SOLEY, J. C. The naval brigade. 6 : 271.
—— On a proposed type of cruiser. 4 : 127.

WHITE, H. C. Cruise of the Tigress. **1** : 39.

WHITE, U. S. G. Tides: their cause. **8** : 467.

—— The want of docking facilities in our navy-yards. **6** : 265.

WHITEHOUSE, F. C. Mœris, the great reservoir of middle Egypt.
11 : 325.

WHITHAM, J. M. Cylinder diameters of marine compound engines.
10 : 497.

—— Surface condensers. **9** : 303.

WILSON, T. D. Experimental determination of the center of gravity
of the U. S. S. Shawmut. **1** : 149.

WINDS. Meigs, J. F. The velocity of. **2** : 17.

—— Nelson, T. Revolving storms. **5** : 231.

—— Tanner, Z. L. The navigation of the China seas. **5** : 181.

WINTERHALTER, A. G. Portuguese apparatus for lowering boats.
(P. N.) **10** : 519.

WIRE-GUNS. Alger, P. R. On the tension of winding. **11** : 141.

—— —— Tension of winding. (P. N.) **9** : 793.

—— Woodbridge, W. E. Wire-wound guns. (P. N.) **11** : 149.

WISSER, J. P., *translator.* Gun-cotton, its military applications.
By M. von Foerster. (P. N.) **15** : 373.

WITU, The march to, by the landing party of the German corvette
Gneisenau. W. H. Beehler. **12** : 625.

WOLCOTT, C. C. Report on hoisting the 100-ton derrick, Mare
Island navy-yard. (P. N.) **13** : 685.

WOOD, Preservation of. C. E. Munroe. **3** : 73.

WOODBRIDGE, W. E. Wire-wound guns. (P. N.) **11** : 149.

WORTHINGTON, W. F. Baird's automatic steam trap. (P. N.)
13 : 278.

—— Baird's deep-sea sounding machine. (P. N.) **14** : 603.

—— Kimball's report on the Panama canal, 1885. (P. N.)
13 : 679.

—— On banked fires. (P. N.) **12** : 648.

—— The Wheeler surface condenser. (P. N.) **13** : 479.

—— *translator.* Combustible fossils, refractory materials and the
iron industry at the Exposition of Turin, 1884. By L. Adami.
(P. N.) **13** : 494.

YARD to the meter, Relation of the. (P. N.) **15** : 524.

ZALINSKI, E. L. The naval uses of the pneumatic torpedo gun.
14 : 9.